S0-AKY-192

QUANTUM EVOLUTION

QUANTUM EVOLUTION

An Introduction to Time-Dependent Quantum Mechanics

JAMES E. BAYFIELD
Department of Physics and Astronomy
University of Pittsburgh

A Wiley-Interscience Publication
JOHN WILEY & SONS, INC.
New York • Chichester • Weinheim • Brisbane • Singapore • Toronto

This book is printed on acid-free paper. ♾

Copyright © 1999 by John Wiley & Sons, Inc. All rights reserved.

Published simultaneously in Canada.

No part of this publication may be reproduced, stored in a retrieval system or transmitted in any form or by any means, electronic, mechanical, photocopying, recording, scanning or otherwise, except as permitted under Sections 107 or 108 of the 1976 United States Copyright Act, without either the prior written permission of the Publisher, or authorization through payment of the appropriate per-copy fee to the Copyright Clearance Center, 222 Rosewood Drive, Danvers, MA 01923, (978) 750-8400, fax (978) 750-4744. Requests to the Publisher for permission should be addressed to the Permissions Department, John Wiley & Sons, Inc., 605 Third Avenue, New York, NY 10158-0012, (212) 850-6011, fax (212) 850–6008, E-Mail: PERMREQ@WILEY.COM.

For ordering and customer sevice, call 1-800-CALL-WILEY.

Library of Congress Cataloging-in-Publication Data:

Bayfield, James E., 1938–
Quantum evolution : an introduction to time-dependent quantum mechanics / by James E. Bayfield
p. cm.
Includes bibliographical references and index.
ISBN 0-471-18174-9 (alk. paper)
1. Quantum theory. I. Title.
QC174. 12.B379 1999
530. 12--dc21 98-53673

Printed in the United States of America.

10 9 8 7 6 5 4 3 2 1

Q C
174
.12
B379
1999
PHYS

CONTENTS

* Indicates an advanced topic.

PREFACE

With the advent of femtosecond lasers, nanostructure fabrication, and other laboratory advances, new research in time-dependent quantum mechanics has been pursued within various fields—atomic and optical physics, physical chemistry, solid state physics, nanoscience, and nanotechnology.

A common feature of this research is preparation of a nonstationary initial state of a quantum system; the time interval for preparation can be short in comparison to the system's excited-state decoherence and relaxation times. The system possesses many energy eigenstates, and more than two of these play a role in the subsequent quantum evolution. When the initial state is prepared with an increased number of important excited eigenstates, its time evolution becomes more semiclassical. In the semiclassical regime, the quantum evolution depends explicitly on properties of the evolution of its classical analog. Thus many quantum systems of interest are related to a nonlinear classical dynamics.

For most quantum systems, neither classical nor quantum equations of motion have exact analytical solutions. Yet great progress has been made investigating time-dependent quantum problems, thanks in part to increasingly powerful computers and new numerical methods. Thus the field of quantum dynamics has grown steadily, and continues to do so.

This book offers an introduction to both the concepts and present status of the field of quantum dynamics. Quantum evolution is discussed for both the strongly quantum and semiclassical regimes. The book is intended, in part, as a bridge between the quantum mechanics of stationary quantum systems and a number of recent advanced theoretical treatises on various aspects of quantum dynamics. These treatises include those works written by Walter Dittrich and Martin Reuter, Martin Gutzwiller, Fritz Haake, Alfredo Ozorio de Almeida, Linda Reichl, and Michael Tabor that are cited in the bibliography. Thus the reader will find many sections in this book that either introduce or survey theoretical topics. The reader, however, will also find representative up-to-date examples of concrete quantum dynamics problems encountered in physics, chemistry, and nanotechnology—examples where much of the knowledge has been acquired with the aid of numerical computation or laboratory experiment.

Chapter 1 reviews some basic quantum concepts, integrated with basic ideas from nonlinear classical dynamics. This first chapter provides a vantage point for exploring short-time quantum evolution. Chapter 2 features studies of the role of quantum wave interference. Chapter 3 further develops needed classical physics, and Chapters 4 and 5 discuss key aspects of classically integrable and classically nonintegrable quantum systems, with specific examples drawn from chemical physics. Chapter 6 covers the important quantum phenomenon of tunneling through classically forbidden regions, both in coordinate space and in phase space; the ideas of the path integral formulation of quantum mechanics are first used here. Chapter 7 addresses periodic orbit theory, which is a recent application of the path integral theory; this chapter also addresses application of periodic orbit theory to important experimental problems in atomic physics, solid state physics, and nanotechnology. Chapters 8 and 9 are concerned with quantum systems that are created by applying time-dependent external forces to otherwise isolated systems—a topic of great interest in many areas of the quantum sciences. Two appendixes discuss computational methods and are often referred to in the text.

Much of this book is at the level of a second-term textbook on quantum mechanics. The reader should be familiar with the foundations of quantum mechanics and with the theory of stationary quantum systems. Previous study of nonlinear dynamics is not necessary, as relevant topics in this area are introduced as needed. If students have a working knowledge of the first ten chapters of the book *Principles of Quantum Mechanics* by Hans Ohanian (1990), or the first twelve chapters of *Quantum Mechanics—An Introduction* by Walter Greiner (1993), or an equivalent, they would be quite prepared for this book. A semester course could cover Chapters 1 through 5 plus the first two sections of Chapter 6, if the surveys of recent developments and advanced topics are not rigorously pursued.

Although many major aspects of theoretical quantum dynamics are introduced, only limited detail can be included. Therefore references are given whenever highly technical or highly mathematical material is omitted. Nonetheless, students should readily be able to prove many of the statements made in the text. This book is written to encourage self-practice of the simpler theoretical and numerical techniques, as well as investigation of the advanced topics.

Due to space limitations, it is not possible to include a chapter on quantum maps. These discrete-time models of quantum evolution have proved important tools for qualitatively exploring new ideas. Persons interested in being at the cutting edge of quantum dynamics should pursue this additional topic. Again due to space limitations, the new field of quantum computing is not included. This active field also merits special attention, for it involves new and interesting quantum time evolution.

The author thanks numerous colleagues all over the world who contributed to the collection of figures. A grant of sabbatical leave by the University of Pittsburgh made rapid progress on this book possible, for which I am very grateful. Professor Rob Coalson read each chapter of the book, and I thank him for his many helpful suggestions. Special thanks to my wife, Janet Bayfield, whose wonderful editing skills and word-processing efforts over the past two years made publication of this book possible.

JAMES E. BAYFIELD

The University of Pittsburgh
Pittsburgh, Pennsylvania
Spring 1999

1

BASIC CONCEPTS

1.1. HAMILTONIAN EVOLUTION

In the macroscopic world we live in, most interesting phenomena involve change. In the microscopic world ruled by quantum mechanics, the situation is similar. Whether at the frontiers of modern physics, chemistry, or biology, we need to understand change—time evolution—at the microscopic level.

Time dependence, however, has received little attention in textbooks about quantum mechanics. A key reason for this has origins in classical physics. Except for model classical systems that involve motion in just one spatial coordinate, and for a few other unusual systems living in more spatial dimensions, the basic equations of classical evolution cannot be solved analytically. Indeed these equations are often devilishly difficult to solve—even approximately—with confidence. Only recently, with the development of modern computers, can we make detailed investigations of many "nonintegrable" classical time-dependent problems of interest. Computers have revolutionized the study of classical physics and the study of quantum mechanics as well.

The physics of thirty years ago, however, cannot be set aside—it remains essential for our understanding. In this first chapter we address equations that form the foundation of Hamiltonian time evolution, starting with Schrödinger's equation in quantum mechanics and Hamilton's equations in classical physics. We also consider a number of important ideas, old and new, for uncovering specific features of the various time evolutions hidden in these equations. Computer-based considerations and numerical results are included where appropriate.

1.1a. Schrödinger's Equation—A Mini-review

Physical systems evolve in time within the usual three-dimensional (3-d) coordinate space. Let a point in space-time be located by $(x, y, z, t) = (\mathbf{r}, t)$ relative to the origin of a reference coordinate system. **(All vectors in this book are in bold type** or in column matrix notation, such as $\mathbf{R} = \bar{\text{R}}$.) At each time t, our system exists in a unique physical state. This state can be described by spatial distributions of physical properties such as mass, charge, and intrinsic particle spin.

By definition, when a physical system is quantum mechanical, wave–particle duality plays a significant role. To describe time evolution under the influence of this duality, we must use the time-dependent Schrödinger equation

$$H\Psi = i\hbar \frac{\partial \Psi}{\partial t}. \tag{1.1}$$

Here we see our first quantum operator—the Hamiltonian operator H. (Whenever confusion with classical quantities might occur, *quantum operators in this book are italicized*.) Our challenge is to explore, characterize, and understand the solutions of this seemingly simple and well-established Schrödinger equation. For particular initial conditions of interest, we shall seek features of time evolution that are common to members of various classes of Hamiltonians H. In this chapter we primarily consider those Hamiltonians that do *not* explicitly depend on time.

All the accessible information about a quantum system is contained in the wavefunction $\Psi(\mathbf{r}, t)$. This $\Psi(\mathbf{r}, t)$ resides in a mathematical "wavefunction space" called Hilbert space; Hilbert space is often (but not always) an infinite-dimensional vector space. The Hamiltonian quantum operator H defines which system we have; H can often (but not always) be obtained from the Hamiltonian H of an appropriate system in classical physics.

As an example where knowledge of H is useful, consider the motion of a particle under the action of a time-independent force $\mathbf{F} = -\nabla V$ determined by a real potential $V(\mathbf{r})$. We have $H = \mathbf{p}^2/2m + V(\mathbf{r})$. Quantum mechanics provides a procedure for finding H for this problem. We replace the momentum $\mathbf{p}$ by $i\hbar\nabla$, where $\nabla = \partial/\partial\mathbf{r}$ is the spatial gradient operator in Cartesian coordinates. Equation (1.1) then becomes

$$\left(\frac{\boldsymbol{p}^2}{2m} + V\right)\Psi = -\frac{\hbar^2}{2m}\nabla^2\Psi + V\Psi = i\hbar\frac{\partial \Psi}{\partial t}. \tag{1.2}$$

In quantum mechanics, relationships between Ψ and those quantities observed by experimentalists are fundamental. Conceptually simplest of these relationships is the connection $P(\mathbf{r}, t) = \Psi^*\Psi$ that relates the wavefunction to the particle mass probability distribution P. Another simple relationship is the connection $\mathbf{J}(\mathbf{r}, t) = (\hbar/2mi)(\Psi^* \, \nabla\Psi - \Psi \, \nabla\Psi^*)$ that relates the wavefunction to the particle mass-probability current distribution $\mathbf{J}$. Noting that V is always real, and using Eq. (1.2) to eliminate time derivatives of Ψ and Ψ^*, we can easily show that $\partial P/\partial t + \nabla \cdot \mathbf{J} = 0$. This is the *equation of continuity*. Although we derived this equation for a particular set of systems, this equation is quite general for real potentials. This equation mathematically states the local conservation of particle mass probability in space, just as a similar differential equation in electromagnetic theory holds for the local conservation of electrical charge.

Among the many possible solutions of Eq. (1.2) there are the special and important stationary solutions $\Psi_E(\mathbf{r}, t)$ for which P is constant in time. We will see that time-evolving quantum wavepackets are often easily constructed of linear superpositions of stationary states. For these stationary states, the time and spatial variables in the wavefunction separate, as follows:

$$\Psi_E(\mathbf{r}, t) = \psi_E(\mathbf{r}) \exp\left(-\frac{iEt}{\hbar}\right). \tag{1.3}$$

Inserting this wavefunction into Eq. (1.2) yields the time-independent Schrödinger equation $H\psi_E = E\psi_E$. The eigenfunction $\psi_E(\mathbf{r})$ of the Hamiltonian belongs to the energy eigenvalue E. For many problems, there are an infinite number of allowable values of E. These values form a set of energies that, in general, includes both discrete and continuous parts.

The set of states $\Psi_E(\mathbf{r}, t)$ is complete. Thus any other complete set of states $\Psi_n(\mathbf{r}, t)$ describing the same system can be expanded as a linear combination of the Ψ_E, with the expansion coefficients C_{En} being constants:

$$\Psi_n(\mathbf{r}, t) = \sum_E C_{En} \psi_E(\mathbf{r}) \exp\left(-\frac{iEt}{\hbar}\right). \tag{1.4}$$

If both sets of states are orthonormal, then the C_{En} are elements of a unitary matrix, and the transformation from the set of states Ψ_E to the set $\Psi_n(\mathbf{r}, t)$ is unitary as well.

Given the stationary states Ψ_E, Eq. (1.4) gives us a connection between each $\Psi_n(\mathbf{r}, t)$ and a set of constants C_{En}. A special case is when the connection constitutes a transformation from the usual Schrödinger representation (where the wavefunctions Ψ_n depend on time and the Hamiltonian does not) to the Heisenberg representation (where wavefunctions are represented by sets of constants $\{C_{En}\}$ and the Hamiltonian must contain all the time dependence). (For details about this transformation, see Merzbacher (1961), Ohanian (1990), or Greiner (1992).) The transformation between these two representations is a rotation in the wavefunction (Hilbert) space, from a "frame" with a stationary set of axes to a frame that rotates with the wavefunction vector. In the latter frame, the wavefunction appears constant in time. Operators that are constant in the Schrödinger representation, such as $\boldsymbol{r}$ and $\boldsymbol{p}$, must become functions of time in the rotating system. Thus, in the Heisenberg representation, these operators contain the time evolution.

We know that the general quantum equations for time derivatives of operators can lead to useful expressions for the time derivatives of $\boldsymbol{r}$ and $\boldsymbol{p}$. For our example of a particle in a potential, these expressions are

$$\frac{d\boldsymbol{r}}{dt} = \frac{\boldsymbol{p}}{m}, \qquad \frac{d\boldsymbol{p}}{dt} = -\boldsymbol{\nabla} V,$$

(see, e.g., Kramers, 1957; Greiner, 1992). For any wavefunction ψ of the system, we can compute mean values (diagonal matrix elements) of the operators, such as $\langle\psi|\boldsymbol{r}|\psi\rangle \equiv \langle\boldsymbol{r}\rangle$. We then see that *Ehrenfest's theorem* must hold,

$$\frac{d\langle\boldsymbol{r}\rangle}{dt} = \frac{\langle\boldsymbol{p}\rangle}{m}, \qquad \frac{d\langle\boldsymbol{p}\rangle}{dt} = -\langle\boldsymbol{\nabla} V\rangle. \tag{1.5}$$

These familiar looking relationships are easy to interpret. We see that mean values for a quantum particle evolve according to the relevant Newtonian classical equations. These classical equations are a special case of Hamilton's equations, which we discuss after an important brief remark.

1.1b. Hamilton's Equations

Hamilton's equations (1.7) below are crucial for us. In the case of many quantum systems, Hamilton's equations determine the time evolution of a closely related and highly relevant classical system. Beginning with Figs. 1.9, 1.11, and 1.12, we'll see that understanding the relevant classical system and its evolution often proves helpful in interpreting the quantum evolution. In the semiclassical regime discussed in Section 2.1c (see Fig. 2.7) and elsewhere in this book, classical evolutions can be used as the "backbone " for accurate calculations of the quantum evolution.

The classical Newtonian physics of a particle in a 3-d potential is contained in the definition of momentum $\mathbf{p} = m(d\mathbf{r}/dt)$ and in Newton's second law $d\mathbf{p}/dt = \mathbf{F}$; compare these equations with Eq. (1.5). For a moment, let us specialize to motion in 1-d along a spatial coordinate q. Then the Hamiltonian has the form $H = H(p, q)$. Letting $\mathbf{Z}$ be the two-component vector $\{q, p\}$, our two equations collapse into one $d\mathbf{Z}/dt = \{p/m,\ F(q)\}$. Since in 1-d problems, a conservative force F can always be written as $-\partial V/\partial q$ and as $p/m = \partial(p^2/2m)/\partial p$, it immediately follows that

$$\frac{d\mathbf{Z}}{dt} \equiv \left(\frac{dq}{dt}, \frac{dp}{dt}\right) = \left(\frac{\partial H}{\partial p}, -\frac{\partial H}{\partial q}\right). \tag{1.6}$$

We define the set of 2-D Hamiltonian dynamical systems as all systems where one can find a coordinate q and a related momentum p, such that Eq. (1.6) is satisfied. (Note that in this book we are using "n-D" to denote a system with an n-dimensional phase space, while "N-d" denotes an N-dimensional coordinate space.)

We stop and introduce a very general definition of a *dynamical system* of order n. Such a system is represented by an n-D real vector variable $\mathbf{Z} = [Z_1, Z_2, Z_3 \ldots, Z_n]$ that satisfies $d\mathbf{Z}/dt = \mathbf{v}(\mathbf{Z}, t)$ for a reasonably well-behaved vector function $\mathbf{v}$. The Z_i can be considered as "coordinates" of an abstract n-D space that is called *phase space*. We call $\mathbf{v}$ the *phase-space velocity function*. Any finite set of coupled ordinary differential equations that describes the time evolution of a system can be expressed as a dynamical system, by defining all derivatives above the first as new dependent variables. The number of independent first-order ordinary differential equations obtained this way becomes the order of the dynamical system.

Our primary concern is a particular class of dynamical systems—the Hamiltonian systems that satisfy a generalization of Eq. (1.6). This generalization can be written in compact form as $d\mathbf{Z}/dt = \{\partial H/\partial \mathbf{p},\ -\partial H/\partial \mathbf{q}\}$, or more explicitly as

$$\frac{dq_i}{dt} = \frac{\partial H}{\partial p_i}, \qquad \frac{dp_i}{dt} = -\frac{\partial H}{\partial q_i}, \qquad i = 1,\ 2, \ldots,\ N. \tag{1.7}$$

Equations (1.7) are *Hamilton's equations.* We see that these equations constitute a dynamical system of order $n = 2N$, where N is the number of degrees of freedom. Equations (1.7) can be written in a still more compact form as $d\mathbf{Z}/dt = \bar{\bar{J}}\,\nabla_Z H(\mathbf{Z})$, where the Poisson matrix $\bar{\bar{J}}$ is the $2N \times 2N$ antisymmetric matrix.

$$\bar{\bar{J}} \equiv \begin{bmatrix} \bar{\bar{0}} & \bar{\bar{I}}_N \\ -\bar{\bar{I}}_N & \bar{\bar{0}} \end{bmatrix}, \tag{1.8}$$

and $\bar{\bar{I}}_N$ is the $N \times N$ identity matrix. It is easy to see that Hamilton's equations can also be written as $d\mathbf{Z}/dt = \{\mathbf{Z}\, , H\}$ where $\{,\}$ denotes the Poisson bracket. Between any two functions $f(\mathbf{Z})$ and $g(\mathbf{Z})$, this bracket is defined as

$$\{f, g\} \equiv \sum_{i,j}^{2N} \left(\frac{\partial f}{\partial Z_i}\right) J_{ij} \left(\frac{\partial g}{\partial Z_j}\right) = \sum_{i=1}^{N} \left[\left(\frac{\partial f}{\partial q_i}\right)\left(\frac{\partial g}{\partial p_i}\right) - \left(\frac{\partial g}{\partial q_i}\right)\left(\frac{\partial f}{\partial p_i}\right)\right]. \tag{1.9}$$

The Poisson bracket form of Hamilton's equations in classical physics corresponds to a similar formula in quantum mechanics—the formula that expresses a time derivative of an operator in terms of the operator's commutator with the Hamiltonian operator. This is one of many important correspondences between related quantum and classical systems, a topic further discussed in Section 1.5.

1.1c. Hamiltonian Flows and Liouville's Theorem

Let us now examine the all-important phase-space velocity function $\mathbf{v}(\mathbf{Z}, t)$ and some of its properties. We begin with some definitions:

- The value of $\mathbf{v}(\mathbf{Z}, t)$ for a particular $\mathbf{Z}$, t is called the *phase-space velocity* at the point located by $\mathbf{Z}$ in phase space.
- If $\mathbf{v}(\mathbf{Z}, t)$ is independent of t, we then say the system is *autonomous.*
- For a given Hamiltonian, a particular evolution $\mathbf{Z}(t)$ is determined by a particular set of initial conditions, and this evolution traces out a continuous curve in phase space called a *phase-space trajectory.*
- The *phase-space flow* is the set of all these particular evolutions.

Out of the phase-space flow we can select a small compact region (a "disk") in phase space that is very thin in one dimension. Along its thin "edge" passes a set of phase-space trajectories that constitute the boundary of a *a tube of phase-space flow.* Figure 1.1 shows an example of such a tube.

In the case of Hamiltonian systems, Liouville's theorem states that the flow within such a tube is volume preserving. This places a constraint on the evolution of the (lower-dimensional) boundary of the tube, as suggested in Fig. 1.2. A rigorous geometrical proof of Liouville's theorem and its lower-dimensional analogs requires the theory of differential forms (Tabor, 1989, Appendix 2.2; Gutzwiller, 1990, Chapter 7).

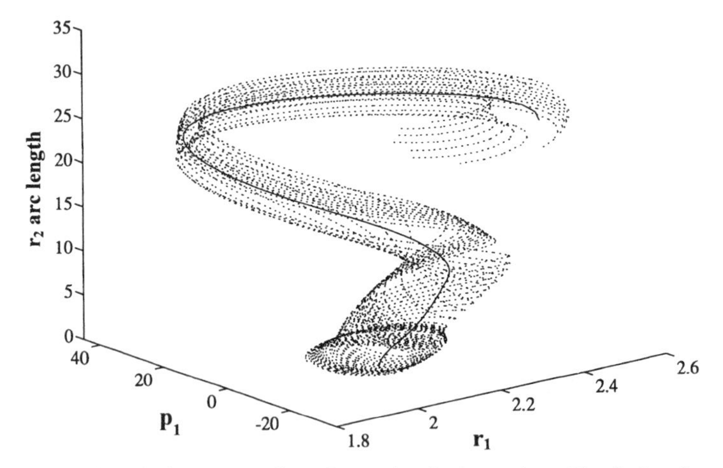

Figure 1.1. A tube of phase-space flow for a chemical reaction. Classical trajectories are plotted versus two spatial variables r_1 and r_2 and a momentum p_1. Several perimeter trajectories of the reaction cylinder are shown as dotted lines. The solid line is a central trajectory at a particular value of total energy. Reprinted from Hutchinson (1996, p.584), by courtesy of Marcel Dekker, Inc.

To see the volume preservation, consider an autonomous system where the Hamiltonian does not explicitly depend on time. It is easy to show that $dH/dt = \mathbf{v} \cdot \nabla_Z H = 0$ for such a system, by using the chain rule and Hamilton's equations. Each phase-space trajectory is then confined to a $(2N - 1)$-dimensional hypersurface on which H is a constant. (For Newtonian systems, where Newton's laws of motion are satisfied, this is a surface of constant energy.) In addition, the phase-space velocity $\mathbf{v}$ is both equal in magnitude and perpendicular in direction to the gradient $\nabla_Z H$ of the Hamiltonian.

Consider a small volume element within the disk that we used while defining our tube of phase-space flow. We can define this volume element as lying between two neighboring surfaces of slightly different constant energy. The size of the element in the direction from one energy surface to the other is inversely proportional to $|\nabla H|$

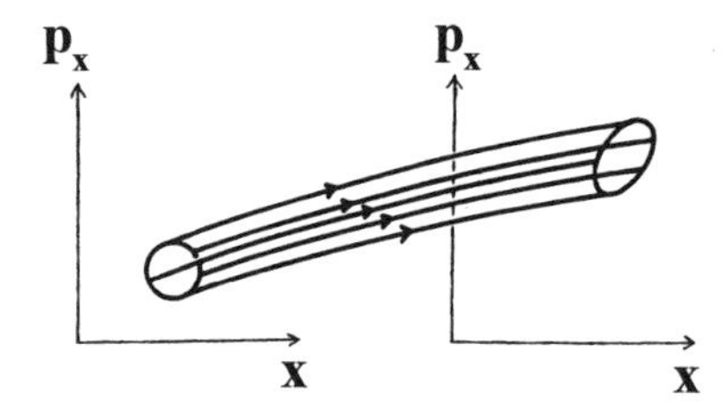

Figure 1.2. A tube of Hamiltonian phase-space flow twice intersecting the (x, p_x) plane in phase space. While the shapes of the boundaries of the two intersections are different, the enclosed areas are the same. Reprinted with permission from Berry (1978). Copyright 1978 American Institute of Physics.

and thus inversely proportional to the local value of $|\mathbf{v}|$. The size of the element in the direction of $\mathbf{v}$—a direction tangential to the energy surface—should be proportional to $|\mathbf{v}|$. In other directions that are orthogonal to the two directions already considered, dimensions of the element should not depend on $|\mathbf{v}|$. Thus the size of the volume element is independent of $|\mathbf{v}|$.

Let us now move along the tube of flow, watching what happens to a volume element in our "disk." The changes in the local properties of the tube are determined entirely by the vector field $\mathbf{v}$. But we have seen that the size of the volume element does not depend on this quantity. Thus the size of the volume element should be conserved as we move along the tube. The volume of the entire disk can be broken down into the volumes of such elements and thus should also be conserved as we move along the tube.

The near symmetry of Hamilton's equations (1.7)—which is reflected in entity (1.8)—immediately leads to the equation of continuity: $\nabla_Z \cdot \mathbf{v} = 0$. In other words, the phase-space velocity function is irrotational, having no sources. If there were such sources, the flow would not be volume preserving. Thus Hamiltonian phase-space flows that satisfy an equation of continuity are quite consistent with Liouville's theorem.

1.1d. Classical Hamiltonians

There are some differences between the autonomous classical Hamiltonians introduced above and *nonautonomous Hamiltonians*—those other Hamiltonians that explicitly depend on time. We know the energy of a nonautonomous Newtonian system is not constant, indicating that the system is interacting somehow with its environment. Because of this interaction, it is sometimes said that a nonautonomous system is "open," while an autonomous system is "closed."

Differences between autonomous and nonautonomous systems, however, are not as great as they may seem. The concept of phase space is very general. One may extend our phase space from 2N to $2(N+1)$ dimensions by defining $p_{N+1} = -H(t)$ and taking $q_{N+1} = t$ itself. As discussed in Section 8.1c, a new Hamiltonian can then be constructed that is constant and that satisfies a version of Hamilton's equations in the extended phase space. It follows that the evolution of a system with a time-dependent Hamiltonian is mathematically equivalent to the evolution that is due to some time-independent Hamiltonian possessing an additional degree of freedom.

Briefly turning to quantum mechanics, from Eq. (1.1) and the quantization equations for the momenta, we see that the full set of parameters of a quantum mechanics problem is simply the set of parameters of the quantum operator H. For Newtonian problems, this full set of parameters is identical to that of the classical Hamiltonian H, appended by the modified Planck's constant $\hbar \equiv h/2\pi$. This Planck's constant enters two ways—through the quantization to produce operators and from the right side of Eq. (1.1).

Speaking of Hamiltonians, there are a set of special classical Hamiltonians that exhibit what is called "scaling." Such Hamiltonians can have their effective number of parameters reduced by at least one. We achieve this reduction by introducing new unitless variables ($\mathbf{p}'$, $\mathbf{q}'$, t') that are scaled by appropriate combinations of the original parameters. This automatically leads to a new, similarly scaled set of parameters, with at least one parameter becoming unity. A further consequence is the production of a unitless *scaled Hamiltonian*.

As an example of such scaling, consider the harmonic oscillator. The energy of the oscillator is given by $E = p^2/2m + m\omega^2 x^2/2$. Dividing by E, we have an equation for an ellipse in phase space, $x^2/A^2 + p^2/B^2 = 1$, $A = (2E/m\omega^2)^{1/2}$, $B = (2mE)^{1/2}$. Clearly we can introduce unitless scaled variables $(x', p') = (x/A_0, p/B_0)$ and replace E by some reasonable value E_0; then the unitless scaled Hamiltonian is simply $H' = H/E_0 = (p')^2 + (x')^2$. For a fixed particle mass m, the parameters have been reduced from E and ω to just the scaled energy $E' \equiv E/E_0$. The area of our ellipse is $\pi AB = 2\pi E/\omega$, which has units of angular momentum or action. Thus a unitless quantum parameter is the scaled Planck's constant $h' = \hbar\omega/E_0$.

When Hamiltonians exhibit scaling, classical phase-space trajectories must divide up into families. In both analytical and numerical work, each family can be treated as a single entity. In Section 7.lb, we will see that scaled experiments have even been carried out. If a quantum system is semiclassical, working with scaled quantities greatly reduces the effort required for a full exploration of the system's evolution. We will also see that if the system is not very semiclassical, the use of scaled quantities readily reveals the fully quantum deviations.

An important problem possessing scaling properties is that of free motion of a particle inside a region with hard boundaries; see Section 7.4. Some other scalable systems have special interaction terms in the Hamiltonian, terms that are homogeneous polynomials in the spatial coordinates; see Sections 4.4, 5.3, and 7.1.

Finally, we should emphasize that particles in potentials are not the only important type of Hamiltonian system. Fields are another. Citing one useful example, we can easily find the Hamiltonian for a uniform and monochromatic classical electromagnetic field. From electromagnetic theory, the cycle-averaged energy stored in a volume V of space is

$$\langle E\rangle = \frac{1}{2}\iiint_V \left[\varepsilon_0\langle \mathbf{E}^2\rangle + \frac{1}{\mu_0}\langle \mathbf{B}^2\rangle\right] dV,$$

where **E** and **B** are the electric and magnetic fields. In phasor notation, the fields are

$$\begin{aligned}\mathbf{E}(\mathbf{r}, t) &= i\omega[\mathbf{A}\exp(-i\omega t + i\mathbf{k}\cdot\mathbf{r}) - \mathbf{A}^*\exp(i\omega t - i\mathbf{k}\cdot\mathbf{r})],\\ \mathbf{B}(\mathbf{r}, t) &= \left(\frac{1}{\omega}\right)\mathbf{k}\times\mathbf{E}(\mathbf{r}, t).\end{aligned} \tag{1.10}$$

Here **A** is a complex amplitude, ω is the angular frequency, and **k** is the propagation vector. Imposing periodic boundary conditions and taking $\mathbf{k}\cdot\mathbf{A} = 0$, we obtain $\langle E\rangle = 2\varepsilon_0 V\omega^2\, \mathbf{A}\cdot\mathbf{A}^*$. If we define $\mathbf{A} \equiv (4\varepsilon_0 V\omega^2)^{-1/2}(\omega Q + iP)\hat{\mathbf{A}}$, where $\hat{\mathbf{A}}$ is the unit vector along **A**, then

$$\langle E\rangle = \tfrac{1}{2}(P^2 + \omega^2 Q^2). \tag{1.11}$$

Mathematically, this equation is the Hamiltonian of a mechanical harmonic oscillator, with Q and P being canonical (conjugate) variables that satisfy Hamilton's equations (1.7).

1.1e. Quantization of Hamiltonians

Earlier, we used the familiar quantization rule $\mathbf{p} \to i\hbar\mathbf{\nabla}$, in Cartesian coordinates when we have more than one degree of freedom. We used this rule to rewrite Schrödinger's equation (1.1) as the partial differential equation (1.2) in the Schrödinger representation. This procedure is an example of what is called *first quantization* of a classical Hamiltonian, a procedure useful when we wish to ultimately solve for the energy eigenvalues E_n and eigenstates ψ_E.

We sometimes wish to discuss very precisely the quantum mechanical interaction of an ensemble of particles (such as those in an atom) with an environment that is represented by one or more fields. When we need to quantize these fields, we use a procedure called *second quantization* (also called the factorization method—see Ohanian, 1990, Chapter 6).

Not only can we use the second-quantization procedure to quantize the electromagnetic field—where the quantum particles involved are massless bosons with intrinsic spin zero—but we can also use the procedure to second-quantize Hamiltonians for any type of fermion or boson, or for mixed ensembles of these elementary particles. This general subject is discussed in textbooks on quantum field theory.

We now demonstrate the second-quantization procedure for the case of the electromagnetic field. This procedure proves useful for the development of time-dependent theory for the interaction of bound electrons with coherent laser light, a topic discussed in Chapter 2. We begin by assuming that the classical Hamiltonian equation (1.11) is first-quantized following the usual first-quantization procedure for a harmonic oscillator, which now has unit mass

$$H = \tfrac{1}{2}(P^2 + \omega^2 Q^2), \qquad [Q, P] \equiv QP - PQ = i\hbar. \tag{1.12}$$

The energy eigenvalues of this problem form an equally spaced ladder of energy levels, $E_n = (n + \frac{1}{2})\hbar\omega$, $n = 0, 1, 2, 3, \ldots$. We recall that the level spacing $\hbar\omega$ is also the energy of one photon in the field. Therefore the integer n labels both the oscillator excitation energy, in units of $\hbar\omega$, and the number of photons in the field above some initial number. To second-quantize the field, we go beyond first quantization and introduce quantum operators corresponding to the amplitudes $\mathbf{A}$, $\mathbf{A}^*$ that appear in Eq. (1.10):

$$a,\ a^\dagger \equiv (2\hbar\omega)^{1/2}(\omega Q \pm iP). \tag{1.13}$$

It is important to consider the *photon-number operator n*, which is defined as the product $aa^\dagger$. After inverting Eq. (1.13) to obtain Q, P and inserting the results into Eq. (1.12), we have the second-quantized Hamiltonian

$$H = (aa^\dagger + \tfrac{1}{2})\hbar\omega = \hbar\omega(n + \tfrac{1}{2}), \qquad [a\ , a^\dagger] = 1. \tag{1.14}$$

Thus the energy eigenstates are also the eigenstates $|n\rangle$ of n (the *photon-number states)*. The eigenvalues of n are the nonnegative integers. From Eq. (1.14) we see that the name "photon-number operator" is quite appropriate for n.

A key observation at this point is that expressions (1.14) imply that

$$Ha^{\dagger}|n\rangle = (E_n + \hbar\omega)a^{\dagger}|n\rangle. \tag{1.15}$$

Thus $a^{\dagger}|n\rangle$ must be $|n+1\rangle$, to within a normalization factor since we have not normalized the $|n\rangle$. The operator $a^{\dagger}$ converts a photon-number eigenstate into another photon-number eigenstate, with the energy increasing by that of one photon. The (photon) *creation operator* $a^{\dagger}$ is also called the "oscillator excitation operator." Similarly, the *destruction operator* a converts $|n\rangle$ to give a result proportional to $|n-1\rangle$, except that $a|0\rangle \equiv 0$. These properties of $(a^{\dagger}, a)$—properties that move the system up and down the energy ladder—are very useful in analytical work. (Sections 2.3 and 2.4 will treat second-quantized systems in more detail.)

1.2. STATIONARY QUANTUM STATES

1.2a. Diagonalization of Hamiltonian Matrices

To find the stationary states of a quantum system, we must solve the time-independent Schrödinger equation $H\psi_E = E\psi_E$. The time-independent Schrödinger equation cannot be solved analytically for most problems, so we need approaches that are at least partially numerical. The basic computational technique is the *basis-set expansion method*, which we now address.

A first step is to separate the Hamiltonian H into the sum of two parts, $H = H_0 + V$. There are an infinite number of ways to do this separation, yet it can often seem like none of them are just what we need. To carry out an accurate basis-set expansion calculation, a number of requirements—often conflicting—must be more or less satisfied at the same time.

First and foremost, one of the partial Hamiltonians—which we call H_0—should have an analytically solvable time-independent Schrödinger equation, $H_0\phi = E^0\phi$. (An equation is an *analytically solvable equation* when its solution can be explicitly written in terms of a numerically convergent power series.) Formulas for the eigenvalues E_k^0 and eigenstates ϕ_k are then available. It is important that the ϕ_k be a complete set spanning the Hilbert space, which is defined by H_0. It is also helpful if the ϕ_k form an orthonormal set. In practice, a second major requirement for an easy basis-set expansion calculation is that analytical formulas exist for the matrix elements $\langle\phi_j|V|\phi_k\rangle$.

Our next step is to expand ψ_E as a linear combination $\psi_E = \sum_k c_k\phi_k$, and to find a set of coupled algebraic equations that can be solved numerically. Inserting the expansion into the full time-independent Schrödinger equation gives

$$(H_0 + V)\sum_k c_k\phi_k = E\sum_k c_k\phi_k.$$

We multiply this equation by the complex conjugate ϕ_j^* of one of the basis states ϕ_j. Then we integrate over the set τ of variables on which the ϕ_k depend, obtaining

$$\sum_k c_k \int \phi_j^* H_0 \phi_k \, d\tau + \sum_k c_k \int \phi_j^* V \phi_k \, d\tau = E \sum_k c_k \int \phi_j^* \phi_k \, d\tau.$$

If the ϕ_k are orthonormal, we can express this equation more compactly as $c_j H_j^0 + \Sigma_k c_k V_{jk} = c_j E$. Considering all possible values of j results in the matrix equation $(\bar{\bar{H}}_0 + \bar{\bar{V}} - E\bar{\bar{I}})\bar{C} = \bar{0}$ where $\bar{\bar{I}}$ is the identity matrix, $\bar{\bar{H}}_0$ is diagonal and known, $\bar{\bar{V}}$ is called the coupling matrix and is known, and $\bar{C}$ is the column matrix of unknown coefficients. The solution of our problem has been reduced to finding these coefficients. To find the coefficients, we diagonalize the Hamiltonian matrix $\bar{\bar{H}} = \bar{\bar{H}}_0 + \bar{\bar{V}}$, thereby obtaining its matrix eignevalues E_i and eigenvectors $\bar{C}_i$. (The computational methods are discussed in Section B.2b.) The vector $\bar{C}_i$ determines a state ψ_E of interest and is associated with the energy $E = E_i$.

In practice, the usually infinite basis set $\{\phi_k\}$ has to be truncated to a finite set because we have a finite amount of computer time and memory available. However, we must verify that changing the size of the basis does not significantly change our results. For some parameter values in some problems, even truncation to two states is an acceptable approximation. In such a case, the computer may be turned off and analytical work carried out instead. (We will investigate examples of two-level systems in Sections 2.2, 9.2a, and 9.3a.)

1.2b. Eikonal Theory for Semiclassical Systems

Over the past few decades, basis-set expansion calculations have been carried out for a wide variety of systems. Due to limitations in computer time, most studies have considered only small sets of parameter values. A great deal of work is required to fully determine the dependence of the numerical results on the parameters. Once the dependence is established, we need to understand the physical origins of those results.

Since the beginnings of quantum mechanics, classical physics has greatly helped us understand quantum systems. We need only recall the Bohr model of the hydrogen atom, which utilizes a circular classical periodic electron orbit. In the real hydrogen atom—even when the electron's wavefunction doesn't behave like a classical particle in a single orbit—there are powerful conceptual connections with ensembles of orbits. To understand quantum time evolution, such conceptual connections must be developed as far as possible. To develop these connections, we will pursue the ideas of what is called "semiclassical quantum theory"—one major focus of this book. As might be expected, the approximations made in this theory are analogous to approximations made in certain theories of classical wave phenomena.

Among other things, the field of optics is concerned with the propagation of electromagnetic waves through spatially inhomogeneous optical media. We know that geometrical optics can give a good approximation to exact optical propagation.

This occurs when the wavelength λ is short compared to the distances x over which the medium's properties change significantly. As a light wave propagates a small distance δx through a large medium, its change $\delta\lambda$ in wavelength must be much smaller than δx, meaning that

$$\left|\frac{\partial\lambda}{\partial x}\right| \ll 1. \tag{1.16}$$

Within this limitation and away from abrupt boundaries, light waves can be well characterized by rays, which are also called trajectories.

Let us explore extending this idea of rays to the matter waves of quantum mechanics. We continue to consider our example system of a particle in a 1-d potential V(x). Our particle is also a matter wave. Its local de Broglie wavelength is $\lambda(x) = h/p(x) = h/mv(x)$ at those points x in space where the local particle momentum is p(x). It follows that

$$\left|\frac{\partial\lambda}{\partial x}\right| = \left|\frac{h}{p^2}\frac{\partial p}{\partial x}\right|.$$

Using $p^2 = 2m[E - V(x)]$, we obtain

$$\frac{hm\left|\dfrac{\partial V}{\partial x}\right|}{[2m(E - V)]^{3/2}} < 1. \tag{1.17}$$

Only when Eq. (1.17) is satisfied can we use an approximation for matter waves that is analogous to geometrical optics. If we want our approximation to be both useful and accurate, then the inequality (1.17) will have to be strongly satisfied. We see this occurs when V varies sufficiently slowly with position, and/or when $E - V$ is not too small. The particle's kinetic energy becomes zero at the turning points x_0 of classical motion, where $p(x_0) = 0$. Thus $E - V$ becomes zero at such turning points. Our approximation will not be justified near these points where the de Broglie wavelength $\lambda(x_0)$ becomes infinite.

As in the single-eikonal theory for a classical wave, we now assume that a matter wavefunction $\psi(\mathbf{r}, t)$ is complex, having the form of a complex number. That is, we take $\psi(\mathbf{r}, t)$ to have a real amplitude $R(\mathbf{r}, t)$ and a real phase $S(\mathbf{r}, t)/\hbar$, so that the *eikonal ansatz* becomes

$$\psi(\mathbf{r}, t) = R(\mathbf{r}, t)\exp\left(i\frac{S(\mathbf{r}, t)}{\hbar}\right). \tag{1.18}$$

The explicit dependence of the phase on $\hbar$, shown in Eq. (1.18), is guided by the free-particle behavior $\psi \sim \exp(ipx/\hbar)$. Just as in electromagnetic theory, we will need to use linear combinations of functions of the form (1.18) to approximately solve most matter-wave problems. Assuming Eq. (1.18), the particle mass probability density is clearly $P(\mathbf{r}, t) = R^2(\mathbf{r}, t)$. Let us insert ansatz (1.18) into the time-dependent Schrödinger equation (1.2) and separate the real and imaginary parts:

$$\frac{\partial R^2}{\partial t} + \nabla \cdot \left(R^2 \frac{\nabla S}{m} \right) = 0, \tag{1.19}$$

$$\frac{\partial S}{\partial t} + \frac{(\nabla S)^2}{2m} + V - \frac{\hbar^2}{2m} \frac{\nabla^2 R}{R} = 0. \tag{1.20}$$

Here $\hbar$ explicitly appears only in the last term of Eq. (1.20). Of course, both R and S may also depend on $\hbar$.

Comparing Eq. (1.19) with the equation of continuity $\partial P/\partial t + \nabla \cdot \mathbf{J} = 0$, we see they are the same if the mass probability current is taken to be $\mathbf{J} = R^2\, \nabla S/m = P\, \nabla S/m$. We identify $\nabla S/m$ as a velocity $\mathbf{v}$ that we call the *probability flow velocity in coordinate space*. (This quantity $\mathbf{v}$ should not be confused with the classical phase-space velocity introduced in Section 1.1c, where by convention we used the same symbol.)

Note that if we were to ignore the last term in Eq. (1.20) and replace ∇S with $m\mathbf{v}$, this would give us $\partial S/\partial t + mv^2/2 + V = 0$. This looks similar to $\partial S/\partial t + H = 0$, where H is the classical Hamiltonian. Indeed, ignoring the last term in Eq. (1.20) gives us the Hamilton–Jacobi equation of classical mechanics, where S in Hamilton's principal function. This function is a classical action integral (Goldstein, 1980). (We return to this subject in Section 3.2b.)

These similarities of eikonal and classical equations suggest that we somehow expand the quantum equations (1.19) and (1.20) about their classical counterparts. A reasonable approach is to expand R and S in powers of the quantum parameter $\hbar$, since $\hbar$ is a rather small number when we use units that characterize the space and time scales of mesoscopic and macroscopic systems. So we write

$$S(x) = \sum_{k=0}^{\infty} \left(\frac{\hbar}{i}\right)^k S_k(x) = S_0 + \left(\frac{\hbar}{i}\right) S_1 + \left(\frac{\hbar}{i}\right)^2 S_2 + \cdots, \tag{1.21}$$

where the first few terms will be a good approximation only if their magnitudes are increasingly small.

1.2c. The Wentzel–Kramers–Brillouin (WKB) Approximation and Bohr–Sommerfeld Quantization

To demonstrate what happens when we pursue Eq. (1.21), let us limit our discussion to the time-independent Schrödinger equation (TISE) and to the particle in a 1-d potential. Then we are discussing energy eigenstates, and the quantities in Eq. (1.18) will not depend on time. At this point it is convenient to absorb R into S by making S a complex phase, writing $\psi = \exp(iS/\hbar)$. If we insert this expression into the TISE, we obtain the differential equation

$$\frac{1}{2m}\left(\frac{\partial S}{\partial x}\right)^2 + (V - E) - \frac{i\hbar}{2m}\left(\frac{\partial^2 S}{\partial x^2}\right) = 0, \qquad \psi \equiv \exp\left(i\frac{S}{\hbar}\right). \tag{1.22}$$

Using Eq. (1.21) in Eq. (1.22), we obtain to second order in $\hbar$

$$\left[\left(\frac{\partial S_0}{\partial x}\right)^2 + 2m(V - E)\right] + \left(\frac{\hbar}{i}\right)\left[2i\frac{\partial S_0}{\partial x}\frac{\partial S_1}{\partial x} + \frac{\partial^2 S_0}{\partial x^2}\right]$$
$$-\left(\frac{\hbar}{i}\right)^2\left[\frac{\partial S_0}{\partial x}\frac{\partial S_2}{\partial x} + \left(\frac{\partial S_1}{\partial x}\right)^2 + i\frac{\partial^2 S_1}{\partial x^2}\right] \approx 0. \tag{1.23}$$

As a mathematics problem, $\hbar$ is just a parameter, and the solutions of Schrödinger's partial differential equation (the TISE) should make sense, independent of the particular value of $\hbar$. Additionally, for those systems (Section 1.1d) that exhibit scaling of the Hamiltonian, the unitless scaled parameters include $\hbar$ divided by a physical scaling quantity that will have units of action. Then changing $\hbar$ is equivalent to changing that scaling quantity, a procedure that does have a physical interpretation. In view of these facts, we require that the coefficients of each power of $\hbar$ in the expansion (1.23) be zero. This produces a set of equations to be satisfied simultaneously. We note that the first coefficient involves only S_0, so we can solve the first equation for this quantity. Because each following coefficient in Eq. (1.23) involves only one new S_k, the whole set of equations can be solved by iteration to obtain the following results (for the case $E > V$) :

$$S_0 = \pm\int_{x_0}^{x} \sqrt{2m(E - V)}\,dx,$$

$$S_1 = \frac{i}{2}\ln\left(\frac{\partial S_0}{\partial x}\right) \quad \text{or} \quad \exp(iS_1) = [2m(E - V)]^{-1/4},$$

$$S_2 = -\frac{1}{2}\frac{m(\partial V/\partial x)}{[2m(E - V)]^{3/2}} + \frac{1}{4}\int\frac{m^2(\partial V/\partial x)^2\,dx}{[2m(E - V)]^{5/2}}.$$

In this last equation, we note that the two terms on the right-hand side are of the same order of magnitude. We take $|\hbar S_2| \ll 1$ as the criterion for applicability of expansion (1.21). This criterion then becomes

$$\left|\frac{\hbar m(\partial V/\partial x)}{[2m(E - V)]^{3/2}}\right| \ll 1,$$

in agreement with Eq. (1.17).

To obtain the stationary-state *WKB wavefunction* for our particle in a 1-d potential, we retain only S_0 and S_1. Consider the special case of bounded motion of the particle in a potential well; see Fig. 1.3. Allowing for both possible signs of S_0, we arrive at the general form of the wavefunction:

$$\psi(x) = \frac{A}{\sqrt{|p(x)|}} \exp\left(\frac{i}{\hbar}\int_{x_0}^{x} p(x')\,dx'\right) + \frac{B}{\sqrt{|p(x)|}} \exp\left(-\frac{i}{\hbar}\int_{x_0}^{x} p(x')\,dx'\right). \quad (1.24)$$

In this two-branch sum, the constants A, B $\neq 0$ and $E > V(x)$ (or $p(x) > 0$) for the classical motion inside the well. There are two turning points for our problem. At those points, p(x) changes sign, and the behavior of $\psi(x)$ switches from oscillatory to exponential, or vice versa. Clearly Eq. (1.24) will not be valid near these turning points, where p(x) becomes zero.

The WKB solution for all x is obtained following the usual matching procedure for finding connection formulas. We begin by finding fully quantum local solutions ψ_{local} for ψ near the turning points. If a linear local approximation is made for V(x) near a turning point x_0, then the local solution is an Airy function (Berry and Mount, 1972). We then make ψ and $\partial\psi/\partial x$ continuous, as the value of x passes into and out of the local regions containing each of the turning points x_0; see, for instance, Sections 19 and 20 of Landau and Lifshitz (1958) or Section 23 of Davydov (1976). This matching procedure also applies for types of 1-d potentials other than wells. In any case, on the classically forbidden side of a turning point there is a monotonically

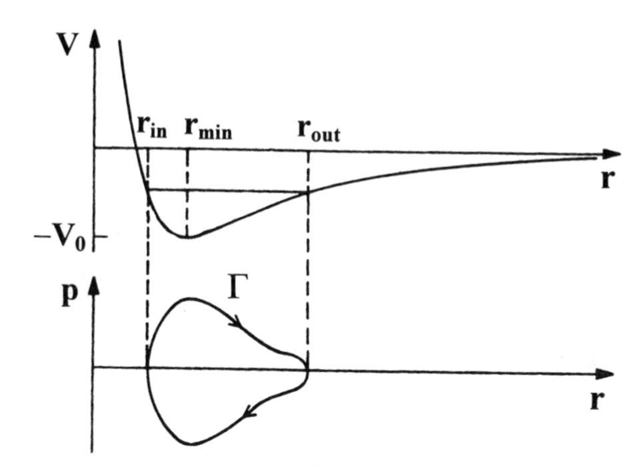

Figure 1.3. Comparison of the classical motion inside the 1-d potential well of a diatomic molecule with its trajectory in phase space. The quantities r_{in} and r_{out} are turning points of the classical motion, for a particular negative value of the energy. The momentum p is zero at the turning points and has a maximum at $r = r_{min}$, where the potential has a minimum. From *Computational Physics—Fortran Version* by Steven Koonin and Dawn Meredith. Copyright 1990 by Addison Wesley Publishing Company. Reprinted by permission of Perseus Books Publishers, a member of Perseus Books, L.L.C.

decreasing real wavefunction. Since there must be an oscillatory wavefunction on the classically allowed side of the turning point, we have the switchover

$$\frac{1}{\sqrt{|p(x)|}} \exp\left(-\frac{1}{\hbar}\left|\int_{x_0}^{x} p(x')\,dx'\right|\right)$$
$$\leftarrow \psi_{\text{local}}(x) \rightarrow \frac{2}{\sqrt{|p(x)|}} \cos\left(\frac{1}{\hbar}\left|\int_{x_0}^{x} p(x')\,dx'\right| - \frac{\phi}{2}\right). \tag{1.25}$$

This notation suggests the following situation: central is the region of the classical singularity where $\sqrt{|p(x)|} \rightarrow 0$; in this region we need a reasonably accurate quantum solution ψ_{local}, which we obtain by approximating the potential V(x) as a linear (or better, quadratic) function of x. On either side of the region of the turning point, we expect the WKB wavefunction to be applicable. The wavefunction and its first spatial derivative must be continuous everywhere, which determines the relative amplitudes and phases of the inner and outer (right and left) WKB wavefunctions shown in Eq. (1.25). The value of the constant phase ϕ is $\pi/2$ in the limit of short de Broglie wavelengths (Landau and Lifshitz, 1958).

We note in passing that the most general expression for the connection formula for a single turning point has been derived in Berry and Mount (1972). Let us call the action integral $(1/\hbar)\int_{x_0}^{x} p(x')\,dx'$ in Eq. (1.25) the quantity w. Then the general expression is

$$\frac{2\sin\mu}{\sqrt{|p|}} e^{|w|} - \frac{\sin(\mu-\gamma)}{\sqrt{|p|}\sin\gamma} e^{-|w|} \leftarrow \psi_{\text{local}}(x) \rightarrow \frac{2\cos[|w| - (\pi/4) + \mu]}{\sqrt{|p|}}$$

The constant γ is undetermined by WKB theory, but it can be shown to be equal to $-\pi/2$. If there is to be pure exponential damping in the classically forbidden region, then we must have $\mu = 0$, the dependence on γ drops out, and we retrieve Eq. (1.25).

To complete the matching procedure for our particle in a potential, we use the remaining boundary conditions at $x = \pm\infty$ to determine A and B in Eq. (1.24) (Merzbacher, 1961, Chapter 7). As a final result of the matching procedure, the matter wave of the particle must "fit" into the potential well, much like a resonant sound wave must "fit" into an organ pipe. The matter-wave fitting is expressed by the well known *Bohr–Sommerfeld quantization condition*

$$\oint_{\Gamma} p(x')\,dx' = 2\pi\hbar(n + \tfrac{1}{2}). \tag{1.26}$$

Here the quantum number n is a nonnegative integer. The quantity Γ is a path of integration. It is a 1-D trajectory in phase space (necessarily a periodic orbit) that starts at one classical turning point, travels to the other point, and returns. Looking again at Fig. 1.3, we see that the left-hand side of Eq. (1.26)—which is the accumulated action around the periodic orbit Γ—is also the phase space area enclosed by that orbit. Since the momentum p depends on the energy E, we can invert Eq. (1.26) to obtain the quantum mechanically allowed values E_n of the energy; that is, we obtain the energy eigenvalues. Using one of these eigenvalues in the matched version of Eq. (1.24) gives us the corresponding energy eigenstate.

After completing the connection procedure, we have a WKB wavefunction $\psi^{WKB}(x)$ that, except for very low-lying energy eigenstates, agrees well with numerical exact wavefunctions for values of x away from the classical turning points. The case of a particular 1-d anharmonic oscillator with the potential $V(x) = -\lambda x^4$ is shown in Fig. 1.4. The upper left-hand corner of the figure shows the energy levels of the ground state ψ_a, of the first excited state ψ_b, and of the fifth excited state ψ_c that has even symmetry under the reflection $x \to -x$. In the other panels, the WKB wavefunctions for these states are compared with precise results (McWeeny and Coulson, 1948). We see little similarity for the ground state ψ_a, yet for ψ_b there is already excellent agreement for most values of x. In the region near the turning points we can greatly reduce the remaining discrepancies by replacing the displayed WKB wavefunction with the special local wavefunction used in the WKB wavefunction matching procedure.

For the harmonic oscillator problem, both the exact and WKB energy eigenvalues are $(n + \frac{1}{2})\hbar\omega$. For other problems, the two sets of eigenenergies are usually different.

Most often, the inversion of Eq. (1.26)—which is needed to find the values of an energy allowed by quantum mechanics—must be done numerically. The first step is to find the turning points, which are the zeros of p(x). (This step can be carried out using the Newton–Raphson method discussed in Section A.4.) In the second step, we must numerically evaluate the integral within a range of energies E, as part of a search routine that converges to an eigenvalue E_n. (A useful Fortran routine for the

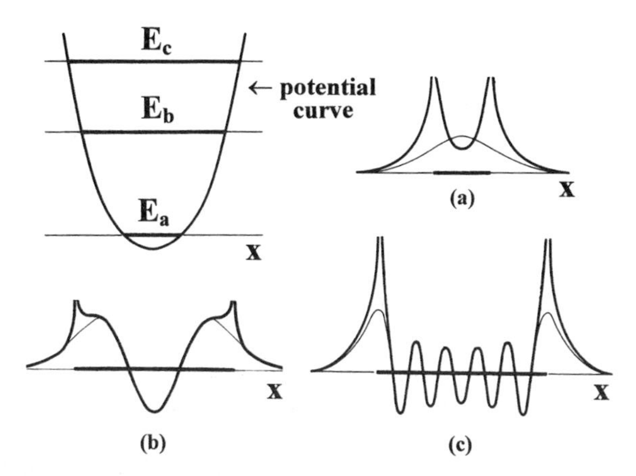

Figure 1.4. Comparison between exact wavefunctions (thin lines) and WKB wavefunctions (heavy lines) for some even states of an anharmonic oscillator. Darkened line segments on the horizontal x-axis denote the regions of classically allowed motion. From Park (1974). Copyright 1974 McGraw-Hill.

search is discussed by Koonin and Meredith (1990) and is available over the Internet.)

In closing this subsection we should recall a scattering picture that is related to the energy eigenstates. The solutions of the time-independent problems we have just discussed are the spatial wavefunctions $\psi(x)$ that enter into the solution $\Psi(x, t) = \psi(x)\exp(-iEt/\hbar)$ of the time-dependent Schrödinger equation. The wavefunction inside the potential well—see the right-hand side of Eq. (1.25)—can be decomposed using $2\cos(|w| - \phi/2) = \exp(i|w| - i\phi/2) + \exp(-i|w| + i\phi/2)$. This decomposition leads to terms in $\Psi(x, t)$ proportional to $\exp[(\pm i|w(x)| - Et/\hbar) \mp i\phi/2]$ that have the form of traveling waves in the $\pm x$ directions. If $p(x)$ were constant inside the well, $w(x)$ would be $p(x - x_0)/\hbar \equiv k(x - x_0)$. Thus textbooks and research papers often utilize scattering terminology when discussing the spatial wavefunctions $\psi(x)$ in classically allowed regions. For instance, the right-hand side of Eq. (1.25) is called a superposition of waves that are incident on and reflected from the wall of the potential at the turning point x_0. The amplitudes of these two travelling waves are clearly equal, so we have elastic scattering. The phase difference between the two waves is $\phi/2 - (-\phi/2) = \phi$ and is called the *phase loss* due to reflection at an isolated turning point. When the potential $V(x)$ near x_0 rises rapidly in comparison to the local de Broglie wavelength, then the phase loss ϕ becomes close to $\pi/2$.

1.2d. Improved Semiclassical Approximations (Advanced Topic)

It is interesting to find out whether the WKB wavefunction (1.24) can be improved to arbitrary accuracy by including more terms in the expansion (1.21). When we include more terms, Eq. (1.23) leads to the recursion relations for all integers $\ell \geq 1$:

$$\left(\frac{\partial S_0}{\partial x}\right)^2 = 2[E - V(x)], \qquad \sum_{k=0}^{\ell}\left[\frac{\partial S_k}{\partial x}\frac{\partial S_{\ell-k}}{\partial x} + \frac{\partial^2 S_{\ell-1}}{\partial x^2}\right] = 0.$$

We can generalize the quantization condition (1.26) to become a uniqueness requirement for the wavefunction; this requirement can be expressed in terms of $S(x)$:

$$\oint_\Gamma dS = \sum_{k=0}^{\infty}\left(\frac{\hbar}{i}\right)^k \oint_\Gamma dS_k = 2\pi\hbar n.$$

If we retain only the $k = 0$ zero-order term, we retrieve Eq. (1.26). If we also include the $k = 1$ first-order correction, then we produce the phase loss $\phi = \pi/2$, which—for two turning points—is counted twice to produce the correction $n \to n + \frac{1}{2}$ included in Eq. (1.26). Handling these terms to arbitrary order has not been done for the general case. In the special case of $V(x) = V_0/[\cos^2(\alpha x)]$, however, a WKB series for the energies can be derived and shown to converge to the exact quantum mechanical energies (Robnik and Salasnich, 1997).

In specific applications we can considerably improve the standard WKB approximation. We may make adjustments forcing the asymptotic (large x) behavior of the WKB wavefunction to agree better with the corresponding behavior of the exact wavefunction. Consider the case of the centrifugal potential $(\hbar^2/2m)(\gamma/r^2)$ with $\gamma = \ell(\ell+1)$. Using an old and commonly used procedure to create the agreement, we alter this potential according to the Langer modification $\gamma \to \gamma' = (\ell + \frac{1}{2})^2 = \gamma + \frac{1}{4}$ (Langer, 1937). (To derive this, Langer had the idea of changing the variable $r \equiv x$ to the variable $y = \ell nx$, and then applying the WKB approximation to the resulting equation.) Altering the potential this way forces asymptotic agreement to order $1/kx$, where $k = 2\pi/\lambda$; see the upper curve in Fig. 1.5. A better approximation is made by leaving the potential alone and replacing the WKB asymptotic phase with the asymptotic phase for the fully quantum problem. This means taking $\phi = \pi/2 + [(\gamma + \frac{1}{4})^{1/2} - \gamma^{1/2}]\pi$ (Friedrich and Trost, 1996). This approach appears to be generally useful in constructing good WKB wavefunctions. The improvement is seen in Fig. 1.5. For radial WKB wavefunctions, unlike those for a true 1-d problem, the ratio of the numerical coefficients on the left and right sides of Eq. (1.25) generally should not be $\frac{1}{2}$ but instead should be a function of the angular momentum ℓ (Popov et al., 1996).

For a particle oscillating between two turning points (see Fig. 1.3), the WKB quantization condition can be written

$$\left(\frac{1}{\hbar}\right)\oint_{\Gamma} p(x)\,dx - \phi_1 - \phi_2 = 2\pi n.$$

Here ϕ_1 is defined as the phase loss due to reflection at the turning point x_1, and ϕ_2 is the phase loss at x_2. This equation leads to the expression

$$\int_{x_1}^{x_2} p(x)\,dx = \left(n + \frac{\mu}{4}\right)\pi\hbar, \qquad \mu = \frac{\phi_1 + \phi_2}{\pi/2}. \tag{1.27}$$

Here we introduce the *Maslov index* μ, which is the total phase loss during one period in units of $\pi/2$. In improved WKB approximations, this index becomes nonintegral. There are sound theoretical procedures for finding Maslov indices, which we discuss in Sections 4.2 and 8.1f.

Semiclassical theory can be developed as an expansion in powers of a dimensionless effective Planck's constant $\hbar_{\text{eff}}$. For example, in the case of the 1-d quartic oscillator, if we write the Hamiltonian as $H(x,p) = (p^2/2m) - (m\omega^2 x^2/2) + bx^4$, then $\hbar_{\text{eff}}$ can be taken as $\hbar b/2m^2\omega^3$; see Section 4.4. For this system, Bohigas et al. (1993) gave an expansion for the energy. They compared the semiclassical energies to fourth order with numerical values of the energy obtained from diagonalization of a Hamiltonian matrix formed using 360 harmonic oscillator basis states. The comparison is shown in Fig. 1.6. Except for the lowest few states, the WKB results are quite accurate up to the quantum number value 260; above this value the

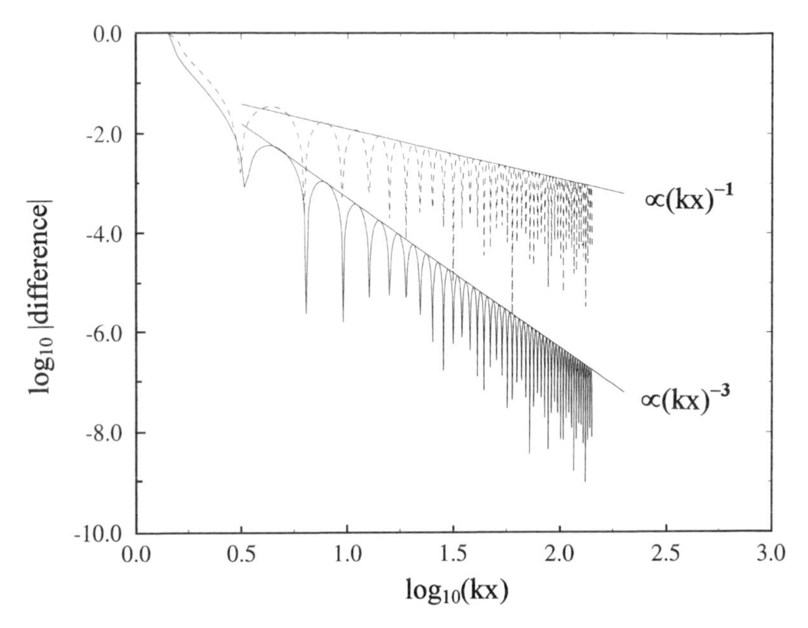

Figure 1.5. Accuracy of WKB wavefunctions in the classically allowed region of particle motion in the radial centrifugal potential. Plotted is the absolute value of the difference between the $\gamma = 2$ WKB wavefunction and a precise numerical wavefunction, as a function of kx. The upper (dashed) curve is for a reflective phase loss $\pi/2$ due to reflection and for the Langer-modified potential, with asymptotic accuracy of order $(kx)^{-1}$. The solid curve is for the exact asymptotic phase and no Langer modification, with improved asymptotic accuracy proportional to $(kx)^{-3}$. From Friedrich and Trost (1996). Copyright 1996 by the American Physical Society.

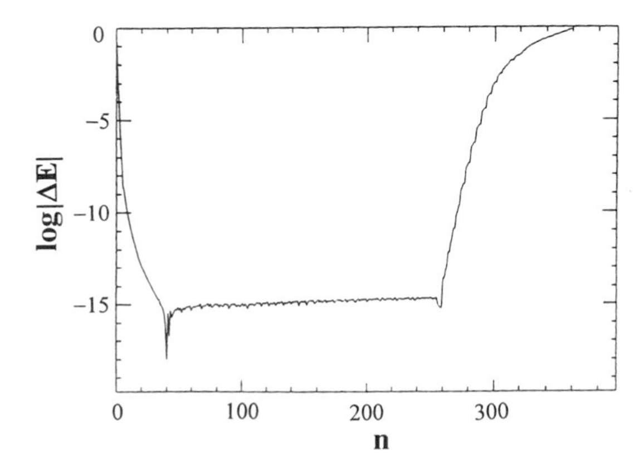

Figure 1.6. Accuracy of fourth-order semiclassical energies of the 1-d quartic oscillator. The absolute value of the difference from precise numerical results is plotted versus the quantum number n. Between $n = 0$ and $n = 40$, the difference is due to inaccuracy in the semiclassical energies. On up to $n = 260$, these numerical results continued to converge and the accuracy is good to 15 decimal places. From Bohigas et al. (1993). Copyright 1993, with permission from Elsevier Science.

numerical energies failed to converge. We see that modern semiclassical theory can give accurate results, except for the very lowest energy eigenstates.

1.3. PROBABILITY DISTRIBUTIONS IN PHASE SPACE

1.3a. Classical Phase-Space Portraits

After seeing the phase-space formulation of Hamilton's equations (Section 1.1), it is fair to ask why we need it. To answer this, we recall that time evolution of a specified 1-d classical system is not uniquely determined by the initial value of the coordinate x. An initial value of the coordinate's time derivative—essentially the momentum p—is also required. Referring to the lower panel of Fig. 1.3, we don't know whether or not we are uniquely following a desired trajectory in position space unless we also know one point on that trajectory's corresponding phase-space trajectory. That is, we need to know simultaneous values of x, p for some moment t. Thus ensembles of complete initial conditions evolve in phase space, not in coordinate space. Such ensembles are often used in studies that seek to summarize the overall behavior of a dynamical system.

We also note that quantum systems require initial states that have spreads in both x and p, in accord with Heisenberg's uncertainty principle. This fundamental restriction on the quantum phase-space distribution makes quantum studies of evolution in phase space even more essential. As we will see, autonomous 2-d systems are prototypes for studying classes of both classical and quantum evolution. Looking at projections of the probability distributions within the 4-D phase space of these systems reveals much about what can happen. But first we need to know how to construct the distributions and projections.

Let us consider the classical motion of a particle in a 2-d potential V(x, y). The full phase space is (x, y, p_x, p_y). Since the system is autonomous, the energy E is conserved, and evolution is constrained to some 3-D energy surface selected by the value of E. Because complicated 3-D motions are hard to display, we construct one or more *Poincaré sections*, also known as "phase-space portraits."

To begin an example of the construction, we define a cut through phase space by the constraints $x = 0$ and $H(x, y, p_x, p_y) = E$. We consider a 2-D subspace S_E—called the 2-D (y, p_y) Poincaré section—as the set of all (y, p_y) that satisfy both $H(0, y, p_x, p_y) = E$ and $p_x > 0$. The subspace S_E then lies within the (y, p_y) plane. We can find a set of points within S_E by following one trajectory of the system at energy E. We do this by numerically integrating Hamilton's equations (1.7), using the techniques discussed in Section A.2. As we iterate along in time, we watch for points where $x = 0$ with $p_x > 0$. When we find ourselves near such a point, we stop for a moment and numerically locate it to the desired accuracy. These points are $p_x > 0$ intersections of the trajectory with the (y, p_y) plane; see Fig. 1.7a. If the x and y motions are periodic with incommensurate periods, plotting these points will gradually fill in a curve within S_E; see Fig. 1.7b. Such curves are called *invariant curves*: once a trajectory produces one point on an invariant curve, it subsequently produces points only on the same curve. In the literature, invariant curves are often

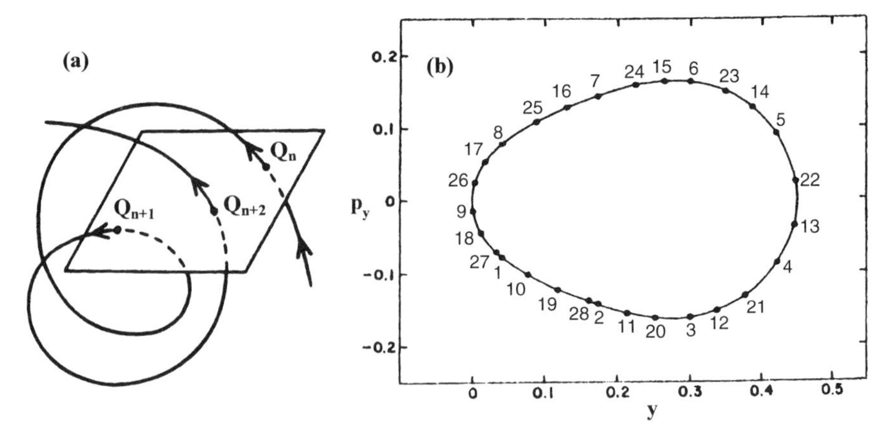

Figure 1.7. Starting the construction of a Poincaré section. (a) Trajectory intersections Q_n with a chosen surface of section are found numerically and are ordered in time by the integer n. (b) Some accumulated points of intersection are obtained for the Hénon–Heiles problem. From Hénon and Heiles (1964). Copyright 1964 American Astronomical Society.

called *phase curves*. Choosing another trajectory will produce another such curve. The uniqueness of classical trajectories requires that the two curves be different and not intersect.

Now let us consider the special, much-studied problem of the Hénon–Heiles potential $V(x, y) = \frac{1}{2}(x^2 + y^2) + x^2y - \frac{1}{3}y^3$, which is discussed further in Section 5.3. Figure 1.8 shows a few invariant curves in a Poincaré section for this system, taken at a low value of the energy. At the points marked "E" and "H," the periods for the x and y motions are commensurate. The commensurate condition causes sets of discrete points, rather than a curve, to appear in S_E.

If we were to follow all orbits from one crossing of the (y, p_y) plane to the next, we would map the Poincaré section onto itself. This discrete-time map is called a *Poincaré (first return) map*. Much can be learned about the continuous evolution of the system from the properties of such maps, especially if the map can be made explicit. For instance, the statistical averages over phase space obtained for the full evolution have several connections to the averages over the Poincaré section that can be obtained using the map. In addition, measures of the stability of orbits for the full problem can be related to measures of stability for the points the orbits produce in S_E, such as the points "E" and "H" of Fig. 1.8.

The special procedure for Poincaré sections we just used might not be optimal for displaying the phase-space structure of the evolutions of a different $N = 2$ system. For instance, to produce a bounded S_E, a change of coordinates can sometimes be useful. Other times, a generalization of the Poincaré procedure produces improvement. To perform this generalization, we first define a 3-D strobing surface that cuts suitably through the 4-D phase space. Let's call this strobe a 3-D *surface of section*. For now, let us assume the existence of a second constant of motion, recognizing that this is often not the case. By itself this second constant of motion defines a one-

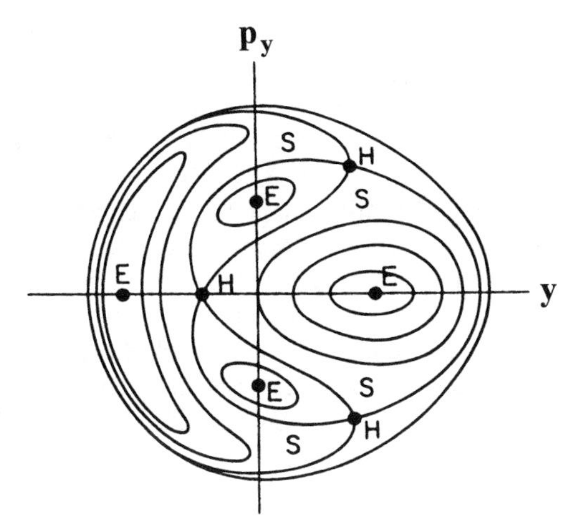

Figure 1.8. Some features of a (y, p_y) Poincaré section of phase space for the Hénon–Heiles problem, at a low particle energy. Stable fixed points are labeled by the letter "E" and unstable ones by "H." The letter "S" labels the separatrix curves between regions of qualitatively different classical evolution. From *Computational Physics—Fortran Version* by Steven Koonin and Dawn Meredith. Copyright 1990 by Addison Wesley Publishing Company. Reprinted by permission of Perseus Books Publishers, a member of Perseus Books, L.L.C.

parameter family of 3-D subspaces in phase space. The intersection of the 3-D energy surface—jointly with the surface of section and with the subspaces—results in a set of 1-D curves. In order to visualize them, these curves in the full phase space can be projected onto one or more planes. If we have relatively simple topologies for the energy surface, the surface of section, and the subspaces, the resulting curves do not cross. Otherwise the curves can cross.

1.3b. Initial Quantum States and Their Classical Analogs

A particular time evolution of a quantum system (having Hamiltonian *H*) depends greatly on the phase-space probability distribution of the initial quantum state. We often want to know a corresponding evolution of the classical system (having Hamiltonian H) to help us understand the quantum evolution. How should we choose the initial classical state?

Sometimes an answer might seem remote. In phase-space representations of quantum mechanics, the wavefunction and the Hamiltonian may be functions of new momentum and coordinate operators P, Q, where P, Q are not the physical quantized classical variables p, q. For instance, we can satisfy most of the mathematical requirements of a quantum mechanical representation by choosing

$$P = \frac{p}{2} - i\hbar\frac{\partial}{\partial q} \quad \text{and} \quad Q = \frac{q}{2} + i\hbar\frac{\partial}{\partial p}; \tag{1.28}$$

see the discussion in Section 1.3d and Torres-Vega and Frederick (1993). Independent of the possibility of introducing such operators, a probability distribution $\rho(\mathbf{q}, \mathbf{p}, t)$ should physically exist in phase space. This distribution always satisfies a differential equation called the quantum Liouville equation (Wilkie and Brumer, 1997a,b).

In general, the quantum Liouville equation is difficult to solve directly, either analytically or numerically; hence we do not pursue this equation here. Nevertheless, this equation is noteworthy because it has an analog in classical physics, not surprisingly called the classical Liouville equation. This classical equation describes the evolution of *ensembles of initial conditions* in phase space. Such an ensemble can be chosen to have the same phase-space probability density distribution function $\rho_{c\ell}(\mathbf{q}, \mathbf{p}, t)$ as a chosen particular quantum mechanical distribution $\rho_{QM}(\mathbf{q}, \mathbf{p}, t)$. Thus we see there is a good theoretical basis for comparing evolutions of quantum and classical phase-space distributions that have the same initial probability distribution functions. Initially at least, the classical evolution contains the effects of the uncertainty principle, partially incorporated in the widths $\Delta\mathbf{q}$ and $\Delta\mathbf{p}$ chosen for $\rho_{c\ell}(\mathbf{q}, \mathbf{p}, t)$, widths that are matched to those of $\rho_{QM}(\mathbf{q}, \mathbf{p}, t)$. As time goes on, of course, the classical evolution will not display the manifestly quantum effects of interference and tunneling. Indeed, exploring these particular effects is one purpose of comparing quantum and classical evolutions.

Alternatively, if we want a quantum–classical comparison to display *all* the quantum–classical differences, then we should choose a classical initial distribution function that includes *only* the classical constraints. Such comparisons are often made. In such a case, a complete matching of classical and quantum initial distributions is not carried out. For instance, if an initial quantum state happens to be an energy eigenstate and there are no other conserved quantities, then the corresponding classical state would involve a set of initial conditions that are uniformly distributed on the infinitely sharp energy surface in phase space. The classical distribution is taken to have no "thickness" locally perpendicular to the energy surface, even though the quantum distribution does have such a thickness. This approach to quantum–classical comparisons works with the full "sharpness" of classical features in phase space. See Fig. 8.10 for an example of this type of classical trajectory ensemble evolution.

A quantum system has an enormous number of possible initial states that can be evolved by using Schrödinger's equation. In the end we want initial states that can be prepared in the laboratory. There are, however, several types of initial states that have particularly simple evolutions. While hard for experimenters to produce, these states are very useful in theoretical work. Among these "simple" initial states are the *coherent states* of a system (Klauder and Skagerstam, 1985; Perelomov, 1986). In one sense or another, these initial states are most close to classical.

Another type of simple initial state is the Gaussian wavefunction. This initial quantum state has been used in the study of many systems (see Sections 2.1c, 2.4b–d, 5.3d, 6.1b, 7.4d, and 9.2b). We note that in the case of the harmonic oscillator problem, the Gaussian wavefunction is a coherent state.

Let us now investigate the evolution of a single Gaussian "wavepacket," a wavefunction given this name because of its relatively tight (and adjustable) localizations in **q** and **p**. Such packets satisfy Heisenberg's uncertainty expression $\Delta q_i \, \Delta p_i \geq h$, with the equality condition satisfied usually at only one moment in time. Often they are called minimal uncertainty states.

1.3c. Evolutions of Wavepackets and Classical Ensembles

It is instructive to see how a 1-d wavepacket evolves under the influence of various Hamiltonians. We will construct a Gaussian wavepacket (in the Schrödinger representation) in terms of the solutions of the free-particle problem. In 1-d, these wavefunctions are

$$\psi_p(x, t) = \frac{1}{\sqrt{2\pi\hbar}} \exp\left(-\frac{i}{\hbar}(Et - px)\right), \qquad E = \frac{p^2}{2m}. \tag{1.29}$$

Clearly these waves have the phase velocity $v_p = E/p = p/2m$. Linear superpositions of such waves also satisfy the free-particle Schrödinger equation, including the continuous sum

$$\psi(x, t) = \int_{-\infty}^{\infty} f(p)\psi_p(x - x_0, t)\, dp. \tag{1.30}$$

We see that this sum is determined by the *spectral function* f(p) that weights the different moment p to determine the composition of the packet. The name "spectral function" comes from the fact that f(p) is the Fourier transform of $\psi(x, 0)$,

$$f(p) = \frac{1}{\sqrt{2\pi}} \int_{-\infty}^{\infty} \psi(x, 0) \exp\left(-i\frac{px}{\hbar}\right) dx,$$

just as energy spectra come from Fourier transforms of time evolutions; see Section A.5a.

We choose f(p) to be the normalized Gaussian function of p, with mean momentum $\langle p \rangle$ and momentum width σ_p,

$$f(p) = \frac{1}{(2\pi)^{1/4}\sqrt{p}} \exp\left[-\frac{(p - \langle p \rangle)^2}{4\sigma_p^2}\right]. \tag{1.31}$$

Using Eq. (1.31), we can analytically evaluate integral (1.30) to obtain a *Gaussian wavepacket*

$$
\begin{aligned}
\psi(x, t) &= A(x, t) \exp[i\phi(x, t)], \\
A(x, t) &= \frac{1}{(2\pi)^{1/4}\sqrt{\sigma_x(t)}} \exp\left(-\frac{(x - x_0 - v_0 t)^2}{4\sigma_x^2(t)}\right), \\
\phi(x, t) &= \frac{1}{\hbar}\left[\langle p\rangle + \frac{\sigma_p^2}{\sigma_x^2(t)}\left(\frac{t}{2m}\right)(x - x_0 - v_0 t)\right](x - x_0 - v_0 t) \\
&\quad + \frac{\langle p\rangle}{2\hbar} v_0 t + \frac{\alpha(t)}{2}.
\end{aligned}
\tag{1.32}
$$

Here $\alpha(t) = \tan^{-1}(2\sigma_p^2 t/m\hbar)$. The group velocity of the packet is $v_0 \equiv \langle p\rangle/m$, and its spatial size grows with time as follows:

$$
\sigma_x(t) - \frac{\hbar}{2\sigma_p}\sqrt{1 + \frac{4\sigma_p^4}{\hbar^2}\left(\frac{t^2}{m^2}\right)}.
$$

From Eq. (1.32) we see that the packet's probability density is

$$
P(x, t) = A^2(x, t) = \frac{1}{\sqrt{2\pi}\sigma_p} \exp\left(-\frac{[x - (x_0 + v_0 t)]^2}{2\sigma_x^2(t)}\right).
$$

Note that $\psi(x, t)$ is properly normalized:

$$
\int_{-\infty}^{\infty} P(x, t)\, dx = 1.
$$

Our Gaussian wavepacket has a minimum uncertainty product only at the initial time $t = 0$, where $\sigma_x(0)\sigma_p = \hbar/2$. Note also that at large enough t, the probability at a fixed spatial point x decays exponentially with time. All in all, looking at Eq. (1.32), we see that the spatial evolution of this Gaussian wavepacket is not completely trivial, even though the expectation values $\langle x(t)\rangle$, $\langle p(t)\rangle$ are classical. And this evolution is under the influence of the simplest classical Hamiltonian, the constant $\langle p\rangle^2/2m$.

How do we find the wavefunction of an initially Gaussian 1-d wavepacket under the influence of a spatially varying, time-independent Hamiltonian *H*? First we solve $H\psi_E = E\psi_E$ analytically, or numerically if need be. We can then use an expansion similar to Eq. (1.4) to evolve our system, once the constant weighting coefficients C_{En} are known:

$$
\Psi(x, t) = \sum_E c_E \psi_E(x) \exp\left(-i\frac{Et}{\hbar}\right).
\tag{1.33}
$$

For convenience, we have dropped the subscript "n." Putting $t = 0$ into Eq. (1.33) must produce our initial Gaussian wavepacket, which then requires that

$$\psi(x, 0) \equiv \frac{1}{(2\pi)^{1/4}\sigma_{x0}} \exp\left(-\frac{(x - x_0)^2}{4\sigma_{x0}^2}\right) \equiv \sum_E c_E \psi_E(x). \tag{1.34}$$

Equation (1.34) expands the initial packet in terms of the time-independent energy eigenstates of our system, whatever that system may be. Let us assume the ψ_E are an orthonormal basis set, multiply Eq. (1.34) by ψ_E^*, and integrate over space. We then obtain an expression for computing the weighting coefficients:

$$c_E = \int_{-\infty}^{\infty} \psi_E^*(x)\psi(x, 0)\, dx. \tag{1.35}$$

These particular coefficients are projections of the initial Gaussian wavepacket onto the time-independent part of the energy eigenstates. With these coefficients, evaluation of Eq. (1.33) gives the desired time evolution, analytically or numerically if necessary.

As a simple example of this procedure to determine Gaussian wavepacket evolution, we follow Brandt and Dahmen (1994) and consider the harmonic oscillator, where $H = p^2/2m + m\omega^2x^2/2$. The expression $E = (n + \frac{1}{2})\hbar\omega$ then relates the energy E to the quantum number n. The needed harmonic oscillator energy eigenfunctions are given by

$$\psi_n(x) = \frac{(\omega m/\pi\hbar)^{1/4}}{(2^n n!)^{1/2}} \exp\left[-\left(\frac{\omega m}{\hbar}\right)\left(\frac{x^2}{2}\right)\right] h_n\left(\sqrt{\frac{\omega m}{\hbar}}x\right).$$

Here the functions h_n are Hermite polynomials, which are finite polynomials of order n. These functions satisfy the recurrence relation $2yh_n(y) = 2nh_{n-1}(y) + h_{n+1}(y)$ and also the identity

$$\exp(-z^2 + 2zy) = \sum_{n=0}^{\infty} h_n(y)\frac{z^n}{n!}.$$

With the help of this last expression, Eq. (1.33) can be evaluated to give the wavefunction of our Gaussian wavepacket under the influence of the harmonic oscillator potential. When this wavefunction is squared, it gives the probability density

$$P(x, t) = \frac{1}{\sqrt{2\pi}\sigma(t)} \exp\left(-\frac{(x - x_0 \cos\omega t)^2}{2\sigma^2(t)}\right).$$

This is a Gaussian wavepacket whose mean location moves along the reference trajectory $x = x_0 \cos \omega t$ of the classical motion. In units of the oscillator ground state width $\sqrt{\hbar/m\omega}$, this packet's width is

$$\sigma(t) = \frac{[4r^4 + 1 + (4r^4 - 1)\cos(2\omega t)]^{1/2}}{2\sqrt{2}},$$

where r is the initial packet width σ_{x0} expressed in the same oscillator ground state units. We see that the width of the oscillator wavepacket oscillates at angular frequency 2ω, rather than growing in time; see Fig. 1.9. This nondispersive (nonspreading) wavepacket evolution is a special property of the harmonic oscillator problem. (In some more complicated systems, however, conditions exist where nondispersive evolution also occurs.) When $r = 1/\sqrt{2}$, the width σ is independent of time—a special case that constitutes an example of a coherent state.

For other 1-d systems, Gaussian wavepacket evolution exhibits spreading of the packet and other interesting effects. The reader might enjoy running the computer programs described by Koonin and Meredith (1990) and by Brandt and Dahmen (1994).

1.3d. Quantum Phase-Space Distributions

In classical mechanics, we've seen that the dynamics of an autonomous system is embodied in a phase-space flow. In principle, this flow can be determined using the phase-space theory of Hamiltonian systems, but numerical computations are easier

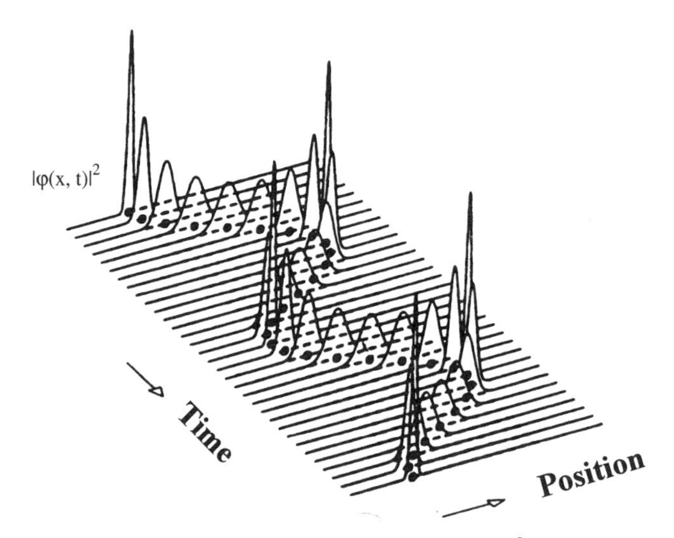

Figure 1.9. Time evolution of the probability distribution $|\psi(x, t)|^2$ for a quantum wavepacket in a 1-d harmonic oscillator potential. From Brandt and Dahmen (1994). Copyright 1994 Springer-Verlag. Also see Brandt and Dahmen (1995).

to carry out. As already noted, for 2-d systems this flow is nicely probed by the numerical construction of Poincaré sections. It is natural to seek quantum mechanical analogs for these flows and sections. The oldest approach is to construct the *Wigner quasiprobability distribution function* $W_\psi(\mathbf{q}, \mathbf{p})$ from the N-dimensional spatial wavefunction $\psi(\mathbf{q})$, using the definition

$$W_\psi(\mathbf{q}, \mathbf{p}) \equiv (2\pi\hbar)^N \int \psi\left(\mathbf{q} - \frac{\mathbf{r}}{2}\right)\psi^*\left(\mathbf{q} + \frac{\mathbf{r}}{2}\right) \exp\left(i\frac{\mathbf{p}\cdot\mathbf{r}}{\hbar}\right) d^N r. \tag{1.36}$$

We recognize this as the Fourier transform of a particular spatial correlation function. The Wigner function W_ψ can be shown to have desirable analytical properties, for its projections into both coordinate and momentum space give the correct quantum momentum and position distributions:

$$\int d\mathbf{q} W_\psi(\mathbf{q}, \mathbf{p}) = |\phi(\mathbf{p})|^2, \qquad \int d\mathbf{p}\, W_\psi(\mathbf{q}, \mathbf{p}) = |\psi(\mathbf{q})|^2$$

(Hilliary et al., 1984). However, $W_\psi(\mathbf{q}, \mathbf{p})$ itself exhibits oscillations that often dip into ranges of negative values, something that is unacceptable for a probability density distribution function. For examples of such regions of negative values see Takahashi (1989). We can smooth out these negative regions by averaging over a "coarse graining" function f. Thus we obtain a *Huisimi distribution function*:

$$H_\psi(\mathbf{q}, \mathbf{p}) = \int W_\psi(\mathbf{q}', \mathbf{p}') f(\mathbf{q}, \mathbf{p}; \mathbf{q}', \mathbf{p}')\, d^N p'\, d^N q'. \tag{1.37}$$

One pays a price for positive definiteness here by losing the uniqueness desired in rigorous theory. At least to date, there is no theory for a "correct" function f. Nevertheless, the calculation of Huisimi functions has proved most useful in revealing quantum structure in phase space. Usually the function f is taken to be the harmonic oscillator coherent state (which we have noted represents a minimal-uncertainty Gaussian wavepacket), now constructed in phase space:

$$f(\mathbf{q}, \mathbf{p}; \mathbf{q}', \mathbf{p}') = (\pi\hbar)^{-N} \exp\left(\frac{-(\mathbf{q} - \mathbf{q}')^2/\sigma_q^2 + \sigma_q^2(\mathbf{p} - \mathbf{p}')^2}{\hbar}\right). \tag{1.38}$$

Here σ_q is the Gaussian spatial width parameter, which is free to be chosen and which determines the relative resolution in **p**-space versus **q**-space.

Let us look at an example of a Huisimi function for an energy eigenfunction of a 1-d model hydrogen atom, shown in (x, p_x) space in Fig. 1.10. The potential for such an atom is defined by $V(x) = (-1/x)$, $x > 0$ and by $V(x) = \infty$, $x \leq 0$. We see the expected ridges. On average, the sum of kinetic and potential energies on the ridges approximately gives the total energy. Oscillations along the ridges reflect oscillations we know occur in the spatial wavefunction. Also as expected, the oscillations in the

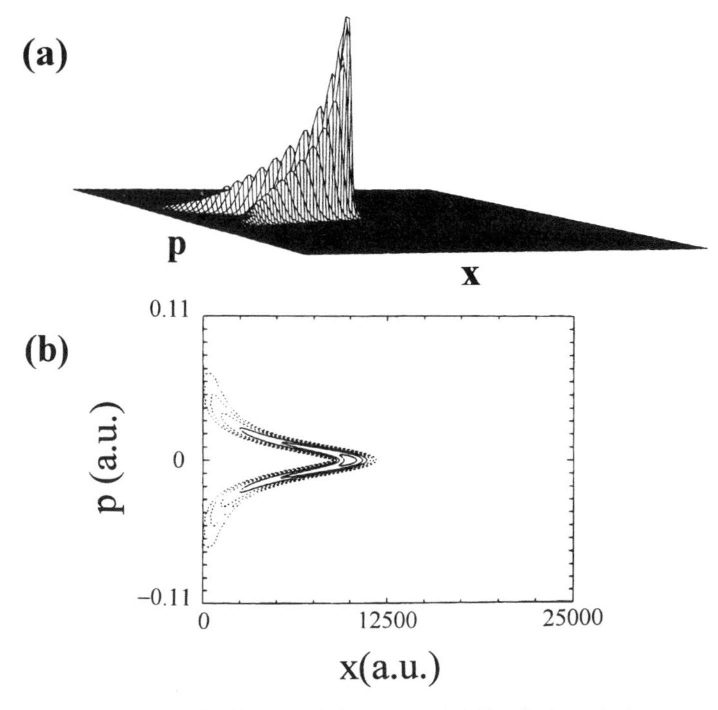

Figure 1.10. The Huisimi distribution of electron probability in (x, p_x) phase space, for the energy eigenstate of the unperturbed 1-d hydrogen atom having quantum number $n = 72$. From Chu and Jiang (1991). Copyright 1991, with permission from Elsevier Science.

figure are largest near the classical outer turning point, where $p_x = 0$, in the region where the orbiting electron spends most of its time.

For useful values of the packet width σ_q, Huisimi probability densities for stationary 1-d problems generally are concentrated close to the classical trajectory in phase space. In the semiclassical regime, each of the high-density regions can be approximated by a Gaussian wavefunction having its mean location on the trajectory. This wavefunction's width is proportional to the square root of the quantum parameter $\sim\hbar$ of the system, and the wavefunction's maximum probability density is inversely proportional to the classical phase-space velocity defined in Section 1.1 Additionally, in this regime, the zeros of H_ψ lie on simple curves, with mean spacing proportional to the quantum parameter of the system (Leboeuf and Voros, 1990, Leboeuf, 1991). There is a phase-space generalization of the counting theorem for the zeros of 1-d wavefunctions in coordinate space (Kurchan et al., 1989). The generalized theorem gives the number of zeros enclosed by the classical trajectory. Numerical studies of the zeros for systems with spatial dimensions larger than one have been reported (Arranz et al., 1996).

With the help of the Huisimi function (1.37), we can also construct a *quantum Poincaré section*. Classically we saw that a Poincaré section for an autonomous system is constructed by eliminating one variable on account of energy conservation and by taking a fixed value of its canonical conjugate variable. The dimension of phase space is lowered by two, and the sections are labeled by the energy. We can obtain a similar quantum mechanical construction by integrating the Huisimi distribution over one variable, while fixing another one that defines the surface of section. For example, in the 2-d Hénon–Heiles problem mentioned earlier, a quantum x Poincaré section would be a probability distribution in the variables (y, p_y) given by $\int H_\psi(x = \text{constant}, p_x, y, p_y)\, dp_x$. An alternative construction of a quantum Poincaré section is to initially restrict the distribution to the energy surface to obtain simply $H_\psi[x, p_x(E, x, y, p_y), y, p_y]$, again with $x = \text{constant}$. When a system is strongly semiclassical, the Huisimi function is sharply peaked on the energy shell, and the two constructions give very similar results. We'll find quantum Poincaré sections to be quite revealing (see Fig. 5.25, for instance).

However, Eqs. (1.36) and (1.37) are still useful when there is time dependence. To this point we've discussed quantum phase-space distributions, primarily with time-independent problems in mind. We now turn to our main topic—time-dependent problems. We begin here by considering the time evolution *in phase space* of initial Gaussian wavepackets. We can use the recently proposed phase-space representation $P = p/2 - i\hbar\, \partial/\partial q$ and $Q = q/2 + i\hbar\, \partial/\partial p$ of Eq. (1.28) and employ a phase-space version of the FFT split propagation operator numerical technique described in Section B.1b.2. As in the coordinate or momentum Schrödinger representations, the time evolution is governed by a quantum propagation operator $\exp(-itH/\hbar)$, which is a formal solution of a phase-space version of Schrödinger's equation (1.1). In the case of the particle in a 1-d potential, a phase-space representation of this operator is

$$\exp\left\{-\frac{i}{\hbar}t\left[\frac{1}{2m}\left(\frac{p}{2} - i\hbar\frac{\partial}{\partial q}\right)^2 + V\left(\frac{q}{2} + i\hbar\frac{\partial}{\partial p}\right)\right]\right\}. \tag{1.39}$$

The potential V is now a function of the quantum operator $q/2 + i\hbar\, \partial/\partial p$. As shown in Torres-Vega and Frederick (1993), the operator (1.39) arises from the formal phase-space Schrödinger equation

$$i\hbar\frac{\partial}{\partial t}\Psi(p, q, t) = \left[\frac{1}{2m}\left(\frac{p}{2} - i\hbar\frac{\partial}{\partial q}\right)^2 + V\left(\frac{q}{2} + i\hbar\frac{\partial}{\partial p}\right)\right]\Psi(p, q, t),$$

where $\Psi(p, q, t)$ resides in the space of square integrable functions of two variables rather than one variable. The choice of the propagation operator (1.39) and the definitions of Eq. (1.28) produce a phase-space representation where (1) the commutator relation $[Q, P] = i\hbar I$ is satisfied; (2) the diagonal matrix elements of the quantum Liouville equation correspond to the classical Liouville equation; and (3) the operators Q, P are "symmetric," with the coefficients of q and p both equal to $\frac{1}{2}$ and the magnitudes of the coefficients of $\partial/\partial q$ and $\partial/\partial p$ both equal to $i\hbar$. This last condition is to place q and p on equal footing, as is the case for q and p for classical phase space. The usual coordinate- and momentum-space Schrödinger equations can be recovered by Fourier projections (Torres-Vega and Frederick, 1993).

If we apply a split-operator approximation to expression (1.39) for a small timestep Δt (see Section B.1b.2 and Feit et al., 1982), an expression can be produced that is very convenient for numerical work (Torres-Vega and Frederick, 1991). Figure 1.11 shows some results for the case of V(q) taken to be the step potential. During the time that the packet "collides" with the step, we see how the wavepacket in phase space splits into the expected incident, transmitted, and reflected parts. There is a part of the initial wavepacket whose momentum is near zero, and thus is slow moving. A second part smoothly changes its direction as it is reflected from the step, where the second part combines with the first part. A third part has enough energy to overcome the step-potential barrier and continues moving past the step $(q = 0)$ with reduced positive momentum p. Note that the quantum phase-space distribution contains a great deal of clarifying detail. These details are not as easily detectable in either the coordinate or momentum distributions alone; these distribu-

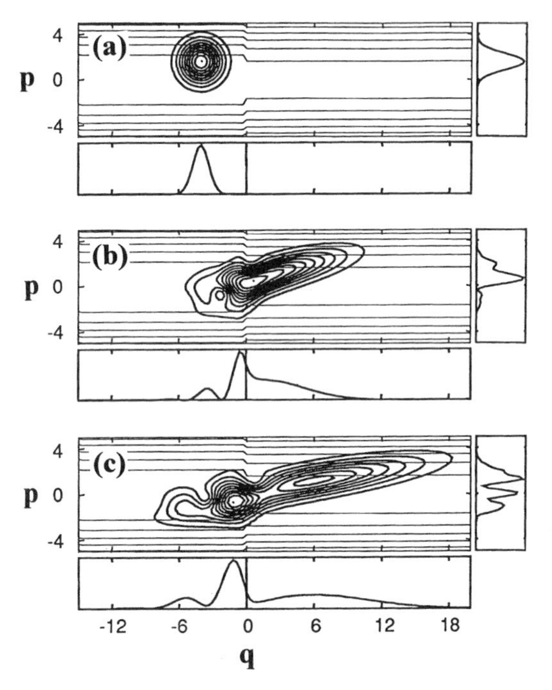

Figure 1.11. Snapshots of quantum wavepacket scattering from a 1-d step potential at position $q = 0$. Dimensionless scaled quantities are used. The mean energy of the initial Gaussian packet is 1.26 times the step height. The thin horizontal line segments are contour plots of the energy surface. Three snapshots of the evolution show the square magnitude of the phase-space wavefunction: (a) before any significant scattering, (b) during the middle of the scattering, and (c) toward the end of the scattering. Square magnitudes of coordinate and momentum wavefunctions are obtained from projections of the wavefunction onto the coordinate and momentum spaces, respectively; for each panel, these projections are shown below and to the right. From Torres-Vega and Frederick (1991). Copyright 1991 by the American Physical Society.

tions are the square magnitudes of the appropriate Fourier projections of the phase-space wavefunction $\Psi(p, q, t)$.

For a comparison, Fig. 1.12 shows snapshots in phase space of 10,000 classical trajectories having initial conditions distributed in phase space according to a Gaussian density. This density is given by the initial quantum wavepacket probability distribution used for Fig. 1.11. The similarities in the quantum and classical evolutions are remarkable, even though they differ because of quantum interference and tunneling effects. As we will see in forthcoming chapters, such early-time similarities in phase-space evolution can occur for numerous systems. Such quantum–classical similarities also occur for stationary problems—for example, in the phase-space distributions of energy eigenstates. These similarities can occur quite far down toward the bottom of the energy spectrum (see Fig. 4.4 for an example).

At present, just as for the Wigner and Huisimi phase-space distributions, fundamental ambiguities exist that concern the physical interpretation of quantities obtained using expression (1.39) and its phase-space Schrödinger equation. The

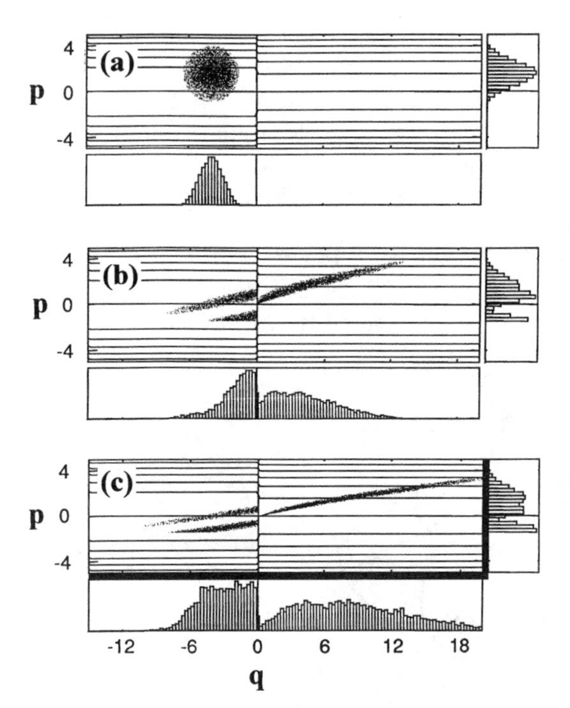

Figure 1.12. Classical scattering of a set of 10^4 particles from the step potential of Fig. 1.11. The initial phase-space distribution was Gaussian, like the wavepacket case. Below and to the right of each snapshot are plotted the coordinate and momentum densities, that is, the averages over the momentum and over the coordinate, respectively. Note the similarities and differences with the quantum distributions and wavefunctions shown in Fig. 1.11. From Torres-Vega and Frederick (1991). Copyright 1991 by the American Physical Society.

difficulty here is that the solution to this phase-space equation is not unique. Although expectation values of quantum operators are unique, transition matrix elements are not (Harriman, 1994; Møller et al., 1997). This issue has generated an interesting newly proposed formulation of quantum mechanics in phase space—a "relative-state" formulation where a reference system is introduced that depends on what we would like to know about our physical quantum system (Ban, 1996).

In this book we will almost always use Huisimi distributions to display probability distributions in phase space, but occasionally a Wigner distribution will be used.

1.4. CLASSICAL DYNAMICS ON TORI AND SEPARATRICES

1.4a. Periodic and Quasiperiodic Orbits on Tori

In the field of nonlinear dynamics, three basic kinds of evolution occur: periodic, quasiperiodic, and chaotic. (Discussion of chaotic evolution is postponed to Section 5.2, as we only begin to need it in Chapter 5.)

If we admit the possibility of an infinite period, the trajectories of 1-d spatially bounded autonomous systems are all periodic (see the example trajectory in Fig. 1.3). To obtain quasiperiodic evolution, we must go beyond these simplest systems to something that has a phase space of at least three dimensions. For many of these systems, a good number of the orbits are no longer periodic. Although some periodic orbits will still appear, now the typical orbit is quasiperiodic whenever there is no widespread chaos.

We again consider 2-d autonomous systems. If a second constant of the motion exists, phase space is filled with 2-D surfaces that contain the possible trajectories, as we will see in Section 3.3. These surfaces have the topology of a torus; see Fig. 1.13.

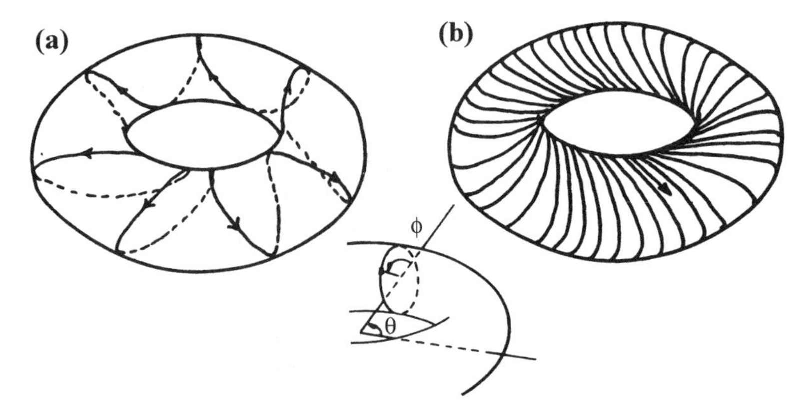

Figure 1.13. Classical trajectories on a torus: (a) a periodic orbit and (b) a quasiperiodic orbit. Reprinted with permission from Berry (1978). Copyright 1978 American Institute of Physics. (The inset shows the two angles θ and ϕ that locate points on the torus.)

We note in passing that for systems with N degrees of freedom, there can be N-dimensional tori in phase space, *N-D tori*, labeled by N constants of motion.

Referring to the inset in Fig. 1.13, suppose that an orbit evolves on the surface of the torus according to $\theta = 2\pi\nu_\theta t$ and $\phi = 2\pi\nu_\phi t$. The quantities ν_θ and ν_ϕ are the linear frequencies for the θ and ϕ evolutions, respectively. The orbit returns to the plane $\theta = 0$ (taken to be the surface of section) whenever $2\pi\nu_\theta t = 2n\pi$, with n a positive integer. At the return time $t = n/\nu_\theta$, the second angle variable has the value $\phi = 2\pi\nu_\phi(n/\nu_\theta)$. Based on these observations, the Poincaré return map is given by

$$\phi_{n+1} = \phi_n + 2\pi\frac{\nu_\phi}{\nu_\theta}.$$

Clearly the orbit will repeat itself, meaning it will be a *periodic orbit*, if and only if the ratio ν_ϕ/ν_θ is a rational number. A periodic orbit produces a finite set of fixed points in the Poincaré section. The frequency ratio ν_ϕ/ν_θ is called the *orbit winding number*, which for the moment we write as q/p, where q and p are coprime integers. (Let us not confuse these integers with canonical variables, which have the same symbols.) For a rational winding number, the orbit winds p times around the torus in the θ way and q times in the ϕ way, before it returns to the starting point (see Fig. 1.13a).

If the winding number is irrational, then our orbit is said to be a *quasiperiodic orbit*, having the two fundamental incommensurate frequencies (ν_θ, ν_ϕ). In this case the orbit never returns to the starting point—although one can show that if we wait long enough, the orbit does get as close as we want to returning at some point in time. As shown in Fig. 1.13b, the trend is to a full covering of the torus by a single quasiperiodic orbit. This necessarily produces the filling-in of a continuous curve in the Poincaré section, as indicated in Figs. 1.7b and 1.8. Indeed, the curves shown in Fig. 1.8 are of this continuous type, except for the indicated fixed points marked "E" and "H."

We should remember that our 2-D torus is embedded in a 4-D phase space, unlike an ordinary 2-d torus lying in 3-d position space. Because of this difference, the projection of our torus onto a plane, such as the 2-d position space (x, y), can look surprising. While projections (a), (b), and (c) in Fig. 1.14 could be expected,

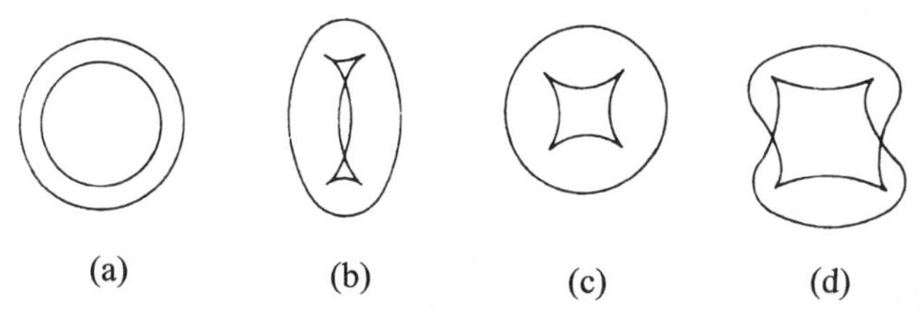

Figure 1.14. Boundaries of possible projections onto a plane, for a 2-D torus residing in a 4-D phase space. From Ozorio de Almeida and Hannay (1982). Copyright 1982 Academic Press, Inc.

projection (d) is not. Just as we expect for an ordinary torus, all these projections fill in portions of the plane, to form a bounded region with projected "layers" on it. The boundary of this bounded region is comprised of portions of *caustic curves*, which are the edges of the layers. These caustics are what is actually shown in Fig. 1.14. (For a simple example of a caustic, turn to Fig. 7.20.) The projection of the torus onto position space is a density distribution within that space, for the set of points on the torus. The number of layers contributing to the density distribution will change as a caustic curve is crossed. Caustics are also called *turning surfaces*, because they are generalizations of the turning points of 1-d systems.

1.4b. The Separatrix

A *separatrix* curve or surface in phase space is what its name suggests—it divides its surrounding phase space into parts that contain qualitatively different kinds of evolution. Fortunately, we can discuss the separatrix while working in 1-d.

We choose an important system as an example: the vertical pendulum in a uniform gravitational field. This system is often called the "classical rotor." (This should not be confused with free or horizontal rotors.) Let L be the length of the pendulum, and let θ be the angle the pendulum makes relative to the downward vertical, the direction of the force due to gravity. The Hamiltonian for the pendulum is $H = T + V = p_\theta^2/2mL^2 - mgL\cos\theta$, where $p_\theta = mL^2\, d\theta/dt$. Hamilton's equations then become

$$\begin{aligned} \frac{d\theta}{dt} &= \frac{p_\theta}{mL^2}, \\ \frac{dp_\theta}{dt} &= -mgL\cos\theta. \end{aligned} \tag{1.40}$$

Equations (1.40) are easily integrated—using the numerical techniques of Section A.2b—to obtain points (θ, p_θ) on the invariant curves for various values of the energy E. Some results are shown in Fig. 1.15b. At low energies one observes near-ellipses centered at $\theta = 0$ and at other values $\theta = \pm 2n\pi$. The duplications of the near-ellipses are the result of the periodicity of the potential V. Many of these nearly elliptical invariant curves span relatively small ranges in θ. These smaller ellipses are approximately those of the simple vibrating (librating) pendulum, where $\sin\theta_{max}$ can be approximated by θ_{max}.

On the other hand, at large enough total energies we expect the kinetic energy to be positive, even when the pendulum is upside down at $\theta = \pi$. At such energies the pendulum rotates continuously in one direction or the other, following the θ-spanning oscillating curves located at large $|p_\theta|$ in Fig. 1.15b, and following the near-circles on the cylinder in Fig. 1.15c. The pendulum is now a rotor. So we see that the phase space of the vertical pendulum contains regions of three different types of motion: clockwise rotational, librational, and counterclockwise rotational.

The separatrices between the regions of different motion pass through points of unstable equilibrium, for instance, at $\theta = \pi$ with $p_\theta = 0$. Because we've chosen the zero of potential energy to be $\theta = 0$ when the pendulum hangs straight down, the

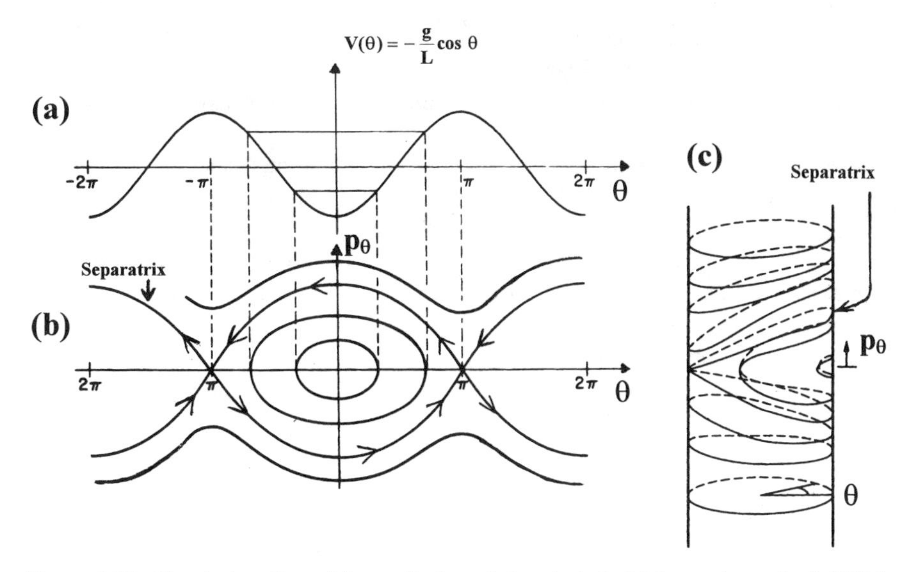

Figure 1.15. Classical motion of the vertical pendulum (rotor): (a) the cosine potential V(x), (b) a few invariant phase-space curves, and (c) invariant phase-space curves drawn on the surface of the (θ, p_θ) cylinder. Panel (a) from Tabor (1989). Copyright 1989 John Wiley & Sons, Inc. Panels (b) and (c) from Rasband (1990). Copyright 1990 John Wiley & Sons, Inc. Reprinted by permission of John Wiley & Sons, Inc.

total energy at such an unstable point is $E = mgL$. The $E = mgL$ is an equation for all the points on the separatrix. As the separatrix is an invariant curve of constant energy, it must be a trajectory. Figures 1.15b and 1.15c show the separatrices, with their points explicitly determined by $p_\theta^2/(2mL^2) - mgL(1 + \cos\theta) = 0$.

The pendulum problem can be addressed analytically with the help of special functions, namely, elliptic integrals and Jacobi elliptic functions. Some details can be found in Tabor (1989) and in Lichtenberg and Lieberman (1992). Included is a result for the time evolution on the separatrix, $\theta(t) = 2\sin^{-1}\{\tanh[t(g/L)^{1/2}]\}$. This result suggests a situation that is general: the approach along the separatrix to the unstable fixed point is exponentially slow, and it takes an infinite time to get there. Needless to say, we should be careful with numerical work close to points of unstable equilibrium.

1.5. INTRODUCTION TO QUANTUM–CLASSICAL CORRESPONDENCE (ADVANCED TOPIC)

Earlier in this chapter we noted specific examples where connections exist between the time evolution of a quantum system and the time evolution of a corresponding classical system. Since an understanding of classical evolution can greatly elucidate quantum evolution, quantum–classical correspondence will reappear almost countless times in subsequent chapters. Let us summarize some things known about such correspondences.

First, there is the obvious connection between a quantum system and a classical system through the set {A} of classical dynamical variables, and the corresponding set $\{A\}$ of quantum mechanical operators. The set $\{A\}$ is obtained from {A} by means of the quantization rules. For a particle in a 1-d potential, {A} includes the quantities x, p, E, kinetic energy T, potential energy V, the derived force F, all powers of x and of p, and so on.

In quantum mechanics, classical conjugate variables—which appear in pairs such as (x, p) or (t, E)—become pairwise complementary, opposing, and interwoven. The members of each pair are "complementary" both because they are classically conjugate and because the uncertainty principle is concerned with them as pairs. Pair members are "opposing" because of the trade-off in their uncertainties that is contained in the uncertainty principle. They are "interwoven" because, for example, phase relations in p-space determine the quantum system's x-distribution, and phase relations in x-space likewise determine the p-distribution. While forming the Gaussian wavepacket, we have explicitly seen these connections between phase relationships and distributions. We can summarize this paragraph by saying that the dynamical aspects of a quantum system, such as p (or E), are complementary to their corresponding space-time aspects, x (or t).

In addition, *complementarity* exists at higher levels in quantum mechanics, as was noted by N. Bohr (Bohr, 1961). We are familiar with complementarity in the nature of matter, where its wave aspects and particle aspects are again interwoven and opposing potentialities. A similar complementarity exists in time evolution itself, between continuous and discontinuous aspects. On the one hand, the wavefunction obtained from first-quantization (and those quantities derived from it) are continuous functions of time. Thus quantum time evolution seems predictable. To a good approximation, it can be controlled. Yet, when the same quantum system is second-quantized, its evolution clearly takes on discontinuous aspects. Now change involves quantum jumps on energy ladders, with the absorption and/or emission of quanta of one type and/or another. These jumps are unpredictable and cannot be controlled. They are absent in classical physics.

In the face of all this complementarity in quantum mechanics, we need to distinguish three classes of simple quantum evolutions that can occur over a *finite* interval of time. Note that the length of this interval is often limited by a decay process, such as radiative decay of the initial energy excitation, neglected quantum tunneling, or interaction with a condensed-matter environment. The three classes of simple quantum evolutions are:

(1) Classically describable (short-time) evolutions.
(2) Essentially quantum evolutions
(3) The host of "semiclassical" (short-time) evolutions that lie between these limiting cases.

In a *classically describable evolution*, the effects of individual quanta are negligible, and their combined effects can be well approximated by a causal description. An *essentially quantum system* is, by definition, the opposite extreme—where a description using second-quantization is essential for any explanation of the time evolution. (This concept of an essentially quantum system must be distinguished from that of an *intrinsically quantum effect*, such as tunneling, radiative decay of excited states, particle spin, and identical particle statistics.)

It is reasonable to define a *semiclassical evolution* as one whose interesting features do not critically depend on the transfer of a few quanta. Note this semiquantitative definition will vary with the system, the observable of interest, and the desired accuracy of a description. A more modern definition of semiclassical evolution is evolution "guided" by a workable number of classical reference trajectories; see Sections 2.1c and 7.2.

At present, a single rigorous mathematical procedure for taking the classical limit of quantum mechanics does not appear to be available. We know of no physical classical system that isn't comprised of a multitude of microscopic bodies. We can describe such many-body systems only in the vernacular of quantum statistical mechanics, which is a theory that is difficult to apply to the problem of quantum time evolution. Two contending "classical limits" are to take the number of bodies very large, or to take the quantum parameter (proportional to a power of $\hbar$) very small. In the case of certain model many-body systems, these two individual mathematical limits do not always appear to give the same results. In addition, if we take the quantum parameter small, it appears that the resulting limit does not commute with a third relevant limit, that of very large times. Thus a major issue remains unsettled—how to rigorously obtain unique classical limits of quantum systems.

Several quantum–classical correspondences have been developed for systems that evolve classically on sets of nested N-D tori. One correspondence is based on the belief that as a quantum system becomes more semiclassical, the appropriately smoothed frequency spectrum of the quantum system will get closer to the spectrum of the classical system. This idea leads to an association of the set of Fourier components $\mathbf{A_k(n)}$ of the classical evolution with the set of displaced quantum matrix elements $\langle \mathbf{n} - \frac{1}{2}\mathbf{k}|A|\mathbf{n} + \frac{1}{2}\mathbf{k}\rangle$ (Greenberg et al., 1995). Hence A is a real Hermitian operator corresponding to the classical quantity $\mathbf{A}$, and $\{n_i, i = 1 \text{ to } N\}$ is the classical set of integers labeling the fundamental frequencies $\{\omega_i\}$ in the quasiperiodic evolution. Quantum mechanically, $\{n_i\}$ labels the members of the ladder of energy eigenvalues.

In another connection between the classical and quantum spectra, values of the *classical frequency vector* $\boldsymbol{\omega}$ itself are associated with sets of displaced energy-level differences. Each component of $\boldsymbol{\omega}$ adheres to

$$\omega_i(\mathbf{n}) \Leftrightarrow E(n_1, \ldots, n_i + \tfrac{1}{2}, \ldots) - E(n_1, \ldots, n_i - \tfrac{1}{2}, \ldots), \quad i = 1, 2, \ldots, N.$$

We know that when the principal quantum number n is large, this association works for the energy levels E(n) of the hydrogen atom. The Kepler orbit frequency is almost equal to the energy separation of adjacent energy levels, divided by $\hbar$.

Furthermore, we can improve the correspondence by using the quantum number displacement procedure $n_i \to n_i \pm \frac{1}{2}$ indicated in the expression above. Such connections between quantum and classical spectra are further supported by experiment and by numerical work; see Chapters 7 and 8.

In the time domain, a powerful correspondence principle is suggested by similarities in the evolutions of classical trajectory ensembles and quantum wavepackets. An example of such similarity appears in Figs. 1.11 and 1.12, and we will see more of such quantum–classical comparisons. In brief, a tube or other appropriate set of classical trajectories that satisfy Hamilton's equations can be associated with a time-dependent wavefunction that at least approximately satisfies Schrödinger's wave equation. As the quantum system becomes more semiclassical through changes in its quantum parameter, we expect this association to improve; see Section 7.2. Sections 5.3, 6.4, 8.2, and 8.3 also offer examples of correspondence between sets of trajectories and wavefunctions. As we shall see, such correspondence can be the basis for accurate calculations, but the rigorous theory for the ab initio reliability of such calculations is still being developed.

2

QUANTUM COLLAPSE AND REVIVALS

2.1. BOUND-ELECTRON WAVEPACKET EVOLUTION

We know that an initial wavepacket, such as that shown in Fig. 1.11a, has a time-dependent wavefunction that is a linear combination of matter waves with different momenta or energies. The packet's spatial probability density is

$$P(\mathbf{r}, t) \equiv |\psi(\mathbf{r}, t)|^2 = \left[\sum_{E} c_E^* \psi_E^*(\mathbf{r}) \exp\left(-i\frac{Et}{\hbar}\right)\right]\left[\sum_{E'} c_{E'} \psi_{E'}(\mathbf{r}) \exp\left(+i\frac{E't}{\hbar}\right)\right].$$

The cross terms in this expression produce wave interference—both constructive and destructive—that is time dependent. This wave interference is a manifestly quantal effect, and it may not be obvious whether long-time evolution will exhibit any signatures of the classical evolution. This question arises even when the initial quantum state is constructed such that the early evolution corresponds to that of a classical probability distribution, as in Figs. 1.11 and 1.12. Let us investigate this issue for the special case of electrons orbiting within atoms.

2.1a. Classical Electron Evolution Within the Hydrogen Atom

First we need to recall some aspects of classical electron motion within the hydrogen atom. We know that radiative decay does not occur for the quantum mechanical lowest-energy eigenstate. Thus we will disregard the classical radiative decay of the accelerating electron. The Kepler orbit of the electron is an ellipse, which lies in some plane within 3-d coordinate space; see Fig. 2.1a. We take this plane to be fixed in space, so that the Euler angles specifying the orientation of the orbit are time independent. In the orbital plane, the orbit can be described using polar coordinates

(r', ϕ') with the origin at one of the focal points of the ellipse; see Fig. 2.1b. We know the equation for the ellipse is

$$r' = \frac{a(1-\varepsilon^2)}{1+\varepsilon\cos\phi'}, \tag{2.1}$$

where ε is the eccentricity of the ellipse, and where the semiaxes of the principal axis are $a(1 \pm \varepsilon)$. Referring to Fig. 2.1b, as point P on the orbit is moved along the orbit, point P′ traces a circle of radius a. The center O of this circle lies at the center of the ellipse. The circle touches the orbit at the two turning points. As ϕ' covers the domain $(0, 2\pi)$, so does the eccentric anomaly angle ξ, which is defined in the figure. Thus either angle can be used for locating a point on the orbit. As is clear from the figure, we can rewrite Eq. (2.1) in terms of this new angle:

$$r' = a(1 - \varepsilon\cos\xi). \tag{2.2}$$

We also see from the figure that

$$x' = a(\cos\xi - \varepsilon), \qquad y' = a\sqrt{1-\varepsilon^2}\sin\xi. \tag{2.3}$$

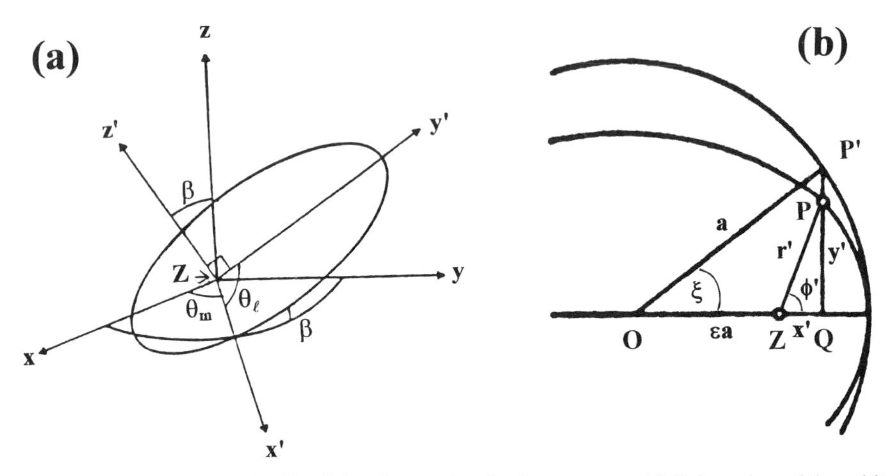

Figure 2.1. The classical orbit of the electron in a hydrogen atom. (a) Orientation of the orbit plane relative to a laboratory reference frame (x, y, z) having its origin at the atomic nucleus Z. The orbit is in the (x', y') plane in the orbit-fixed reference frame. The Euler angles describing the relative orientation of the two frames are θ_m, θ_ℓ, and β. The electron's orbital angular momentum vector lies along the z'-axis. (b) Spatial variables and characteristic distances in the orbit-fixed reference frame. The rectangular and polar coordinates that locate the electron at point P are $(x', y') \equiv (\overline{ZQ}, \overline{QP})$ and (r', ϕ'), respectively. Point O is the orbit center; $a = \overline{OP'}$ is the maximum distance of the orbit from O; ε is the orbit eccentricity; and ξ is the eccentric anomaly angle. After Born (1960).

If the hydrogen atom is in a time-oscillating electric field having a constant direction, then we can sometimes ignore the motion perpendicular to that direction, at least for a short time (see Section 8.3). In such a case, a 1-d model of the hydrogen atom is useful. This model is obtained by letting $\varepsilon \to 1$ in Eqs. (2.3). Taking $z \equiv -x'$, we then have

$$z = a(1 - \cos \xi). \tag{2.4}$$

The electron then moves on one side of the atomic nucleus, between $z = 0$ and an outer turning point at $z = a$. (In this model, the potential is taken equal to $+\infty$ for $z \leq 0$.)

We also know that in spherical coordinates $\{q_i\} = (r, \theta, \phi)$, the 3-d classical Kepler problem can be separated and solved. The Hamiltonian $H = p^2/2m - Ze^2/r$ and the angular momentum vector $\mathbf{L} = \mathbf{r} \times \mathbf{p}$ are both constants of the motion. Alternatively, we can take the set of constants of the motion to include three 1-d action integrals, with the integrations being once around the orbit Γ, $I_i = \oint_\Gamma p_i \, dq_i$. Finding $\mathbf{p}$ and then $\mathbf{L}$ in spherical coordinates, and then doing the integrals, these actions are

$$\begin{aligned} I_\phi &= 2\pi L_z, \\ I_\theta &= 2\pi(L - L_z), \\ I_r &= -2\pi\left[L - \frac{1}{\sqrt{-2H}}\right], \end{aligned} \tag{2.5}$$

where $L \equiv |\mathbf{L}|$. From Eqs. (2.5) it immediately follows that

$$H = -\frac{2\pi^2}{(I_r + I_\theta + I_\phi)^2} \equiv -\frac{2\pi^2}{I^2}, \tag{2.6}$$

where I is the constant total action. We note that the classical frequencies $\omega_i = \partial H/\partial I_i$ of the motions in the individual coordinates q_i are all the same. They are equal to the Kepler frequency $\omega_I \equiv 4\pi^2/I^3$. This is a special situation arising from the special symmetry of the Kepler problem.

Since the Hamiltonian can be written in terms of the single action variable I, a conjugate coordinate is cyclic and must be an angle variable that we call θ_I. This angle evolves uniformly in time according to $\theta_I = \omega_I t + \text{constant}$. An important connection exists between θ_I and the eccentric anomaly angle ξ, namely,

$$\theta_I = \xi - \varepsilon \sin \xi. \tag{2.7}$$

This relationship is proved in a detailed discussion of the classical hydrogen atom (Born, 1960). For a brief but more extensive summary of this atom, see Section IIIA in Delos et al. (1983a).

There is a natural set of units for the hydrogen atom (and other atoms) called *atomic units*. Atomic units are defined by setting $\hbar$, m and e each equal to unity. Other dynamical quantities then become scaled. Lengths become scaled to the Bohr radius $\hbar^2/me^2$ of the ground-state nominal orbit, velocities become scaled to the electron velocity $e^2/\hbar$ of that orbit, and so forth (Bethe and Salpeter, 1977).

When we work with excited atomic states, it is often advantageous to introduce further scaling. This scaling will account, approximately, for the dependence of atomic properties on the principal quantum number n. The mean radius of an n, $\ell = 0$ quantum nS energy eigenstate of hydrogen scales as n^2. The corresponding velocity scales as the square root of the energy, that is, as n^{-1}. These scalings determine the n-scaling of other dynamical quantities in *scaled atomic units* so that, for instance, time is scaled by $n^2/n^{-1} = n^3$. The result is the removal of the gross dependence on n of all the dynamical quantities. Some orders of magnitude for $n = 1$, 50, and 500 are given in Table 2.1.

A scaled quantity often used in Chapters 8 and 9 is that for the magnitude of an electric field F. The expectation value of the Coulomb electric field e^2/r^2 within the hydrogen atom is $n^{-3}(\ell + 1/2)^{-1}$ atomic units (Bethe and Salpeter, 1977, page 17). When we work with states having a typical mean angular momentum $\ell \sim n/2$, the scaling goes as n^{-4} and the characteristic scaled electric field is $F_0 \equiv n^4 F$ atomic units, as indicated in Table 2.1.

2.1b. Rydberg-Atom Wavepacket Collapse and Revivals

We now want to search for quantum–classical correspondence in the orbital evolution of the bound electron in a hydrogen-like excited atom, also called a Rydberg atom. The Rydberg electron is taken to be located well outside the rest of the atom. To find the correspondence we must, as usual, choose the initial quantum and classical states. Various choices have been pursued in the literature. These choices range from the generalized coherent states of the hydrogen atom (Nauenberg, 1989) to the states that can be produced by short-pulse laser excitation. Such excitation begins with the ground state or some low-lying excited state (Parker and

TABLE 2.1. Gross Properties of the Hydrogen Atom[a]

	Scales as	$n = 1$	$n = 50$	$n = 500$
Atom diameter	n^2	0.1 nm	250 nm	25 μm
Classical orbit period	n^3	0.15 fs	20 ps	20 ns
Orbital frequency	$1/n^3$	7 THz	60 GHz	60 MHz
Electron velocity	$1/n$	2×10^6 m/s	4×10^4 m/s	4×10^3 m/s
Electron energy	$1/n^2$	13.6 eV	5 meV	50 μeV
Ionizing electric field	$1/n^4$	10^9 V/cm	200 V/cm	20 mV/cm
Critical magnetic field	$1/n^3$	2×10^5 T	2 T	2×10^{-3} T
Radiative lifetime	n^3	∞	100 μs	0.1 s

[a] The quantity n is the principal quantum number.

Stroud, 1986). In any case we will need to carry out basis-set expansion computations (Section 1.2a).

To evolve a particle-like quantum atomic electron, we use the wavepacket approach introduced in Section 1.3c. We will take the initial quantum state to be an appropriate Gaussian (and hence minimally uncertain) wavepacket ψ_0^G. The initial classical state will again be a corresponding set of initial conditions for trajectories in phase space. These classical initial conditions and the initial wavepacket will have the same phase-space probability distribution. We let $(\mathbf{r}_0, \mathbf{p}_0)$ be the expectation values for the initial electron packet in the 6-D phase space. We then choose

$$\psi_0^G(\mathbf{r}, \mathbf{r}_0, \mathbf{p}_0) = \frac{1}{(2\pi)^{3/4}\sigma^{3/2}} \exp\left[-\frac{(\mathbf{r}-\mathbf{r}_0)^2}{4\sigma^2} - i\frac{\mathbf{p}_0 \cdot \mathbf{r}}{\hbar}\right], \tag{2.8}$$

where σ is the packet's spatial width. This particular initial packet exhibits no correlation between $\boldsymbol{r}$ and $\boldsymbol{p}_0$; in other words, the expectation values of these operators satisfy $\frac{1}{2}[\langle x_i p_i\rangle + \langle p_i x_i\rangle] - \langle x_i\rangle\langle p_i\rangle = 0$. This lack of correlation allows us to connect expectation values to the corresponding values of a single central classical orbit. Both the quantum expectation values and classical central-orbit values satisfy the same equation for conservation of energy, as well as the same equation for conservation of momentum.

Since we will need to expand expression (2.8) in terms of a hydrogen atom basis set, we will work in the angular momentum representation. Thus we introduce spherical coordinates determined by the direction of the expectation value of the initial angular momentum $\mathbf{L}_0 = \mathbf{r}_0 \times \mathbf{p}_0$. We then choose $\mathbf{r}_0 \equiv (x_0, b, 0)$ and $\mathbf{p}_0 \equiv (-p_0, 0, 0)$. Hence, when written in Cartesian coordinates, $\mathbf{L}_0 = (0, 0, bp_0)$.

The next step is to perform the initial wavepacket expansion in the hydrogen atom energy eigenfunctions

$$\psi_0^G = \sum_{n\ell m} a_{n\ell m}\psi_{n\ell m}(\mathbf{r}), \qquad \psi_{n\ell m}(\mathbf{r}) = R_{n\ell}(r)Y_{\ell m}(\theta, \phi). \tag{2.9}$$

Here the $R_{n\ell}$ are the usual hydrogen atom radial wavefunctions, and $Y_{\ell m}$ are spherical harmonics (Bethe and Salpeter, 1977). If we follow our procedure in Section 1.3a for finding expansion coefficients, then we get an analytical result for the $a_{n\ell m}(\mathbf{r}_0, \mathbf{p}_0)$. This result is written in terms of a radial integral involving spherical Bessel functions in r, and of spherical harmonics in θ, ϕ (Boris et al., 1993). The time evolution of the bound electron wavepacket is then described—now in atomic units—by

$$\psi^G(\mathbf{r}, t, \mathbf{r}_0, \mathbf{p}_0) = \sum_{n=1}^{\infty} \exp\left(-i\frac{t}{2n^2}\right)\sum_{\ell=0}^{n-1} R_{n\ell}(r) \sum_{m=-\ell}^{\ell} a_{n\ell m}Y_{\ell m}(\theta, \phi). \tag{2.10}$$

This equation is evaluated numerically.

Since we have a 6-D phase space, it's difficult to construct useful quantum or classical Poincaré sections in order to visualize the evolution. We choose instead to look directly at snapshots of plots of the probability density $P_{xy}(x, y, 0, t)$ in the plane $z = 0$. We choose $(\mathbf{r}_0, \mathbf{p}_0)$ so that the mean value $\langle n \rangle$ of the principal quantum number of the initial packet is quite large: 324. As an example, we take the case of Boris et al. (1993), namely, $\varepsilon = 0.9$, $b = 0$, $p_0 = 7.1 \times 10^{-4}$ a.u., and $\sigma_n / \langle n \rangle = 0.015$ for the fractional width in the principal quantum number n.

Figure 2.2 compares quantum and classical packet evolution during the time interval T_K of the first Kepler period. We see very similar spreading and contracting of both packets, as they follow the orbit of a classical particle (solid line). This particle orbit has the chosen initial condition $(\mathbf{r}_0, \mathbf{p}_0)$.

Following the evolutions further (not shown), this spreading and contracting continues for a number of periods, with an overall packet-spreading from period to period. This overall spreading from period to period is called the *collapse* of the wavepacket. During the collapse, the uncertainty principle product $\Delta p_x \, \Delta_x$ increases.

So far, the quantum–classical correspondence appears very strong. However, as time continues further, we begin to observe deviations of the quantum probability distribution P_{xy} from the classical distribution. The quantum distribution develops lumps along the orbital path, whereas the classical one does not. For times near $T_r \equiv (\langle n \rangle / 3) T_K$, there is a strong primary *revival* of the localization of the quantum wavepacket; see Fig. 2.3. The revival is located near the atomic nucleus rather than at the original packet location x_0 (the outer turning point). The quantum uncertainty principle product drops back near the initial value.

In addition, at times considerably earlier than T_r, there already occurred *fractional revivals* of order M/N, where M and N are positive integers, here with $N > M$. These revivals occur near times $(M/N)T_r$. See the results for P_{xy} in Fig. 2.4 for some examples. For this figure, $M = 1$ and $N = 5, 4, 3, 2$. Note that these revivals involve the temporary development of N lumps in P_{xy}, lumps that actually are equally distributed in time along the orbit. The origins of these fractional revivals are understood in detail (Leichtle et al. 1996).

Thus we see that—as we might have expected—the quantum evolution does retain some aspects of the classical evolution. For a fairly long time, the packet stays localized along the central orbit. At early times the packet evolution is almost classical. On the other hand, the packet does spread along the orbit. Eventually, quantum interference of the packet with itself produces the primary and fractional packet revivals. Close to early primary revival times, the quantum packet is back close to the classical distribution.

In the example just discussed, we don't have particularly good conditions for sharp revivals. To see why, we should expand the phase factor $\exp[(-i/2n^2)t]$ in each term of the expansion (2.10). We first expand in n about the mean value $\langle n \rangle$, and second in t about a revival time T_r. The requirement that residual phases be small in order to escape destructive interference then leads us to $\sigma_n^3 \ll \frac{3}{2} \langle n \rangle$ (Boris et al., 1993, Appendix A). For the parameters chosen for Figs. 2.2–2.4, this inequality is only weakly satisfied, as the two quantities differ by a factor of four.

Using lower values of $\langle n \rangle$ and smaller σ_n, laboratory experiments have been carried out to observe collapse and revival. Many of these experiments involve packet evolution in only the radial spatial coordinate of the electron. However, a proposal for 3-d electron localization for arbitrary orbit eccentricity ε has been

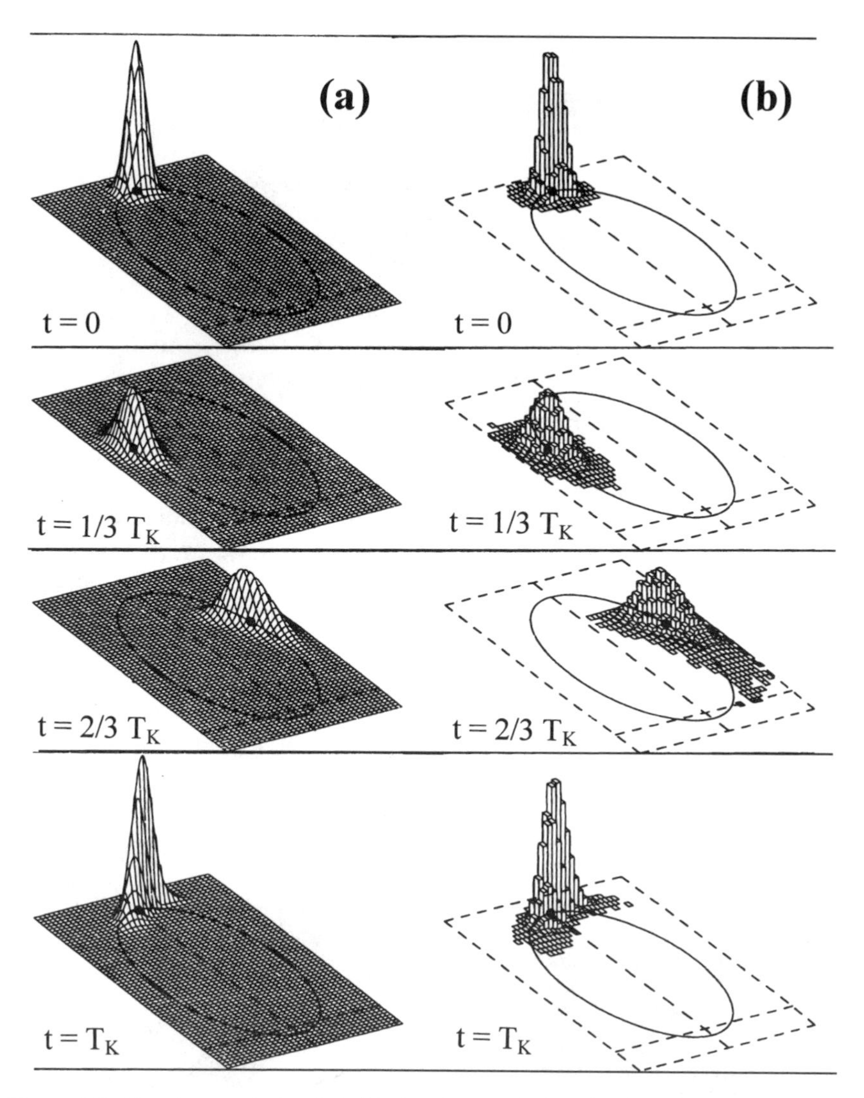

Figure 2.2. Quantum and classical Kepler time evolution in the (x, y) orbit plane. (a) Snapshots of an initially Gaussian wavepacket, during the first classical Kepler period $T_K = 2\pi\langle n\rangle^3$ a.u. (b) Corresponding snapshots of a phase-space distribution of classical trajectories having the same distribution of initial conditions. The mean initial quantum number (or classical action) is high, $\langle n\rangle = 324$. From Boris et al. (1993). Copyright 1993 by the American Physical Society. Also see Brandt and Dahmen (1995).

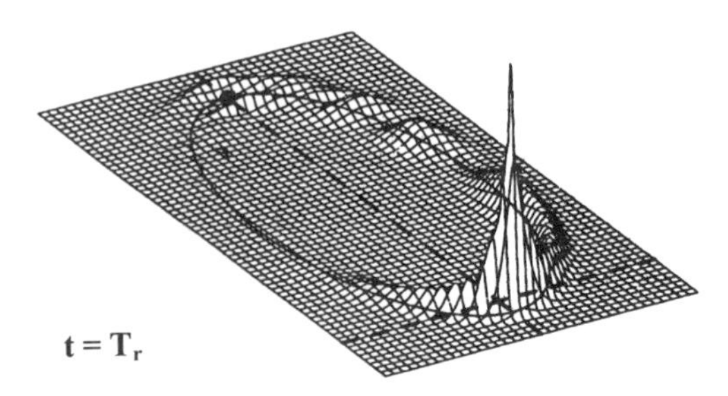

Figure 2.3. The Gaussian wavepacket of Fig. 2.2, after an elapsed time equal to the packet first revival time T_r. From Boris et al. (1993). Copyright 1993 by the American Physical Society. Also see Brandt and Dahmen (1995).

made—see Gaeta et al. (1994). Among their findings, non-Gaussian initial packets can produce relatively sharp revival signals.

In an experiment using a pulsed laser to photoexcite Rydberg atoms, the band of states initially populated by the laser excitation is determined by two factors: the laser bandwidth and, less importantly, the relative amplitudes for excitation of the states within the band. We can numerically calculate the initial quantum wavefunction produced by a short (fs or ps) laser pulse, by using a time-dependent extension of our basis-set expansion approach (Parker and Stroud, 1986). In order to qualitatively predict the results of experiments, the numerical amplitudes in the wavefunction are used as the $a_{n\ell m}$ in Eq. (2.10). For a quantitative treatment,

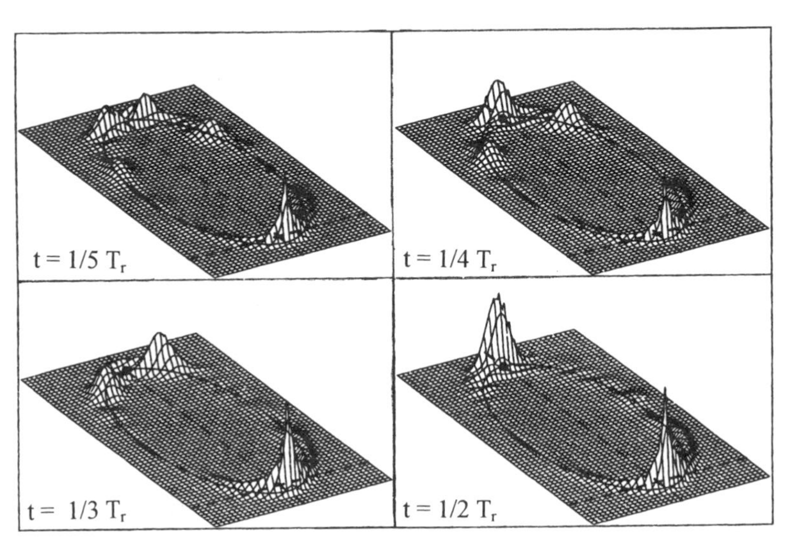

Figure 2.4. Fractional revivals of the wavepacket of Fig. 2.2 that first occur at times $(M/N)T_r$ with $M/N = \frac{1}{5}, \frac{1}{4}, \frac{1}{3}$ and $\frac{1}{2}$. From Boris et al. (1993). Copyright 1993 by the American Physical Society. Also see Brandt and Dahmen (1995).

however, we must also include the effects of the experimental process for probing the time evolution for an ensemble of atoms, of course, not just one atom.

Conceptually, the simplest experimental approach would be to observe the evolution of the spontaneous radiative decay of the atoms. In practice, it is easier to use a second laser pulse as a probe. This probe photoionizes some of the Rydberg atoms, and the product ions are collected. In what is called a *laser pump-probe experiment*, one varies the delay time between the excitation pulse and the probe (detection) pulse. Figure 2.5a shows some laboratory results.

Both the spontaneous decay and photoionization detection schemes are based on the fact that radiative processes are associated with charged-particle acceleration. When the electron wavepacket is near the atomic nucleus, its acceleration is large, and hence the detected signal is large. The signal is much smaller at the outer turning point of the orbit. Thus as the electron goes around its orbit, the signal should oscillate. Figure 2.5 shows reasonable accord between an experimental photoionization signal and numerical predictions. Figure 2.6 shows some predicted effects of fractional revivals. In recent years, many experiments have been carried out on atomic-electron wavepacket collapse and revival. The atoms are often in external static electric and/or magnetic fields.

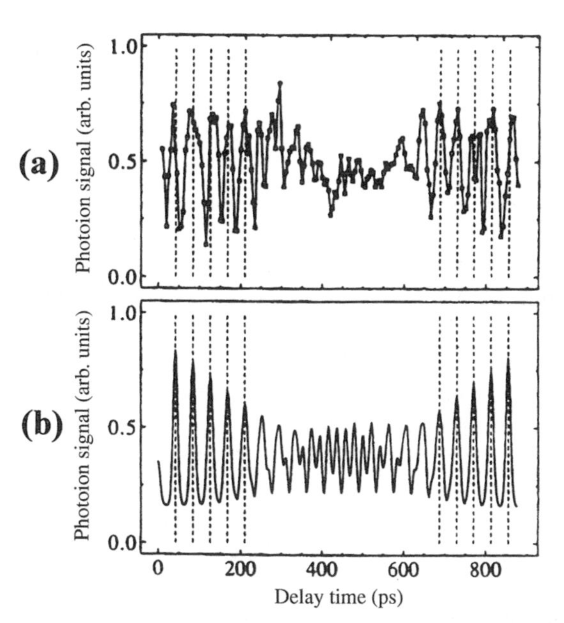

Figure 2.5. Time evolution of a radial Rydberg-electron wavepacket with $\langle n \rangle = 65.2$: (a) experimental laser photoionization signal as a function of the time delay of a probe laser pulse; and (b) corresponding results of quantum numerical calculations. The vertical dashed lines are separated by the orbit period T_K. The early strong oscillations become damped, but then revive because of constructive matter-wave interference. From Yeazell et al. (1990). Copyright 1990 by the American Physical Society.

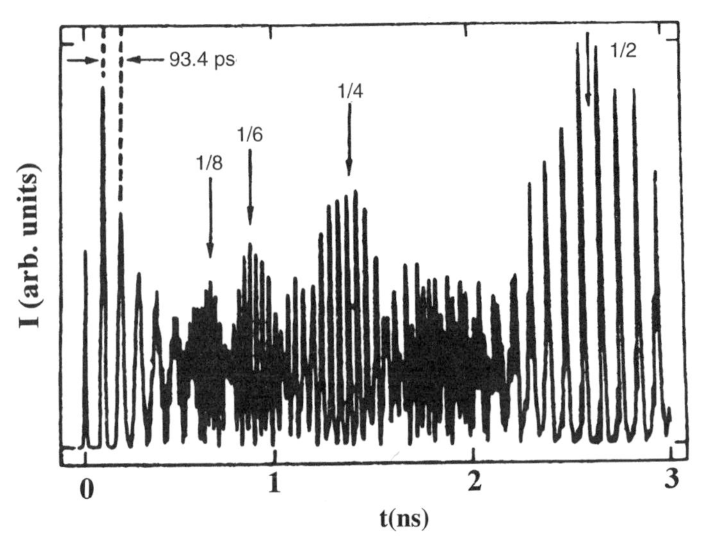

Figure 2.6. Calculated intensity of spontaneous light emission by an $\langle n \rangle = 85$ Rydberg–atom wavepacket excited by a short laser pulse. The times of fractional wavepacket revivals are indicated by the arrows and their values of M/N. Reprinted with permission from Averbukh and Perel'man (1990). Copyright 1990 American Institute of Physics.

2.1c. Semiclassical Propagation for the Coulomb Potential (Advanced Topic)

The numerical evidence to date clearly shows that atomic wavepackets often can be guided by a central classical periodic orbit. The expectation values of the packet's location in phase space are connected with the evolution of the central orbit. So far, we've dealt theoretically with a purely quantum mechanical coupling of an appropriate set of basis states. What we'll now see is that, for the semiclassical regime touched on in Sections 1.2, 1.3, and 1.5, the packet can be guided by a *set* of classical reference trajectories that control the full quantum evolution of the *entire* packet. Observable quantities will then be expressed as sums over reference trajectories. The sums contain parameters that are evaluated only with classical information. All the quantum interference of the quantum primary and fractional revivals can accurately be described by this procedure. The general theory justifying this *semiclassical reference-trajectory wavepacket propagation method* will be discussed in Section 7.2.

We can adequately demonstrate the new ideas behind semiclassical propagation by considering a 1-d problem. In atomic units, Schrödinger's equation (1.2) becomes

$$i\frac{\partial\psi(x,t)}{\partial t} = -\frac{1}{2}\frac{\partial^2\psi(x,t)}{\partial x^2} + V(x)\psi(x,t). \tag{2.11}$$

For the 1-d hydrogen atom satisfying the trajectory equation (2.4), with z now being written as x,

$$V(x) = -\frac{1}{x}, \qquad x > 0, \qquad V(x) = \infty, \qquad x \leq 0. \tag{2.12}$$

A particular reference trajectory can be specified by that trajectory's collection of phase-space points $\{x_r, p_r\}$ for all times. To find a solution of Eq. (2.11) that is best suited for the neighborhood of a reference trajectory, we first expand V(x) in a Taylor series about x_r through second order in $(x - x_r)$. It is important to keep in mind that this Taylor series expansion is continuously evolving in time, so that both x_l and all the expansion parameters are time dependent. Since we have kept expansion terms in the potential only up to second order, our evolving local solution of Eq. (2.11) is that of a quantum harmonic oscillator with both time-varying frequency and time-varying central position.

Suppose that at $t = 0$, the wavefunction is a Gaussian $\psi_\beta^G(x, 0)$, where β is the set of parameters describing the packet. In Section 1.3c we saw how a (now local) harmonic oscillator potential influenced an initial Gaussian wavepacket. Under the potential's influence, the wavepacket (now local in both space and time) evolved with no change in its phase-space probability distribution. We conclude that a sufficiently localized initial Gaussian will evolve along a reference trajectory. It will retain its Gaussian form but distorted by the time-dependent parameters. Thus the (local) solution of Eq. (2.11) can be written

$$\psi_r(x, t) = A \exp\{i[B_r(x - x_r) + C_r(x - x_r)^2 + D_r]\}. \tag{2.13}$$

Here B_r, C_r, and D_r are complex functions of time and depend on the packet initial conditions β. If we had kept only the linear term in our expansion of the potential, then these packet initial conditions would have been its initial mean values—x_0, p_0; then $x_r(t)$ would be the time-evolving mean packet position prescribed by the classical trajectory $\{x_r(t), p_r(t)\}$. But we have expanded further, so this constraint is relaxed.

Substituting Eq. (2.13) and the expanded potential into Eq. (2.11), and equating the coefficients of like powers of $(x - x_r)$, we again find a dynamical system (Section 1.1b). This system is comprised of the Newtonian form of Hamilton's equations for the variables $\{x_r, p_r\}$, and of first-order ordinary differential equations for the time derivatives of B_r, C_r, and D_r. These equations are solved in Suárez Barnes et al. (1994), where they investigated a bound orbit of the 1-d hydrogen atom starting at (x_0, p_0) and having energy $E = p_0^2/2 - 1/x_0$.

A theoretical quantity displaying the revivals observed in experiment should involve the overlap of the time-evolving wavefunction at time t, with some wavefunction localized near the atomic nucleus. We choose the *autocorrelation function*, which is relatively easy to evaluate:

$$\mathbb{C}_\beta(t) \equiv \int dx\, \psi_\beta^{G*}(x,0)\psi_\beta^G(x,t) \equiv \langle \beta | \beta(t) \rangle. \tag{2.14}$$

Clearly, trajectories that can effect $\mathbb{C}_\beta(t)$ are those that (nearly) return to the nucleus.

A particular reference trajectory is only useful within its neighborhood in phase space, a neighborhood determined by small perturbations. Therefore we expect to need more than one reference trajectory to fully describe $\mathbb{C}_\beta(t)$. Letting the label on the needed set of trajectories be j, Eq. (2.14) becomes

$$\mathbb{C}_\beta(t) = \sum_j \langle \beta | \beta(t) \rangle_j. \tag{2.15}$$

Since x_r, p_r, B_r, C_r, and D_r are known functions of time for any and all reference trajectories, and since the initial wavepacket parameters can be chosen appropriately, the sum (2.15) can now be evaluated, once a suitable set of reference trajectories are found.

It is crucial to realize that the potentially important reference trajectories *at each time* t are simply those trajectories that start at some point along the x-axis at time $t=0$, and return to that same point at the later time t. These trajectories give strong constructive interference; that is, they give large contributions to Eq. (2.15). As our system evolves, the reference trajectories that give these contributions will change. The search for returning trajectories, however, is strictly a classical physics problem, one that can be dealt with analytically in the present case. To obtain exact trajectory closure at time t, we only need tune the energy.

The above procedure for semiclassical propagation has been carried out (Tomsovic, 1995). Let the wavepacket start at the outer turning point (x_0, $p_0=0$). Then the result is

$$\mathbb{C}_\beta(t) = \sum_{j=1}^{\infty} \left(\frac{t'}{t' + \mu_\beta j^2} \right)^{1/2} \exp\left(i \frac{3\tau_\beta}{x_\beta} (j^2 t')^{1/3} + i \frac{2C_0 \mu_\beta x_\beta^2 j^2 [(t'/j)^{2/3} - 1]^2}{(t' + \mu_\beta j^2)} \right). \tag{2.16}$$

Here $t' \equiv t/\tau_\beta$ is the time in units of the Kepler period τ_β of the central trajectory, $x_\beta = x_0$ is the mean location of the initial wavepacket started at the outer turning point, and $\mu_\beta \equiv 3\tau_\beta / 8C_0 x_\beta^3$. The quantity C_0 is C_r evaluated for the central classical orbit.

Equation (2.16) only holds for times $t > \tau_\beta/2$. This is because the initial collapse of the packet occurs at earlier times and must be treated separately. In Fig. 2.7 we compare $|\mathbb{C}_\beta(t)|$ obtained from expression (2.16) with $|\mathbb{C}_\beta(t)|$ obtained from precise basis-set expansion calculations. We see that the semiclassical reference-trajectory propagation approach works very well.

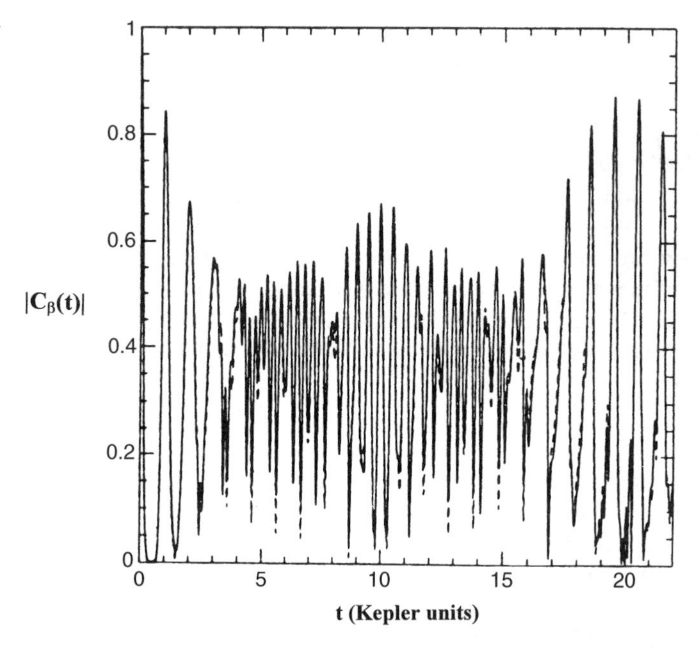

Figure 2.7. Semiclassical and precise numerical collapse and revivals of an initially Gaussian Rydberg-atom wavepacket with $\langle n \rangle = 60$. The semiclassical results (solid curve) were obtained by guiding a set of wavepackets along an appropriate set of reference electron trajectories. The packet autocorrelation function $|\mathbb{C}_\beta(t)|$ is plotted versus time in units of the Kepler period of the initial central electron orbit. The exact quantum revivals of the packet (dashed curve) are accurately described semiclassically, particularly for $M/N = \frac{1}{3}$ near $t = 20$. From Suárez Barnes et al. (1993). Copyright 1993 by the American Physical Society.

2.1d. Wavepacket Evolution Within Solid State Systems

For some time there has been considerable interest in the time evolution of atomic particles in small structures within solid materials. When the structure has one or more spatial dimensions smaller than one micron, it is called a *nanostructure*. The utilization of nanostructures is part of a field called *nanotechnology*. When the spatial size is considerably less than one micron, the particle-structure system is quantum mechanical. One example of such a system is electrons in semiconductor nanostructures. Electron transport in semiconductor nanostructures is a subject under active investigation; see Ferry and Goodnick (1997).

Quantum mechanics allows materials to have bands of energy levels. These bands are a consequence of the quantization of the energies of the constituent atoms in the material. The bands are essentially continuous because there are so many individual atoms. Materials are classified as insulators, semiconductors, or conductors. Their classification depends on two factors: whether or not electrons fill the highest potentially active energy band, and the size of the energy gap above that band.

Special *charge carriers* can be placed within materials by introducing impurity atoms. These carriers can be either unbound electrons or the positively charged unbound absences of electrons, called *holes*. An electron and a hole can be bound together to form a (moving) *exciton*, a neutral quantum mechanical "particle" analogous to the positronium atom (e^-e^+).

Solid state structures that are "doped" by impurities can have one or more significant spatial dimensions. Some of these dimensions can be microscopic, meaning they are comparable to, or only somewhat larger than, the de Broglie wavelength of the charge-carrier particle or exciton. When the microscopic dimensions number 1, 2, or 3, we have, respectively, a 2-d nanostructure (a *quantum well*), a 1-d nanostructure (a *quantum wire*) or a 0-d nanostructure (a *quantum dot*). For these three structures there will be respectively, a 1-d, a 2-d, or a 3-d quantum evolution of particles within the nanostructure. Because some of the dimensions are microscopic, there will be discrete charge-carrier energy-level structure combined with the material's band structure.

When there is a low concentration of carriers in a structure, on time average the individual carriers interact mostly with their environment, which is the undoped base material. In the simplest model of the effects of this environment, the mass and charge of the carriers are modified. The procedure is similar in spirit to that for introducing permittivity, permeability, and conductivity into the electromagnetic theory for linear media. In a solid material, the *effective mass* and *effective charge* of a charge carrier are usually very different from its mass and charge in a vacuum. Since many different materials and geometries are available for the construction of solid state structures, the quantum energy bands and levels can be adjusted considerably.

To be more specific, let us consider a quantum well with a carrier interaction potential V(x) along the microscopic narrow dimension x. The shape of V(x) can be tailored by varying the concentration of some constituent atom in the material, as we proceed across the region in x defining the well. All sorts of well shapes are possible. These shapes include the square well potential, as well as the parabolic well potential where the carrier motion is that of a harmonic oscillator. (In Section 1.3c we discussed wavepacket evolution for the harmonic oscillator). Packet evolution in the infinite square well is as easy to study as the Coulomb potential discussed earlier in this section; see Brandt and Dahmen (1995) for some details. More applicable to multiple quantum well nanostructures are cases where V(x) consists of some number of finite steps; these steps can go up and down by different and adjustable amounts. We can also numerically investigate wavepacket evolution in these multistep systems, using the computer code supplied by Brandt and Dahmen (1994).

The atoms of solid materials vibrate. In a nanostructure these vibrations couple to the degrees of freedom of the charge carriers. The vibrations limit the lifetimes of the carrier energy eigenstates, with the energies therefore becoming complex quantities. Under these conditions, we generally do not expect quantum evolution of individual carriers to persist for long times. Indeed, this evolution is often cut off in less than a picosecond by energy relaxation to the carrier's environment. However, we can prepare packets of quantum bound carrier states by femtosecond pulsed-laser excitation. Just as in the isolated atom experiments discussed above, the pulse duration may be as short as 5–10 fs. Fairly common now are laser pump–probe studies of carrier evolution in quantum solid state structures. The focus of such studies, however, is often on the nature of the relaxation processes.

Laser spectroscopy easily reveals the existence of quantum energy levels in quantum wells. One interesting type of experiment investigates the spectrum as a function of well width. Such studies can be facilitated by preparing a *lateral heterostructure* consisting of parallel layers of quantum wells, in this case wells of different widths where the wells are sandwiched together; see the inset of Fig. 2.8a. The dimensions of the indicated layers are large both in the vertical direction and in the direction perpendicular to the page. In Fig. 2.8a, the photoluminescence spectra of all the different wells are superimposed to form the observed spectrum. The locations of the different spectral peaks can be analyzed using the stationary-state quantum mechanics of finite square wells, taking into account the different well widths. The results can be excellent (see Fig. 2.8b), proving that quantized carrier states exist in the wells. To reduce the broadening of the spectral lines due to relaxation, these CW laser experiments were carried out with the solid state structures held below room temperature.

Recent experiments have investigated the evolution of a quantum wavepacket within a single quantum well at room temperature, using 12 fs laser pulses for the charge-carrier wavepacket preparation (Bonvalet et al., 1996). To enhance the packet evolution signal, they used a lateral heterostructure of 100 nominally identical wells very spatially isolated from one another, hence a *multi-quantum-well* (*MQW*) *structure*. After each laser pulse, they used a very fast infrared (IR) detector to

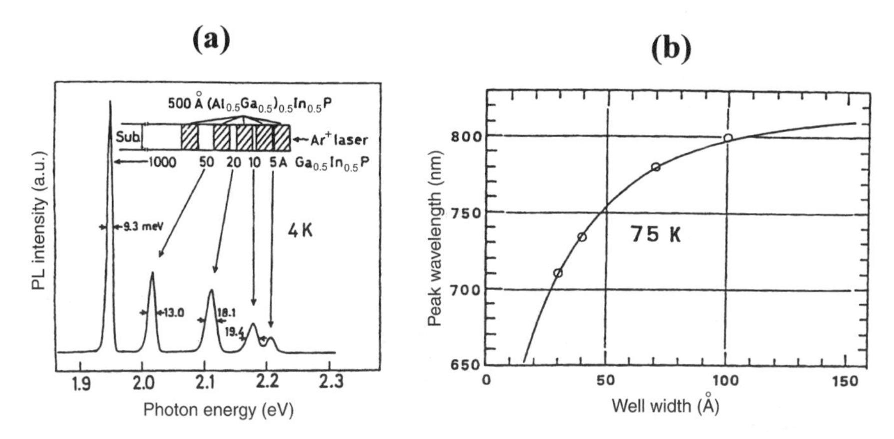

Figure 2.8. (a) Photoluminescence of an array of GaInP/(AlGa)InP quantum wells having the indicated different well widths, from 5 Å to 1000 Å. An argon ion laser excites the first excited state in each well, which spontaneously decays while emitting resonant light with one of the photon energies indicated by the spectral peaks. (b) The wavelength of the photoluminescence peak is shown versus the well width of the quantum well, for GaAs sandwiched by $Al_{0.54}Ga_{0.46}As$. The circles are experimental data and the curve is calculated. From Watanabe et al. (1987). Copyright 1987 Plenum Publishing.

monitor IR emission by the charge carriers in the wells. The laser pulse produced the simplest of wavepackets—a packet comprised of just two coherently populated energy eigenstates. In Fig. 2.9 we see packet evolution having oscillations of period 33 fs, which corresponds to an expected intersubband transition energy of 126 meV and to the emitted IR wavelength of 9.8 μm. The oscillations show no collapses or revivals, of course, because there are only two component states in the packet. The rise in the envelope of the oscillations is due to the response time of the IR detector, and the fall time of 180 fs is believed due to the energy relaxation to the environment.

Other experiments have investigated wavepacket evolution *within* excitons present in semiconductor quantum-well structures. The time–evolving excitonic quantum states were initiated by 110 fs laser pulses, which were shorter than the electron round trip time within the exciton orbital. The laser bandwidth and exciton energy-level structure were such that the exciton 1s and 2s states were excited, along with higher-lying states believed to be in the continuum. The observed experimental time evolution—a signal generated by a nonlinear optics technique—is shown in Fig. 2.10b. Here the center of the laser frequency band is tuned onto the energy level of the 2s state. In the figure we also see the corresponding results of a 1s–2s–continuum theoretical model of the packet evolution. The dashed line shows the effect of omitting the discretized continuum above the 2s state. Clearly, the "continuum" states played a significant role in the evolution.

In Figs. 2.9 and 2.10, the basic period of the oscillations reflects matter-wave interference arising from a prepared linear combination of two energy eigenstates. These temporal oscillations are similar to the two–amplitude spatial interference pattern of Thomas Young's two-slit optics experiment.

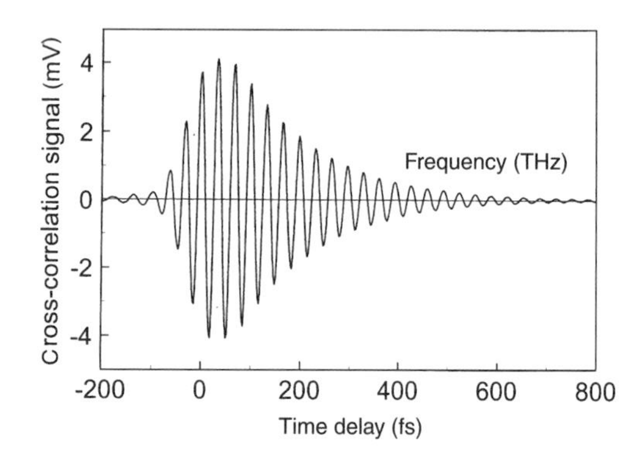

Figure 2.9. Time evolution of mid-infrared emission from a multiple quantum well structure. The MQW structure was made up of 100 undoped compositionally asymmetric [$Ga_{0.9}In_{0.1}As$(4.6 nm)/GaAs(5.4 nm)]AlAs barriers and was operated at room temperature. A 12 fs laser pulse was used to excite both the structure and a [110] grown GaAs sample as a reference. The data show the correlation between the electric field radiated by the structure and the reference electric field, as a function of the time delay between the detected emission and the laser pulse. The dephasing time is seen to be about 180 fs. From Bonvalet et al. (1996). Copyright 1996 by the American Physical Society.

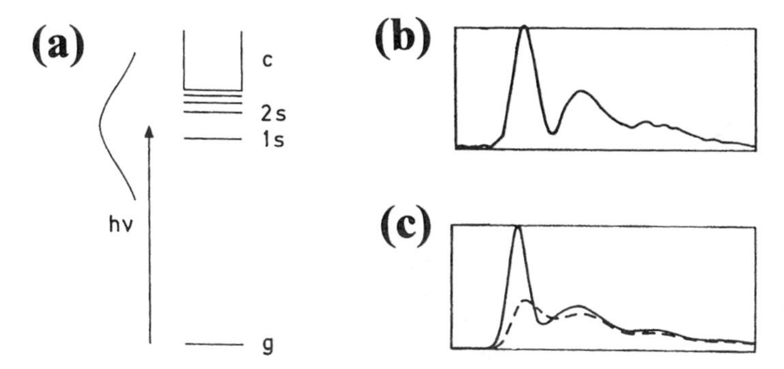

Figure 2.10. Excitation and decay of excitons in a MQW structure at a temperature of 5 K. Shown are (a) the exciton energy levels compared with the photon energy spectrum of a particular 110 fs laser pulse (g = ground state, c = continuum states); (b) the time evolution of the experimental nonlinear optics (transient four-wave mixing) signal, for zero mean detuning of the g → (1s, 2s) exciton transition; and (c) theoretically predicted signals with (solid line) and without (dashed line) coupling to a discretized continuous part "c" of the exciton spectrum of panel (a). From Feldmann et al. (1993). Copyright 1993 by the American Physical Society.

2.2. RABI OSCILLATIONS

In Section 2.4 we will conclude this chapter with a discussion of wavepacket collapse and revival in a particular driven system—the nonautonomous system that consists of an atom (assumed to have just two active energy levels) and of an applied monochromatic electromagnetic field. Here we begin our discussion of two-level atoms in an electromagnetic field by showing I. I. Rabi's "flopping" of probability back and forth between the two atomic levels. This flopping originates in the initial condition where the atom starts out at $t = 0$ with unit probability amplitude in just one of its two unperturbed energy eigenstates. Note that this initial state is not an eigenstate of the atom–field combined system, a fact that must lead to time evolution of the occupation probabilities of both energy eigenstates. The initial condition is also that for the *sudden* application of an interaction to the atom, a limiting case of the more general problem of the switching-on of a perturbation over some period in time (Section 9.3). Rabi flopping is almost all that occurs when the applied field is relatively weak (Rabi, 1932). An understanding of Rabi oscillations will help prepare us for collapse and revivals at higher field strengths.

The Hamiltonian for our present problem is $H = H_0 + V$, where H_0 is the Hamiltonian for the two-level atom by itself and V is the atom's interaction with the applied electromagnetic field. From electromagnetic theory we know that the interaction of the atomic electron with the external field can be written as a multipole expansion. We know too that the leading term is the electric dipole interaction, which classically is $-\mathbf{D} \cdot \mathbf{F}$ with $|\mathbf{F}| = \mathrm{F}_0 \cos \omega t$. We take the electric field direction along the z-axis, $\mathbf{F} = \mathrm{F} \cos(\omega t + \phi)\hat{\mathbf{z}}$. We expand the wavefunction of our two-level atom in the orthonormal (two-state) basis of the two energy eigenstates $\psi_1(\mathbf{r})$ and $\psi_2(\mathbf{r})$, where these eigenstates correspond to the energies E_1, $E_2 > E_1$:

$$\begin{aligned} \Psi(\mathbf{r},t) &= c_1(t)\psi_1(\mathbf{r},t) + c_2(t)\psi_2(\mathbf{r},t), \\ \psi_i(\mathbf{r},t) &\equiv \exp[-i\zeta_i(t)]\psi_i(\mathbf{r}), \quad i = 1,\ 2. \end{aligned} \tag{2.17}$$

We have introduced phase factors $\exp[-i\zeta_i(t)]$ into the basis $\psi_i(\mathbf{r},t)$, factors that otherwise could be in the expansion coefficients $c_i(t)$. These phase factors add flexibility that is helpful in addressing our problem, while not affecting the state occupation probabilities $P_i(t) = |c_i(t)|^2 = |\langle\psi_i(\mathbf{r},t)|\Psi(\mathbf{r},t)\rangle|^2$.

Since isolated atoms cannot have permanent electric dipole moments because of invariance under angular rotations, V has null diagonal matrix elements for energy eigenstates. Of course H_0 is diagonal for such states. If we insert ansatz (2.17) into the time-dependent Schrödinger equation, multiply the result by each $\psi_i^*(\mathbf{r},t)$ to obtain two equations, and then integrate each equation over all space, we have

$$\hbar\frac{d}{dt}\begin{pmatrix} c_1(t) \\ c_2(t) \end{pmatrix} = -i\begin{bmatrix} E_1 - \hbar\dot{\zeta}_1(t) & V_{12}(t)\exp(i\zeta_2 - i\zeta_1) \\ V_{21}(t)\exp(-i\zeta_2 + i\zeta_1) & E_2 - \hbar\dot{\zeta}_2(t) \end{bmatrix}\begin{pmatrix} c_1(t) \\ c_2(t) \end{pmatrix}. \tag{2.18}$$

Note that taking the time-dependent phases $\zeta_i(t)$ unequal to constants will add time-derivative terms $-\hbar\dot{\zeta}_i(t) \equiv -\hbar d\zeta_i(t)/dt$ to the diagonal matrix elements of $H(t)$ and will modify the phase of the off-diagonal elements $V_{ij}(t) \equiv V_{ij}\cos(\omega t + \phi)$ through the multiplicative factors $\exp[i\zeta_j(t) - i\zeta_i(t)]$. We now define $\hbar\omega_0 \equiv E_2 - E_1 > 0$ and make a particular phase choice:

$$\begin{aligned} \zeta_2(t) &= \zeta_1(t) + \omega t + \zeta_0 \\ \hbar\dot{\zeta}_1(t) &= E_1 + \tfrac{1}{2}(E_1 + E_2 - \hbar\omega) = E_1 + \tfrac{1}{2}\hbar(\omega_0 - \omega) \\ \hbar\dot{\zeta}_2(t) &= E_2 + \tfrac{1}{2}(E_1 + E_2 + \hbar\omega) = E_2 + \tfrac{1}{2}\hbar(\omega_0 + \omega). \end{aligned}$$

We also define $\hbar U^* \equiv V_{21}\exp(-i\phi + i\zeta_0)$. With our phase choice, Eq. (2.18) becomes

$$\frac{d}{dt}\begin{pmatrix} c_1(t) \\ c_2(t) \end{pmatrix} = -\frac{i}{2}\begin{bmatrix} -(\omega_0 - \omega) & U(1 + e^{-2i\omega t - 2i\phi}) \\ U^*(1 + e^{2i\omega t + 2i\phi}) & \omega_0 - \omega \end{bmatrix}\begin{pmatrix} c_1(t) \\ c_2(t) \end{pmatrix}.$$

No simple exact analytic solutions of this equation are known. However, there are a number of useful approximate solutions (Shore, 1990). One useful solution is a near-resonance, weak-perturbation approximation called the *rotating wave approximation* (RWA). The assumption made here is that the part of time evolution that is of interest is slow, occurring over many periods $T = 2\pi/\omega$ of the time-oscillating perturbation. We then time-average the two ordinary differential equations equivalent to the matrix equation just above; the averaging over one period T introduces the cycle-averaged quantities

$$\bar{c}_i(t) = \frac{1}{T}\int_t^{t+T} dt'\, c_i(t'), \qquad i = 1, 2$$
$$\frac{d}{dt}\bar{c}_i(t) = \frac{1}{T}\int_t^{t+T} dt' \frac{d}{dt'} c_i(t').$$

By our assumption, the interesting part of $c_i(t)$ changes very little in time T, so we can replace the cycle average of $\exp(2i\omega t)c_i(t)$ by its average value

$$\frac{1}{T}\int_t^{t+T} dt' \exp(2i\omega t')c_i(t') \cong \bar{c}_i(t)\int_t^{t+T} dt' \exp(2i\omega t') = 0.$$

Then our matrix equation reduces to

$$\frac{d}{dt}\begin{pmatrix} \bar{c}_1(t) \\ \bar{c}_2(t) \end{pmatrix} = -\frac{i}{2}\begin{bmatrix} -(\omega_0 - \omega) & U \\ U^* & \omega_0 - \omega \end{bmatrix}\begin{pmatrix} \bar{c}_1(t) \\ \bar{c}_2(t) \end{pmatrix}. \tag{2.19}$$

Defining $\omega_1, \omega_2 \equiv \pm(\omega_0 - \omega)$, Eq. (2.19) can be written out as

$$\frac{1}{i}\left(\frac{d\bar{c}_1}{dt}\right) + \omega_1\bar{c}_1 + U\bar{c}_2 = 0$$
$$\frac{1}{i}\left(\frac{d\bar{c}_2}{dt}\right) + \omega_2\bar{c}_2 + U^*\bar{c}_1 = \frac{i\gamma}{2}\bar{c}_2.$$

The only thing we have added to these last equations is a radiative decay term for the excited state ψ_2, a term proportional to the decay rate γ. We sometimes need to include this term. The decay term can be derived from the full second-quantized theory of atom interactions with electromagnetic fields (Loudon, 1983).

To proceed, we introduce new amplitudes

$$a_1(t) \equiv \bar{c}_1(t)\exp(-i\omega_1 t),$$
$$a_2(t) \equiv \bar{c}_2(t)\exp(-i\omega_2 t - \gamma t/2),$$

which then satisfy

$$\begin{aligned} &\frac{da_1}{dt}\exp(-i\chi t) + iUa_2 = 0, \\ &\frac{da_2}{dt}\exp(i\chi t) + iU^* a_1 = 0. \end{aligned} \tag{2.20}$$

Here $\chi \equiv \Delta\omega - i\gamma/2$ is the complex detuning. Taking the time derivative of the first of Eqs. (2.20), and then using the second of Eqs. (2.20) to eliminate da_2/dt, we have

$$\frac{d^2 a_1}{dt^2} - i\chi\frac{da_1}{dt} + |U|^2 a_1 = 0. \tag{2.21}$$

Similarly, the other amplitude a_2 satisfies Eq. (2.21), with the sign of the second term changed. We can solve the uncoupled second-order ordinary differential equation (2.21) by using either the method of characteristic exponents or the method of Laplace transforms. Assuming the atom is initially in the excited state ψ_2, the results are

$$\begin{aligned} a_1(t) &= \exp\left(i\frac{\chi}{2}t\right)\left[-\frac{i|U|}{\omega_p}\sin\frac{\omega_p}{2}t\right], \\ a_2(t) &= \exp\left(-i\frac{\chi}{2}t\right)\left[i\frac{\chi}{\omega_p}\sin\frac{\omega_p}{2}t + \cos\frac{\omega_p}{2}t\right]. \end{aligned} \tag{2.22}$$

Here $\omega_p^2 = \chi^2 + |U|^2$. For the case of no damping, $\gamma = 0$, and the time evolution of the *occupation probability* for state 1 becomes

$$P_1(t) = |a_1(t)|^2 = \frac{\omega_R^2}{(\Delta\omega)^2 + \omega_R^2}\sin^2\tfrac{1}{2}[(\Delta\omega)^2 + \omega_R^2]^{1/2}t. \tag{2.23}$$

The quantity $\omega_R \equiv eF_0\langle 1|z|2\rangle = |U|$ is proportional to the strength of the electromagnetic field and is called the *Rabi frequency.* Equation (2.23) exhibits oscillating time evolution, called Rabi oscillations, at the *Rabi-flopping frequency* $\Omega \equiv [(\Delta\omega)^2 + \omega_R^2]^{1/2}$. On resonance, $\Delta\omega = 0$; then the oscillations have the frequency ω_R and vary between zero and one. Far off the resonance, the Rabi flopping is very rapid and very small in amplitude. Then most of the occupation probability remains in the initial state, in spite of the driving electromagnetic field. We can parameterize Eq. (2.23) by defining $\tan\alpha \equiv \omega_R/\Delta\omega$ to obtain $P_1(t) = \frac{1}{2}\sin^2\alpha\ (1 - \cos\Omega t)$. The population inversion $P_2(t) - P_1(t)$ then becomes $1 - \sin^2\alpha(1 - \cos\Omega t)$.

Precision Rydberg atom experiments have been carried out that confirm the theory of Rabi flopping (Gentile et al., 1989). It is possible, but not easy, to find and couple only two levels of an atom that are sufficiently isolated from the rest of the levels. To avoid electric dipole radiative decay of the upper level down to the lower level, one can utilize a resonant two-photon transition between two levels having the same angular momentum quantum number. The extension of our theory to the two-photon case is discussed in Gentile et al. (1989). Some experimental results are compared with expectations in Fig. 2.11.

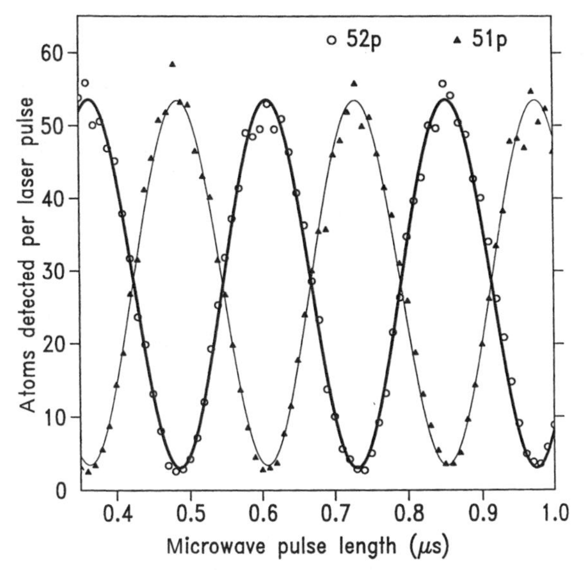

Figure 2.11. Time evolution of microwave two-photon Rabi oscillations for calcium atoms optically excited to the 52p state and then exposed to a pulse of microwave electric field. The frequency of the microwaves is two-photon resonant with the 52p → 51p transition. The microwaves are switched on at time 0.35 μs, and the microwave pulse length is varied. The measured final relative 52p and 51p occupation probabilities (○ and ▲ symbols) are shown as a function of microwave pulse length. The solid curves are a fit of the data to a sinusoidal function of time. From Gentile et al. (1989). Copyright 1989 by the American Physical Society.

2.3. ATOM INTERACTION WITH QUANTIZED RADIATION FIELDS

As we will see, the evolution of an atom in the presence of a quantized electromagnetic field can exhibit collapse and revivals, even when only two states are involved for the atom. Yet in Section 2.2, the results for a classical field exhibited only simple Rabi flopping. In order to obtain revivals we will prepare nonclassical states of the field. We will second-quantize both the field and atom, along the lines of Section 1.1e. From that section we conclude that the second-quantized vector potential for a monochromatic electromagnetic field in a large cavity of volume V is

$$\boldsymbol{A}_{\mathrm{k}}(\mathbf{r}, \mathrm{t}) = \sqrt{\hbar/(2\varepsilon_0 \mathrm{V}\omega)}[a\mathrm{e}^{-\mathrm{i}\omega\mathrm{t}+\mathrm{i}\mathbf{k}\cdot\mathbf{r}} + a^{\dagger}\mathrm{e}^{\mathrm{i}\omega\mathrm{t}-\mathrm{i}\mathbf{k}\cdot\mathbf{r}}]\,\hat{\boldsymbol{\varepsilon}}_{\mathrm{k}}.$$

Here $\hat{\boldsymbol{\varepsilon}}_{\mathrm{k}}$ is a unit vector parallel to $\boldsymbol{A}_{\mathrm{k}}$ and is called the *polarization vector.* The second-quantized electric field is $\boldsymbol{E}_{\mathrm{k}}(\mathbf{r}, \mathrm{t}) = \mathrm{i}\omega\boldsymbol{A}_{\mathrm{k}}(\mathbf{r}, \mathrm{t})$, and the interaction Hamiltonian is

$$H' = -\boldsymbol{D}\cdot\boldsymbol{E}_{\mathrm{k}}(0, \mathrm{t}) = -\mathrm{i}\omega\boldsymbol{D}\cdot\boldsymbol{A}_{\mathrm{k}}(0, \mathrm{t}), \tag{2.24}$$

where $\boldsymbol{D}$ is the dipole moment operator for the atom. We have temporarily written the field as $\boldsymbol{E}_{\mathrm{k}}(0, \mathrm{t})$ to denote our assumption that the field is uniform over the spatial size of the atom—that is, the field's wavelength is not much above optical.

We formally second-quantize the atom by introducing *atom transition operators* that are outer products of orthonormal energy eigenstates. For a two-state atom, these operators are

$$S_{12} \equiv |1\rangle\langle 2|, \qquad S_{21} \equiv |2\rangle\langle 1|,$$

and they indeed switch atomic states, for example, $S_{12}|2\rangle = |1\rangle\langle 2|2\rangle = |1\rangle$. These operators clearly satisfy $S_{21}S_{12} = |2\rangle\langle 1|1\rangle\langle 2| = |2\rangle\langle 2|$, $S_{12}S_{21} = |1\rangle\langle 1|$, as well as $S_{12}S_{12} = S_{21}S_{21} = 0$.

It is possible to write the Hamiltonian for the atom as

$$H_{\mathrm{a}} = \mathrm{E}_1|1\rangle\langle 1| + \mathrm{E}_2|2\rangle\langle 2| \equiv \mathrm{E}_1 S_{11} + \mathrm{E}_2 S_{22},$$

where E_1 and E_2 are the two possible energies of the atom. This expression is justified, since correct results are obtained when H_{a} operates on each element of the entire basis set, $|1\rangle$ and $|2\rangle$. Using Eq. (1.14) as the Hamiltonian for the field and using H_{a} for the atom, we write the full second-quantized Hamiltonian for the atom–field system as

$$H = \hbar\omega a^{\dagger}a + \mathrm{E}_1 S_{11} + \mathrm{E}_2 S_{22} + \hbar \mathrm{g}(a^{\dagger} + a)(S_{12} + S_{21}). \tag{2.25}$$

Here g is called the atom–field coupling constant. The term proportional to g is the atom–field interaction: it includes all the possible products of field and atom transition operators that are in accord with the linearity of Eq. (2.24). Two of these products, $a^{\dagger}S_{21}$ and aS_{12}, represent simultaneous atom and field excitation and de-excitation, respectively. The terms with these products are not energy-conserving—thus these two products represent "virtual" processes. To be in accord with the uncertainty principle, the results of the actions of the two energy-nonconserving terms must be accompanied by additional transitions within very short times. Adequate transition rates for this require relatively high field strengths, that is, high values of $\mathrm{g} \gtrsim 0.01\omega_0$. Omitting the energy-nonconserving terms defines the *Jaynes–Cummings model* (JCM).

In the literature it is common to choose the energy levels to be $\pm\frac{1}{2}\hbar\omega_0$. In practice it is also common to define atomic operators σ_{x}, σ_{y}, σ_{z}, that have the mathematical structure of the Pauli spin matrices, as follows:

$$\begin{aligned}
\sigma_{\mathrm{x}} &\equiv S_{12} + S_{21}, \\
\sigma_{\mathrm{y}} &\equiv \mathrm{i}(S_{12} - S_{21}), \\
\sigma_{\mathrm{z}} &\equiv S_{22} - S_{11}, \\
\sigma_{+,-} &\equiv \sigma_{\mathrm{x}} \pm \mathrm{i}\sigma_{\mathrm{y}}.
\end{aligned}$$

Then the *full* Hamiltonian can be written as $H = \hbar\omega a^\dagger a + \frac{1}{2}\hbar\omega_0\sigma_z - \hbar g(a - a^\dagger)(\sigma_+ + \sigma_-)$. The Hamiltonian for the Jaynes–Cummings model then becomes

$$H^{\mathrm{JCM}} = \hbar\omega a^\dagger a + \tfrac{1}{2}\hbar\omega_0\sigma_z - i\hbar g(a\sigma_+ - a^\dagger\sigma_-). \tag{2.26}$$

Several different complete sets of conserved quantities for the *full* Hamiltonian exist, as can be verified by computing the appropriate commutators. One set contains $a^\dagger a$ and σ_z; another set is H and the parity operator $P = \exp[i\pi(a^\dagger a + \frac{1}{2}\sigma_z + \frac{1}{2})]$.

For the approximate Hamiltonian H^{JCM}, the operator $K \equiv a^\dagger a + \frac{1}{2}\sigma_z$ is also conserved. Then the JCM problem can be solved (Jaynes and Cummings, 1963). The problem reduces to that of an infinite set of uncoupled two-state Schrödinger equations, where each pair of states is identified by the number of photons present when the atom lies in its *lower* energy state. In other words, the general solution is the sum of the solutions of an infinite set of independently evolving Rabi-flopping problems! For each specified photon number n, the atom–field state becomes

$$\Psi(n, t) = \exp(-in\omega t - E_1 t/\hbar)[c_1(n, t)\phi_1(n) + c_2(n, t)\phi_2(n)],$$

where the basis states are the atom–field product states

$$\phi_1(n) = |n\rangle\psi_1, \qquad \phi_2(n) = |n-1\rangle\psi_2.$$

The Hamiltonian matrix representation of Eq. (2.26) then takes the form

$$\bar{\bar{H}}(n) = \frac{\hbar}{2}\begin{bmatrix} 0 & \Omega(n) \\ \Omega(n) & 2\Delta \end{bmatrix},$$

where Δ is the frequency detuning and $\Omega(n) = n^{1/2}g$ is called the *n-photon Rabi frequency*. The factor $n^{1/2}$ stems from the raising or lowering of the photon-number basis states $|n\rangle$. Our general solution of the JCM problem can then be written as

$$\Psi(t) = \sum_n f_n\Psi(n, t) = \sum_{n,m} f_n c_m(n, t)|n - m + 1\rangle\psi_m, \tag{2.27}$$

with $m = 1, 2$. By selecting the coefficients f_n appropriately, we can form various wavepackets and study their time evolution, much as in Section 2.1b. The difference is that we now have packets of field states rather than packets of atomic states. The dependence of $\Omega(n)$ on n will combine with the nonlinear dependence of the Rabi-flopping frequency $\omega_p(n)$ on $\Omega(n)$ to produce superpositions of states with incommensurate ω_p that must collapse.

2.4. EVOLUTION OF FULLY QUANTIZED DRIVEN TWO-LEVEL ATOMS

2.4a. Coherent Photon States

One of the most interesting initial states we can use to investigate the evolution of a quantum system is the coherent state. By definition, coherent states evolve almost classically when the level of energy excitation is high. The coherent state of a monochromatic (single cavity mode) second-quantized electromagnetic (EM) radiation field—which is relatively easy to construct and work with analytically—is chosen as the initial field state in this section.

In analytical quantum optics work, we usually use the photon-number states $|\mathrm{n}\rangle$ as the basis set because the action of the operators a, $a^\dagger$ on this set is so simple. Let $\alpha \equiv |\alpha|e^{i\theta}$ be the polar form of an arbitrary complex number; this complex number will determine the degree of energy excitation of our field state. We then define the *coherent photon (EM-field) state* $|\alpha\rangle$ by

$$|\alpha\rangle \equiv \left[\exp(-\tfrac{1}{2}|\alpha|^2)\right] \sum_{n=0}^{\infty} \frac{\alpha^n}{\sqrt{n!}} |n\rangle \equiv \sum_{n=0}^{\infty} f_n(\alpha)|n\rangle. \tag{2.28}$$

It is easy to show that this state is normalized, $\langle\alpha|\alpha\rangle = 1$. However, the set of coherent (photon) states is not orthogonal, $|\langle\alpha|\beta\rangle|^2 = \exp(-|\alpha - \beta|^2)$. The action of a on $|\alpha\rangle$ is

$$a|\alpha\rangle = \left[\exp(-\tfrac{1}{2}|\alpha|^2)\right] \sum_n \frac{\alpha^n}{\sqrt{n!}} \sqrt{n}|n-1\rangle = \alpha|\alpha\rangle,$$

where the last step requires redefining the summation index as $n' \equiv n - 1$. Thus $|\alpha\rangle$ is the eigenfunction of the destruction operator (but *not* also of the creation operator $a^\dagger$). The mean number of photons of an EM cavity mode in a coherent state is the expectation value

$$\langle\alpha|n|\alpha\rangle = [\exp(-|\alpha|^2)] \sum_n \frac{(\alpha^*)^n}{\sqrt{n!}} \frac{\alpha^n}{\sqrt{n!}} \langle n|n|n\rangle = [\exp(-|\alpha|^2)]|\alpha|^2 \sum_n \frac{(|\alpha|^2)^{n-1}}{(n-1)!} = |\alpha|^2,$$

and similarly $\langle\alpha|n^2|\alpha\rangle = |\alpha|^4 + |\alpha|^2$. This means that the root mean square (r.m.s.) spread $\Delta n \equiv \sqrt{(\langle n^2\rangle - \langle n\rangle^2)} = |\alpha|$. Thus the fractional width in n of a coherent state is $1/|\alpha|$, which becomes increasingly small as $|\alpha|$ becomes large. In other words, our coherent state is really a wavepacket of photon-number states, the energy eigenstates of the field. The probability distribution of the number states $|n\rangle$ contained in the packet is

$$P(n) = |\langle n|\alpha\rangle| = [\exp(-|\alpha|^2)] \frac{|\alpha|^{2n}}{n!}, \tag{2.29}$$

which is a Poisson distribution about the mean value $|\alpha|^2$.

It is possible to define an operator conjugate to n and to show, with effort, that $|\alpha\rangle$ has the minimum uncertainty product (Loudon, 1983). Equally interesting is the expectation value of the magnitude of the electric field

$$\begin{aligned}\langle\alpha|\boldsymbol{E}_k|\alpha\rangle &= i\sqrt{\frac{\hbar\omega}{2\varepsilon_0 V}}[\langle\alpha|a\exp(-i\omega t + i\mathbf{k}\cdot\mathbf{r})|\alpha\rangle - \langle\alpha|a^\dagger\exp(i\omega t - i\mathbf{k}\cdot\mathbf{r})|\alpha\rangle] \\ &= \mathcal{E}_0\sin[(\mathbf{k}\cdot\mathbf{r}) - \omega t + \theta]. \end{aligned} \tag{2.30}$$

Here $\mathcal{E}_0 \equiv \sqrt{(\hbar\omega/2\varepsilon_0 V)}|\alpha|$; we have used $\langle\alpha|a^\dagger|\alpha\rangle = \alpha^*$, and the phase shift is the quantity θ present in the definition of α. Thus the electric field of a coherent state evolves classically for *all* α, strongly supporting the idea that a coherent quantum state should be the most nearly classical quantum state. One can similarly compute the r.m.s. spread $\Delta\mathcal{E}_k$ for $\mathcal{E}_k$, which is $\sqrt{(\hbar\omega/2\varepsilon_0 V)}$, independent of α. So the field resolution $\Delta\mathcal{E}_k/\mathcal{E}_0$ of the coherent states again varies as $1/|\alpha|$; this resolution is small at large excitation $|\alpha|$.

2.4b. The Phase-Space Q-Function of an Interacting Coherent Photon State (Advanced Topic)

Our coherent state depends solely on the location of the complex number α in the complex plane. For a single electromagnetic cavity mode, the complex plane is therefore the state's phase space. We can expand any second-quantized quantum state of the electromagnetic field in terms of the coherent states of different α and $\mathbf{k}$, and then watch how that state evolves in this phase space. To display the evolution, the phase-space probability density distribution (see Section 1.3d) is usually taken to be the Q-function—which is a particular projection of the field state onto the coherent-state basis set. Without going into the mathematical formalism for a rigorous derivation (Eiselt and Risken, 1991), let us just be aware of the following. For our JCM model of a two-level atom in our quantized electromagnetic field, we can construct the Q-function for an initial coherent photon (field) state in terms of two sets of quantities: the photon-number expansion coefficients $f_n(\alpha)$ for the field and the expansion coefficients $C_m(n, t)$ for the atomic states. The result of the construction is

$$Q(\alpha, t) = \sum_n\sum_m f_{n-m+1}(\alpha) f_n(\alpha) C_m(n, t). \tag{2.31}$$

Here $C_1(n, t)$ and $C_2(n, t)$ are the two atomic probability amplitudes, and $f_n(\alpha) \equiv \langle n|\alpha\rangle$ are coherent-state expansion coefficients defined by Eq. (2.28). As shown in Fig. 2.12, $Q(\alpha, t)$ initially appears as a Gaussian distribution in the complex α plane. For the choice of α real, this distribution has its mean location on the horizontal axis at a distance $\sqrt{\langle n\rangle}$ from the origin. As time increases, the distribution bifurcates into two peaks that move in opposite directions around a circular path with radius $\sqrt{\langle n\rangle}$. As the peaks meet at the opposite side of the circle from the start, we

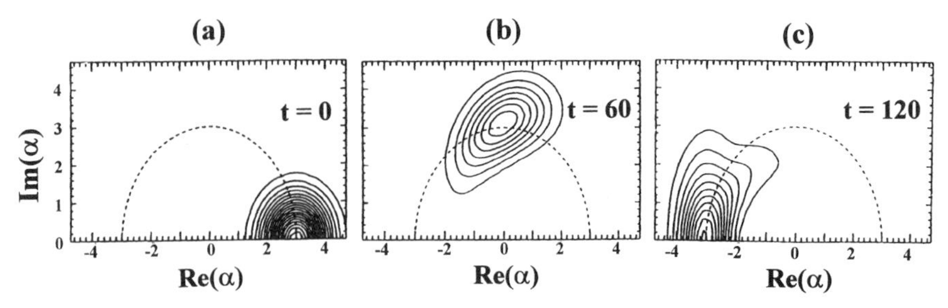

Figure 2.12. Snapshots of the time evolution of the photon quasiprobability Q-function, for an initially excited single two-level atom interacting with an electromagnetic field that is initially in a coherent state. The field has an initial mean photon number $\langle n \rangle = 9$. The unshown lower half of the complex α plane is a mirror image of the upper half shown. From Shore and Knight (1993). Copyright 1993 Taylor & Francis.

would expect some sort of revival. Further revivals should occur as the peaks repeatedly pass each other along the circular track. The situation appears similar to that of revivals of electron wavepackets orbiting in the hydrogen atom, except that now we have photon wavepackets orbiting in their phase space, the complex α plane.

2.4c. Collapse and Revivals of a Two-Level Atom in a Coherent-State Electromagnetic Field

Using the JCM model, we now wish to compute the time evolution of the state probabilities of a two-level atom that is exposed to a single coherent-state photon wavepacket. We proceed as in Sections 1.3c and 2.1b. We express the initial field as a superposition of photon-number states and, for convenience, take the atom to be initially *excited*. The initial state is then known, and from Eq. (2.27) this state must be equal to

$$\Psi(0) \equiv \sum_{n=0}^{\infty} f_n \Psi(n, 0).$$

Since the JCM has the property that each state $\Psi(n, t)$ in Eq. (2.27) evolves independently, the probability for observing the atom in state m at time t is

$$P_m(t) = \langle S_{mm} \rangle = \sum_{n=0}^{\infty} |f_n c_m(n, t)|^2. \tag{2.32}$$

Figure 2.13 shows the evolution of the atomic population inversion; again this is $P_2(t) - P_1(t)$, now for *initial* coherent field states with $f_n = f_n(\alpha)$ and shown for particular different values of $\langle n \rangle = |\alpha|^2$. Since the atom is initially excited and the excitation energy can pass from the atom to the field, for $\langle n \rangle = 0$ there is simple Rabi flopping at the Rabi frequency ω_R; see the end of Section 2.2. However, at

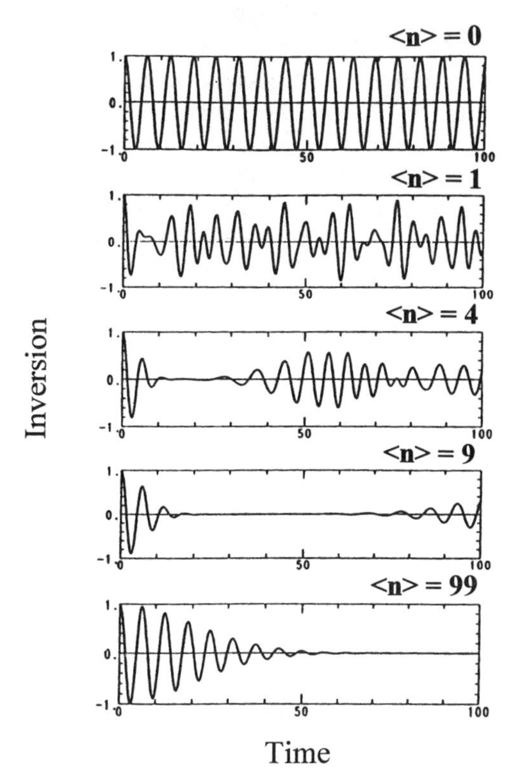

Figure 2.13. Time evolution of the population inversion $P_2(t) - P_1(t)$ of an initially *excited* two-level atom in an electromagnetic cavity field that is initially in a coherent state. The mean photon numbers of the *initial* field vary from $\langle n \rangle = |\alpha|^2 = 0$ (top) to $\langle n \rangle = 99$ (bottom). The times are in units of the inverse of the mean (angular) Rabi frequency. At the higher values of $\langle n \rangle$, the Rabi oscillations completely collapse for awhile. From Shore and Knight (1993). Copyright 1993 Taylor & Francis.

"large" $|\alpha|$ (even the value 3), there is a collapse and then a revival at the revival time

$$T_r = 4\pi \frac{\sqrt{\langle n \rangle}}{\omega_R}.$$

The revival time and the collapse time $2/\omega_R$ are well understood (Eberly et al., 1980; Narozhny et al., 1981). It can be shown that the revival time is essentially the inverse of the separation between neighboring distinct Rabi-flopping frequencies $\Omega(n)$. We should note that the "dormant" period seen between revivals in the population inversion is a bit misleading. Both the atomic dipole moment vector and the field continue to evolve during these time intervals. Such evolution is also readily seen in the expectation values of other operators of our problem, such as $\langle S_{21} \rangle$ (Narozhny et al., 1981).

2.4d. Strong Fields (Advanced Topic)

The JCM has proved quite useful for exploring quantum time evolution, because of its abundant generalizations that include further physical effects. These generalizations include varying the initial conditions; including models of dissipation and damping; treating more than one atom in the system; including more than two atomic levels; using multimode fields; and treating multiphoton interactions rather than the original single-photon electric dipole interaction.

Yet one basic question remains—it concerns the approximation that makes the JCM solvable and involves the neglect of the energy-nonconserving terms in Eq. (2.25). This approximation is actually a quantized-field version of the RWA (Section 2.2), indicating that the approximation will break down at strong field strengths. As mentioned earlier, the quantity $K = aa^{\dagger} + \frac{1}{2}\sigma_3$ will then no longer be conserved, the problem is not analytically solvable, and computational explorations are then necessary.

Precise basis-set expansion calculations for the two-level atom and a coherent field state—without the RWA—have been carried out in the (*H*, *P*) representation mentioned in Section 2.3. As many as 400 basis states were included in the calculations performed by Graham and Höhnerbach (1984). As we see in Fig. 2.14, differences from the RWA evolution already appear when the coupling constant g is $0.01\omega_0$. Within the "dormant" region between the revivals (Fig. 2.13), there now

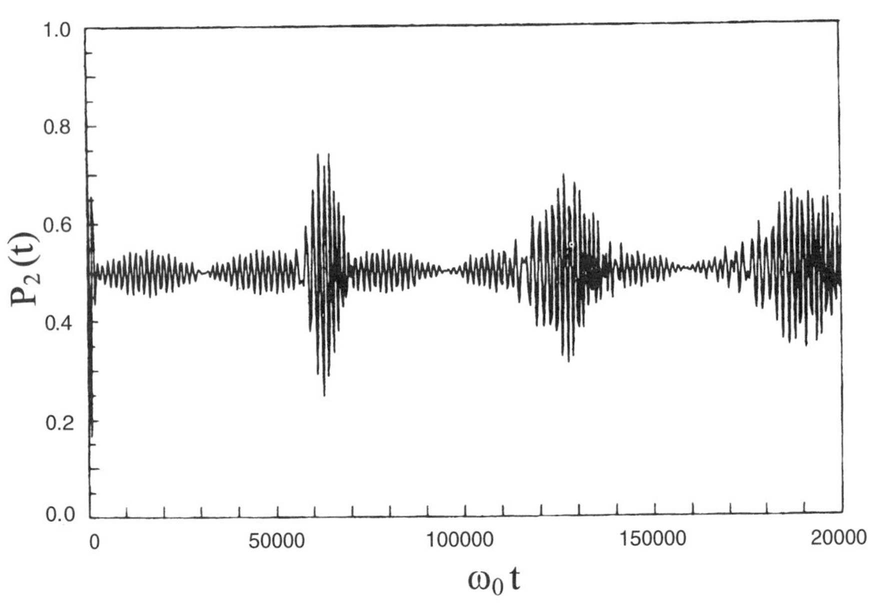

Figure 2.14. Precise numerical results for the time evolution of the occupation probability $P_2(t)$ for a two-level atom in a resonant electromagnetic field, $g/\omega_0 = 0.01$ and $\alpha = 10$. Initially the atom is prepared in the energy eigenstate 2 having the upper energy E_2, while the field is prepared in a coherent state, Eq. (2.28). From Graham and Höhnerbach (1984). Copyright 1984 Springer-Verlag.

appear Rabi oscillations of considerable amplitude. There is total destructive interference only in the middle between revivals. This destructive interference is due to cancellation between contributions of the states of positive and negative parity P. These definite-parity states taken by themselves would each have maxima at the midpoint between revivals. Thus the change of the RWA constant of motion K over to the exact one P leads to a reduction of the destructive interference in the population inversion. This reduction is from a whole time interval in-between the revivals to a single point in time.

At the higher value $g = 0.05\omega_0$, "revivals" are still discernible but no longer prominent. The Rabi oscillations in the intermediate time intervals have about the same amplitude as the "revivals." At $g \geq 0.1\omega_0$, the evolution depends strongly on g/ω_0 and is complex, displaying all kinds of different behavior as parameters are varied. Beginning with Chapter 5, chaos will demand a great deal of our attention. If the field and spin are taken to be classical, then there is large-scale chaotic behavior when $4g/(\omega\omega_0)^{1/2} \gtrsim 1$. This is the generic situation for an analytically unsolvable quantum problem, with at least some chaos appearing in the corresponding classical problem.

2.4e. Two Atoms (Advanced Topic)

Another critical area of study concerns the many-body limit—meaning we take the number of atoms N that are simultaneously in the electromagnetic field to be large, rather than equal to one. The many-body limit again tends to produce chaos in the classical system. The parameter for breakdown of the RWA generalizes to

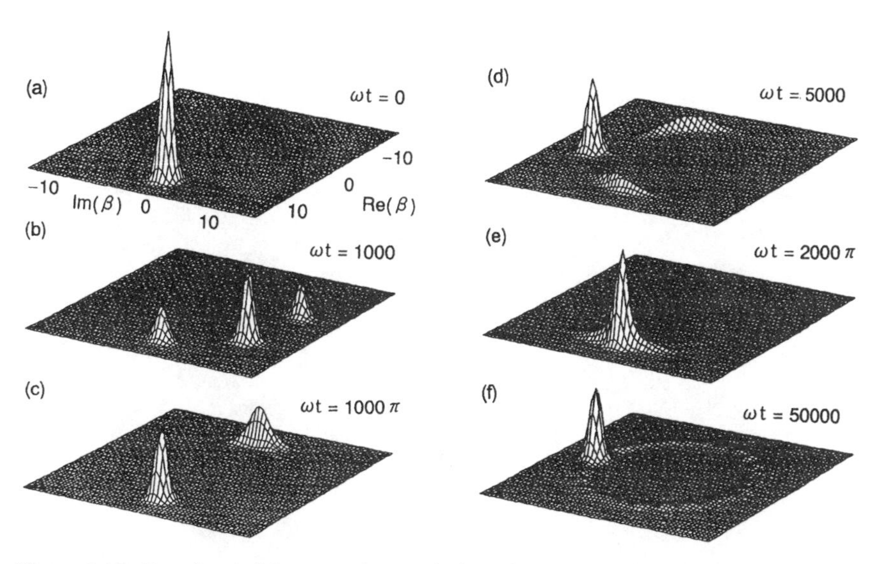

Figure 2.15. Snapshots of the exact time evolution of the quasiprobability Q-function for *two* initially excited two-level atoms in a resonant coherent electromagnetic field, $g/\omega_0 = 0.01$ and $\alpha = 10$. From Ueda et al. (1996). Copyright 1996 by the American Physical Society.

$4g(N/\omega\omega_0)^{1/2}$. Exploration of the fully quantized system of N two-level atoms in a monochromatic electromagnetic field holds the promise, as yet unfulfilled, of resolving the issues of the two "classical" limits of N large and of $\hbar$ small, as well as the issue of the limit of large times t. A recent small step in this direction is the discovery of an exact analytical solution for $N=2$, for the special case where both atoms are initially excited and the field is initially a coherent photon state (Ueda et. al., 1996). There are three distinct wavepackets in the evolution seen in the Q-function; see Fig. 2.15. This contrasts with the two wavepackets seen for one atom; see Fig. 2.12. Ueda and co-workers find that oscillations in the atomic dipole fluctuations and in measures of the photon statistics are synchronized. Experiments would require both a single-mode cavity field and a long-term coherent atom–field interaction, with tens of thousands of field oscillations. It soon may be possible to create these conditions, either by using very cold (laser-cooled) atoms, or by using diluted molecules in special solid state crystals that exhibit very narrow (10 MHz) spectral linewidths.

3

FURTHER TOPICS IN CLASSICAL THEORY

3.1. LINEAR STABILITY ANALYSIS

As we've seen, geometric structures in phase space play an important role in the time evolution of both quantum and classical systems. Stability is a key property of such geometric structures, determining what role each structure will play. If we start a system's (mean) initial conditions located on a structure, one kind of time evolution results. If instead we start somewhat displaced off the structure, then what will the evolution be?

We can imagine several possibilities, including (1) a very different kind of evolution, (2) no qualitative difference, or (3) a return back onto the phase-space structure. In the first case, we would be inclined to call the structure unstable with respect to changes in initial conditions, whereas the other two cases would be stable in some sense. However, the third case does not occur in classical Hamiltonian systems, as dissipation would be required (Lichtenberg and Lieberman, 1992).

In this section, we will investigate some concrete notions of stability in Hamiltonian classical physics. The ideas will be developed for two simple structures that are central to dynamical systems: the equilibrium (fixed) point and the periodic orbit. For the periodic orbit, we will introduce the monodromy matrix, Eq. (3.4), which describes the stability of the orbit. Monodromy matrices are not only important for classical evolution; they also appear in those expressions for quantum evolution that are given by semiclassical periodic orbit theory, Eqs. (7.16) and (7.22). Procedures for numerical determinations of monodromy matrices are discussed in Section A.4.

3.1a. Stability of Fixed Points

In Sections 1.1b and 1.1c we considered a classical trajectory $\mathbf{Z}(t)$ lying within the phase space of a dynamical system. A trajectory is a solution of the differential equations of evolution $d\mathbf{Z}/dt = \mathbf{v}(\mathbf{Z})$ for some initial condition $\mathbf{Z}(0) \equiv \mathbf{Z}_0$. Recall that $\mathbf{v}(\mathbf{Z})$ is the phase-space velocity function; this function determines the local evolution at any and all points located by values of $\mathbf{Z}$ in phase space. A *fixed point* in phase space is a point $\mathbf{Z}^{FP}$ where $\mathbf{v}(\mathbf{Z}^{FP}) = 0$. Such a point is also quite accurately called an *equilibrium point.* It is clear that a fixed point should be a zero of $\mathbf{v}(\mathbf{Z})$—for if $\mathbf{Z}_0 = \mathbf{Z}^{FP}$ and $\mathbf{v}(\mathbf{Z}^{FP}) = 0$ at time $t = 0$, then $\mathbf{Z}(t) = \mathbf{Z}^{FP}$ for all times $t > 0$. For a Hamiltonian system at a fixed point, we also have $v_i = (\partial H/\partial p_i, -\partial H/\partial q_i) = (0, 0)$ and hence also $\nabla_Z H = (\partial H/\partial \mathbf{q}, \partial H/\partial \mathbf{p}) = 0$.

We define a *stable fixed point* $\mathbf{Z}^{FP}$ by the following requirement: for all sufficiently small $\delta > 0$, there exist $\epsilon(\delta)$ such that $|\mathbf{Z}(t) - \mathbf{Z}^{FP})| < \delta$ for all t, whenever $\mathbf{Z}_0$ satisfies $|\mathbf{Z}_0 - \mathbf{Z}^{FP}| < \epsilon$. In the neighborhood of $\mathbf{Z}^{FP}$, we then can make a Taylor series expansion of $\mathbf{v}(\mathbf{Z})$ about $\mathbf{Z}^{FP}$ and keep only the first partial derivatives. This procedure is called *linearization* of the phase-space flow. For the example of a 2-D dynamical system with phase-space variables x, p,

$$\begin{pmatrix} v_x(x,p) \\ v_p(x,p) \end{pmatrix} = \begin{bmatrix} \dfrac{\partial v_x}{\partial x} & \dfrac{\partial v_x}{\partial p} \\[2ex] \dfrac{\partial v_p}{\partial x} & \dfrac{\partial v_p}{\partial p} \end{bmatrix}_{\mathbf{Z}=\mathbf{Z}^{FP}} \begin{pmatrix} x - x^{FP} \\ p - p^{FP} \end{pmatrix} + \mathcal{O}(\mathbf{Z} - \mathbf{Z}^{FP})^2. \tag{3.1}$$

Put in matrix notation, $\bar{v} = \bar{\bar{M}}(\bar{Z} - \bar{Z}^{FP})$. The Jacobian matrix $\bar{\bar{M}}$ of first partial derivatives is called the *stability matrix*. Usually Eq. (3.1) is written with $\mathbf{Z}^{FP} = \mathbf{0}$. We can do this because a translation $\mathbf{Z}' = \mathbf{Z} - \mathbf{Z}^{FP}$ means that

$$\frac{d\mathbf{Z}'}{dt} = \frac{d\mathbf{Z}}{dt} = \mathbf{v}(\mathbf{Z}) = \mathbf{v}(\mathbf{Z}^{FP} + \mathbf{Z}') = \mathbf{v}'(\mathbf{Z}')$$

for some function $\mathbf{v}'(\mathbf{Z}')$ that has $\mathbf{v}'(\mathbf{0}) = \mathbf{v}(\mathbf{Z}^{FP}) = \mathbf{0}$.

Generalizing Eq. (3.1), we have the set of n equations of evolution for a linearized classical Hamiltonian system near a fixed point $\mathbf{Z}^{FP}$:

$$d\mathbf{Z}/dt = \mathbf{MZ}. \tag{3.2}$$

Here the stability matrix $\bar{\bar{M}} \equiv \mathbf{M}$ is a constant matrix, having $[\partial v_i/\partial Z_j]_{\mathbf{Z}=\mathbf{Z}^{FP}}$ in its ith row and jth column. (Note we temporarily omit the notation indicating that $\mathbf{M}$ is a matrix.) To solve the set (3.2) of ordinary differential equations with constant coefficients, we know that we can use the method of characteristic exponents, also called the method of normal modes. We take $\mathbf{Z}(t) = \exp(st)\mathbf{u}$, where $\mathbf{Mu} = s\mathbf{u}$, so that the exponent s is the eigenvalue of $\mathbf{M}$ belonging to the eigenvector $\mathbf{u}$. The eigenvalues $(s_1, \ldots, s_n)$ are the zeros of the characteristic polynomial $\det|\mathbf{M} - s\mathbf{I}|$.

Note that $\mathbf{M}$ is real. Usually the set of eigenvectors $\{\mathbf{u}_1, \ldots, \mathbf{u}_n\}$ spans the space of $\mathbf{u}$. If so, there exist constants $\xi_1, \xi_2, \ldots, \xi_n$ such that

$$\mathbf{Z}(0) = \sum_{j=1}^{n} \xi_j \mathbf{u}_j.$$

Then the solution of Eq. (3.2) is

$$\mathbf{Z}(t) = \sum_{j=1}^{n} \xi_j \exp(s_j t)\mathbf{u}_j, \qquad \text{for } t \geq 0. \tag{3.3}$$

This result reminds us of the evolution equation (1.33) of an initial wavepacket, where the equation was written in terms of an initial packet expansion in the eigenvectors of the Hamiltonian operator that was responsible for the evolution.

In the present case, since $\mathbf{M}$ and $\mathbf{Z}(0)$ are real, each set $(\xi_j, s_j, \mathbf{u}_j)$ is either a real triplet or one of a conjugate pair of triplets. Each eigenvalue s_j corresponds to a normal mode of the system; that is, each eigenvalue s_j corresponds to a possible independent evolution. If an $\mathrm{Re}(s_j)$ is positive (negative), then we have an exponentially growing (decaying) evolution that is unstable (stable). If $\mathrm{Re}(s_j) = 0$ for all j, then we have oscillatory evolution, which is stable. In that case we call $\mathbf{Z}^{FP}$ an *elliptic fixed point*. One example is the vertical pendulum hanging straight down and not moving. The name "elliptic fixed point" comes from the fact that, in a 2-D phase space, evolution of a nearby initial condition traces out an ellipse; see Fig. 1.15b.

For a simple example of how the stability analysis of fixed points proceeds, let us consider the vertical pendulum (Fig. 1.15), choosing the range of the pendulum angle variable θ to be $(-\infty, \infty)$. To find the fixed points, we set the right-hand sides of Eqs. (1.40) equal to zero. We immediately obtain an infinite set of fixed points $(\theta_n, p_{\theta_n}) = (\pm n\pi, 0)$, $n = 0, 1, 2, \ldots$. The linearized equation (3.2) that is needed for the stability analysis becomes

$$\frac{d}{dt}\begin{bmatrix} \delta\theta \\ \delta p_\theta \end{bmatrix} = \begin{bmatrix} 0 & 1 \\ -\frac{g}{L}\cos\theta_n & 0 \end{bmatrix} \begin{bmatrix} \delta\theta \\ \delta p_\theta \end{bmatrix}.$$

We explicitly indicate here that we are considering differential displacements $(\delta\theta, \delta p_\theta)$ away from the fixed point; see Eq. (3.1). The eigenvalues of the stability matrix $\bar{\bar{M}}$ becomes $s_{1,2} = \pm i[(g/L)\cos\theta_n]^{1/2}$. At the bottom of one of the cosine potential wells, $\theta_n = 2n\pi$ for some n, $s_{1,2} = \pm i\sqrt{(g/L)}$, and we have a stable fixed point. When we are at the top of one of the barriers, $\theta_n = (2n + 1)\pi$ for some n, $s_{1,2} = \pm\sqrt{(g/L)}$, and the fixed point is unstable.

We noted in Section 1.4b that, for the pendulum, the time evolution on the separatrix trajectory is easily calculated. The unstable fixed points are "end points" of this evolution. So we might expect connections between the parameters of the separatrix evolution and the parameters $s_{1,2}$ of the fixed points. Indeed, these parameters are the same—namely, $\sqrt{(g/L)}$.

Any one of the unstable fixed points is an end point for four "branches" of the separatrix. These branches are 1-D examples of what more generally are multi-dimensional manifolds in phase space. The arrows in Fig. 1.15b indicate the directions of the evolution along each of these manifolds. We call each of the two manifolds having direction toward the unstable fixed point (i.e., away from the region of a neighboring stable fixed point) an ingoing or *stable manifold*. Similarly, each of the other two manifolds we call an outgoing or *unstable manifold*. The approach to the unstable fixed point along a stable manifold varies approximately as $\exp(st) = \exp[-\sqrt{(g/L)}t]$. Similarly, the departure from the unstable fixed point along an unstable manifold evolves according to $\exp(st) = \exp[+\sqrt{(g/L)}t]$. In this pendulum example, the *exponentiation rates* for the evolutions along the separatrix stable/unstable manifolds are $\pm\sqrt{(g/L)}$.

3.1b. Stability of Periodic Orbits

Next to fixed points—which are 0-D structures in phase space—the next least-complicated structures are the trajectories themselves, which are 1-D in phase space. We are interested in whether or not a trajectory is *orbitally stable*. If it is stable, neighboring trajectories in phase space will remain close to it. The situation is different from the stability for fixed points, discussed above. Now we are not concerned with the system's location on a given orbit at a given time. To see the difference, consider uniform evolution on a circle of radius r_0 in coordinate space. The individual points on the circular orbit are unstable. This is so because the evolution for such a point separates from the evolution for a nearby point located on a neighboring concentric circle of radius $r_0 + \delta r$, according to $|\mathbf{R}(t) - \mathbf{R}_0(t)| = [(r_0 + \delta r)^2 - 2r_0(r_0 + \delta r)\cos[(\delta r\ t) + r_0^2]^{1/2}$. This separation of points will not remain small, no matter how small δr is taken. Yet the orbits traced by $\mathbf{R}(t)$ and $\mathbf{R}_0(t)$ in the plane remain close at all times; they are concentric circles, separated by the small parameter δr. It is only the polar angles—that is, the phases of $\mathbf{R}(t)$ and $\mathbf{R}_0(t)$—that do not remain close. The evolution of $\mathbf{R}_0(t)$ is orbitally stable.

We want to introduce a quantitative measure of the stability of a trajectory, with respect to small changes in the trajectory's initial values. For a phase-space trajectory $\mathbf{Z}(t)$ with initial conditions $\mathbf{Z}(0)$, we define the *monodromy matrix* $\bar{\bar{U}} \equiv \mathbf{U}$ (Arnold and Avez, 1968) by

$$U_{ij}(t) = \frac{\partial Z_i(t)}{\partial Z_j(0)}. \tag{3.4}$$

(For simplicity, we temporarily omit the notation that reminds us that $\mathbf{U}$ is a matrix.) Consider now two trajectories $\mathbf{Z}(t)$ and $\mathbf{Z}_0(t)$ with infinitesimal initial separation $\mathbf{z}(0) \equiv \mathbf{Z}(0) - \mathbf{Z}_0(0)$. We define the *trajectory difference function* $\mathbf{z}(t) \equiv \mathbf{Z}(t) - \mathbf{Z}_0(t)$. Notice that $\mathbf{Z}_0$ is now a trajectory, whereas in the previous subsection, it located a point and was equal to $\mathbf{Z}(0)$. As long as $\mathbf{z}(t)$ remains small, a Taylor series expansion and definition (3.4) give

$$\mathbf{z}(t) = \mathbf{U}(t)\mathbf{z}(0), \qquad \mathbf{U}(0) = \mathbf{1}. \tag{3.5}$$

Thus $\mathbf{U}(t)$ is a linearized time evolution operator that determines the evolution of differences of the coordinates and momenta of neighboring trajectories, in terms of the initial values of those differences. If $\mathbf{Z}_0(t)$ is called the *reference trajectory*, we can say that $\mathbf{U}(t)$ determines the evolution of the deviations of the coordinates and momenta from the reference trajectory, starting from specified initial deviations. To determine whether or not a trajectory is orbitally stable, we will have to investigate the growth of $\mathbf{U}(t)$ with time t. Using the chain rule and Hamilton's equations $d\mathbf{Z}/dt = \mathbf{J}\cdot\nabla_Z H(\mathbf{Z})$—see Eq. (1.8) and just above it—we obtain the differential equation that determines the evolution of $\mathbf{U}(t)$:

$$\frac{d\mathbf{U}(t)}{dt} = \mathbf{J}(\nabla_Z\nabla_Z H)|_{Z=Z_0}\mathbf{U}(t). \tag{3.6}$$

Here $\nabla_Z\nabla_Z H$ is the symmetric matrix of second derivatives $\partial^2 H/\partial Z_i \partial Z_j$ of H; and the reference trajectory $\mathbf{Z}_0(t)$ is an exact solution of Hamilton's equations for all times $t \geq 0$.

The matrix $\mathbf{U}(t)$ satisfies the matrix equation

$$\mathbf{U}\mathbf{J}\mathbf{U}^{\dagger} = \mathbf{J}, \tag{3.7}$$

which is the definition of a *symplectic matrix*. This fact is a consequence of Hamilton's equations. The sympletic nature of the evolution of Hamiltonian systems distinguishes this evolution from the evolution of all other systems.

Consider now the special case where the reference trajectory $\mathbf{Z}_0(t)$ is a periodic orbit, $\mathbf{Z}_0(T) = \mathbf{Z}_0(0)$, where T is the period of the orbit. The matrix $\mathbf{U}(T)$ is called the *one-period monodromy matrix*, and its eigenvalues are useful for describing the orbital stability of $\mathbf{Z}_0(t)$. $\mathbf{U}(T)$ depends on the parameter T rather than the time variable t. For problems having a coordinate space beyond 2-d, sometimes one can find analytical results for these eigenvalues, but nongeneric system symmetry is usually required. More often, we need to use numerical matrix diagonalization in such problems; see Section B.2b for some methods.

For a 1-d problem, $\mathbf{U}(T)$ is a 2×2 matrix, where the eigenvalues are determined by

$$\begin{vmatrix} U_{11} - s & U_{12} \\ U_{21} & U_{22} - s \end{vmatrix} = 0. \tag{3.8}$$

This is the algebraic equation $s^2 - s(U_{11} + U_{22}) + (U_{11}U_{22} - U_{12}U_{21}) = 0$, which can also be written $s^2 - s(\text{trace}\ \underline{\underline{U}}) + \det \underline{\underline{U}} = 0$. Because Hamiltonian systems are area-preserving (satisfying Liouville's theorem as described in Section 1.1c),

$\det \bar{\bar{\mathrm{U}}} = 1$. Thus the values of s are

$$s_{1,2} = \tfrac{1}{2}\left[\text{trace } \bar{\bar{\mathrm{U}}} \pm \sqrt{(\text{trace } \bar{\bar{\mathrm{U}}})^2 - 4}\right].$$

Except for the special degenerate case $|\text{trace } \bar{\bar{\mathrm{U}}}| = 2$, there are two possibilities. First, when $|\text{trace } \bar{\bar{\mathrm{U}}}| < 2$, the $s_{1,2}$ are a complex conjugate pair lying on the unit circle; this is the case of an (orbitally) *stable periodic orbit*. However, when $|\text{trace } \bar{\bar{\mathrm{U}}}| > 2$, the eigenvalues are real numbers satisfying $s_2 = 1/s_1$, which means that we have an *unstable periodic orbit*. The fact that the eigenvalues come in pairs—with one member of the pair being an inverse of the other—is actually a consequence of the sympletic condition (3.7) (Lichtenberg and Lieberman, 1992, Section 3.3a).

For another special case—that of a periodic orbit of an autonomous 2-d system—we can reduce expression (3.5) to a pair of evolution equations; then there will be an associated eigenvalue equation that has the 1-d form of Eq. (3.8). We make this reduction by introducing the local arc length (the variable along the orbit) and a second coordinate that is the local perpendicular distance away from the orbit (Bogomolny, 1988). Instead of the time, the value of the arc length variable evolves the system, and the stability equations describe the evolution of the perpendicular distance variable and its conjugate momentum. The mathematics of the stability analysis for the 1-d case then apply for a reduced monodromy matrix.

For the case of a stable periodic orbit, numerical results show that nearby orbits are quasiperiodic, meaning that they wind in the full phase space around the periodic orbit upon concentric surfaces that have elliptical cross sections. These surfaces are the nested tori (Section 1.4a) that lie in the immediate neighborhood of a stable periodic orbit. The winding number w of one of the quasiperiodic orbits (Section 1.4a) is related to the *residue* $R \equiv \frac{1}{2}$ trace $\bar{\bar{\mathrm{U}}}$ (Greene, 1968, 1979a), according to

$$R = \sin^2 \pi w. \tag{3.9}$$

3.2. EXCERPTS FROM TRANSFORMATION THEORY

3.2a. Canonical Transformations

In classical mechanics, there are many representations associated with different coordinate systems and with different phase-space variables $(\mathbf{q}, \mathbf{p})$. Transformations are the functional relationships that connect different representations. In Hamiltonian dynamics, certain classes of such transformations can help us find solutions—exact or approximate—of problems.

First, we want to know which transformations preserve the form of Hamilton's equations. Such transformations are particularly helpful since they preserve physically important properties, such as the conservation of phase-space area provided by Liouville's theorem. Second, we are interested in transformations that can simplify a

Hamiltonian—for instance, reducing it to a form independent of the new coordinates $\mathbf{Q}$. If this reduction isn't possible, then at least we might hope to transform away uninteresting but large parts of the evolution, to facilitate the study of what remains; an example of this clarification are transformations to moving coordinate frames. Third, we are interested in finding those representations where numerical calculations can proceed rapidly and accurately.

Recalling Section 1.1b, we define a set of coordinates $\mathbf{q}$ and momenta $\mathbf{p}$ to be a "canonical set" if Hamilton's equations (1.7) are satisfied. To maintain this condition under a transformation $(\mathbf{q}, \mathbf{p}) \to (\mathbf{Q}, \mathbf{P})$, we require that phase-space areas be conserved so that the numerical value of all areas (or volumes) is independent of the representation. The general formula for changing coordinate differentials can be integrated to relate areas (or volumes) in the two representations:

$$\iint_{A_Q} d\mathbf{Q}\, d\mathbf{P} = \iint_{A_q} d\mathbf{q}\, d\mathbf{p} \frac{\partial(\mathbf{Q}, \mathbf{P})}{\partial(\mathbf{q}, \mathbf{p})}.$$

Here the Jacobian determinant is

$$\frac{\partial(\mathbf{Q}, \mathbf{P})}{\partial(\mathbf{q}, \mathbf{p})} = \frac{\partial(Q_1, \ldots, Q_N, P_1, \ldots, P_N)}{\partial(q_1, \ldots, q_N, p_1, \ldots, p_N)}.$$

For one degree of freedom (1-d), this Jacobian determinant becomes

$$\frac{\partial(Q, P)}{\partial(q, p)} = \begin{vmatrix} \dfrac{\partial Q}{\partial q} & \dfrac{\partial P}{\partial q} \\ \dfrac{\partial Q}{\partial p} & \dfrac{\partial P}{\partial p} \end{vmatrix}.$$

Since we want equal areas, $\iint_{A_Q} d\mathbf{Q}\, d\mathbf{P} = \iint_{A_q} d\mathbf{q}\, d\mathbf{p}$, we define a *canonical transformation* as one satisfying

$$\frac{\partial(\mathbf{Q}, \mathbf{P})}{\partial(\mathbf{q}, \mathbf{p})} = 1 \quad \text{and also} \quad \frac{\partial(\mathbf{q}, \mathbf{p})}{\partial(\mathbf{Q}, \mathbf{P})} = 1.$$

Time is held constant while calculating the necessary partial derivatives. One can show that the set of all canonical transformations satisfies the algebraic properties of a group.

Any particular canonical transformation can be specified by a single function, called the *generating function* F. Depending on the choice of independent variables, there are four different types of generating function: $F_1(\mathbf{q}, \mathbf{Q}, t)$, $F_2(\mathbf{q}, \mathbf{P}, t)$, $F_3(\mathbf{Q}, \mathbf{p}, t)$, and $F_4(\mathbf{P}, \mathbf{p}, t)$. Table 3.1 indicates how the remaining dependent variables relate to the various generating functions F_i (Tolman, 1938; Miller, 1974). The F_i are related to one another. For example, $F_2(\mathbf{q}, \mathbf{P}, t) = F_1(\mathbf{q}, \mathbf{Q}, t) + \mathbf{Q} \cdot \mathbf{P}$.

TABLE 3.1. Equations of Transformation for the Four Generating Functions F_1, F_2, F_3, and F_4

$F_1(\mathbf{q}, \mathbf{Q}, t)$	$F_2(\mathbf{q}, \mathbf{P}, t)$	$F_3(\mathbf{Q}, \mathbf{p}, t)$	$F_4(\mathbf{P}, \mathbf{p}, t)$
$\mathbf{p} = \partial F_1/\partial \mathbf{q}$	$\mathbf{p} = \partial F_2/\partial \mathbf{q}$	$\mathbf{p} = -\partial F_3/\partial \mathbf{p}$	$\mathbf{q} = -\partial F_4/\partial \mathbf{q}$
$\mathbf{P} = -\partial F_1/\partial \mathbf{Q}$	$\mathbf{Q} = \partial F_2/\partial \mathbf{P}$	$\mathbf{P} = -\partial F_3/\partial \mathbf{Q}$	$\mathbf{Q} = \partial F_4/\partial \mathbf{P}$

If both the original Hamiltonian H and the canonical transformation are time independent, then the new transformed Hamiltonian K is clearly

$$K(\mathbf{Q}, \mathbf{P}) = H(\mathbf{q}(\mathbf{Q}, \mathbf{P}), \mathbf{p}(\mathbf{Q}, \mathbf{P})).$$

Then it is not hard to show that $\partial K/\partial P = \partial H/\partial P$ and $\partial K/\partial Q = \partial H/\partial Q$, which means that $K = H + \text{constant}$. For time-dependent transformations, the relation between K and H is (Tolman, 1938)

$$K(\mathbf{Q}, \mathbf{P}, t) = H(\mathbf{q}, \mathbf{p}, t) + \partial F/\partial t. \tag{3.10}$$

Here F is any one of the generating functions, since their time derivatives $\partial F_i/\partial t$ are all equal; see also Lichtenberg and Lieberman (1992), Section 1.2. Persons interested in the detailed study of canonical transformations might enjoy Chapter 6 of Percival and Richards (1982).

3.2b. The Hamilton–Jacobi Equation

We will see in Section 3.3 that some classical Hamiltonian systems (beyond 1-d systems) are analytically solvable. To find out whether or not we can solve a system having a particular Hamiltonian H, we introduce here a particular F_2-type canonical transformation that forces the issue. We insist that this transformation produce new coordinates Q_i that are constant, namely, $d\mathbf{Q}/dt = \mathbf{0}$. We wish to interpret the Q_i as the initial values of the original coordinates, $q_i(0)$. Then we are forcing the transformation to provide the solution of the problem, where the solution is written in terms of the generating function.

For the case of a time-dependent Hamiltonian, the procedure is to force $K \equiv 0$. Referring to Eq. (3.10) and Table 3.1, this means

$$H\left(\mathbf{q}, \frac{\partial F_2}{\partial \mathbf{q}}, t\right) + \frac{\partial F_2}{\partial t} = 0, \tag{3.11}$$

where F_2 is then called *Hamilton's principal function*. For the case of no explicit time dependence, we set $K = E$ and have

$$H\left(\mathbf{q}, \frac{\partial F_2}{\partial \mathbf{q}}\right) = E. \tag{3.12}$$

This is called the *Hamilton–Jacobi equation* for *Hamilton's characteristic function* $\mathbf{S} \equiv F_2(\mathbf{q}, \mathbf{P})$.

For the case of $H = \mathbf{p}^2/2m + V(\mathbf{r})$, Eq. (3.12) becomes

$$\frac{(\nabla S)^2}{2m} + V(\mathbf{r}) = E,$$

an equation already mentioned in Section 1.2b.

The Hamilton–Jacobi equation is a good starting point for trying to separate variables (in a selected coordinate system). To begin the separation, we use the ansatz $S \equiv \sum_{j=1}^{n} S_j(q_j, \alpha_1, \ldots, \alpha_N)$. Here the α_j are new, constant momenta associated with N constants of the motion. We will carry the separation through in Section 3.3.

3.2c. The Classical Interaction Representation

Fortunately, the Hamiltonian can often be written as a sum

$$H = H_0 + H_I, \tag{3.13}$$

where the partial Hamiltonian H_0 is special in that we already know something about it. Then the rest of the Hamiltonian—namely, H_I—is called the *interaction Hamiltonian*. We'll soon see that the Hamiltonian breakup shown in Eq. (3.13) also defines an interaction representation. Particularly useful are the cases where the evolution generated by H_0 can be found analytically. Examples are the problems of free-particle motion, the harmonic oscillator, the Morse oscillator, the rigid rotor, the pendulum, the Kepler system, and separable multidimensional combinations of one or more of the above. We note that many interaction representations exist, because we can add any canonical partial Hamiltonian H' to H_0 if we also subtract it from H_I. Some interaction representations can help us apply analytical theory to a particular problem. Others may enhance computational accuracy or reduce computer running time.

We construct a transformation to a *classical interaction representation* (CIR) by first defining three sets of variables, one for each quantity in Eq. (3.13). Let one set be $(\mathbf{q}, \mathbf{p})$ for the Hamiltonian H, another set be $(\mathbf{q}_0, \mathbf{p}_0)$ for H_0, and the set of CIR variables be $(\mathbf{q}_I, \mathbf{p}_I)$ for H_I. We parameterize all these variables in terms of common initial conditions at some time $t = t_0$; for example, we write $\mathbf{q} = \mathbf{q}[\mathbf{q}(t_0), \mathbf{p}(t_0); t_0, t]$. If $t_0 > t$, backward propagation in time is indicated. We now define the CIR by forward propagation from 0 to t using H, followed by backward propagation using H_0 (Skodje, 1984). Thus

$$\begin{aligned} \mathbf{q}_I &\equiv \mathbf{q}_0(\mathbf{q}, \mathbf{p}; t, 0), \\ \mathbf{p}_I &\equiv \mathbf{p}_0(\mathbf{q}, \mathbf{p}; t, 0). \end{aligned} \tag{3.14}$$

These equations state that the CIR variables are results of a backward time propagation from t to 0 using H_0, with "initial" conditions at time t of $(\mathbf{q}, \mathbf{p})$, which is the phase-space point forward-propagated to t by H. It is easy to show that the $(\mathbf{q}_I, \mathbf{p}_I)$ evolve according to Hamilton's equations with Hamiltonian H_I.

Thus transformation (3.14) is canonical. The generating function of the transformation from $(\mathbf{q}, \mathbf{p})$ to $(\mathbf{q}_I, \mathbf{p}_I)$ is the particular Hamilton's principal function S_0 that is appropriate for H_0. When the evolution for H_0 is analytically solvable, then the appropriate Hamilton–Jacobi equation can be solved for S_0, and we have the transformation (3.14). Table 3.2 gives the transformation equations for three specific systems.

If we want to numerically evolve a particle in a potential $V(\mathbf{r})$, it is natural to consider taking H_0 to be the free-particle Hamiltonian, and thus $H_I = V$. Then we would see large time variations in the CIR variables only when V is at least

TABLE 3.2. Transformation Equations from the (q, p) Representation to the Classical Interaction Representation (q_I, p_I), for Three Specific Hamiltonian Systems

The Free Particle

$$H_0 = p^2/2m$$
$$q_I = q - pt/m$$
$$p_I = p$$

The Harmonic Oscillator

$$H_0 = p^2/2m + \tfrac{1}{2} m\omega^2 q^2$$
$$q_I = q\cos\omega t - (p/m\omega)\sin\omega t$$
$$p_I = p\cos\omega t + qm\omega\sin\omega t$$

The Morse Oscillator

$$H_0 = p^2/2m + D(1 - e^{-aq})^2$$
$$q_I = -\frac{1}{\alpha}\ln\left[\frac{1-\delta}{1+\delta^{1/2}\sin(svt+c)}\right]$$
$$p_I = \frac{-mvs\delta^{1/2}}{a}\,\frac{\cos(svt+c)}{1+\delta^{1/2}\sin(svt+c)}$$

where

$$\delta = H_0/D$$
$$v = (1-\delta)^{1/2}(2Da^2/m)^{1/2}$$
$$c = \sin^{-1}\left(\frac{1-\delta-e^{-aq}}{e^{-aq}\delta^{1/2}}\right)$$
$$s = -\mathrm{sign}(p)$$

Source: Skodje (1984). Copyright 1984, with permission from Elsevier Science.

comparable in magnitude to H_0. Depending on what $V(\mathbf{r})$ is—such as how spatially localized it is—the time interval when V is large may be much smaller than the time interval needed to accurately propagate H itself. In the $(\mathbf{r}, \mathbf{p})$ representation, a trajectory is always a strong function of time, requiring a relatively short-time stepsize. In the CIR, such short stepsizes are needed only over the smaller time interval when $V \gtrsim H_0$. We see that computer time will be reduced, sometimes dramatically. Similar advantages occur for the quantum interaction representation, which is therefore useful in numerical quantum time evolution.

3.3. ANALYTICALLY SOLVABLE CLASSICAL PROBLEMS

3.3a. Integrable Classical Systems

So far, most of our specific examples of classical systems have been analytically solvable—solvable in the sense that we have other options beyond using brute-force numerical evolution of differential equations of evolution. On the other hand, the Hénon–Heiles problem mentioned in Section 1.3a is not analytically solvable.

Thus one might have the impression that most problems of interest are analytically solvable, but quite the opposite is true. In dealing with an unsolvable problem, it is very important to know about any related problems that can be solved: these are prospective starting points for approximate analytical solutions that can help explain numerical results. The interaction representation discussed in the last section is based on this idea of utilizing a related solvable problem.

One might be surprised that mathematicians have not yet achieved a complete rigorous description of the set of all analytically solvable Hamiltonian problems. After all, the symmetries of a problem can lead to constants of the motion. Hence enough symmetry could lead to solutions of a problem. However, in practice, these symmetries are not always easy to identify when phase space is larger than 2-D. In addition, it is not always clear to what extent a knowledge of the symmetries leads to explicit solutions of the equations of evolution.

An often-noted example of "hidden symmetry" is the three-particle *Toda lattice problem*. Three particles with identical masses are confined to moving on a circle and they have exponentially decreasing, repulsive two-body forces between them. If the exponentials are expanded up to third order, the potential energy becomes exactly that of the unsolvable Hénon–Heiles problem (Tabor, 1989; Lichtenberg and Lieberman, 1992). Thus it was a surprise when Ford et al. (1973) reported that numerical calculations of trajectories for the Toda lattice problem produced only smooth curves in Poincaré sections, up to arbitrarily high total system energy. The "apparently unsolvable" Toda problem must be solvable after all! Confirming this, an additional highly nontrivial constant of the motion was subsequently found, although it was not related to any obvious conservation law or symmetry. Further numerical calculations showed that when the particle masses are made unequal, the "hidden symmetry" of the Toda lattice is broken, the constant of the motion no longer exists, and the modified problem is analytically unsolvable. But this nonintegrability was not predicted beforehand.

Using a traditional approach, let us define a completely *integrable Hamiltonian system* with N degrees of freedom (N-d) by requiring that there exist N independent integrals of the motion $\alpha_1, \alpha_2, \ldots, \alpha_N$ that are "in evolution;" in other words, all the Poisson brackets (Eq. (1.9)) of these integrals must be zero (Whittaker, 1964, Section 147). Thus the α_k are N constants of the motion. So the essence of an integrable classical Hamiltonian system is that the number of constants of the motion equals the spatial dimensionality of the system.

A subset of the set of integrable problems are the separable problems. The Hamilton–Jacobi equation (3.12) is (completely) *separable*—if and only if Hamilton's characteristic function can be written as a sum of terms, with each term depending on a single independent coordinate q_k of some coordinate system

$$S(\mathbf{q}, \boldsymbol{\alpha}) = \sum_{k=1}^{N} S_k(q_k, \boldsymbol{\alpha}) \tag{3.15}$$

Then each conjugate momentum p_k is a function of only one q_k,

$$p_k = \frac{\partial}{\partial q_k} S_k(q_k, \boldsymbol{\alpha}), \tag{3.16}$$

and therefore $\partial H/\partial p_k$ is some function $f_k(q_k)$ for each value of k. This separation then produces N 1-d problems. Each 1-d problem can be solved by integrating its first Hamilton's equation (1.7) when written as $dt = dq_k/f_k(q_k)$. The solution of the full problem is then found.

3.3b. Action-Angle Variables

If the evolution of an N-d separable system is periodic in each of the q_k, then a set of action variables can be introduced:

$$I_k(\boldsymbol{\alpha}) \equiv \frac{1}{2\pi} \oint_{\Gamma_k} p_k(q_k, \boldsymbol{\alpha})\, dq_k. \tag{3.17}$$

Here Γ_k is the closed path in phase space that is traced out during the periodic evolution in the single spatial variable q_k. Since the α_k are constants of the motion, the $I_k(\boldsymbol{\alpha})$ are also constants of the motion: $dI_k/dt = 0$. Having obtained the I_k and their relation to the α_k, the α_k can be found by involution and then substituted into S to obtain $S = S(\mathbf{q}, \mathbf{I})$. Since S is an F_2-type generating function for a canonical transformation, the angle variable conjugate to I_k must be

$$\theta_k = \frac{\partial S}{\partial I_k} = \sum_{i=1}^{N} \frac{\partial}{\partial I_k} S_i(q_i, \mathbf{I}). \tag{3.18}$$

The transformed Hamiltonian H′ is a straightforward change of variables

$$H'(\mathbf{I}, \theta) = H(\mathbf{p}, (\mathbf{I}, \theta), \mathbf{q}(\mathbf{I}, \theta)).$$

Because the action-angle variables are canonical, $dI_k/dt = -\partial H'/\partial\theta_k$ for each k. Since we have noted that $dI_k/dt = 0$ for each k, we conclude that H′ is independent of each θ_k; that is, $H' = H'(\mathbf{I})$. Thus in action-angle variables, Hamilton's equations become simply

$$\frac{dI_k}{dt} = -\frac{\partial H'(\mathbf{I})}{\partial\theta_k} = 0, \tag{3.19a}$$

$$\frac{d\theta_k}{dt} = \frac{\partial H'(\mathbf{I})}{\partial I_k} \equiv \omega_k(\mathbf{I}). \tag{3.19b}$$

Here ω_k is the frequency of the kth periodic evolution. Clearly Eqs. (3.19) imply that

$$\mathbf{I} = \text{constant} \quad \text{and} \quad \boldsymbol{\theta} = \boldsymbol{\omega}(\mathbf{I})t + \text{constant},$$

where the "constants" make up the set of 2N constants of integration that must appear in the solution of an N-d solvable problem. The quantity $\boldsymbol{\omega}(\mathbf{I})$ is called the *frequency vector* for the periodic integrable system. We note that Eq. (3.18) implies that

$$\frac{\partial\theta_k(\mathbf{q}, \mathbf{I})}{\partial q_k} = \frac{\partial}{\partial q_k}\frac{\partial S_k(q_k, \mathbf{I})}{\partial I_k} = \frac{\partial}{\partial I_k}\frac{\partial S_k}{\partial q_k},$$

from whence it follows that

$$\oint_{\Gamma_k} d\theta_k = \oint_{\Gamma_k}\frac{d\theta_k}{dq_k}\,dq_k = \frac{\partial}{\partial I_k}\oint_{\Gamma_k}\frac{\partial S_k}{\partial q_k}\,dq_k = \frac{\partial}{\partial I_k}\oint_{\Gamma_k} p_k\,dq_k = \frac{\partial}{\partial I_k}(2\pi I_k) = 2\pi.$$

Thus θ_k changes by the amount 2π during a complete period of its evolution, justifying our naming it an angle variable.

A second, more modern approach to defining an integrable system begins by requiring there be a complete set of independent action variables

$$\mathbf{I}_k \equiv \frac{1}{2\pi}\oint_{\Gamma_k}\sum_{i=1}^{N} p_i\,dq_i$$

with their corresponding set of conjugate angle variables

$$\theta_k = \frac{\partial S(\mathbf{q}, \mathbf{I})}{\partial I_k}.$$

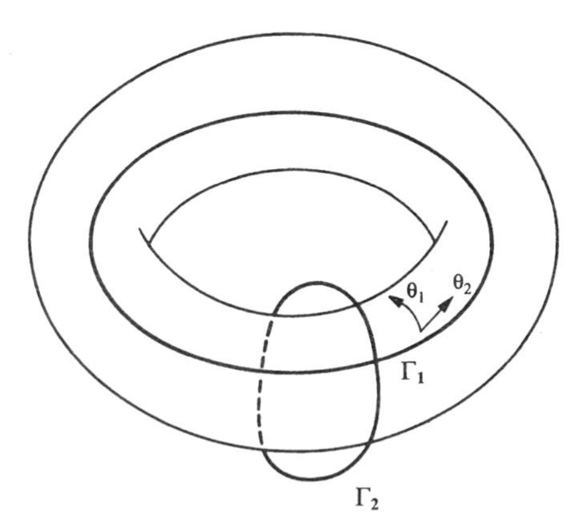

Figure 3.1. Two topologically independent curves on a 2-D torus. From Tabor (1989). Copyright 1989 John Wiley & Sons, Inc. Reprinted by permission of John Wiley & Sons, Inc.

Here $(\mathbf{q}, \mathbf{p})$ is any set of canonical variables for the system, and the Γ_k are topologically independent curves on an N-D torus in phase space, as Fig. 3.1 schematically suggests. Thus the existence of an N-D torus is in essence the starting point for this second approach. For integrable systems with $N \geq 2$, the orbits are *not* all periodic. It is always found that tori are present and that the typical orbit is quasiperiodic; recall Section 1.4a. A good example of this situation for a separable system appears in a recent discussion of the hydrogen atom in a uniform static electric field; see Gao and Delos (1994). We also note that a set of topological theorems has been proved, showing that the torus-based definition of *classical integrability* is equivalent to the traditional definition given earlier in this section in terms of the Poisson brackets. For an outline of the reasoning, see Tabor (1989).

For future reference, we note that the explicit time evolution of the vertical pendulum action-angle variables (I, θ) can be expressed in terms of the Jacobi elliptic functions cn(u) and sn(u, k) (Zaslavsky and Filonenko, 1968). These functions are the inverse functions of elliptic integral functions (Abramowitz and Stegun, 1964). Let $2\kappa^2 \equiv 1 + E/mgL$. Then for librational motion $(\kappa < 1)$ we have

$$\theta(t) = 2\sin^{-1}\left[\kappa \, \mathrm{sn}\left(t\sqrt{(g/L)}, \kappa\right)\right],$$
$$I(t) = 2\sqrt{gL}\kappa \, \mathrm{cn}(\omega_0 t).$$

The frequency of the pendulum is

$$\omega(\kappa) = \tfrac{1}{2}\pi\sqrt{gL}[\mathscr{K}(\kappa)]^{-1},$$

where $\mathscr{K}(\kappa)$ is the complete elliptic integral of the first kind. Similar expressions hold for the other case of rotational motion, $\kappa > 1$.

For the troublesome case of an integrable (but nonseparable) multidimensional Hamiltonian system, there are brute-force numerical techniques for finding action-angle variables; see Skodje and Borondo (1985). Since their approach constitutes a good example of how numerical techniques can take over where analytical theory seems to leave off, let us take a look at their basic ideas.

Our system needs to have a partial Hamiltonian H_0 that is separable:

$$H = H(\mathbf{p}, \mathbf{q}) = H_0 + H_1.$$

If there is no natural separable partial Hamiltonian available, we can construct any artificial one we want and include its negative in H_1; we just need a canonical and solvable partial Hamiltonian to drive the evolution.

A key idea behind the numerical technique will involve the time-dependent Hamiltonian

$$H(t) = H_0 + \lambda(t)H_1, \quad 0 \leq t \leq T. \tag{3.20}$$

Here the switching function $\lambda(t)$ should be chosen to go from zero to one, something like $t/T - [\sin(2\pi t/T)]/2\pi$. The solution of this new time-dependent problem is known at $t = 0$, and hopefully becomes that of our original problem at $t = T$. The basic assumption, borne out numerically even for some applications involving near-integrable systems (Skodje et al., 1985), is that we can switch slowly enough (i.e., make T large enough) so that "good" initial actions will conserve their values. This assumption is based on the existence of adiabatic invariants in classical mechanics; see Section 9.1a. (We note that this basic assumption of the *adiabatic switching method* breaks down even for integrable systems whenever separatrices are crossed; the classical action is discontinuous at a separatrix crossing (Section 4.4). Although the effects of separatrix crossing can be included in a modified ("uniform") adiabatic invariant analysis (Skokje, 1989), the adiabatic switching method ceases to be useful in the case of strongly nonintegrable systems where the separatrix is replaced by a stochastic layer of significant width.)

We know that there are action-angle variables for H_0,

$$\begin{aligned} I_i^0 &= \oint p_i \, dq_i, \\ \theta_i^0 &= \omega_i^0(I_i^0) + \theta_i^0(t = 0), \quad i = 1, \ldots, N. \end{aligned}$$

By the adiabatic hypothesis, we assume there exists a family of variables $(\mathbf{I}^\lambda, \theta^\lambda)$ that are good action-angle variables for each constant value of λ. We then define

$$\boldsymbol{\theta}(t) = \boldsymbol{\theta}^0(0) + \int_0^t \boldsymbol{\omega}(\mathbf{I}^\lambda; \lambda) \, dt, \quad \lambda = \lambda(t), \qquad 0 \leq t \leq T, \tag{3.21}$$

and we begin our numerical work by finding $\boldsymbol{\omega}$ from its definition, using

$$\omega_i \equiv \frac{\partial E(\mathbf{I}^{\lambda}; \lambda)}{\partial I_i^{\lambda}} \equiv \lim_{\delta I_i \to 0} \frac{E(\mathbf{I}^{\lambda} + \delta I_i; \lambda) - E(\mathbf{I}^{\lambda} - \delta I_i; \lambda)}{2\delta I_i}.$$

We just adiabatically propagate two trajectories with slightly different actions—$(I_1, \ldots, I_i + \delta I_i, \ldots, I_N)$ and $(I_1, \ldots, I_i - \delta I_i, \ldots, I_N)$—and monitor the energy difference as we go along. In testing this procedure on 1-d and 2-d problems, we can typically find the frequencies to five or six significant figures by running a small number of trajectories for perhaps 100 periods.

Knowing $\boldsymbol{\omega}(\mathbf{I}^{\lambda}; \lambda)$, we then numerically propagate $(\mathbf{I}^0, \boldsymbol{\theta}^0)$ adiabatically using Hamiltonian (3.20) until $t = T$; we simultaneously numerically integrate Eq. (3.21), which gives us $\boldsymbol{\theta}(T)$. Then we continue propagating the same trajectory using the original time-independent Hamiltonian H, recording the relationship between $(\mathbf{p}, \mathbf{q})$ and $[\mathbf{I} = \mathbf{I}^0,\ \boldsymbol{\theta}(t)]$, where $\boldsymbol{\theta}(t) = \boldsymbol{\theta}(T) + \boldsymbol{\omega}(\mathbf{I};\ \lambda = 1)(t - T)$. Repeating this for different values of $\mathbf{I}^0$ will numerically build up a discrete numerical representation of the transformation $(\mathbf{p}, \mathbf{q}) \to (\mathbf{I}, \boldsymbol{\theta})$, something useful when analytic approaches are cumbersome.

4

CLASSICALLY INTEGRABLE QUANTUM SYSTEMS

4.1. QUANTUM INTEGRABILITY (ADVANCED TOPIC)

In quantum mechanics, unlike classical physics, observables are of various types. These range from observables with only two discrete eigenvalues (such as the familiar spin $\frac{1}{2}$) to those with a continuous range of eigenvalues (such as a spatial coordinate). Because some types of quantum observables are not found in classical physics, the Hamiltonian of a quantum system may or may not have a corresponding classical Hamiltonian. Some quantum observables are described by quantum operators that can be derived from classical physics using the procedures of first-quantization. We call these *canonical quantum observables*.

Some essentially quantum systems (defined in Section 1.5) do not have any canonical observables. Other essentially quantum systems have mixed types of observables, some canonical and some not; an example are electrons in an atom that possess both spatial coordinates and spin variables.

When a quantum system has a classical counterpart that is both integrable and separable, then Schrödinger's equation usually separates and becomes analytically solvable, meaning that no numerical solution of differential equations is required. In this case there is a carryover of the classical definition of integrability expressed by the existence of independent constants of motion equal in number to the system's number N of degrees of freedom. The Poisson brackets of classical physics become commutators in quantum mechanics. Thus *canonical quantum systems* have a complete set of commuting observables (CSCO), also N in number. The number of quantum numbers describing these systems is N. These systems are included among those that we call quantum integrable.

When we move to the systems with noncanonical observables or with mixed quantum observables, the notion of quantum integrability is not so developed. Such systems may have various CSCOs that differ in their number of observables. Thus one proposed definition of quantum integrability (Graham and Höhnenbach, 1984) is as follows. Consider the CSCO provided only by the observables of the constituent particles and fields within the quantum system. Usually this set does not contain the Hamiltonian H itself. We will then say that a quantum system is integrable only if there also exists a larger CSCO that has two attributes: the larger CSCO includes H, and it has types for its observables that are in one-to-one correspondence with the types of observables of the original CSCO. Under these conditions, the quantum system will usually be separable.

Finally, there are those quantum systems that derive from a classical Hamiltonian—but from a Hamiltonian for a nonintegrable classical problem. As we will see in Chapter 5, for such Hamiltonians there are at least some chaotic classical trajectories in phase space. Not all trajectories lie on the N-D tori characteristic of classically integrable systems. It is believed that quantum systems that are classically nonintegrable will also be quantum nonintegrable. Given the present status of rigorous theory, the *a priori* nature of the quantum evolution is not yet predictable. Numerical computational approaches are still available to assist us, however, and these can show us the nature of the quantum evolution.

4.2. EBK QUANTIZATION

We are now prepared to develop a major subject—the semiclassical theory for the energy eigenstates and eigenvalues of an integrable quantum system having N degrees of freedom (N-d). Named *EBK quantization* after A. Einstein, L. Brillouin, and J. B. Keller, this theory is based on the realization that classical actions should be defined on an (invariant) N-D torus in phase space (Einstein, 1917). (Otherwise, for those quantum systems separable in more than one coordinate system, the results of quantization would not be unique. There are such systems; for instance, the hydrogen atom problem separates in both spherical and parabolic coordinates.) Thus we want to further develop the ideas behind both the 1-d WKB approximation and Bohr–Sommerfeld quantization, extending them to N-dimensional quantum states supported by N-D tori rather than by periodic orbits. Now we have quasiperiodic orbits as well as periodic orbits. This extension is not only called EBK quantization but also "semiclassical quantization" and "torus quantization."

Before constructing our theory, let us summarize those elements already at hand. For the quantum mechanics, the 1-d WKB wavefunction was developed in Section 1.2c—see Eq. (1.24); crucial to its construction were the connection formulas—like Eq. (1.25)—that bridge the wavefunction over the classical turning points, where the momentum p is zero and where the WKB amplitude $\sim p^{-1/2}$ diverges. By "fitting the WKB wavefunction into the potential well," we obtained Eq. (1.26) for the quantization of the action and the related energy. The allowed energy eigenvalues depended not only on an integer quantum number but also on the Maslov index μ. The Maslov index μ is the total "additional" phase change accumulated around the periodic orbit during one period, in units of $\pi/2$; see Eq. (1.27).

As for the N-d classical mechanics, we know about the N-D tori (see Fig. 1.13) and about the topologically independent closed loops lying on these tori (see Fig. 3.1). The existence of these tori and loops is intimately connected with the existence of the action-angle variables (see Eqs. 3.17–3.19). In addition, the time-dependent and time-independent Hamilton–Jacobi equations (3.11) and (3.12) have already been written in terms of a set of canonical variables $(\mathbf{q}, \mathbf{p})$ for an N-d system.

The underpinning of our 1-d semiclassical theory is the eikonal theory applied to matter waves. We considered this subject (Section 1.2b) for the case of a particle in a 1-d potential. We need to generalize Eqs. (1.18)–(1.20) to any integrable canonical system having N degrees of freedom. In the Schrödinger coordinate representation, we write the eikonal trial solution as

$$\psi(\mathbf{q}, t) = A(\mathbf{q}, t)\exp\left(i\frac{S(\mathbf{q}, t)}{\hbar}\right). \tag{4.1}$$

When this equation is inserted into Schrödinger's equation (1.1) it produces

$$i\hbar\frac{\partial A}{\partial t} - A\frac{\partial S}{\partial t} = \exp\left(-i\frac{S}{\hbar}\right)H(\boldsymbol{p}, \mathbf{q})\exp\left(i\frac{S}{\hbar}\right)A. \tag{4.2}$$

We note that $\exp[-iS(\boldsymbol{q}, t)\hbar]$ is a unitary operator for which

$$\exp\left(-i\frac{S(\boldsymbol{q}, t)}{\hbar}\right)\boldsymbol{p}\exp\left(i\frac{S(\boldsymbol{q}, t)}{\hbar}\right) = \boldsymbol{p} + \frac{\partial S}{\partial \boldsymbol{q}}, \tag{4.3}$$

and that algebraic relations are preserved by unitary transformations. It follows that the right-hand side of Eq. (4.2) contains the coordinate representation of the operator

$$\exp\left(-i\frac{S(\boldsymbol{q}, t)}{\hbar}\right)H(\boldsymbol{p}, \boldsymbol{q})\exp\left(i\frac{S(\boldsymbol{q}, t)}{\hbar}\right) = H\left(\boldsymbol{p} + \frac{\partial S}{\partial \boldsymbol{q}}, \boldsymbol{q}\right). \tag{4.4}$$

Thus Schrödinger's equation can be written exactly as

$$i\hbar\frac{\partial A}{\partial t} - A\frac{\partial S}{\partial t} = H\left(\boldsymbol{p} + \frac{\partial S}{\partial \boldsymbol{q}}, \boldsymbol{q}\right)A. \tag{4.5}$$

We now recall the replacement $\boldsymbol{p} \rightarrow i\hbar(\partial/\partial\mathbf{q})$. When we make this replacement and then neglect everything in Eq. (4.5) that explicitly involves $\hbar$, we again get Eq. (3.11), with F_2 replaced by S. As before, we can then obtain Eq. (3.12) for the time-independent case that we presently wish to pursue. Since our system is integrable, canonical action-angle coordinates $(\mathbf{I}, \boldsymbol{\theta})$ exist such that

$$S(\mathbf{q}, \mathbf{I}) = \int_{\mathbf{q}_0}^{\mathbf{q}} \mathbf{p}(\mathbf{q}', \mathbf{I})\, d\mathbf{q}. \tag{4.6}$$

Here $\mathbf{q}_0$ locates some arbitrary initial point in coordinate space. Since S is an F_2-type generating function, we know that

$$\theta = \nabla_I S(\mathbf{q}, \mathbf{I}), \qquad \mathbf{p} = \nabla_{\mathbf{q}} S(\mathbf{q}, \mathbf{I}). \tag{4.7}$$

Equation (4.6) gives us the function in the exponential factor of our N-d eikonal matter wave (4.1). We also need the wave's amplitude $A(\mathbf{q}, t)$. Van Fleck (1928) derived this amplitude from the next term in the expansion of Eq. (4.5) in powers of $\hbar$; this derivation parallels our approach in Section 1.2b. However, for autonomous systems Van Fleck also obtained $A(\mathbf{q}, t)$ from an argument based on the uncertainty principle, which goes as follows. Since $S(\mathbf{q}, \mathbf{I})$ generates a canonical transformation, we can take $\mathbf{I}, \boldsymbol{\theta}$ to correspond to quantum operators $\boldsymbol{I}, \boldsymbol{\theta}$ that satisfy the commutator relationships

$$[I_i, I_j] = [\theta_i, \theta_j] = 0, \qquad [\theta_i, I_j] = i\hbar. \tag{4.8}$$

However, $\mathbf{I}$ is a constant of the motion, having no uncertainty. So in the $\boldsymbol{\theta}$ representation, the wavefunction must have a constant amplitude—the amplitude can contain no information about $\boldsymbol{\theta}$. From the conservation of probability under a change of spatial variables, the amplitude A of the wavefunction in the $\mathbf{q}$ representation will differ from the constant amplitude in the $\boldsymbol{\theta}$ representation by a factor involving the Jacobian determinant of the transformation. So we have

$$A^2 \propto \left|\det \frac{\partial \boldsymbol{\theta}}{\partial \mathbf{q}}\right| = \left|\det \frac{\partial^2 S}{\partial \mathbf{q}\, \partial \mathbf{I}}\right|.$$

With this equation, we have $\psi_{\mathbf{I}}(\mathbf{q})$ to within a constant phase factor $\exp(i\alpha)$ and to within a normalization constant c:

$$\psi_{\mathbf{I}}(\mathbf{q}) = c\left|\det \frac{\partial^2 S}{\partial \mathbf{q}\, \partial \mathbf{I}}\right|^{1/2} \exp\left(i\frac{S(\mathbf{q}, \mathbf{I})}{\hbar} + i\alpha\right). \tag{4.9}$$

It is crucial to emphasize that wavefunction (4.9) should *not* really be a function of the phase-space variables as the right-hand side suggests, but only of the position-space variables, as indicated on the left-hand side. However, Einstein suggested that actions to be quantized should be defined on the N-D torus in phase space. Then we would have a situation that can parallel that for 1-d problems (Section 1.3d): phase-space probability distributions are projected onto position space to obtain position-space probability distributions. Some such projections were displayed in Fig. 1.11.

It was Keller (1958a) who decided to put the eikonal WKB wavefunction on the torus and to connect this phase-space wavefunction with the projection implied by Eq. (4.9). In discussing Fig. 1.14 we noted that projections of a torus generally have overlapping pieces, called layers or "leaves." Therefore we now say that $S(\mathbf{q}, \mathbf{I})$ has a number of branches $S_j(\mathbf{q}, \mathbf{I})$; that is, S is a multivalued function of $\mathbf{q}$. The general solution of a linear differential equation is the sum over all distinct possible solutions. Thus in accord with the superposition principle for matter waves, we write a sum over the branches:

$$\psi_{\mathbf{I}}(\mathbf{q}) = c \sum_j \left| \det \frac{\partial^2 S_j}{\partial \mathbf{q}\, \partial \mathbf{I}} \right|^{1/2} \exp\left(i \frac{S_j(\mathbf{q}, \mathbf{I})}{\hbar} + i\alpha_j \right). \tag{4.10}$$

The equation of continuity guarantees the continuity of the phase-space velocity function $\mathbf{v} = d\mathbf{Z}/dt$ for trajectories on the torus. These trajectories are now the paths of integration for specific evaluations of the action function (4.6). However, for the projections contributing to Eq. (4.10), there are singularities in the second and higher derivatives of $S_j(\mathbf{q}, \mathbf{I})$ along the boundaries of the layers. That is, singularities exist at the folds that produce the layers. We can demonstrate these singularities by rewriting the expression for the squared amplitude A^2 given above as

$$\det \frac{\partial^2 S}{\partial \mathbf{q}\, \partial \mathbf{I}} = \det \frac{\partial \boldsymbol{\theta}}{\partial \mathbf{q}} = \det \frac{\partial \boldsymbol{\theta}}{\partial \mathbf{p}} \det \frac{\partial \mathbf{p}}{\partial \mathbf{q}}. \tag{4.11}$$

At a caustic—that is, a fold in $\psi_{\mathbf{I}}(\mathbf{q})$—the last determinant in Eq. (4.11) becomes infinite.

At the caustics, the normal component of the momentum $\mathbf{p}$ is zero. As we pass through a caustic, this component changes signs. This situation is analogous to the turning points in the 1-d WKB problem (Section 1.2c), where connection formulas had to be found. Now we must make connections across the caustics instead. To do this, the wavefunction in the neighborhood of the caustic should be separated locally into pieces that correspond to classical evolution that is normal and tangential to the caustic. Then, through the development of connection formulas, one can find approximations for the normal part over each of the caustics.

It was V. Maslov's idea to switch to momentum space when $\mathbf{q}$ approaches a caustic, and then to replace the eikonal ansatz (4.1)—and all the equations that follow from it—by a similar ansatz and similar equations in momentum space. If we move along members Γ_i of the set of irreducible independent closed curves on the N-D torus, it can be shown that the caustics in momentum space do not coincide with those in position space. Thus the Fourier transforms of the local wavefunctions in momentum space can be used to produce the needed connection formulas across the caustics in position space.

Maslov's procedure for connecting up special coordinate- and momentum-space wavefunctions (outlined just above) is quite abstract (Maslov and Fedoriuk, 1981). To pursue this topic further, one should first read the arguments in Berry (1981) and in Section 7.2 of the long chapter on torus quantization in Ozorio de Almeida (1988).

Stitching together the $\mathbf{q}$-space and $\mathbf{p}$-space wavefunctions in their caustic-free regions produces the global semiclassical wavefunction that we want. The total phase increment accumulated around the loop Γ_i becomes

$$\frac{\Delta S_i}{\hbar} = \frac{1}{\hbar} \oint_{\Gamma_i} \mathbf{p} \cdot d\mathbf{q} - \frac{\pi}{2} \mu_i = \frac{2\pi I_i}{h} - \frac{\pi}{2} \mu_i. \tag{4.12}$$

This increment must be a multiple n_i (2π) for the wavefunction to be single valued. Equation (4.12) for the increment then becomes the 1-d result, Eq. (1.27), extended to each of the loops Γ_i on the N-D torus. We remember that each of these special loops involves one angle variable θ_i changing continuously from 0 to 2π, with the conjugate action I_i staying constant. It's not surprising then that Eq. (4.12) implies that EBK energy eigenvalues E are obtained by first setting the vector $\mathbf{I}$ of constant actions equal to the sum of two constant vectors

$$\mathbf{I}(E) = h\left(\mathbf{n} + \frac{\boldsymbol{\mu}}{4}\right), \tag{4.13}$$

where $\mathbf{n}$ is a vector of integer quantum numbers, and $\boldsymbol{\mu}$ is the vector of Maslov indices that is a characteristic of the particular integrable system under consideration. We next need to find $\boldsymbol{\mu}$. The Maslov index μ_i for the loop Γ_i is the number of passages made past the caustics as the evolution proceeds once around Γ_i in phase space (Arnold, 1967). (The caustics are produced by projecting Γ_i onto coordinate space.) The index is the number of "reflections" made in coordinate space at the boundaries of the layers of the projection of the torus—and the index also labels the accumulated number of wavefunction phase changes for all the caustics passed by.

Knowing $\boldsymbol{\mu}$, we can invert Eq. (4.13) and find all the EBK semiclassical energy levels:

$$E_{\mathbf{n}}(\boldsymbol{\mu}) = H\left[I = h\left(\mathbf{n} + \frac{\boldsymbol{\mu}}{4}\right)\right]. \tag{4.14}$$

In most cases these approximate values for the quantum energy levels are accurate for sufficiently high degrees of excitation, that is, for large quantum numbers n_i. The accuracy of the EBK energies is often surprisingly good for low energies as well, and in exceptional cases, the results are exact. For the N-d harmonic oscillator, all the different basic evolutions are librational, involving two turning points (i.e., caustics) so that all the $\mu_i = 2$. Thus

$$E_{\mathbf{n}} = \sum_i \hbar\omega_i(n_i + \tfrac{1}{2}),$$

which gives the correct zero-point energy. Two other exact results can be obtained. One is for the N-d infinite square well, which is relevant to quantum well studies (see Section 7.3). The other exact result is for the hydrogen atom, where $\mu_r = \mu_\theta = 2$ and $\mu_\phi = 0$. Using Eq. (2.6), we find that the EBK energies for the atom satisfy the Bohr formula $E = -1/2n^2$ a.u., with the principal quantum number $n = n_r + n_\theta + n_\phi + 1$, which is exact.

4.3. THE VERTICAL PENDULUM: QUANTUM ENERGIES VERSUS CLASSICAL MOTION

In Section 1.4b we considered the classical physics of the vertical pendulum. The solutions of Hamilton's equations (1.40) were discussed and interpreted in terms of

Fig. 1.15. The two principal modes of time evolution—librational and rotational—were easy to visualize. In addition, with some effort, we can obtain an analytical result for the evolution on the separatrix trajectory. Given this happy situation, it is natural to look at how the quantization of the pendulum is carried out. Of particular interest are the roles that the three different motions (librational, rotational, and separatrix) might play in the quantum system.

There is a further special reason for studying the quantum pendulum. As we will see in Section 5.6 and beyond, generic nonintegrable classical systems exhibit nonlinear resonance zones in phase space. A general approximate theory of the evolution within one of these zones leads to the equations of motion of the pendulum. These equations are written in terms of parameters and canonical variables different from those of the actual pendulum, of course. Thus a knowledge of pendulum problems, both classical and quantum, has broad applicability.

We may write the Hamiltonian for the vertical pendulum as

$$H = \frac{1}{2m} p_\theta^2 - E_{sep} \cos\theta. \tag{4.15}$$

Here $p_\theta = mL(d\theta/dt)$, and the energy for evolution on the separatrix is $E_{sep} = mgL$. In this section we are interested in the stationary quantum states, that is, the energy eigenstates and eigenvalues obtained by solving the time-independent Schrödinger equation $H\psi = E\psi$. Since θ is an angle variable, we quantize the system according to

$$p_\theta = -i\hbar \frac{\partial}{\partial\theta}. \tag{4.16}$$

Inserting Eq. (4.16) into the time-independent Schrödinger equation and selecting units such that $m = \hbar = 1$, we obtain

$$-\frac{d^2\psi(\theta)}{d\theta^2} - 2E_{sep} \cos\theta\, \psi(\theta) = 2E\psi(\theta).$$

To rewrite this ordinary differential equation in a standard form, we introduce the half-angle $\alpha \equiv \theta/2$ to get

$$\frac{d^2\psi(\alpha)}{d\alpha^2} + [8E - 8E_{sep} \cos(2\alpha)]\psi(\alpha) = 0. \tag{4.17}$$

The Mathieu equation is usually written as

$$\frac{d^2\psi(\alpha)}{d\alpha^2} + [a - 2q \cos(2\alpha)]\psi(\alpha) = 0, \tag{4.18}$$

where a and q are constants. Thus we see that the stationary states of the quantum pendulum are just the well-known π-periodic Mathieu functions $\psi_k(\alpha; q)$, where the constant "a" has characteristic eigenvalues $a_k(q)$, which arise from the periodic boundary conditions. The integer $k = 0, 1, 2, \ldots$ is called the *Mathieu index*. The $a_k(q)$ determine the energy eigenvalues $E_k(E_{sep})$. The standard reference on Mathieu functions is Abramowitz and Stegun (1964).

If the classical motion is librational ($E < E_{sep}$ or $a_k < 2q$), then k is an action quantum number. To see this, we first need the action-angle variables (J, θ_J) given in Lichtenberg and Lieberman (1992) by

$$J = \frac{8}{\pi}\sqrt{E_{sep}}[\mathbb{E}(\kappa) - (1 - \kappa^2)\mathbb{K}(\kappa)] \tag{4.19}$$

$$\phi_J = [\mathbb{K}(\kappa)]^{-1}\mathbb{F}(\eta, \kappa). \tag{4.20}$$

Here $2\kappa^2 \equiv 1 + E/E_{sep}$, $\kappa \sin\eta \equiv \sin\alpha$, and $\kappa < 1$ for librations. The functions $\mathbb{E}(\kappa)$ and $\mathbb{K}(\kappa)$ are complete elliptic integrals. The function $\mathbb{F}(\eta, \kappa)$ is an incomplete elliptic integral of the first kind, a function that depends on the intermediate angle variable η. For rotational motion, there are similar but different equations for the action-angle variables J and ϕ_J (Lichtenberg and Lieberman, 1992).

For the rotational case, when $\kappa \gg 1$, a WKB approximation suitable for the angular coordinate ϕ_J was developed and applied in Reichl and Haoming (1990). At sufficiently high values of the WKB quantum number r, the WKB energy eigenvalues become proportional to r^2. This occurs when $E_{sep}/r^2 \ll 1$, where the energy is much larger than E_{sep} and the rotor is essentially free. In this regime, the WKB and exact energies closely agree, enabling the identification of r with the Mathieu index k. Thus it's well established that the pendulum states at large κ, or large enough k, are identified as "rotational states." In the other limit of $\kappa \ll 1$, or small enough k, we similarly have "librational states."

There is a range of κ, near unity, where the wavefunctions are not easily described. These wavefunctions are influenced by the separatrix trajectory of the classical problem and hence are called *separatrix wavefunctions*. These wavefunctions are partially supported by both the librational and rotational portions of the classical phase-space flow near the separatrix.

Numerical values for the pendulum's exact quantum energies are shown in Fig. 4.1. Curves are drawn as a function of $q = 4E_{sep}$ for values of the Mathieu index $k = 0$–5. We also see a dashed line representing the separatrix location condition $E = E_{sep}$, and this dashed line passes close to the inflection points of the different energy curves. As E_{sep} is increased above zero, an initial rise of the quantum energies is associated with rotational classical motion. Likewise, the nearly linear fall of the energies at sufficiently large E_{sep} (not shown for $k \geq 2$) is associated with librational classical motion.

As one crosses the separatrix, the classical action J changes discontinuously, with the rotational value being one-half the librational value. This change follows directly from the analytical formulas for J (Lichtenberg and Lieberman, 1992). Accompany-

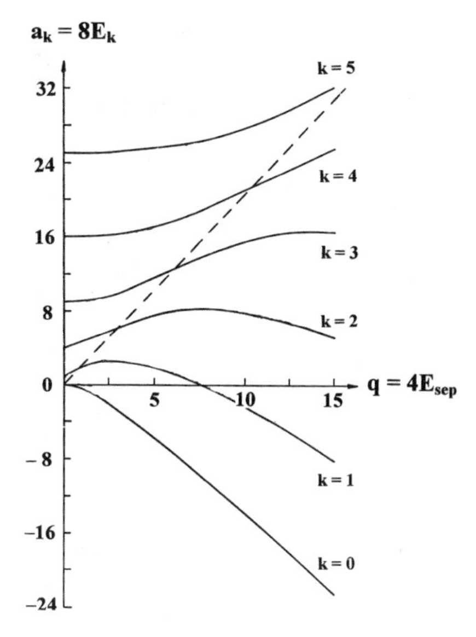

Figure 4.1. The k = 0 to k = 5 energy levels for the even π-periodic states of the quantum pendulum, as a function of the Mathieu parameter $q = 4E_{sep}$, where E_{sep} is the scaled separatrix energy. The vertical axis is eight times the scaled energy; see Eqs. (4.17) and (4.18). The dashed straight line is given by the separatrix location condition $E = E_{sep}$. Well above this line, the classical pendulum motion supporting the quantum states is rotational, while well below this line the support is librational. From Abramowitz and Stegun (1964).

ing this change are changes in orbit-averaged quantities such as $\langle p_\theta \rangle$ and $\langle p_\theta^2 \rangle$, averages that are obtained by integrating angle variables over their ranges $[0, 2\pi)$. The corresponding quantum expectation values reflect these changes in the classical averages; see Fig. 4.2 for an example obtained using the Mathieu wavefunctions. For the chosen values of the parameters, there are just two separatrix quantum states. These states have expectation values that don't follow the behavior characteristic of either the rotational or librational states.

What we have learned so far is that even in a 1-d quantum system, situations do arise where the applicability of the usual EBK quantization procedure is not evident. In particular, further considerations are needed for energies near those of classical separatrices in phase space.

4.4. QUANTUM MANIFESTATIONS OF SEPARATRICES: THE DOUBLE-WELL POTENTIAL

Many important problems can be modeled by particle-like evolution due to the action of a force $\mathbf{F} = -\nabla V$, with the potential V(x) having two minima separated by a finite barrier region. Such problems include (1) the transfer of hydrogen atoms along chemical bonds in molecules (Skinner and Trommsdorff, 1988; Oppenländer

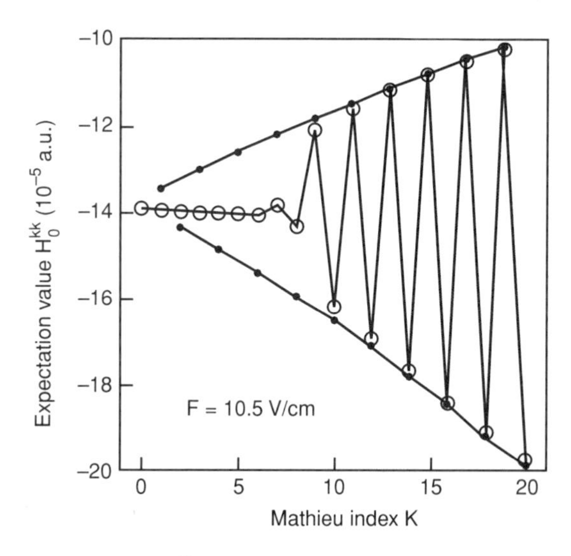

Figure 4.2. Expectation values H_0^{kk} of the free-atom Hamiltonian H_0 for the 1-d hydrogen atom in a time-oscillating field; see Section 8.3. The nonlinear resonance approximation reduces this nonintegrable problem to an integrable quantum pendulum problem; see the procedure in Section 5.6. The values shown as large open circles are computed using Mathieu functions analogous to the wavefunctions $\psi_k(\theta)$ of the vertical pendulum with Hamiltonian (4.15). The smaller solid dots are computed using the energy eigenstates of H_0. The lines connecting points were drawn only to guide the eye. For the chosen value of $E_{sep} \approx 7$, the expectation values obtained with the two different sets of wavefunctions are almost the same when the Mathieu index $k \geq 9$; this similarity indicates that these pendulum states are almost free rotor states, and that the classical motion of the pendulum is rotational. For $k \leq 6$, the H_0^{kk} are almost constant, having the value for the central librational state $k = 0$. The two states $k = 7$ and 8 are in the transition region between librational and rotational classical motion; they are called "separatrix" states. Adapted from Sirko and Koch (1995). Copyright 1995 Springer-Verlag.

et al., 1989), (2) the transport of hydrogen isotopes between interstitial sites in metals (Grabert and Wipf, 1990), and (3) macroscopic coherence phenomena in SQUID devices (Tesche, 1990).

In 1-d models of the chemical reaction $I + HI \rightarrow IH + I$, the light hydrogen atom moves in an effective double-well potential whose time development is governed by the slow evolution of the iodine atoms. The top of the barrier (see Fig. 4.3) can then pass the system's energy level during a first stage of the reaction. During a second stage there will be a second passage in the opposite direction. Thus chemists are interested in time-modulated double-well problems. A prototype system having a double well is the *quartic oscillator*, defined by the Hamiltonian

$$H(q, p, t) = \frac{1}{2m} p^2 - \tfrac{1}{2} m\omega^2 q^2 + bq^4, \tag{4.21}$$

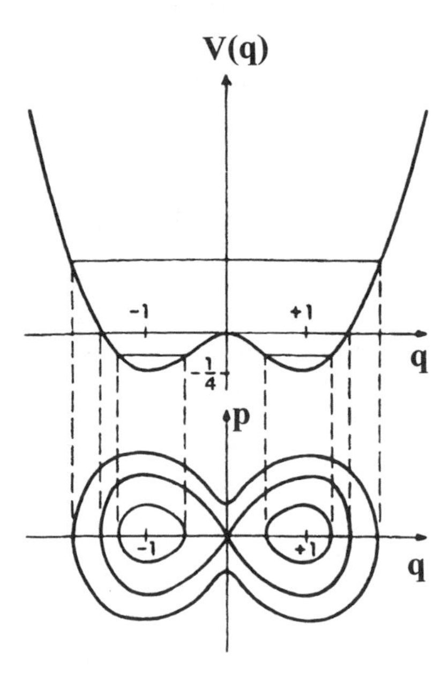

Figure 4.3. The matching of invariant phase-space curves with classical motion in the quartic-oscillator double-well potential, with scaled parameter $\gamma = 0.25$. Compare this with Figs. 1.3 and 1.15. From Tabor (1989). Copyright 1989 John Wiley & Sons, Inc. Reprinted by permission of John Wiley & Sons, Inc.

where b and/or ω depend on the time. In Chapter 9 we will consider problems where parameters in the Hamiltonian become explicitly time dependent.

In this section, however, we consider ω and b to be constant, and we will investigate—for different values of b—the region of energies close to the top of the barrier, where there is clearly a point of unstable equilibrium. The top of the barrier is located at $q = 0$, where $V(0) = 0$. For future reference, the bottoms of the two wells are located at $q_{min} = \pm\omega(m/4b)^{1/2}$, where $V(q_{min}) = -m^2\omega^4/16b$. Thus the barrier height $E_b = V(0) - V(q_{min}) = m^2\omega^4/16b$. We mention in passing that the classical quartic double-well problem has some mathematical similarities with the pendulum problem of Section 4.3. In particular, the actions and frequencies may be written in terms of complete elliptic integrals, which is useful at times.

For Hamiltonian (4.21) there are several different sets of dimensionless scaled variables and parameters used in the literature. One set is obtained by defining $q' \equiv (m\omega^2/E_0)^{1/2}q$, $p' \equiv p/(mE_0)^{1/2}$, and $\gamma \equiv E_0 b/m^2\omega^4$, where E_0 is some characteristic energy. Equation (4.21) becomes

$$H' \equiv \frac{H}{E_0} = \tfrac{1}{2}(p')^2 - \tfrac{1}{2}(q')^2 + \gamma(q')^4. \tag{4.22}$$

Other sets of variables utilize different constant factors between q' and q, and between p' and p. This can change the coefficients in front of the $(p')^2$ and $(q')^2$

terms, also of course redefining γ. Classically E_0 could be chosen to be the barrier height E_b; in quantum work either $\hbar\omega$ or $\hbar\omega_0$ has been chosen, where ω_0 is the harmonic oscillator frequency near the bottom of either one of the wells. In the case of these quantal choices for E_0, the parameter γ becomes proportional to $\hbar$ and thus is the chosen quantum parameter. In order of magnitude, these choices of the quantum parameter are the same and are inversely proportional to the dimensionless barrier height $D \equiv E_b/\hbar\omega_{(0)}$.

Since a particle in the double-well potential is so important in molecular science, many interesting studies have focused on this system. We consider a few studies that pertain to the role of separatrices in quantum mechanics. Our agenda for understanding this system parallels that for previous sections. Figure 4.3 shows the matching of phase-space trajectories with classical energy levels in the double-well potential. Unlike the pendulum (Fig. 1.15), there now exist energies E where the classical particle motion is only in one or the other of the two wells. At such energies, there are a total of four classical turning points, rather than two. The separatrix is a figure-of-eight curve in (q, p) phase space, with two lobes that touch at the unstable fixed point $(q, p) = (0, 0)$. In unscaled variables the area A_{sep} inside the two lobes of the separatrix is equal to $32E_b/3\omega$; this is another characteristic classical action of the double-well system that can be used to define a scaled Planck's constant $\hbar/A_{sep}$. The separatrix energy E_{sep} is equal to the value of the potential V(q) at the top of the barrier in the middle of the double well, that is, $E_{sep} = 0$. At E_{sep} the inner two turning points coalesce. These turning points completely disappear for $E > E_{sep}$, where the motion is that of a particle in a single well with an unusual shape for the well bottom.

As for the quantum mechanics, one can pursue the energy eigenstates and eigenvalues using the ideas we developed for the quantum pendulum. We can also obtain WKB wavefunctions. Perhaps easiest, we can carry out numerical calculations for the energy eigenstates and then compute phase-space probability distributions (Section 1.3d). For $\gamma = 0.06$, the first and third eigenstates are shown in Fig. 4.4a, along with their q-space and p-space projections. For the chosen values of the parameters, the locations of the bottoms of the two wells are near $q = 2$, and the well-depth energy is near -1.0. We see that the lowest-energy eigenstate is well localized symmetrically in the two wells; the third eigenstate, with energy above the top of the barrier, fits inside the larger well and looks more like a wave with three lobes. Using a construction similar to that for Fig. 1.12, Fig. 4.4b shows the classical analogs of the two quantum distributions in Fig. 4.4a. Even for these two quantum eigenstates with small quantum numbers, the classical physics plays a noticeable role, as do quantum interference effects.

We can also do calculations of the propagation of an initial Gaussian wavepacket in phase space (Section 1.3c). Such calculations have been done with γ reduced to 0.01. Now eight quantum energy eigenstates exist, with energies below the top of the barrier. We start off our packet within the phase-space region of the left-hand well, as shown in the first quantum Poincaré section in Fig. 4.5. The scaled mean energy of the packet is 0.2, near $E_{sep} = 0$. We see that the packet evolves under the guidance of trajectories near the separatrix, with one part of the packet passing over the region

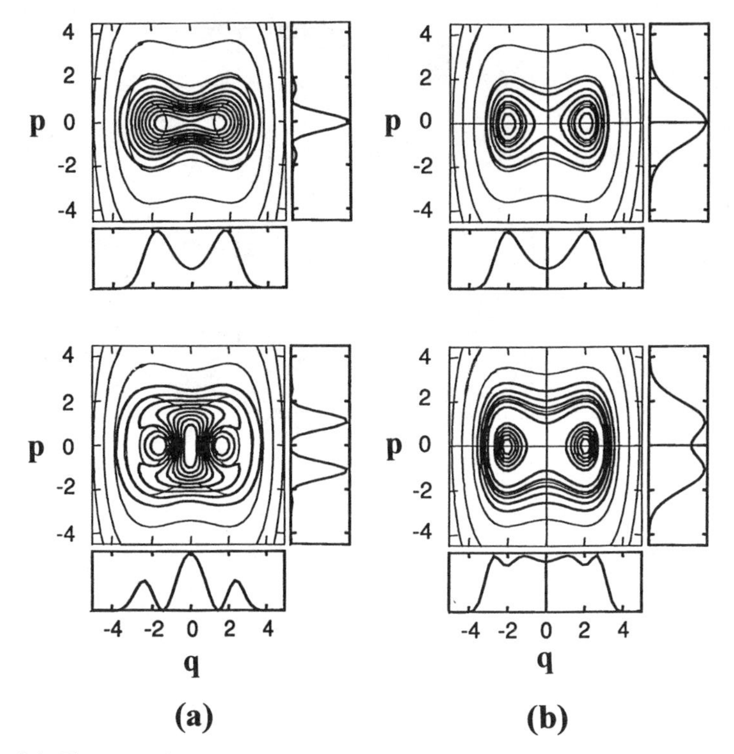

Figure 4.4. The ground and second excited quantum energy eigenstates for a particle in the quartic-oscillator double-well potential (top and bottom left), compared with their classical analogs on the right. As in Figs. 1.11 and 1.12, phase-space probability distributions are shown, as well as their projections onto the coordinate and momentum spaces. The quantum phase-space distributions are Huisimi functions; see Section 1.3. The scaled double-well quantum parameter is $\gamma = 0.06$. From Torres-Vega and Frederick (1991). Copyright 1991 by the American Physical Society.

near the unstable fixed point. Part of the packet will follow the separatrix, enter the region of the second well, and stay there awhile (not shown). This type of evolution is called *quantum separatrix crossing*. It results in a temporary split-up of the packet into portions localized in each of the wells. We see that, at least initially, classical evolution guides the quantum packet evolution. As time goes on, one expects collapse and revival effects to occur, like those in Sections 2.1b and 2.1c.

So far we've looked at the impact of the separatrix, choosing some values of the parameter γ that produce a relatively shallow double-well structure. Such wells support only a few quantum states that have energies below the top of the potential barrier. Let us now decrease γ dramatically, to $\gamma = 1.77 \times 10^{-8}$. The barrier height $E_b \sim \gamma^{-1/2}$ is much larger than before, and there are about 6×10^6 states with energies below the top of the barrier.

Now we ask a question: For this new value of γ, what regions of phase space—if any—are not clearly described by unmodified semiclassical theory? We address this

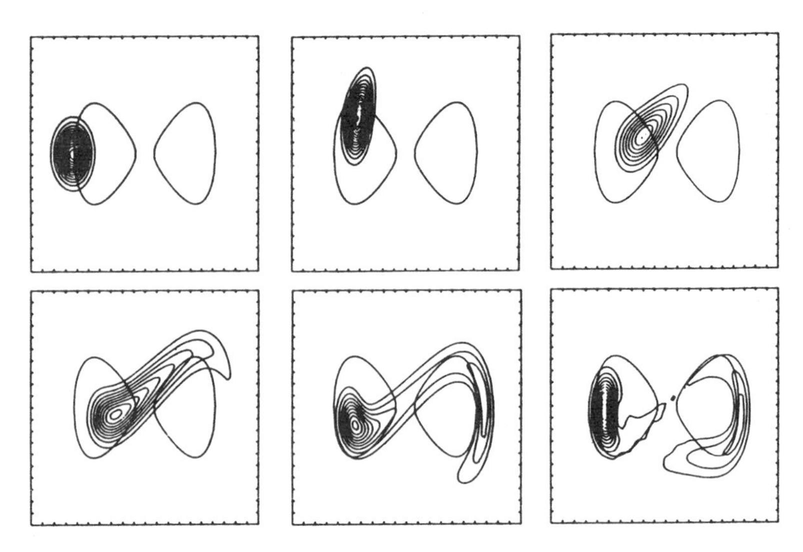

Figure 4.5. Time evolution of a wavepacket partially crossing over a separatrix, for a particle in the quartic-oscillator double-well potential. From top left to bottom right, the six snapshots of (q, p) phase space are for early moments, 0.0, 0.1, 0.2, ..., 0.5 times the vibrational period $T_{vib} = 2\pi/\omega_0$ for the particle near the bottom of either single well. The scaled quantum barrier-height parameter is $\gamma = 0.01$. The partial crossing is occurring at $t = 0.3\ T_{vib}$ and $t = 0.4\ T_{vib}$ and is completed at $t = 0.6\ T_{vib}$ (not shown). From Skodje et al. (1989). Copyright 1989 by the American Physical Society.

by checking a correspondence mentioned in Section 1.5. If a quantum system is semiclassical, then the separations of nearest-neighbor energy levels should be close to the local classical orbit frequency, multiplied by $\hbar$. Following Cary et al. (1987), we will find that the correspondence principle breaks down for quantum states with energies in a very narrow band close to the separatrix energy E_{sep}. As we approach E_{sep} within this band, the quantum level separations do not vanish. The classical frequency, however, does vanish because the period of the evolution on the separatrix is infinite (Section 1.4b).

Near the top of the barrier, the potential energy must be parabolic:

$$V(q) \approx V(0) - \tfrac{1}{2} m\omega^2 q^2. \tag{4.23}$$

Using the approach to stability analysis given in Section 3.1a, we find that ω is the exponentiation rate that is associated with evolution on the separatrix near the unstable fixed point in phase space. As already noted, the energy $\hbar\omega$ sets an energy scale for the quantum problem. We define the scaled energy shift δ away from E_{sep} by $\delta \equiv (E - E_{sep})/\hbar\omega$. Connor (1969) considered the WKB problem for a parabolic barrier and derived the connection formulas. For E near E_{sep}, while keeping $\delta \gg 1$,

Cary et al. (1987) found that the equation determining the WKB energy eigenvalues is

$$\tan\left(\frac{\pi I_a}{h} - \frac{\pi}{2} - \frac{\phi}{2}\right)\tan\left(\frac{\pi I_b}{h} - \frac{\pi}{2} - \frac{\phi}{2}\right) = e^{2\pi\delta}\left[1 + \left(1 + e^{2\pi\delta}\right)^{1/2}\right]^{-2}. \tag{4.24}$$

Here $\phi = \delta - \delta\ \ln|\delta| + \arg[\Gamma(\frac{1}{2} + i\delta)]$, and I_a, I_b are the classical actions for the separate motions in the two wells a, b. Tennyson et al. (1986) considered the classical motion just below the top of the barrier and found formulas for these actions with the form

$$I_{a,b} = Y_{a,b} + \delta\left(1 + \ln\left|\frac{E_{a,b}}{E - E_{sep}}\right|\right). \tag{4.25}$$

Here $Y_{a,b}$ are the separatrix actions—the values of $I_{a,b}$ for $E = E_{sep}$. The quantities $E_{a,b}$ are constants. The orbit period just below the separatrix energy is

$$T_a = \frac{\partial I_a}{\partial E} = \frac{1}{\omega}\ln\left|\frac{E_a}{E - E_{sep}}\right|, \tag{4.26}$$

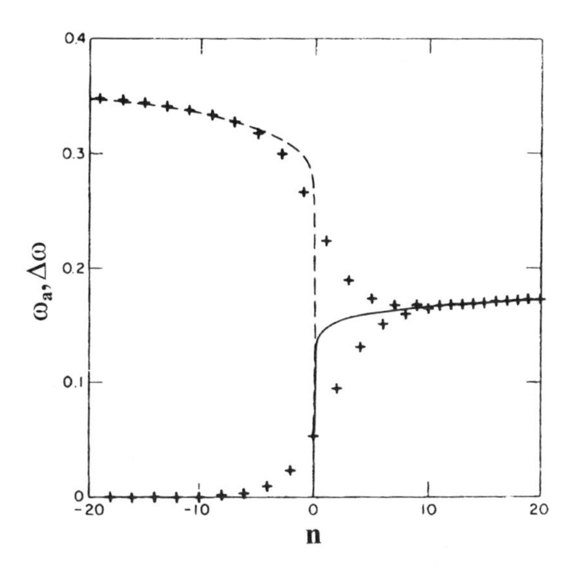

Figure 4.6. Separatrix effects on the energies of quantum eigenstates for a particle in a deep quartic-oscillator double-well potential. The quantum parameter is $\gamma = 1.77 \times 10^{-8}$. The horizontal axis is the quantum number relative to its value closest to the separatrix classical action. The plus signs are the quantum energy-level differences $(E_n - E_{n-1})/\hbar$. The dashed and solid curves are the classical orbit frequencies for motion below and above the top of the barrier, respectively. In this deep semiclassical regime, the nonzero energy-level differences are nonclassical within a very narrow region near the top of the barrier. From Cary et al. (1987). Copyright 1987 by the American Physical Society.

which gives $\omega_a \equiv 2\pi\omega/\ln|\delta_a/\delta|$ for our quartic-oscillator double well. A similar formula gives ω_b.

We finally insert Eq. (4.25) into Eq. (4.24) and numerically solve for the quantum energies. We then compare these quantum results with the classical results we have for ω_a and $\omega_b = \omega_a$. A similar procedure is applied for just above the barrier—for region c, where $\omega_c = \pi/\ln|\delta_c/\delta|$. Our check of the correspondence principle is the comparison of energy-level separations and orbit frequencies ω_a, ω_b, and ω_c shown in Fig. 4.6. The plus signs are the quantity $(E_n - E_{n-1})/\hbar$ plotted versus the energy-level number n, which is set equal to zero for the energy closest to E_{sep}. The zero values of energy-splitting below the top of the barrier, located at $n \lesssim -10$, are due to the quantum degeneracy associated with classical motion in one well or the other. The rise above zero of these energy values, as the separatrix or top of the barrier is approached, is due to quantum tunneling. We discuss this in Section 6.2. The classical frequencies are plotted versus a "classical action quantum number" n, which is given by the action relative to the separatrix action for region c, $n = (\delta/\pi)(1 + \ln|\delta_c/\delta|)$. We indeed see the expected answer to our question: out of the roughly 10^7 quantum states below or near the separatrix, only about 20 do not closely satisfy our correspondence principle! But these 20 intrinsically quantum states that are not "nearly classical" are indeed there. The influence of separatrices on quantum systems is undeniable. Wave interference and quantum tunneling are the origins of the unusual behavior. Of course, a few more intrinsically quantum energy levels are found near the bottoms of the double well.

5

INTRODUCTORY ASPECTS OF NONINTEGRABLE QUANTUM SYSTEMS

5.1. KAM SYSTEMS (ADVANCED TOPIC)

With this chapter, we begin our discussion of quantum systems that are not classically integrable. The simplest of these quantum systems are the KAM systems—slightly perturbed integrable systems that turn out to be nonintegrable. Although one might think a straightforward perturbation theory would prove useful for describing classical KAM systems, in fact this is usually not the case. Indeed, only with the relatively recent work of A. Kolmogorov, of V. Arnold, and of J. Moser has the study of slightly perturbed integrable classical systems been put on a firm mathematical foundation. Their work produced the celebrated "KAM" theorem, which we will discuss. The KAM nonintegrable systems exhibit many important and interesting features that are not present in integrable systems. Thus we must spend a few sections of this book addressing nonintegrable classical systems, before turning to their corresponding quantum systems.

Consider an integrable autonomous system with Hamiltonian $H_0(\mathbf{I})$, where $\mathbf{I}$ stands for the set of action variables $I_1, I_2, \ldots, I_N$, $N \geq 2$. (The case $N = 1$ is excluded; 1-d systems are always integrable if the Hamiltonian does not explicitly depend on time.) We add to H_0 a perturbation, which can be written in terms of the action-angle variables $(\mathbf{I}, \boldsymbol{\theta})$ for H_0. The total Hamiltonian is then taken to have the form

$$H = H_0(\mathbf{I}) + \epsilon H'(\mathbf{I}, \boldsymbol{\theta}), \tag{5.1}$$

where ϵ is a parameter characterizing the strength of the perturbation. We assume that H determines a nonintegrable system, at least for some nonzero values of ϵ. For these nonzero values of ϵ, the (θ_i, I_i) are no longer action-angle variables. However, they do remain canonical variables, which is important.

An interesting question is whether or not the system (5.1) is integrable for small ϵ. What happens to the invariant tori of the integrable system? If, using a computer, we integrate Hamilton's equations (1.7) for enough different conditions and display the results in a Poincaré section (Section 1.3a), then two dramatic new features appear. First, the Poincaré section displays sections of a multitude of new structures that look like—and are—tori of a type different from those in the integrable system. Second, and even more surprising, trajectories are found that do not lie on tori at all! However, there are also many, many tori that look very similar to the original tori, as we would expect when ϵ is small. These "remaining" tori we call *KAM tori.* Figure 5.1b schematically shows a small part of a typical mixture of the three types of trajectories.

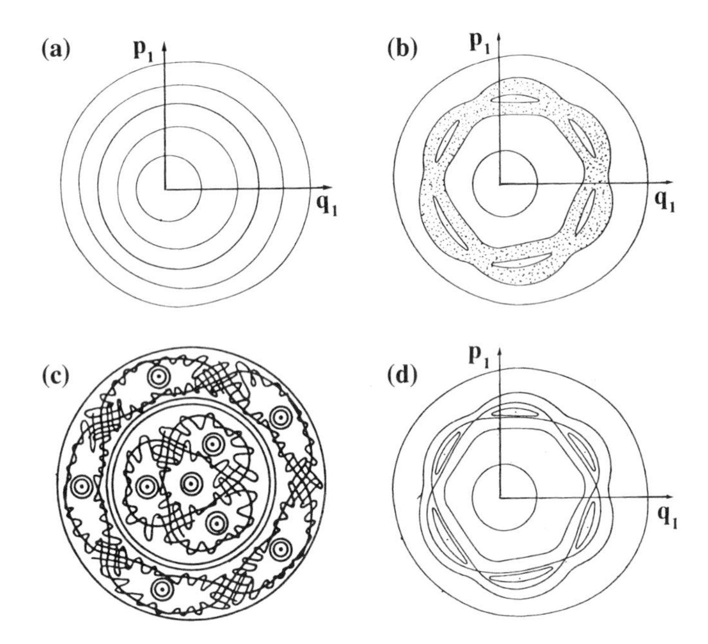

Figure 5.1. The modification of Poincaré sections produced by the nonintegrable perturbation of an integrable classical system. (a) A simple integrable system displays a family of closed invariant curves. (b) The perturbation destroys many of the original invariant curves, replacing them by the closed curves of "island" tori surrounded by chaotic orbits. (c) The chaotic bands around the island tori in (b) contain structures called manifolds that intersect at unstable periodic orbits; see further Fig. 5.9 and the end of Section 5.2. (d) The perturbed system can be approximated by a set of integrable systems; see Section 5.6a. Panels (a), (b), and (d) reprinted from Ozorio de Almeida (1984). Copyright 1984 American Chemical Society. Panel (c) reprinted with permission from Berry (1978). Copyright 1978 American Institute of Physics.

In classical mechanics, standard canonical perturbation theory is based on a search for a particular kind of F_2-type generating function of new momenta and of old coordinates, a function that can be written as a perturbation expansion in powers of ϵ. This expansion is worked out in many standard textbooks (e.g., Rasband, 1990; Lichtenberg and Lieberman, 1992). In the generating function $F_2(\mathbf{q}, \mathbf{I}) = F_1(\mathbf{q}, \boldsymbol{\theta}) + \mathbf{I}\cdot\boldsymbol{\theta}$ for the integrable system, the result for the correction to the term $\mathbf{I}\cdot\boldsymbol{\theta}$ turns out to be $\epsilon S_1(\mathbf{I}, \boldsymbol{\theta})$, where

$$S_1(\mathbf{I}, \boldsymbol{\theta}) = \sum_{\mathbf{m}\neq 0} \frac{iH'_{\mathbf{m}}(\mathbf{I})}{\boldsymbol{\omega}(\mathbf{I})\cdot\mathbf{m}} e^{i\mathbf{m}\cdot\boldsymbol{\theta}}. \tag{5.2}$$

Here $\mathbf{m}$ is an N-d vector of integers, $\boldsymbol{\omega}(\mathbf{I})$ is the set of N classical frequencies of the unperturbed system, and the $H'_{\mathbf{m}}(\mathbf{I})$ are coefficients in an expansion of H' in a Fourier series $\sum_{\mathbf{m}} H'_{\mathbf{m}} \exp(i\mathbf{m}\cdot\boldsymbol{\theta})$. Whenever the denominator of any term in Eq. (5.2) is zero—that is, $\boldsymbol{\omega}\cdot\mathbf{m} = 0$ for some $\mathbf{m}$—then the standard perturbation theory and Eq. (5.2) do not apply.

For each vector $\mathbf{m}$ that causes the frequencies ω_j to be commensurate, $\boldsymbol{\omega}\cdot\mathbf{m} = 0$, we say we have a *nonlinear resonance*. These nonlinear resonances are associated with some of the "new" tori present in the KAM system. (We discuss these resonances in some detail in Section 5.6.) Frequency space becomes fragmented by regions associated with the nonlinear resonances. This fragmentation is reflected back into action space, as suggested in Fig. 5.2.

There is a rigorous theorem that is related to the existence of the nonlinear resonances. Consider the Poincaré section of an unperturbed integrable system. There are smooth curves associated with trajectory evolution on tori in the full phase space. The curves can be "rational" or "irrational," depending on whether or not a point on them is mapped back onto itself after a finite time. The *Poincaré–Birkhoff theorem* states that, upon adding a perturbation, only even numbers of points remain from a rational curve. If the winding number of the rational curve is the ratio of integers p/q, then the remaining numbers of points are 2kq for some $k = 1, 2, 3\ldots$ (Birkhoff, 1927).

On the other hand, there exists a nondenumerable number of values for the actions I_i; most of these values are associated with frequencies $\boldsymbol{\omega}_i(\mathbf{I})$ that are incommensurate. In such cases, $\boldsymbol{\omega}(\mathbf{I})\cdot\mathbf{m} \neq 0$ for all integers m_i. One requirement of the KAM theorem is that we remove from consideration the neighborhoods of all points in the action-angle phase space where this incommensurability condition does *not* hold. That is, we avoid all the nonlinear resonances.

One statement of the *KAM theorem* is given in Arnold and Avez (1968). We paraphrase it as follows. If ϵ is "small enough," then for "almost all" incommensurate $\boldsymbol{\omega}$ there exists an invariant torus $T(\boldsymbol{\omega})$ of the perturbed system, such that $T(\boldsymbol{\omega})$ is "close to" an invariant torus $T_0(\boldsymbol{\omega})$ of the unperturbed system. The set of all such tori $T(\boldsymbol{\omega})$ has positive measure—that is, the set is nondenumerable. The complement of this set has a measure that goes to zero as $\epsilon \to 0$.

Key issues for this KAM theorem involve the three phrases used above—"small enough," "almost all," and "close to." To address the first issue, there generally

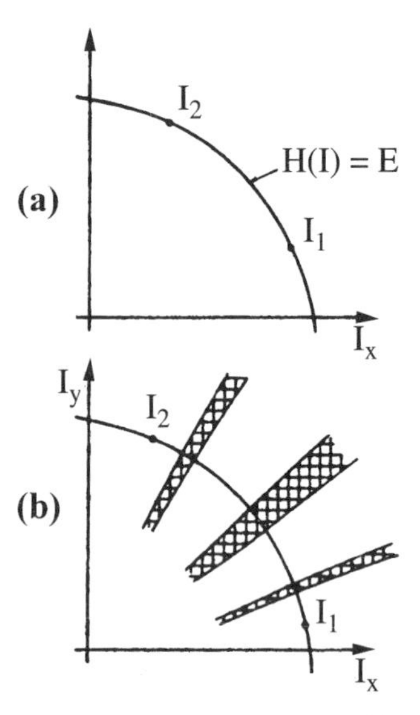

Figure 5.2. Fragmentation of action space. (a) In a 2-d integrable system, a torus having total action I_1 can be smoothly deformed to give a torus with total action I_2. A curve is followed in a 2-D action space. (b) In a KAM nonintegrable system, the action space is fragmented because of new regions in phase space such as those shown in Figs. 5.1b, c. No smooth deformation is possible. From Wilkinson (1986). Copyright 1986, with permission from Elsevier Science.

appears to be no precise estimate of how small "small enough" is. However, the existence of an adequately small ϵ was shown by KAM, by abandoning the conventional ideas underlying Eq. (5.2) and employing "superconvergent" perturbation theory. Standard perturbation theory is based on the selection of an initial value about which to expand, and then consistently using this value out to the desired order in ϵ. In a superconvergent theory, however, the expansion is always about the latest best-known value, step by step as the order is increased. Successive canonical transformations can then be chosen so that the order in ϵ increases by the square of the preceding order, for each step:

$$\epsilon H_1' \to \epsilon^2 H_2' \to \epsilon^4 H_3'. \to \cdots \to \epsilon^{2^{n-1}} H_n'.$$

This is also called *quadratic convergence*. One might be familiar with this approach, as it is used in the Newton–Raphson method for finding roots (zeros) of an equation; see Section A.4. For a direct application of the superconvergence approach to the pendulum problem, see Chirikov (1979).

KAM theory is based on superconvergent perturbation theory taken relative to a "sufficiently irrational" torus $T_0(\boldsymbol{\omega})$. The issues of "almost all" and "close to" mentioned above basically reduce to the question of an irrationality condition for the preserved tori $T(\boldsymbol{\omega})$. This question is answered, in part, by considering the continued-fraction rational approximations of irrational numbers. The KAM result is that the set of missing (unpreserved) tori is included in those tori that satisfy the condition (for $N = 2$)

$$\left|\frac{\omega_1}{\omega_2} - \frac{r}{s}\right| < \frac{K(\epsilon)}{s^{2.5}}, \tag{5.3}$$

where r and s are integers, $\boldsymbol{\omega} \equiv (\omega_1, \omega_2)$ is the frequency vector of the 2-D torus $T_0(\boldsymbol{\omega})$, and the function $K(\epsilon) \to 0$ as $\epsilon \to 0$. Thus the winding numbers ω_1/ω_2 for each of the missing 2-D tori must be close enough to some rational number r/s for the difference to be bounded by the indicated function of s on the right side of Eq. (5.3). For values of r/s within the unit interval, we have $s \geq r$. Consider summing both sides of Eq. (5.3) over all $r \leq s$ and then over all $s \geq 2$. The right-hand side of Eq. (5.3) becomes

$$\sum_{s=2}^{\infty} \frac{K(\epsilon)}{s^{2.5}} s.$$

This quantity will at most equal $K(\epsilon)$, which can be made as small as one wants by taking ϵ sufficiently small. Values of r/s within other unit intervals can be treated similarly, by first removing the integer part of the left-hand side of Eq. (5.3). Thus Eq. (5.3) implies the last sentence in our statement of the KAM theorem—that is, the complement of the set of tori $T(\boldsymbol{\omega})$ has a measure that goes to zero as $\epsilon \to 0$.

We note the KAM theorem says nothing about the nature of the phase-space regions where the slightly perturbed tori are missing. Nor does it pertain to nonintegrable systems that are not slightly perturbed integrable systems. Thus we need to further investigate classical nonlinear dynamics before we can return to quantum time evolution.

5.2. LYAPUNOV EXPONENTS, CHAOS, AND STOCHASTIC LAYERS

5.2a. Lyapunov Exponents for Trajectories in Phase Space

In characterizing the behavior of nonintegrable classical systems in different regions of phase space, it is important to investigate how initially neighboring trajectories separate as time evolves. One possible dramatic event is for some measure of the initial separation of two trajectories to grow exponentially with time. The Lyapunov (also spelled Liapunov) exponent is constructed as a mathematically rigorous measure of any such growth behavior. Our approach for the construction of the exponent uses the general discussion of orbit stability analysis from Section 3.1b, Eqs. (3.4)–(3.7).

We begin by selecting a reference fiducial phase-space trajectory $\mathbf{Z}_0(t)$. We want to evolve this point's neighborhood in phase space, starting at the initial time $t = 0$. To evolve such a neighborhood, we start by choosing an infinitesimal 2N-D sphere of initial conditions centered at the position $\mathbf{Z}_0(0)$ in phase space. Let the sphere radius be ϵ. Since ϵ is infinitesimal, the sphere becomes a 2N-D ellipsoid as time goes on, because of the locally deforming nature of the Hamiltonian phase-space flow. Of course, this time-dependent deformation satisfies the constraints imposed by Liouville's theorem. Figure 5.3 schematically shows this evolution. Each point on the initial 2N-D sphere lies on some trajectory $\mathbf{Z}(t)$, and $\mathbf{Z}(t) - \mathbf{Z}_0(T)$ evolves according to Eq. (3.5).

The 2N-D ellipsoid has principal axes; we denote their directions by $\hat{e}_j$ and their lengths by $r_j \equiv \epsilon L_j$, $j = 1, \ldots, 2N$. The ellipsoid in general will rotate, with the $\hat{e}_j$ being time dependent. However, in a reference frame rotating with the ellipse, the new axes will have constant directions. In this frame, we can define the 2N *Lyapunov functions* by

$$\lambda_j(t) \equiv \frac{1}{t} \log \frac{r_j(t)}{r_j(0)} = \frac{1}{t} \log L_j(t), \tag{5.4}$$

where the all-important dependence on the selection of $\mathbf{Z}_0(t)$ has been suppressed in the notation. If there is a long-term sufficiently smooth behavior for the locally time-averaged evolution of the trajectories involved, then it is reasonable to consider the *Lyapunov exponents* defined by

$$\lambda_j \equiv \lim_{t\to\infty} \lambda_j^{\text{ave}}(t), \tag{5.5}$$

since the limit then could exist. Oseledec (1968) has actually shown that if two initially neighboring trajectories in phase space are initially separated by a phase-space vector $\mathbf{Z}(0) - \mathbf{Z}_0(0) \equiv \epsilon\mathbf{R}$, then

$$\lim_{t\to\infty} \frac{1}{t} \log[\det(\mathbf{U}\mathbf{R})] \equiv \Lambda(\mathbf{R})$$

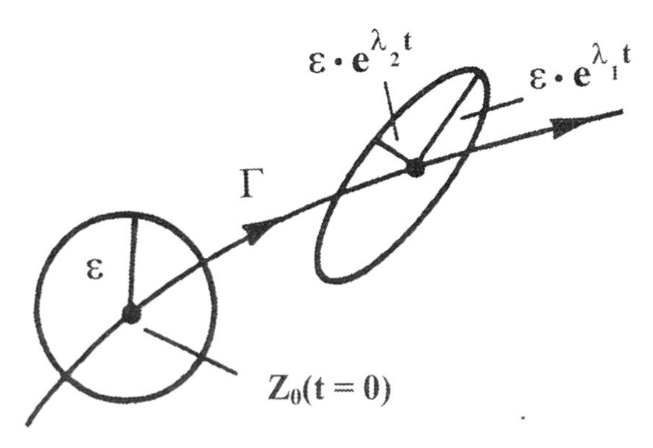

Figure 5.3. Lyapunov exponents describe the time evolution of a small spherical surface of initial conditions taken around a point located by a position $\mathbf{Z}_0(t = 0)$ on a trajectory Γ in phase space. At time t, the sphere of radius ϵ is deformed into an ellipsoid with principal axes $\epsilon \exp(\lambda_1 t)$ and $\epsilon \exp(\lambda_2 t)$, where λ_1 and λ_2 are Lyapunov exponents. From Schuster (1992). Copyright 1992 Springer-Verlag.

does exist for all nonvanishing $\mathbf{R}$. Here $\bar{\bar{U}}$ is the monodromy matrix defined by Eq. (3.4). Oseledec also showed that for every nonzero $\mathbf{R}$, $\Lambda(\mathbf{R})$ takes on one of 2N distinct values λ_j—values that are independent of the particular choice of the generalized coordinates $\mathbf{Z} \equiv (\mathbf{q}, \mathbf{p})$, independent of the units of $\mathbf{q}$ and $\mathbf{p}$, and independent of the energy E. Thus the Lyapunov exponents λ_j simply scale with the unit of time. It can be shown that if there are k independent local "constants of the motion" $G_k(\mathbf{Z}_0)$, then there are at least 2k vanishing Lyapunov exponents. The $G_k(\mathbf{Z}_0)$ need not be the global constants of motion of which we usually think. We only need that $G_k[\mathbf{Z}_0(t)]$ remain constant for the particular reference trajectory $\mathbf{Z}_0(t)$ under investigation, and also that the components of the gradient $\nabla_{\mathbf{Z}_0} G_k(\mathbf{Z}_0)$ exist, be continuous along $\mathbf{Z}_0(t)$, and be linearly independent.

For the special case of an autonomous 2-d Hamiltonian system, the number N of degrees of freedom is two, and there are $2N = 4$ Lyapunov exponents. Since E is a constant of the motion, at most two of the exponents can be nonzero. As was the case for the stability of fixed points (Section 3.1a), Liouville's theorem requires that the remaining two exponents be additive inverses. One possibility is that both exponents are zero; the other is that they be $\pm\lambda$. Thus stretching along one principal axis must be compensated by contraction along another principal axis.

The above situation for $N = 2$ generalizes to any value of N. Liouville's theorem requires that the sum of all the Lyapunov exponents be zero. Actually, the exponents λ_j add up to zero in pairs (Oseledec, 1968). Thus the stretching and contraction along principal axes is pairwise.

For the special case of the integrable systems of Chapter 4, we know there are sets of action-angle variables $(\mathbf{I}, \boldsymbol{\theta})$. When written in terms of these variables, the Hamiltonian H does not depend on $\boldsymbol{\theta}$, and the action vector $\mathbf{I}$ is constant. Therefore the matrix $\bar{\bar{J}}(\nabla_Z \nabla_Z H)$ of Eq. (3.6) becomes independent of the time. The solution of Eq. (3.6) is then

$$\bar{\bar{U}}(t) = \exp[\bar{\bar{J}}(\nabla_Z \nabla_Z H)t] = \bar{\bar{1}} + \bar{\bar{J}}(\nabla_Z \nabla_Z H)t,$$

where the last equality follows from $[\bar{\bar{J}}(\nabla_Z \nabla_Z H)]^2 = 0$, primarily as a consequence of θ-derivatives of $H(\mathbf{I})$ and of $\partial H(\mathbf{I})/\partial \mathbf{I}$ being zero. The linearity in time of this expression implies that all Lyapunov exponents λ_j will be zero for all trajectories of integrable systems.

For any dynamical system, a phase-space trajectory is called a *regular trajectory* if all the λ_j for that trajectory are zero. Integrable systems are "regular" everywhere in phase space. Periodic and quasiperiodic trajectories on N-D tori are usually regular. On the other hand, we have an *irregular trajectory* when there is a positive value for at least one of the λ_j for a trajectory of a nonintegrable system. We have a *chaotic trajectory* if the phase-space trajectory is irregular and not periodic. Unstable periodic orbits are irregular, since all sufficiently close trajectories are chaotic and therefore separate exponentially from the periodic orbit in at least one "direction."

For nonintegrable systems, the set of Lyapunov exponents is generally not trivial to calculate. One must deal with the often complicated, time-dependent rotation of the 2N-D ellipsoid. In addition, since exponentially varying quantities can quickly become susceptible to computer limitations (such as overflow/underflow and limited number of accurate digits), one must repeatedly rescale the lengths of the principal axes and reset the time to a new initial value, as suggested by Fig. 5.4. (An accepted algorithm for calculating a set of Lyapunov exponents is discussed in Section A.3.)

We should also note that we don't have enough computer time to numerically take the limit in Eq. (5.5). Thus to determine if a trajectory is regular or irregular, we calculate the Lyapunov functions (5.4) out to "sufficiently large" finite times and then stop. We look for apparent convergence to the needed accuracy. Figure 5.5 shows some results of such calculations for a regular trajectory and for a chaotic trajectory. There are some dangers in following this apparent-convergence procedure. One is that some trajectories of some Hamiltonian systems exhibit intermittent

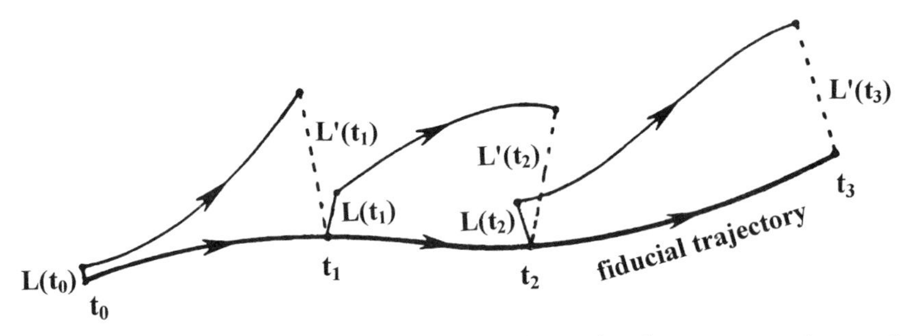

Figure 5.4. Schematic representation of the evolution and replacement procedure used to numerically compute Lyapunov exponents. From Wolf et al. (1985). Copyright 1985, with permission from Elsevier Science.

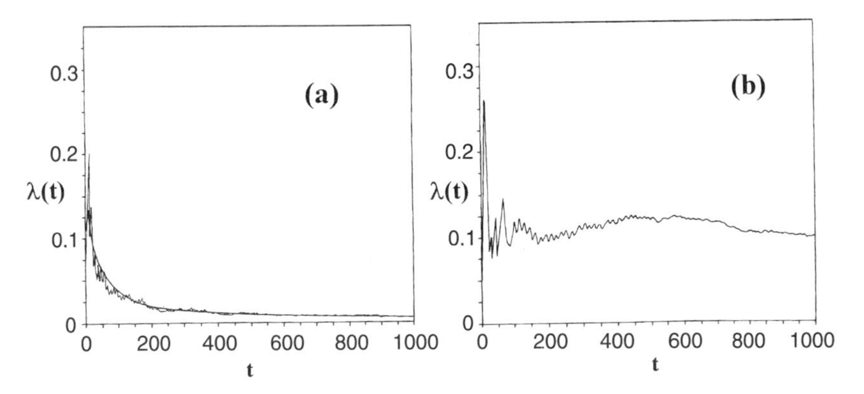

Figure 5.5. Time evolution of regular and chaotic Lyapunov functions. (a) The Lyapunov function $\lambda(t)$ for a regular trajectory goes to zero at large times t. For this particular trajectory, there is a good fit to the function $(2t)^{-1}\log(1+\beta^2t^2)$ shown as a solid line. (b) A Lyapunov function for a chaotic trajectory, which converges to a positive Lyapunov exponent at sufficiently large times. Reprinted with permission from Meyer (1986). Copyright 1986 American Institute of Physics.

behavior, with large delayed changes in behavior occurring in the Lyapunov functions. An example is the set of coupled Bloch–Maxwell equations for an ensemble of two-level atoms in a resonant cavity and where the atoms are driven by an additional amplitude-modulated electromagnetic field (Alekseev and Berman, 1994). Special concepts for transient chaos have been developed and sometimes are needed (Benkadda et al., 1994; Klafter et al., 1995).

In characterizing nonintegrable dynamical systems, an important issue is how the nature of the periodic orbits changes as the system's parameters are varied. The orbits can change from regular to irregular, or vice versa. Figure 5.6 shows some examples of changes in values of a Lyapunov exponent, using a particular $N = 2$ autonomous system that is important for the study of quantum dynamics. Following some ideas in statistical mechanics, we say that *critical points* are those sets of parameter values where a change from zero to a positive Lyapunov exponent occurs, or vice versa. One type of critical point is the bifurcation point of a periodic orbit; see Section 7.1b for an example of the impact of bifurcations on a quantum system.

5.2b. Stochastic Layers and Global Chaos

What is the nature of the regions of missing tori in phase space—the regions implied in the discussion of KAM systems in Section 5.1? We can now explore this issue numerically by calculating the Lyapunov exponents of the trajectories lying in these many regions. Chaotic trajectories occupy parts of these regions, for the trajectories have positive Lyapunov exponents. The larger chaotic subregions are noticeable layers in phase space, with the layers replacing the separatrices of the unperturbed

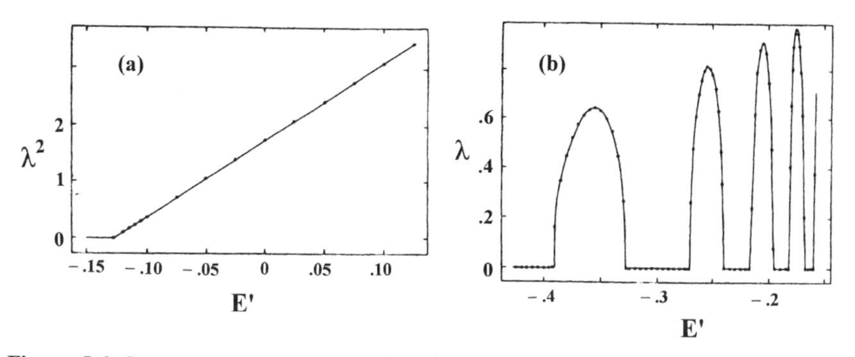

Figure 5.6. Lyapunov exponents as a function of a scaled energy parameter E′, for two different periodic orbits found in the problem of Section 7.1, the hydrogen atom in a static magnetic field. (a) A particular periodic orbit perpendicular to the field is regular up to a threshold for neighboring chaos. At the threshold (bifurcation) point, the stable periodic orbit becomes unstable and new periodic orbits appear in phase space as well (not shown). Above the threshold the Lyapunov exponent is positive and its square is linearly increasing. (b) For the periodic orbit parallel to the magnetic field, the Lyapunov exponent alternates between zero and positive values, indicating a sequence of periodic orbit bifurcations. From Wintgen (1987). Copyright 1987 Institute of Physics Publishing.

integrable system, as suggested in Fig. 5.7a. A single initial condition taken inside one of these subregions starts us on a chaotic trajectory that indefinitely produces, as time evolves, more and more points in a subregion of the Poincaré section. Such subregions are called *stochastic layers*, because the points appear to randomly fill them up.

As we begin to increase the perturbation parameter ϵ, the widths of the stochastic layers increase. For the case of Fig. 5.7a, the changing width is in the action variable. After some further increase, ϵ becomes large enough for the widest stochastic layers to merge, as indicated in Fig. 5.7b. We now have *global chaos*, with chaotic trajectories wandering about in a much larger (perhaps infinite) portion of phase space. A precise *threshold for global chaos* occurs at the exact value of ϵ at which a last N-D torus just disappears (this last N-D torus is of the KAM type, characteristic of the unperturbed integrable system). The threshold parameter value for the removal of a boundary that inhibits global chaos is another kind of critical point.

The qualitative situation suggested by Fig. 5.7 is typical of autonomous $N = 2$ systems. As we will see in Chapter 8, Fig. 5.7 is also typical of $N = 1$ integrable systems driven by sinusoidally time-varying perturbations. For systems of higher dimension, the region of missing KAM tori can be a "stochastic web" that penetrates the entire phase space. In such cases, any nonzero perturbation strength ϵ may produce some global chaos; see Fig. 5.8.

Why do stochastic layers begin to appear at small ϵ? Let us again consider the vertical pendulum, referring back to Fig. 1.15. Far away from the unstable fixed point (point of unstable equilibrium), a small perturbation has little effect. The force that the perturbation produces on the pendulum is small, compared to the active

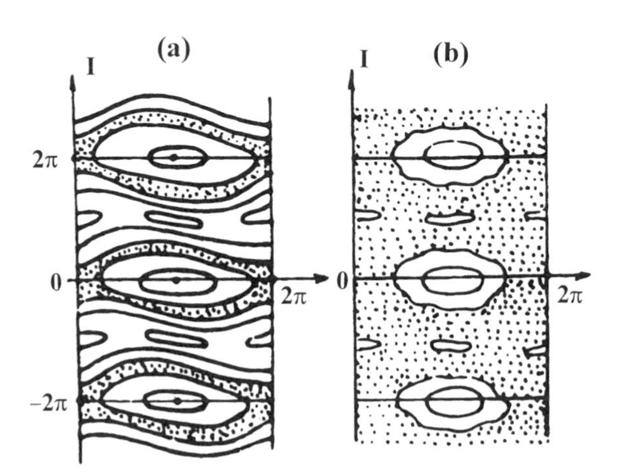

Figure 5.7. A schematic illustration of Poincaré sections showing the change from stochastic layers around island chains characteristic of KAM systems (left panel) to the global chaos of more strongly perturbed systems (right panel). A chaotic trajectory started in a region of truly global chaos is free to roam over the full ranges of both phase-space variables used for the section. From Sagdeev et al. (1988). Copyright 1988 Harwood Academic Publishers.

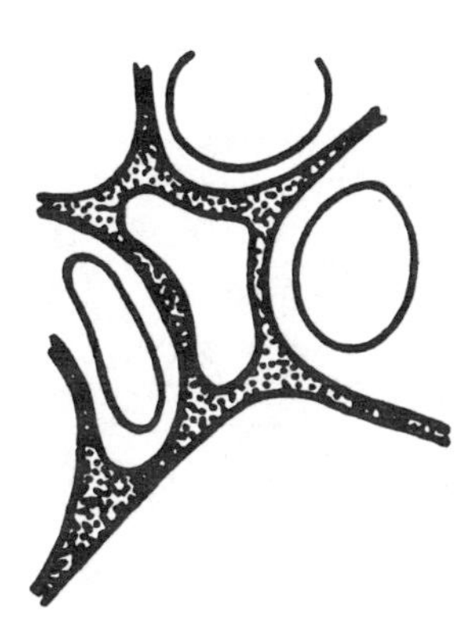

Figure 5.8. In problems with spatial dimension $N \geq 3$, stochastic layers can join to form an Arnold stochastic web that can penetrate all of phase space. From Sagdeev et al. (1988). Copyright 1988 Harwood Academic Publishers.

component of the force due to gravity. Thus even an orbit fairly close to the separatrix will respond primarily to the unperturbed Hamiltonian. However, at the unstable fixed point, the active force due to gravity is zero, so the perturbation is dominant near this point. When we start a trajectory near the separatrix and near the unstable fixed point, the perturbation can switch librations into rotations and back again. Successive switchings will be uncorrelated, because the periods of rotations and librations lying close to the separatrix vary over infinite ranges. (The period of the separatrix is infinite, while a bit away from the separatrix, orbit periods are quite finite.) This is the origin of the chaos in what is often loosely called a "separatrix" stochastic layer.

In the unperturbed pendulum system, the separatrices are the stable and unstable manifolds of the equilibrium (Section 1.4b). The unstable manifolds are those branches of the separatrix trajectories (or, more generally, those manifolds of trajectories) that correspond to growing evolution away from the unstable fixed point. The stable manifolds are branches that correspond to a slowing approach to that fixed point. Interestingly, such stable manifolds continue to exist in the Poincaré section of the irregular perturbed system; see Fig. 5.9. Near the unstable fixed point in a Poincaré section—a point produced by an (orbitally) unstable periodic orbit in the full phase space—the manifolds look similar to the separatrices of the unperturbed system. However, far from those unstable fixed points, the extensions of the manifolds intersect one another almost transversely at discrete points. The topology of the manifolds is indicated in Fig. 5.9. This figure shows a *homoclinic* case, where a trajectory starts out at an unstable point as t goes to minus infinity, and exhibits returns to the region of the same point as t goes to plus infinity. If a trajectory exhibits returns to the region of a different unstable fixed point, then we have a *heteroclinic trajectory* instead of a homoclinic trajectory.

When we further explore the regions of missing tori for a KAM system, we find an enormous amount of additional structure in phase space. There may be some

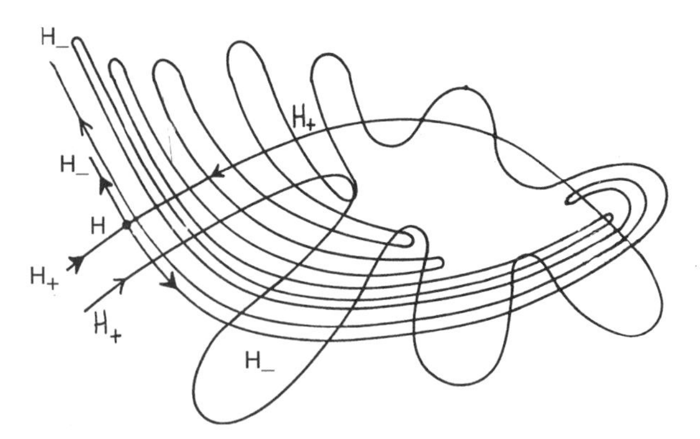

Figure 5.9. A Poincaré section of a stable manifold H_+ and an unstable manifold H_- intersecting at an unstable periodic orbit that produces the unstable fixed point labeled H. Also see Fig. 5.1c. Reprinted with permission from Berry (1978), after Moser. Copyright 1978 American Institute of Physics.

original KAM tori left, but there also are new tori, new manifolds, and new stochastic layers. There are new structures associated with each value of r/s in Eq. (5.3)! One little piece of this additional structure is indicated in Fig. 5.1c. This kind of structure is self-similar, meaning that it repeats itself on ever smaller phase-space length scales. Although this is a fascinating fact, at the present time it does not appear very relevant for physically accessible quantum systems.

5.3. DYNAMICS OF THE HÉNON–HEILES SYSTEM

5.3a. Early History

In Section 1.3a we briefly considered the problem of a particle of mass m in the Hénon–Heiles potential so that we could display a "typical" Poincaré section for an autonomous 2-d problem (Fig. 1.8). In Cartesian coordinates, the Hamiltonian defining the Hénon–Heiles problem is

$$H = \frac{1}{2m}(p_x^2 + p_y^2) + \frac{1}{2}m\omega^2(x^2 + y^2) + m\omega^2\left(\frac{1}{a}\right)\left(x^2 y - \frac{y^3}{3}\right). \tag{5.6}$$

The first two terms form an integrable partial Hamiltonian—for harmonic oscillator evolution in each of the x and y directions. The frequencies of the 1-d oscillators are the same, with the value ω. Thus this partial Hamiltonian is that for a 2-d harmonic oscillator. This oscillator is perturbed by the last term in the Hamiltonian, which has the parameter a that defines a length scale. The second term in the perturbation makes the y-oscillator anharmonic, while the first term x^2y couples the two oscillators.

The study of the Hénon–Heiles problem is historic, both in classical science and in quantum science. Two astronomers—Hénon and Heiles (1964)—did the initial classical work. They were interested in modeling the motion of one star moving through the mean gravitational field of a galaxy. A model of a galaxy should include a central potential well where most of the galaxy's stars are trapped. There should also be an effect modeling the centrifugal barrier, an effect that tends to keep stars away from the exact galactic center. Finally, stars with sufficient energy should be able to ultimately escape the galaxy's attraction. Hamiltonian (5.6) models all these effects. The numerical work of Hénon and Heiles was the first to indicate the coexistence of both ordered (regular) and stochastic (chaotic) regions in the phase space of a physically interesting system.

About a decade later, theoretical chemists realized that the Hénon–Heiles coupled oscillator system could model the evolution of two coupled vibrational modes (bonds) of a triatomic molecule. An exciting chain of studies of the quantized Hénon–Heiles problem followed, addressing the question of what role classical chaos might play in quantum evolution. Thus we have the beginnings of the investigation of physically interesting nonintegrable systems, both classical and quantum.

5.3b. Classical Evolution

Our Hamiltonian (5.6) has a special three-fold (equilateral) spatial symmetry, which can be seen by writing Eq. (5.6) in polar form (taking $m = 1$):

$$H = \frac{1}{2}\left(p_r^2 + \frac{1}{r}p_r + \frac{p_\theta^2}{r^2}\right) + \frac{1}{2}\omega^2 r^2 + \frac{\lambda}{3} r^3 \sin 3\theta,$$

where $\lambda \equiv m\omega^2/a$. The symmetry group is C_{3v} (Tinkham, 1964; Schutte, 1968). Figure 5.10 displays the symmetry seen in the equipotential lines of the potential.

The Hénon–Heiles potential $V(x, y)$ has a minimum value of 0 at the origin. For the $\omega = 1$, $\lambda = 1$ special case shown in Fig. 5.10, the condition $V(x, y) = \frac{1}{6}$ for an equipotential becomes a polynomial that factors according to

$$(y + \tfrac{1}{2})\left(x - \frac{(y-1)}{\sqrt{3}}\right)\left(x + \frac{(y-1)}{\sqrt{3}}\right) = 0.$$

This factorization into products of linear functions results in the special triangular equipotential contour in Fig. 5.10. For particle energies $E < \frac{1}{6}$, the motion is confined to be within the potential well located inside the triangle. Thus the depth of the potential well is $D = \frac{1}{6}$. However, for $E > \frac{1}{6}$, there are three ridges in $V(x, y)$, each ridge opposite a valley on the other side of the well. The vertices of the triangle are saddle points in the (x, y) position space and correspond to unstable fixed points

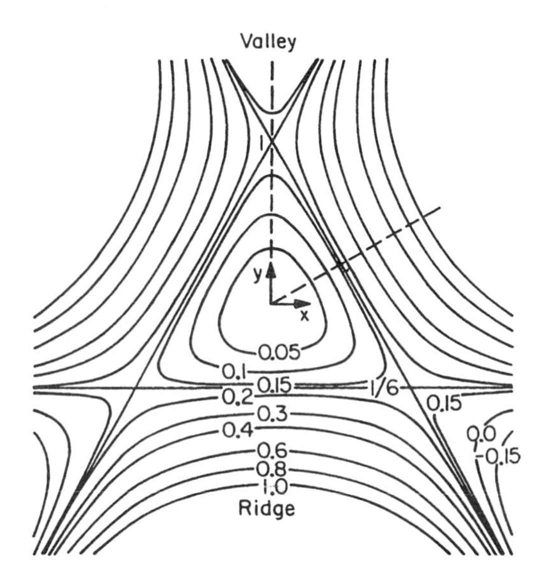

Figure 5.10. Equipotential contours for the Hénon–Heiles potential V(x, y) that is defined by the terms in Eq. (5.6) proportional to ω^2. From *Computational Physics—Fortran Version* by Steven Koonin and Dawn Meredith. Copyright 1990 by Addison Wesley Publishing Company. Reprinted by permission of Perseus Books Publishers, a member of Perseus Books, L.L.C.

in phase space. Still for our special case, if x, y, p_x, and p_y are all divided by $E^{1/2}$, then we have the scaled Hamiltonian $H' \equiv H/E$ given by

$$H' = \tfrac{1}{2}[(p'_x)^2 + (p'_y)^2 + (x')^2 + (y')^2] + \sqrt{E}\left((x')^2 y' - \frac{(y')^3}{3}\right).$$

There there is only one parameter $\sqrt{E}$ that determines the strength of the perturbation of the 2-d oscillator and hence controls how nonintegrable the system will be.

It is particularly easy for us to explore phase space for the Hénon–Heiles problem, since the needed computer code for Poincaré sections is available; see Koonin and Meredith (1990). The (y, p_y) Poincaré section for $E = \frac{1}{12}$ (Fig. 1.8) indicates the existence of seven periodic orbits. Actually, the boundary of this section corresponds to an eighth periodic orbit that is stable, with the spatial evolution of the orbit being up and down the y-axis. On the y-axis of the section there are three fixed points, two stable and one unstable (Greene, 1979b). The boundary periodic orbit and the periodic orbits through the fixed points on the y-axis are "Type A" orbits, meaning that orbits have two points in phase space where x and p_y are both zero. The remaining periodic orbits we call "Type B," for they have vanishing momentum at two turning points, $p_x = p_y = 0$. The boundary Type A periodic orbit is also Type B, so we call it "Type AB."

Now let us increase the particle energy to $E = \frac{1}{8}$ and obtain a new Poincaré section. The result is Fig. 5.11a. We see a large portion of the section where no invariant curves are produced by the intersection of the $x = 0$ surface of section with 2-D tori in the full 4-D phase space. Yet it is apparent that some tori continue to exist, producing curves in the neighborhoods of the stable fixed points shown in

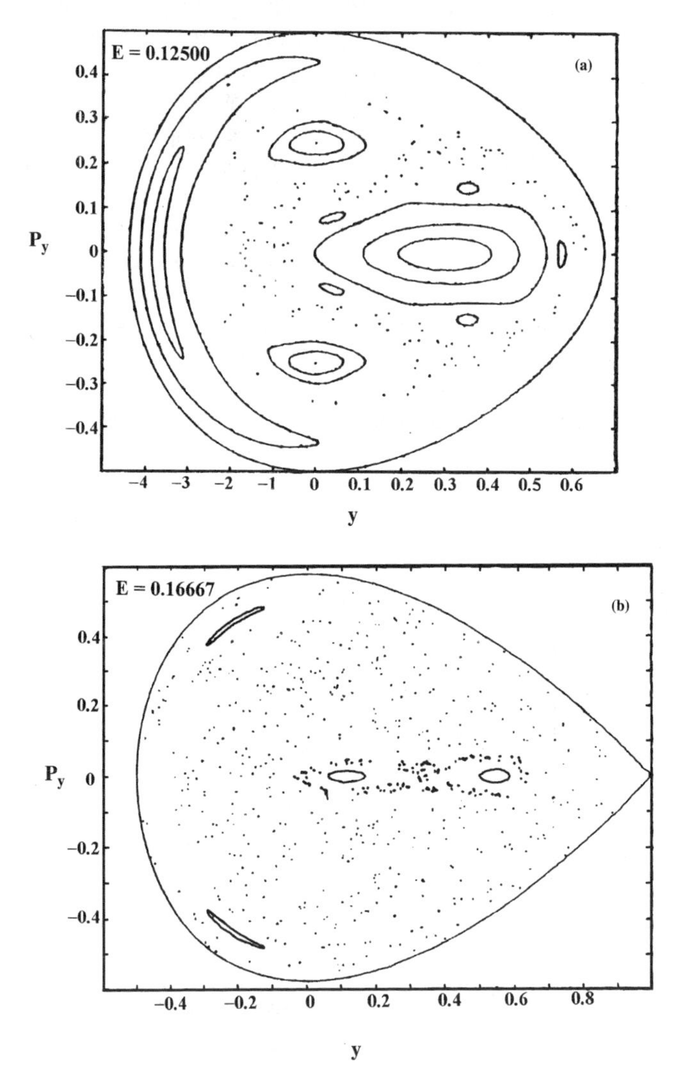

Figure 5.11. Poincaré sections (y, P_y) of phase space, for the Hénon–Heiles problem. (a) For the intermediate energy $E = 0.12500$, there is a mixed phase space containing both large regular and large chaotic regions. (b) For the energy $E = 0.16667$ at the triangular "top edge" of the potential well (see Fig. 5.10), phase space is largely chaotic. From Hénon and Heiles (1964). Copyright 1964 American Astronomical Society.

Fig. 1.8. In the neighborhoods of the unstable fixed points, we find no signs of tori, which is in keeping with our expectations based on remarks above about the perturbed pendulum (Section 5.2b). We suspect we have chaotic trajectories in these neighborhoods. We can verify the chaos by rigorously calculating Lyapunov functions and exponents, along the lines indicated in Sections 5.2a and A.3.

These Lyapunov exponents have been known for some twenty years (Benettin et al., 1976). Following a procedure for discrete-time systems, these authors obtained the largest exponent from the evolution of points in the Poincaré section (a discrete-time mapping we have called a Poincaré map). Although the Lyapunov functions obtained this way differ somewhat from the ones obtained using full continuous trajectory evolution, the Lyapunov exponents do correspond. Benettin and co-workers found that the largest Lyapunov exponent was either zero or positive. The value depended solely on the energy E, not on the particular starting point within a region of zero exponent or within a region of positive exponent.

We call a region of zero Lyapunov exponent a *regular region* and a region of positive Lyapunov exponent a *chaotic region*. (A chaotic region is also called an *irregular region*.) Regular and chaotic regions appear either in phase space or within Poincaré sections. Figure 5.12 shows some Lyapunov functions for the $E = \frac{1}{12}$ Poincaré section, presented in a log–log plot. The three functions for evolution in a regular region are similar at large times and are approaching zero inversely with time. The three functions for trajectories in the irregular region are all going to the

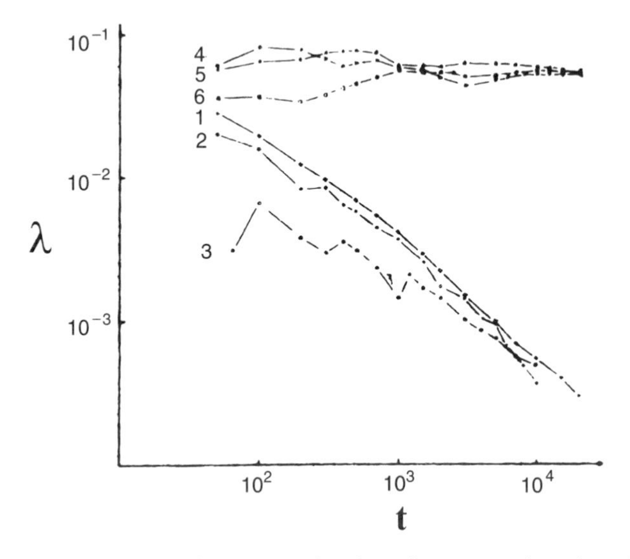

Figure 5.12. Time evolution of Lyapunov functions for some trajectories of the Hénon–Heiles system, at the intermediate energy $E = 0.125$. Curves 1–3 are for trajectories with initial points in regular regions of Fig. 5.11a, while curves 4–6 arise from starting in the chaotic region of that figure. From Benettin et al. (1976). Copyright 1976 by the American Physical Society.

same positive value at large times, showing that we do have three chaotic trajectories.

At $E = \frac{1}{6}$, we are at the top of the Hénon–Heiles potential well, where the three saddle-point vertices in the potential lie. Astronomers call this critical energy the "star escape energy," while chemists call it the "molecule dissociation energy." The Poincaré section is shown in Fig. 5.11b. Now there are only small regular regions; most of the section is irregular. Figure 5.13a shows the fraction of the area of the Poincaré section that is regular, as a function of energy E. At small E this fractional area is known to be *not* exactly one, but it does decrease exponentially with decreasing E, with the exponent proportional to E^{-1}. Figure 5.13b shows the largest positive Lyapunov exponent, again as a function of E. This exponent increases exponentially with E—except for the two points near the "threshold" energy $E_{th} \sim 0.113$ of Fig. 5.13a, where the computations are not as reliable.

5.3c. Regular and Irregular Quantum Eigenstates

We have just seen that the classical physics of nonintegrable systems differs profoundly from that of integrable systems—now there is some chaos. Let us investigate the most interesting question of what, if any, impact this new classical phenomenon has on the corresponding quantum system. We continue to explore the Hénon–Heiles problem.

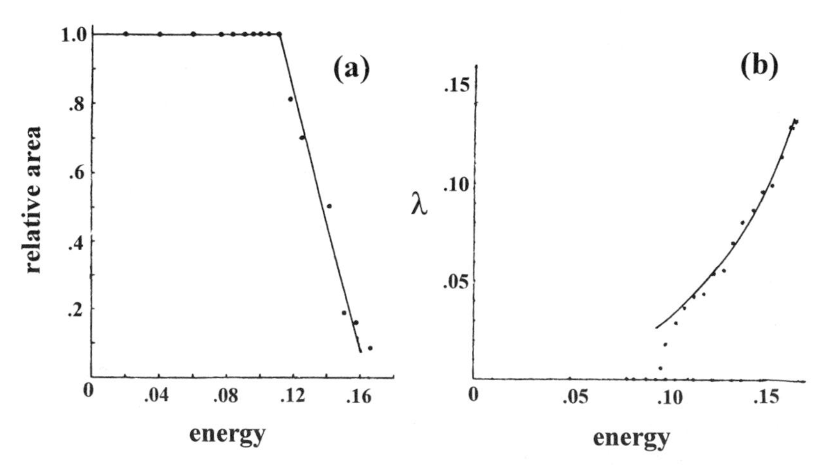

Figure 5.13. The transition to chaos in the Hénon–Heiles problem. (a) The fraction of the area of the Poincaré section that is regular. There is an apparent threshold for chaos near the particle energy $E = 0.11$. (b) The largest positive Lyapunov exponent for chaotic trajectories (points above the energy axis) and vanishing values for regular trajectories (points on the energy axis). Above the threshold energy region of panel (b), the fitted solid line indicates that the positive Lyapunov exponent increases exponentially with energy. The "threshold" for chaos in (b) is lower than in (a), as a careful search for chaotic trajectories located a very few near this threshold. From Benettin et al. (1976). Copyright 1976 by the American Physical Society.

The quantum Hamiltonian in dimensionless coordinates ($m = \hbar\omega = 1$) (Noid and Marcus, 1977) is

$$H = -\frac{1}{2}\left(\frac{\partial^2}{\partial x^2} + \frac{\partial^2}{\partial y^2}\right) + \frac{1}{2}(x^2 + y^2) + \lambda'\left(x^2 y - \frac{y^3}{3}\right) \tag{5.7a}$$

$$= -\frac{1}{2}\left(\frac{\partial^2}{\partial r^2} + \frac{1}{r}\frac{\partial}{\partial r} + \frac{1}{r^2}\frac{\partial^2}{\partial \theta^2}\right) + \frac{1}{2}r^2 + \frac{\lambda' r^3}{3}\sin 3\theta. \tag{5.7b}$$

We note that the symmetry of H includes reflection ($x \to -x$) invariance with respect to the x-axis, which leads to even and odd (E and A) quantum states. For historical reasons, in this subsection we will often set $\lambda' \equiv \lambda/2\pi = 0.111803$, or $\lambda = 0.70$. This is a little different from the value $\lambda = 1$ chosen previously.

When the coupling parameter λ' is small, the Hamiltonian (5.7b) is almost that of a quantum 2-d central-force problem, where the problem of a 2-d harmonic oscillator has the separable Hamiltonian

$$H^{\mathrm{I}} \equiv -\frac{1}{2}\left(\frac{\partial^2}{\partial r^2} + \frac{1}{r}\frac{\partial}{\partial r} + \frac{1}{r^2}\frac{\partial^2}{\partial \theta^2}\right) + \tfrac{1}{2}r^2. \tag{5.8}$$

Thus one set of potentially useful quantum numbers for labeling Hénon–Heiles states is comprised of the effective radial "principal" quantum number n, and of an angular momentum quantum number *l*. For the full Hénon–Heiles problem, *l* will be an approximate quantum number for the component of the particle's angular momentum that is directed along the z-axis. Recall that the z-axis is the symmetry axis of the potential well. In Fig. 5.10 the z-axis is directed perpendicularly out from the page toward the reader.

In a different approximation of the Hamiltonian, if we ignore just the term $\lambda' x^2 y$ in the Hamiltonian (5.7a), then we have another separable Hamiltonian:

$$H^{\mathrm{II}} \equiv -\frac{1}{2}\left(\frac{\partial^2}{\partial x^2} + x^2\right) - \frac{1}{2}\left(\frac{\partial^2}{\partial y^2} + y^2 - \frac{y^3}{3}\right). \tag{5.9}$$

This is the Hamiltonian of two uncoupled 1-d oscillators. The x-oscillator is harmonic and the y-oscillator is anharmonic. Thus we have a second set of potentially useful quantum numbers—n_x and n_y—which are the quantum numbers for these oscillators.

For $\lambda' = 0.1118$, Fig. 5.14 shows all 99 quasibound energy levels for the Hénon–Heiles system (5.7) (Wyatt et al., 1983). These are the levels below the dissociation limit E_{diss} in the units chosen. Basis-set expansion numerical techniques (Section B.2b) were used to compute these levels, with some additional levels at $E > E_{diss}$ also being approximately determined. The basis states were products of harmonic oscillator functions in the x and y coordinates. These energies have been reproduced to within 100 ppm (Noid et al., 1980), although this does not mean that levels near the dissociation limit E_{diss} are very accurate. FFT grid results (Section B.2e) agree to

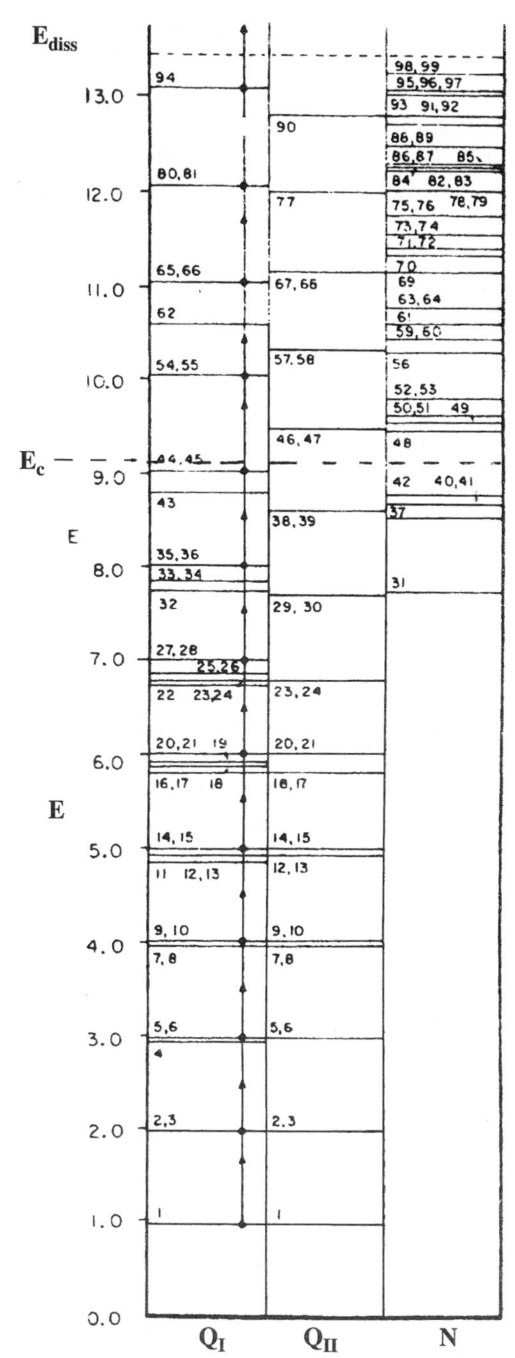

Figure 5.14. An energy-level diagram for the quantum Hénon–Heiles problem, for the parameter value $\lambda' = 0.1118$. All 99 quasibound levels are shown. They are grouped within columns according to their types, which depend on the nature of their dominant classical character; see the text. Some near-degeneracies are not resolved in this diagram. From Wyatt et al. (1983). Copyright 1983 by the American Physical Society.

four significant figures (Feit et al., 1982). Figure 5.14 groups the levels into three columns labeled Q^I, Q^{II}, and N. The criteria for these groupings are empirical yet informative. A Hénon–Heiles quantum energy eigenstate is of Type Q^I if its overlap with the similarly computed states of integrable Hamiltonian (5.8) is greater than 50%. Likewise, a state is of Type Q^{II} if the overlap with states of integrable Hamiltonian (5.9) is greater than 50%. These $Q^{I,II}$ states are called KAM *regular states*, because they are dominated by amplitudes of a physically-relevant approximate integrable system.

A Hénon–Heiles state can be both Types Q^I and Q^{II}. It can also be neither, in which case it is called a Type N state. The energies of the Type N states are shown in the third column of Fig. 5.14. The total number of distinct levels in the figure is only 66, since there are some degenerate E states in the spectrum. The total number of Q^I states is 44, that of Q^{II} states is 22 (including 15 that are simultaneously Q^I), and that of N states is 48. Note that the density of N states increases rapidly above the critical energy E_c—the chaos threshold—where the classical Poincaré section begins to reveal significant stochastic layers (and nonlinear resonance surely occurs; see Section 5.6). Roughly 60% of the 99 energy levels are above E_c, with about 70% of these levels being for Type N states. Below E_c, only 10% of the energy levels are for Type N states.

By definition, both types of Q states have wavefunctions dominated by projections onto wavefunctions of regular subsystems. Both subsystems have a full set of rigorous constants of the motion and have their phase space filled with 2-D tori. Thus with high probability, these Q states act as if they have extra "quasi" constants of the motion, just as is the case for classical evolution on the remaining tori of the nonintegrable system; see Figs. 5.15a,b. (For the Hénon–Heiles system and xy coordinates, it is not easy to visually display the localization of regular quantum states on sets of nested tori in the full phase space, but we display this for other systems in Sections 8.2 and 8.3.)

The energy eigenvalues of the Type Q states are those energy levels that can be approximately found by applying the EBK theory of Section 4.2 (Noid and Marcus, 1977). The Type Q states have $|\psi(x, y)|^2$ localized in those same regions in the xy-plane that contain the projections of the periodic phase-space trajectories used in the EBK semiclassical quanitization (Davis and Heller, 1981a). Figure 5.15 shows an example of a regular state in this position space, along with the projection of a quasiperiodic orbit lying on an associated torus.

A remaining question concerns the character of the Type N states, which constitute almost half the quasibound states of our system. Some of these Type N states have spatial probability distributions that nicely cover regions "filled" with single chaotic trajectories; see, for example, Fig. 5.16 which was calculated by Noid et al. (1979). Other Type N states are more complicated. Quite aptly, we call all these complicated Type N states *irregular states*. Some investigators have been tempted to use the term "chaotic state" for a case such as that seen in Fig. 5.16. However, we need to define just what is chaotic about it, for chaos refers to a specific type of classical time evolution. In the next subsection we start developing an answer to the question "What is quantum chaos?" and we return to this issue in subsequent

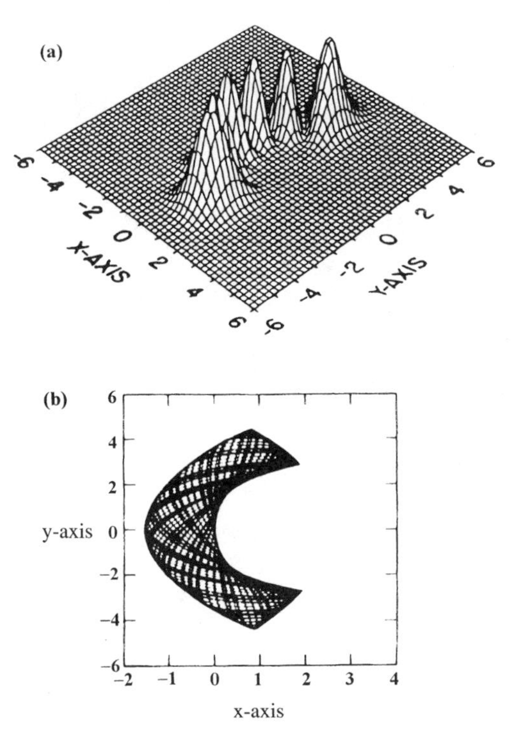

Figure 5.15. A regular Hénon–Heiles energy eigenstate and its classical support. (a) The probability density $|\psi(x, y)|^2$ is plotted for a regular quantum state of the generalized Hénon–Heiles problem, which has the potential $V(x, y) = 0.98x^2 + 0.245y^2 - 0.08(xy^2 - 0.08y^3)$. The energy eigenvalue is $E = 4.265$. (b) A projection onto position space is made of a quasiperiodic orbit lying on an associated EBK torus. Reprinted with permission from Noid et al. (1979). Copyright 1979 American Institute of Physics.

chapters. Yet it is fair to say that a complete definitive answer to this simple-sounding question is not yet available.

An early hint of the differences between regular and irregular quantum energy eigenstates comes from studies of the sensitivity of the energy levels to changes in the parameters. For the Hénon–Heiles Hamiltonian (5.7), this sensitivity was investigated using changes in the parameter λ' (Pomphrey, 1974). Pomphrey carried out numerical calculations of the energy levels for $\lambda' = 0.086, 0.087, \ldots, 0.090$ and then evaluated the second energy-level differences Δ_i defined by

$$\Delta_i \equiv |[E_i(\lambda' + \Delta\lambda') - E_i(\lambda')] - [E_i(\lambda') - E_i(\lambda' - \Delta\lambda')]|, \tag{5.10}$$

where $\Delta\lambda' = 0.001$ was the increment used for changing the value of λ'. A plot of Pomphrey's second differences Δ_i is shown in Fig. 5.17. Up to energies $E \sim 16$, all the differences are small. Above this value, which is close to E_c, some of the second differences become large, indicating the presence of a different kind of energy

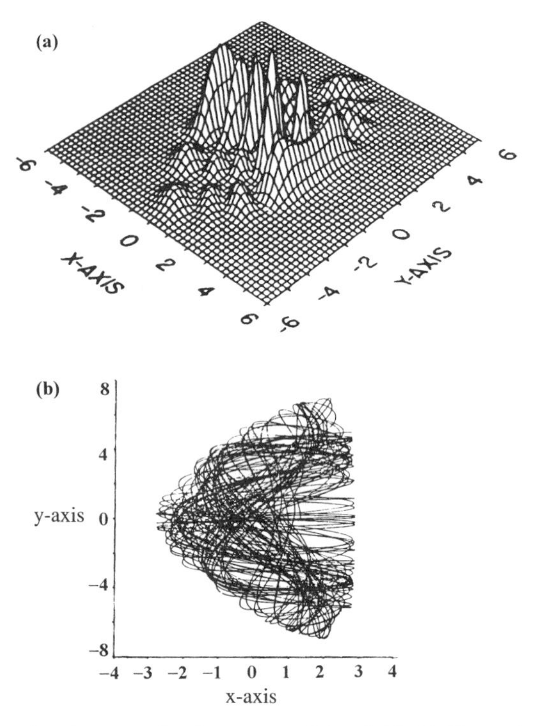

Figure 5.16. An irregular eigenstate for the system of Fig. 5.15. (a) The probability density of a state at the higher energy E = 8.0 is shown. (b) A typical chaotic trajectory at the same energy. The quantum probability is spread over the position space covered by the chaotic trajectory. Reprinted with permission from Noid et al. (1979). Copyright 1979 American Institute of Physics.

eigenstate that we now know as the irregular eigenstate. The breakup of the Δ_i distribution into two subsets indicates that irregular states are more sensitive to parameter values than are regular states. The idea of *sensitivity analysis* introduced here has been applied to a number of other nonintegrable quantum systems. Usually the focus has been on the *curvature distribution* of energy levels, that is, on $d^2E_i(\lambda)/d\lambda^2$. For applications of sensitivity analysis to the problems discussed in Sections 7.1 and 7.4, see Zakrzewski and Delande (1993).

The semiclassical character of many states of nonintegrable quantum systems has potentially important applications. This was already recognized over twenty years ago by those studying the Hénon–Heiles system (Wyatt et al., 1983). Perhaps the holy grail of quantum chemical dynamics is the controlled selective breaking of chemical bonds, using pulsed-laser excitation or other means. Within the set of Q^I energy levels shown in Fig. 5.14, there is a special ladder-like subset of levels that is almost harmonic. This ladder is indicated by the vertical arrows. This ladder will enhance laser-induced multiphoton transitions, which can proceed up just that one ladder. The result should be a special breakup of the Hénon–Heiles system, that is,

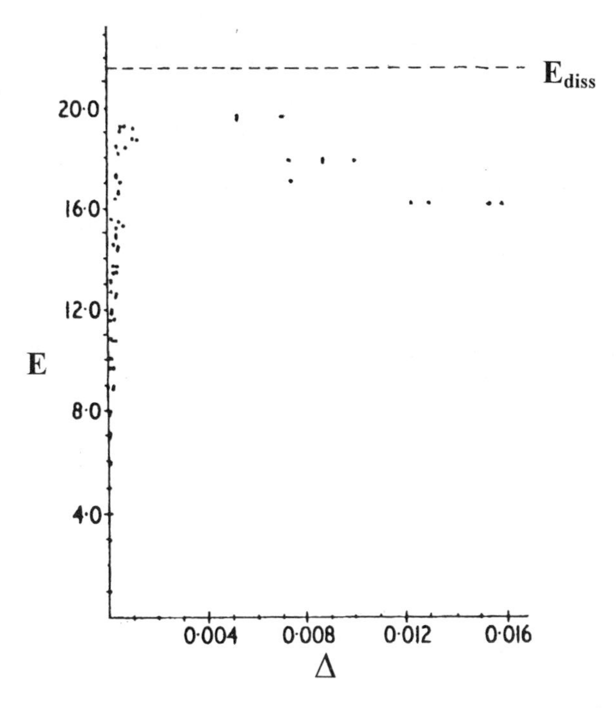

Figure 5.17. Sensitivity analysis of Hénon–Heiles energy eigenstates, changing the parameter λ'. Shown are the quantum second energy-level differences Δ, defined by Eq. (5.10) and based on the parameter values $\lambda' = 0.086, 0.087, \ldots, 0.090$. The vertical axis is the energy level E, and the horizontal axis is Δ. Above the energy $E \sim 16$, some states have large differences Δ because they are irregular states. From Pomphrey (1974). Copyright 1974 Institute of Physics Publishing.

something like the specific dissociation of a polyatomic molecule. This question—whether or not quantum transitions on a set of tori-supported quantum states can give us some useful control of quantum time evolution—is typical of the issues to be discussed in Chapter 9.

5.3d. Wavepacket Evolution

As we begin this subsection, let us briefly recap some of what we know about wavepackets. In Section 1.3c we studied the time evolution of wavepackets for 1-d systems and found that displays in phase space of such evolution were particularly informative. In Section 2.1b, we set up the problem of wavepacket evolution for the 3-d hydrogen atom, another integrable problem. Equation (2.8) was introduced for the initial Gaussian electron wavepacket in the position coordinates $\mathbf{r}$, a packet having initial expectation values $(\mathbf{r}_0, \mathbf{p}_0)$ in the full 6-D phase space. We then discussed recent packet evolution results both for 3-d numerical calculations and for a 1-d model of the hydrogen atom. Our focus was on the imposition of quantum collapse and revivals upon the packet as its evolution was guided by a central classical periodic orbit. Other nearby orbits helped guide the packet as well. In Section 2.4 we found that wavepacket evolution was crucial to the study of an essentially-quantum system, the two-level atom in a monochromatic electromagnetic field. When the field was strong, exotic evolution occurred. This was a result of the system becoming nonintegrable.

Now that we know more about nonintegrability and have some ideas about what this means for classical systems, we can profitably extend our exploration of wavepacket evolution to a large class of nonintegrable systems.

In Davis and Heller (1979), we find the introduction of complex wavepackets of the form Eq. (2.8) into bound-state molecular problems. We are of course free to study the evolution of any initial quantum state we please; the key is to include in this state the right amount of the physics, in order to produce an evolution that will be physically interpretable. When we consider wavepacket molecular vibrations, it is essential to provide freedom to adjust both the average momentum $\mathbf{p}_0$ and the average coordinates $\mathbf{q}_0$ of the initial wavepacket. This is accomplished by the phase factor $\exp(i\,\mathbf{p}_0 \cdot \mathbf{q})$ in Eq. (2.8). One may want even more flexibility, which can be achieved by writing (in 1-d)

$$\psi^G(q; q_0, p_0) = \left(\frac{2\alpha}{\pi}\right)^{1/4} \exp[-\alpha(q - q_0)^2 + ip_0 q] \quad (\hbar = 1), \tag{5.11}$$

where α is a complex number. Complex values of α introduce position–momentum correlation in the packet, as suggested by the bottom panel of Fig. 5.18.

The integral in Eq. (1.36) for the Wigner function can be evaluated for the case of wavepacket (5.11). Thus we obtain

$$W^G(q, p) = \frac{1}{\pi} \exp\left(-\frac{1}{2\,\mathrm{Re}\,\alpha}[4|\alpha|^2(q - q_0)^2 + (p - p_0)^2 + 4(\mathrm{Im}\,\alpha)(p - p_0)(q - q_0)]\right). \tag{5.12}$$

We see that $|\alpha|^2$ controls a trade-off in the widths of the packet in p and in q, much like quantity σ_q did in Eq. (1.38). The quantity Im α adds extra flexibility, when needed. From Eq. (5.12) we can see that the packet has its mean location at (q_0, p_0) and that α determines the orientation and relative magnitude of the major and minor axes of the ellipses that form the lines of constant probability for the packet; see Fig. 5.19. This role of α allows us to "line up" the wavepacket to any desired degree along the local direction of the central trajectory that passes through (q_0, p_0). By introducing linear combinations of such aligned stretched wavepackets, with mean locations distributed along an interesting curve in phase space, we can construct a composite initial quantum state "covering" that curve.

Let us now use such rudimentary complex-number wavepackets to study evolution in the Hénon–Heiles system. We will focus on a comparison of packet evolutions starting in the regular and in the irregular parts of the Poincaré section, as

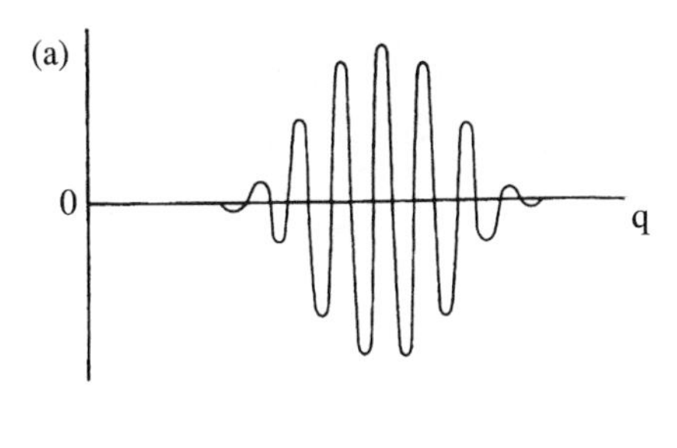

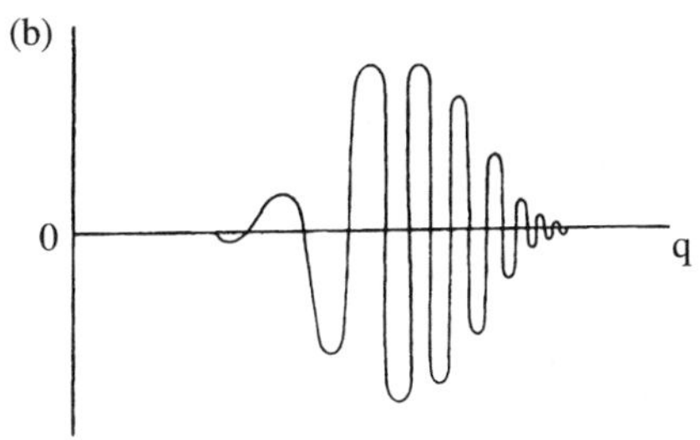

Figure 5.18. Real parts of the spatial dependence of real and complex Gaussian wavepackets. (a) The usual real packet parameter α is used in Eq. (5.11). (b) A complex α generates position–momentum correlation in the packet, so that the de Broglie wavelength varies with the position q. Reprinted with permission from Davis and Heller (1979). Copyright 1979 American Institute of Physics.

indicated in Fig. 5.20. We see that packet 1 has its mean location in the large region of regular motion associated with the fixed point near $(y, p_y) = (2.4, 0)$. This fixed point represents one of the stable periodic orbits discussed earlier in this section. The circle indicating the full width at half maximum (FWHM) of packet 1 lies entirely in a regular part of the Poincaré section. On the other hand, the circle indicating the

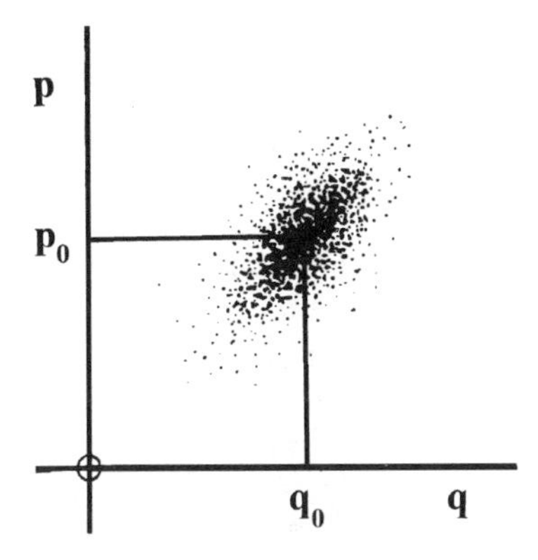

Figure 5.19. The Wigner phase-space distribution of a Gaussian wavepacket with complex α. The choice of α stretches the packet in phase space, along a chosen direction. Reprinted with permission from Davis and Heller (1979). Copyright 1979 American Institute of Physics.

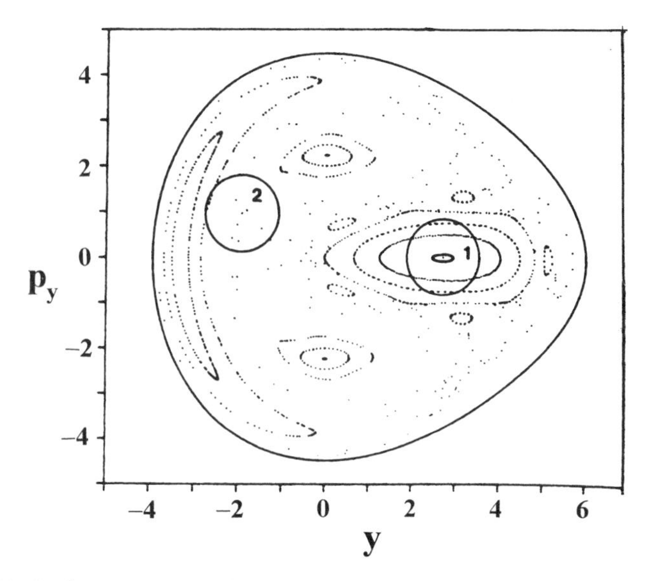

Figure 5.20. Setting up a wavepacket study of quantum regular and irregular time evolution. The locations of two different initial Gaussian wavepackets are shown superimposed on a (y, p_y) Poincaré section for the Hénon–Heiles system at energy $E = 10.0$. Packet 1 is localized in a regular region, while packet 2 is localized in a chaotic region. From Davis et al. (1980). Copyright 1980, with permission from Elsevier Science.

FWHM of packet 2 lies in a region of the section that is primarily chaotic. Thus Fig. 5.20 sets up a direct, controlled comparison of wavepacket evolutions that start in regular and in chaotic regions of phase space.

For the Hénon–Heiles parameter, we choose the value $\lambda = 0.1118$, the same value as for the energy-level spectrum of Fig. 5.14. Following Davis et al. (1980), we take the wavepackets ψ_1^G and ψ_2^G that cover the regions 1 and 2 in Fig. 5.20 to be

$$\psi_1^G = \pi^{-1/2}\exp\left[-\tfrac{1}{2}x^2 + 3.762ix - \tfrac{1}{2}(y - 2.707)^2\right],$$
$$\psi_2^G = \pi^{-1/2}\exp\left[-\tfrac{1}{2}x^2 + 3.976ix - \tfrac{1}{2}(y + 1.914)^2 + 0.961iy\right].$$

The classical average energy of each packet is 10.0, while the quantum average energy of each is 11.0.

Before we evolve our two packets, it is informative to check out which Hénon–Heiles energy eigenstates are contained in each of these initial packets. So we calculate the projections (wavefunction overlap integrals) of each packet on the basis set of exact Hénon–Heiles energy eigenstates. The results are shown in Fig. 5.21. By comparing the energies of the peaks in the upper panel of Fig. 5.21 with the energy levels in the three columns of Fig. 5.14, we see that packet 1 is comprised almost entirely of Type Q^I regular states—primarily those on the energy-level ladder connected by the vertical arrows in the left-hand column of Fig. 5.14. On the

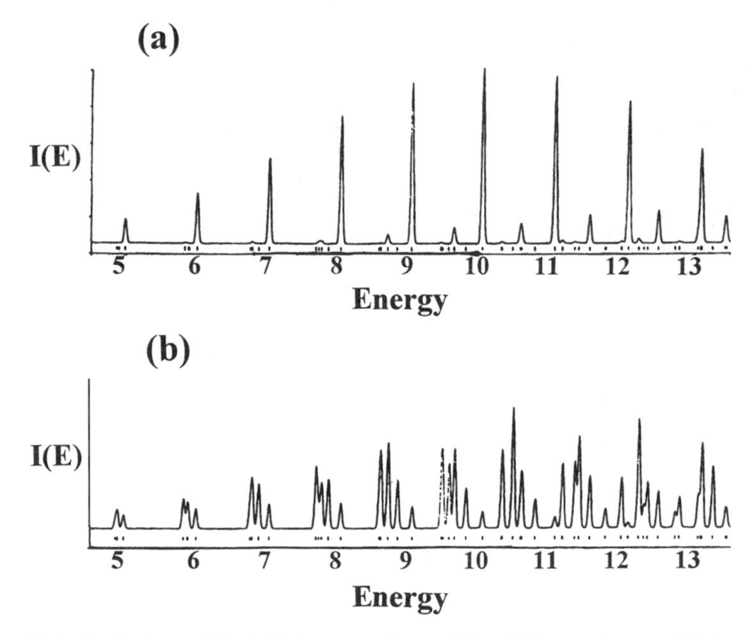

Figure 5.21. Projections of the initial wavepackets of Fig. 5.20 onto the set of Hénon–Heiles energy eigenstates. (a) Packet 1 is comprised of Type Q^{I} regular eigenstates having energies in the left column of Fig. 5.14. (b) Packet 2 is comprised mainly of the Type N irregular eigenstates having energies in the right column of Fig. 5.14. From Davis et al. (1980). Copyright 1980, with permission from Elsevier Science.

other hand, packet 2 is comprised mainly of Type N "irregular" states. This is what we would have expected from the definition of the three types of eigenstates. It is noteworthy just how well the division into these three types works for packets 1 and 2.

Now let us numerically evolve each packet, computing the autocorrelation function—Eq. (2.14)—as we go along. Figure 5.22 shows the results. There are dramatic differences in the two evolutions. In the regular evolution shown in the upper panel, there is a dominant frequency component with a large amplitude. This frequency component is located near the frequency of the primary classical periodic orbit, an orbit that is associated with the fixed point in the selected regular region. On the other hand, the lower panel shows a quick collapse of the autocorrelation function. This collapse indicates that the irregular packet quickly spreads to parts of phase space that were not initially covered. There are no signs of a revival over the time interval shown.

Similar results have been obtained by Moiseyev and Peres (1983). A small wavepacket placed in a regular region of phase space slowly disperses while following a set of almost periodic trajectories. This is typical of quantum evolution whenever the important Lyapunov exponents are zero. On the other hand, a wavepacket placed in a chaotic region disperses much more rapidly. When a

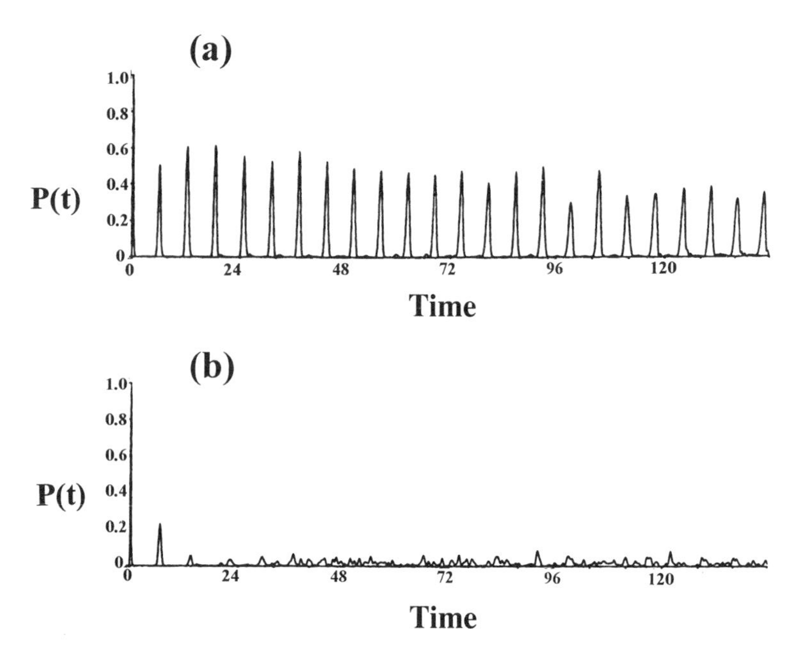

Figure 5.22. Time evolution of the autocorrelation function Eq. (2.14) for the two initial wavepackets shown in Fig. 5.20. The top panel is the evolution of the regular packet 1, while the bottom panel is for the irregular packet 2. The regular packet oscillates strongly near the frequency of the periodic orbit at the center of the regular region of phase space. The irregular packet collapses rapidly, because of a rapid filling of the chaotic region of phase space. From Davis et al. (1980). Copyright 1980, with permission from Elsevier Science.

Lyapunov exponent is positive, this rapid dispersal characterizes the quantum time evolution.

5.4. ORBIT SCARRING OF IRREGULAR QUANTUM EIGENSTATES

The fascinating classical–quantum comparison in Fig. 5.16 suggests a combination of classical orbit effects and quantum wave interference effects. However, this association of chaotic trajectories with irregular quantum time evolution is not the only one we might make. We know that unstable periodic orbits are also present in the nominally chaotic portions of the phase space of a system. We begin investigating the effects of these unstable orbits by considering irregular energy eigenstates, such as our Type N Hénon–Heiles states. It must be strongly emphasized that stationary quantum states are a different matter than short-time quantum wavepacket evolution. Important periodic orbit effects on energy eigenstates do not necessitate dominant similar effects in the early packet evolution. Here we wish to compare the phase-space distributions of individual Type N eigenstates with the location of unstable periodic orbits. Conceivably there could be constructive wave interference if part of the quantum wavefunction were to "fit in the orbit."

The Hénon–Heiles problem is generic in the sense that bifurcations of periodic orbits occur as the energy or the parameter λ' is smoothly changed. When a bifurcation occurs, a periodic orbit is altered or disappears, while new periodic orbits appear. This creates a strong dependence of classical phase-space structure on the values of the parameters of the system, greatly complicating a general study of classical–quantum correspondence. However, there are systems that possess special scaling properties where these orbit bifurcations do not occur as a function of a certain parameter. One such system is the 2-d quartic oscillator, defined by the Hamiltonian

$$H = \tfrac{1}{2}(p_x^2 + p_y^2) + \tfrac{1}{2}x^2y^2 + \beta(x^4 + y^4). \tag{5.13}$$

Hamiltonian (5.13) possesses a special property—the homogeneity of the kinetic energy and the potential energy, where both energies depend solely on even powers of phase-space canonical variables. As a result, there is a scaling property for the classical system. A trajectory $\mathbf{r}(t)$, $\mathbf{p}(t)$ at energy E is determined by another trajectory $\mathbf{r}'(t')$, $\mathbf{p}'(t')$ at energy E', according to

$$t' = \epsilon^{-1/2}t, \qquad \mathbf{r}' = \epsilon^{1/3}\mathbf{r}, \qquad p' = \epsilon^{2/3}\mathbf{p},$$

where the classical scaling parameter is $\epsilon \equiv (E'/E)^{3/4}$. We note that the classical action $S = (1/2\pi)\int \mathbf{p}\cdot d\mathbf{r}$ then scales simply as $S'(E') = \epsilon S(E)$. For fixed β, the scaling means a trivial dependence of periodic orbits on the energy E, without bifurcations occurring when either E or ϵ is changed.

As β approaches zero in this system, Poincaré sections exhibit chaotic fractional areas that approach unity (Carnegie and Percival, 1984; Meyer, 1986). It appears difficult, however, to numerically calculate a large number of quantum eigenstates at $\beta = 0$. Thus very small nonzero values are used in practice, for example, $\beta = 0.004167$ in the work of Waterland et al. (1988), and $\beta = 0.0025$ in that of Eckhardt et al. (1989). Already at $\beta = 0.004167$, only two weakly stable periodic orbits have been found, lying along the coordinate axes. The residues of these two orbits—see Eq. (3.9)—are 0.996, very close to one. This indicates weak stability. These orbits are surrounded by very small regions of tori, barely visible in a Poincaré section. On the other hand, at $\beta = 0.0025$, 50 periodic orbits have been found at an energy $E = \frac{1}{2}$, all having positive Lyapunov exponents and thus being orbitally unstable.

The periodic orbits can be located in phase space by searching along the symmetry lines in Poincaré sections (De Vogelaere, 1958). For Hamiltonian (5.13), some of these lines can be found in (x, y) space. These symmetry lines are the coordinate axes, the diagonals, and the boundary of the classically allowed coordinate space at energy E. For the coordinate axes and the diagonals, one numerically initiates trajectories perpendicular to the symmetry line and searches for perpendicular intersections with the line after a given number of crossings, using interval bisection techniques. For orbits evolved on points along the boundary, one can use the turning-point method of Pollak et al. (1980), which is also implemented with interval bisections. In Fig. 7 of Eckhardt et al. (1989), we find pictures of 36 identified periodic orbits, sorted according to their geometrical symmetry. Employing the linear stability analysis techniques of Section 3.1b, one finds all the orbits discussed in this paragraph to be unstable at $\beta = 0.0025$. Most of these periodic orbits are present but not visible in the ($y = 0$, $p_y > 0$) Poincaré section shown in Fig. 5.23. The darkened area in this figure was produced by iterating one chaotic trajectory to produce 100,000 points. Chaos indeed dominates the section, and we might wonder what could be the impact of the unstable periodic orbits on the quantum eigenstates.

Schrödinger's equation for the 2-d quartic oscillator is

$$\left[-\tfrac{1}{2}\hbar^2\nabla_2^2 + \tfrac{1}{2}x^2y^2 + \beta(x^4 + y^4)\right]\psi = E\psi.$$

We can see a quantum scaling here, in addition to the classical scaling of orbits that differ only in their energies. The quantity $\hbar$ is replaced by unity if we introduce double-primed quantities according to $x = \hbar^{1/3}x''$, $y = \hbar^{1/3}y''$, and $E = \hbar^{4/3}E''$. Note that we now have physical meaning for the classical limit $\hbar \to 0$, which now is mathematically equivalent to $E'' \to \infty$ with E held fixed.

For Hamiltonian (5.13), Eckhardt et al. (1989) carried out basis-set expansion quantum calculations (Sections 1.2a and B.2b). They found quantum energy eigenstates by using harmonic oscillator basis states to diagonalize the Hamiltonian matrix. Their paper displays the probability density $|\psi(x, y)|^2$ for 124 eigenstates.

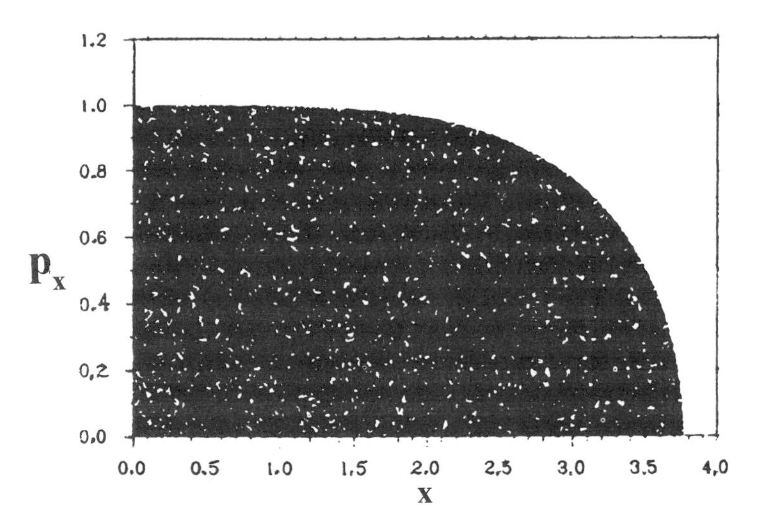

Figure 5.23. An (x, p_x) Poincaré section for a 2-d quartic oscillator having the classical Hamiltonian (5.13). Because of the symmetries, only one quadrant is shown. At $\beta = 0.01$, phase space is almost entirely chaotic. From Provost and Brumer (1995). Copyright 1995 by the American Physical Society.

Comparing projections of the unstable periodic orbits onto the xy-plane with these eigenstates, one sees that in almost all cases, patterns in the eigenstates lie close to patterns produced by the orbits! An example of this phenomenon is shown in Fig. 5.24. Following tradition, we call this matching of patterns the "scarring" of quantum eigenstates by periodic orbits, and we introduce the term *scarred quantum eigenstate*. We conclude that unstable periodic orbits play a major role in the stationary-state quantum physics of classically chaotic systems. This relationship has been confirmed for many systems.

Toward the end of Section 5.2, we noted that unstable periodic orbits are accompanied by stable and unstable manifolds in phase space. These manifolds replaced the separatrices of integrable systems. Let us now look in phase space at one of the quantum eigenstates, following Waterland et al. (1988). They computed the quantum Poincaré section (Section 1.3d) for a particular state, shown in Fig. 5.24a. After overlaying the stable and unstable manifolds of the "diagonal" and "box" orbits of Fig. 5.24b onto the quantum section, the result is Fig. 5.25. We see that the manifolds tend to guide the shapes of the four major ridges of quantum probability density in the quantum section. The probability density remains relatively high along the manifolds, as we proceed away from the unstable fixed points. This structural underpinning of the quantum eigenstate is not apparent in the plot in coordinate space (Fig. 5.24a). This again illustrates the point made in Section 1.3d that distributions in phase space are very useful for the display of the underlying physics of quantum states.

For the particular state shown in Fig. 5.24a, the actions of the scarring orbits are close to quantizing values—that is, a Bohr–Sommerfeld-like quantization seems to work even when there are no tori! Evidence for the partial relevance of such a

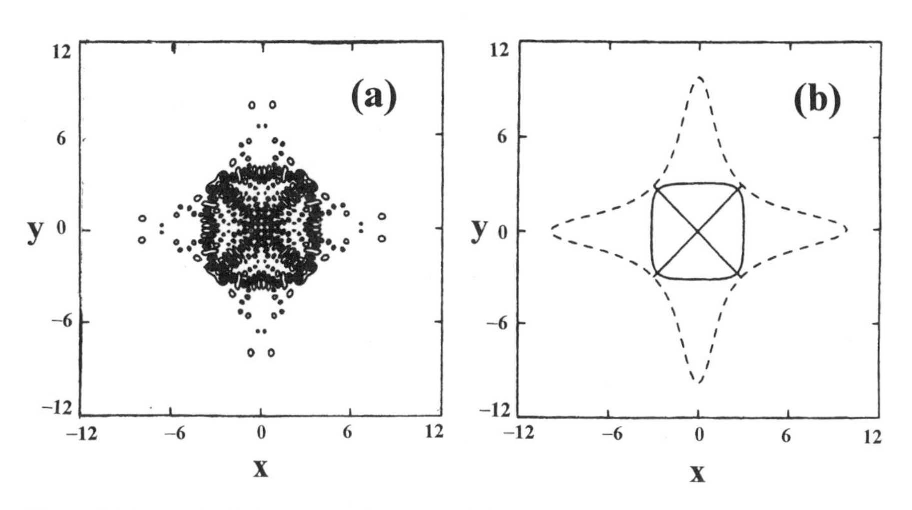

Figure 5.24. Scarring of an energy eigenstate of the 2-d quartic oscillator. (a) The probability density in position space of a high-lying eigenstate. (b) The four unstable periodic orbits that influence ("scar") the eigenstate shown in panel (a). From Waterland et al. (1988). Copyright 1988 by the American Physical Society.

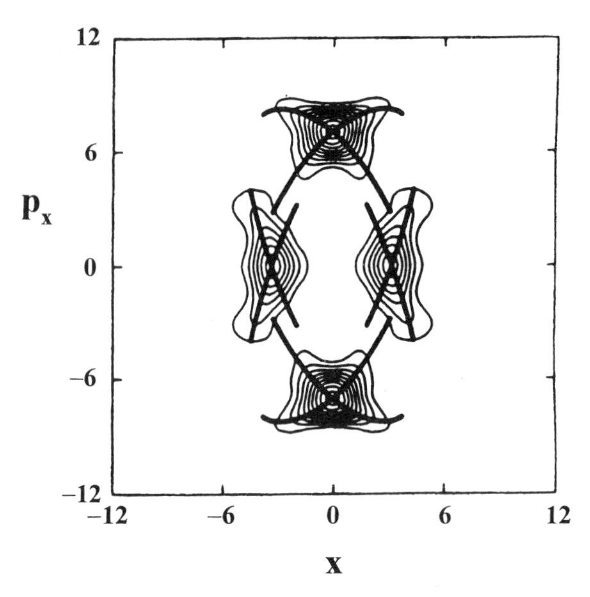

Figure 5.25. Contours of the quantum Poincaré section of the eigenstate shown in Fig. 5.24a, superimposed on the unstable and stable manifolds (heavy curves) associated with the "diagonal" and "box" orbits shown by the solid sections of curves in Fig. 5.24b. The scarred wavefunction is stretched out along the directions of the manifolds. From Waterland et al. (1988). Copyright 1988 by the American Physical Society.

quantization has appeared for other systems. These systems include a chaotic quartic oscillator (Eckhardt et al., 1989); the problem of a hydrogen atom in a magnetic field, to be discussed in Section 7.1; the problem of a quantum well in a tilted magnetic field, to be discussed in Section 7.3; and the stadium billiards problem, to be discussed in Section 7.4 (Agam and Fishman, 1994). Thus some scarred eigenstates show maximal scar strengths near energies that satisfy a 1-d quantization rule for the action S_j of a periodic orbit j:

$$S_j = (\nu + \phi_j)h, \quad \nu = 1, 2, 3, \ldots.$$

This equation gives the number 2ν of antinodes in the scarred wavefunction along a closed classical path, with ϕ_j fixed for a given orbit. Such scarred wavefunctions usually correspond to the eigenvalues closest to the energies predicted by the above equation.

In general, however, the idea of Bohr–Sommerfeld-like quantization for unstable periodic orbits seems too simple. A significant fraction of the quantum eigenstates of Hamiltonian (5.13) are not obviously related to any small set of classical periodic trajectories. Still, we cannot rule out possible connections with possibly large numbers of long-period or highly unstable periodic orbits.

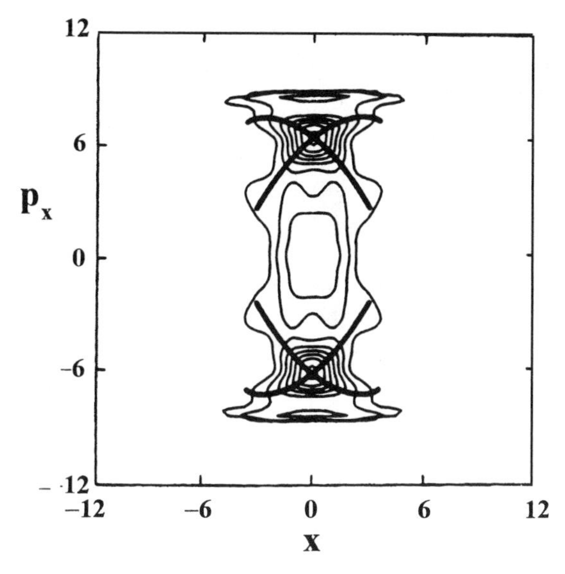

Figure 5.26. The sum of quantum Poincaré sections for all energy eigenstates whose energies fall within an appropriate energy band. The summing builds up the structure that arises from a short-period unstable periodic orbit and its manifolds, the latter shown as the heavy curves. From Waterland et al. (1988). Copyright 1988 by the American Physical Society.

In Section 4.2 we developed a quantum–classical correspondence principle for integrable systems, a principle that states that a regular quantum eigenstate is related to trajectories on one or more tori in phase space. As a prelude to a more detailed discussion in Section 7.2, we now mention that there is a different correspondence principle for irregular quantum eigenstates. This second principle states that an unstable periodic orbit and its manifolds are related to an appropriate set of quantum eigenstates, where the eigenstates have energies in an appropriate range ΔE. This relationship can be demonstrated by summing up the quantum Poincaré sections for that set of eigenstates; see a result in Fig. 5.26. Such sums are strongly influenced by short-period orbits, with periods $T \gtrsim \hbar/\Delta E$. As one sums the quantum sections, the recurrent quantum structure associated with short-period orbits and their invariant manifolds does build up, rather than smooth out. For Hamiltonian (5.13) this buildup is a common phenomenon, occurring in a number of different energy regimes. In Section 7.2 we see this phenomenon can be explained in terms of formal semiclassical theory. Let us now preview this rigorous theory.

5.5. OVERVIEW OF PATH INTEGRAL THEORY FOR IRREGULAR QUANTUM SYSTEMS (ADVANCED TOPIC)

In Section 1.2c we reviewed Bohr–Sommerfeld quantization for 1-d systems, a procedure based on the existence of periodic orbits. In Section 4.2 we returned to the same line of reasoning while treating the EBK quantization of higher-dimension integrable systems. This extension was based on periodic orbits that are loops on N-D tori. In Section 5.4 we empirically found that periodic orbits play a major role in the properties of many energy eigenstates of irregular (classically chaotic) quantum systems.

So it is reasonable to ask several questions: Can we construct a rigorous formal theory for irregular systems, based on the empirical observations? Would such a theory improve our understanding of an irregular system, without the need for numerical calculations? Can such a theory enhance our ability to carry out the necessary numerical calculations of the eigenstates and of the time evolution of nonintegrable systems? The respective answers to these three questions are presently "to a considerable extent," "a little," and "sometimes." This reflects the fact that application of the theory is at present both complicated and difficult. These answers, however, are in transition.

To date, most direct numerical solutions of Schrödinger's equation (1.1) depend on powerful methods that were developed for dealing with differential equations. The past use of these methods has often relied more on experience than on input of a detailed knowledge of the classical structures present in phase space. Until very recently, these structures have not entered in a natural way into most ab initio calculations.

There exists an alternative formulation of the very foundations of quantum mechanics itself—a path integral formulation that is the starting point for the semiclassical theory of irregular quantum systems. At the beginning of this theory, one does not see Schrödinger's differential equation (1.1), but rather its formal inversion into a very general integral equation. This integral equation states that a definite quantum time evolution—which concretely connects two different points in space-time—can be written as a particular sum or integral over *all* the conceivable paths in space-time connecting those two points. There are a great many paths to be summed over, which seems discouraging. However, if a stationary-phase approximation is used to select the most important paths, then progress can be made in reducing the number of paths. In Section 7.2 we discuss the use of this approximation, after setting up the integral formalism during our discussion of quantum tunneling in Section 6.3. The standard textbooks, by necessity, are quite mathematical in character. They include Feynman and Hibbs (1965) and Schulman (1981). Since the general theory contained in these books has been applied more recently to irregular nonintegrable systems, neither of these texts directly discusses our specific application.

In our path integral theory—called *periodic orbit theory*, or more generally *closed orbit theory*—closed orbits will arise naturally as special contributing paths, because they retrace or at least come back on themselves in time. It is a matter of constructive interference of matter waves. The stationary-phase approximation naturally "zeros in" on these special trajectories. When these stationary trajectories appear in our semiclassical approximation, they produce resonances in the expansion of physical quantities over all the paths. Both the locally averaged energy eigenfunctions (Bogomolny, 1988) and the number density of energy eigenvalues, dN/dE (also called the *energy density of states*) (Gutzwiller, 1982), become expressed as a sum over periodic orbits. The contribution of each orbit to the sum has a characteristic form. For each of the eigenfunctions, the spatial probability distribution—averaged over suitable small ranges of energy and of coordinates—has the form

$$\langle|\psi(\mathbf{q})|^2\rangle = \rho_0(\mathbf{q}) + \hbar^{(N-1)/2}\sum_p \operatorname{Im}\left\langle A_p(x)\exp\left(i\frac{S_p}{\hbar} + i\frac{W_p^{km}(x)}{2\hbar}y_k y_m\right)\right\rangle. \quad (5.14)$$

In this equation there are implicit summations over the integers k and m. In the term for each periodic trajectory p, the local variable x is chosen along that trajectory. The y_m-axes, $m = 1, \ldots, N-1$, are locally perpendicular to the trajectory. The integer N is the number of degrees of freedom, $\mathbf{q} \equiv \{x, y_1, \ldots, y_{N-1}\}$, and $S_p = \oint_p \mathbf{p}\cdot d\mathbf{q}$ is a classical action accumulated around the trajectory p. The quantity $\rho_0(\mathbf{q})$ coincides with the projection of the classical (microcanonical) phase-space distribution onto coordinate space. The quantities $A_p(x)$ and $W_p^{km}(x)$ are classical quantities defined through the elements of the monodromy matrix for the trajectory p; see Sections 3.1b and 7.2c. In the case of Eq. (5.14), this matrix $\bar{\bar{M}}$ determines a solution of the linearized classical equations of evolution in the coordinate space perpendicular to the orbit p, according to

$$\begin{bmatrix} y_k(T_p) \\ \dot{y}_k(T_p) \end{bmatrix} = \bar{\bar{M}}\begin{bmatrix} y_k(0) \\ \dot{y}_k(0) \end{bmatrix},$$

where T_p is the period of the orbit p. In practice, the summation in Eq. (5.14) is performed over a finite number of the periodic orbits, a number that depends on the value ΔE chosen for the energy averaging.

In a time-dependent version of the theory, semiclassical expansions of the form (5.14) can also be used in calculations of wavepacket evolution; see Section 7.2e. The approach for this evolution will be similar in spirit to the approach for our semiclassical packet studies of the integrable hydrogen atom problem (Section 2.1c). We will also see that such numerical packet evolutions can produce time-dependence data that—when Fourier transformed into energy space—give good *single* irregular energy eigenstates.

When applying periodic orbit theory, there usually are two primary concerns: (1) the convergence of the infinite sum over the periodic orbits and (2) the existence of ranges of parameter values where, as parameters are changed, there is exponential proliferation of the periodic orbits due to bifurcations. Some progress has been made in developing practical summation schemes. One tries to sort the infinity of orbits into groups, where each group is more easily summed.

Analytical continuation of one or more space-time variables into the complex plane can be another mathematical tool that aids computations based on periodic orbit theory. Imaginary-time techniques have proved useful when working with both the differential equation formulation and the path integral formulation of quantum mechanics. Oscillating and exponential dependencies in the wavefunction factors exp(iEt) are interchanged as t is changed from real to imaginary. We shall use imaginary time in some of our studies of quantum tunneling in Chapter 6. This approach produces a theory that can utilize the imaginary-time solutions of Hamilton's equations, with their different mathematical properties such as the boundary conditions. In path integral theory, the time-oscillating phase factors in Eq. (5.14) then become decaying exponentials; convergence of sums over imaginary-time trajectories is therefore expected to be relatively rapid. At sufficiently long time, only a lowest-energy term contributes to the summation over paths. This term can then be used to produce a good ground energy eigenstate (Weiper et al., 1996). "Subtracting out" this term, one repeats the procedure to get the next most long-lived term and its corresponding eigenstate, and so forth. Although this approach is very new, it at least demonstrates that semiclassical path integral theory provides a growing foundation for the quantum mechanics of nonintegrable systems.

5.6. NONLINEAR RESONANCE APPROXIMATIONS

In this chapter we have been concerned with the quantum evolution of classically nonintegrable systems. First we considered the common situation where a partial Hamiltonian is integrable and the remainder of the Hamiltonian has a small effect. We noted those qualitative classical changes arising from nonintegrability—changes that include new tori that are different in character from the KAM tori and therefore are not just distorted versions of tori of the integrable subsystem. We then focused on a new kind of classical evolution—irregular evolution in regions of classical chaos—and then on some aspects of irregular quantum evolution and irregular quantum eigenstates. We now return to the new tori of different character, discussing their origins in nonlinear resonance and describing a useful and illuminating approximation for them. Both a classical approximation and a corresponding quantum approximation exist.

The new tori arise from a frequency resonance between two interacting subsystems. The subsystems might be parts of an autonomous system—a case we call *internal nonlinear resonance*. An example is the enhanced interaction of two molecular vibrational modes that arises when the frequencies of the two vibrations are commensurate. A second case is *external nonlinear resonance*, where one subsystem produces a sinusoidally oscillating external driving force. External resonance can occur, of course, in many driven systems; see Chapter 8. The theory of external resonance is actually a little simpler than the theory we describe here for internal resonance.

5.6a. Internal Classical Nonlinear Resonance

Nonlinear resonance between two (or more) degrees of freedom is possible when $N \geq 2$. We consider the $N = 2$ case and assume that for each degree of freedom there is a 2-D dynamical system with a Hamiltonian H_{0i}, $i = 1$ or 2, that separates out of the Hamiltonian H of the full nonintegrable system. Introducing the canonical action-angle variables (θ_i, I_i) for each of the two subsystems, we may then write

$$H = H_{01}(I_1) + H_{02}(I_2) + \epsilon V(I_1, I_2, \theta_1, \theta_2). \tag{5.15}$$

Here V contains the phase-space dependence of the interaction Hamiltonian that couples the two degrees of freedom (θ_1, θ_2), and ϵ is a coupling strength parameter that could be made small enough for us to have a KAM system. However, ϵ can often be considerably larger, depending on how accurate we want the following approximations to be. Hamilton's equations become (Section 3.3b).

$$\frac{dI_i}{dt} = -\frac{\partial H}{\partial \theta_i}, \qquad \frac{d\theta_i}{dt} = \frac{\partial H}{\partial I_i}, \qquad i = 1, 2. \tag{5.16}$$

For classical frequency resonance, by definition there must exist integers k_0, l_0 such that

$$k_0 \frac{d\theta_1}{dt} - l_0 \frac{d\theta_2}{dt} \equiv k_0 \omega_1(I_{10}) - l_0 \omega_2(I_{20}) = 0. \tag{5.17}$$

Here $\omega_i(I_{i0})$ are the frequencies of oscillation of the two subsystems, when those subsystems are individually isolated. Note that the resonance condition depends on the two action values I_{i0}, which leads us to call $\mathbf{I}_0 \equiv (I_{10}, I_{20})$ the *resonant action vector.* Since the subsystems are usually nonlinear to begin with, and the coupling V is likewise nonlinear, we say that Eq. (5.17) is the condition for *nonlinear resonance.* The two frequencies $\omega_i(I_{i0})$ generally are nonlinear functions of their action variables. For the special case of a linear system (the harmonic oscillator), these frequencies are constants.

We know that the resonance condition (5.17) is exactly that needed for a certain periodic orbit to exist in the phase space of the nonintegrable system. The properties of such a periodic orbit do depend on the coupling V that creates the nonintegrable system. From numerical studies we have seen that these new periodic orbits come in pairs—one orbit stable and one unstable; see Figs. 1.8 and 5.1c for examples. The unstable orbits have chaotic neighborhoods in phase space (Sections 5.2b and 5.3b). We now focus on one of the stable periodic orbits. By definition, the orbit's neighborhood in phase space is regular, containing nested tori on which the action-angle variables θ_i, I_i are well defined. In general, the evolution of these variables is quasiperiodic evolution.

Two major approximations are made in the theory of nonlinear resonance. The first is that the variations in actions over the nested tori are small compared to the resonance actions: $I_i - I_{i0} \ll I_{i0}$. We can then make Taylor series expansions of the

Hamiltonians and of the frequencies of the subsystems, expansions taken about the resonance values of the actions

$$H_{0i} \cong H_{0i}(I_{i0}) + \omega(I_{i0})(I_i - I_{i0}) + \tfrac{1}{2}\omega_i'(I_i - I_{i0})^2, \tag{5.18}$$

$$\omega_i(I_i) \cong \omega_i(I_{i0}) + \omega_i'(I_i - I_{i0}). \tag{5.19}$$

Here we've used the fact that $\omega = \partial H/\partial I$ and have written $\omega' \equiv \partial\omega/\partial I = \partial^2 H/\partial I^2$.

The second major approximation is that, once a double Fourier expansion of V is made in the angle variables θ_1, θ_2, a single Fourier component dominates over all the others. We begin by writing

$$V = \frac{1}{2}\sum_{k,l} V_{kl}(I_1, I_2)\exp[i(k\theta_1 - l\theta_2)] + \text{c.c.} \tag{5.20}$$

When the coupling ϵ is not too large, the angle variables approximately satisfy $\theta_i = \omega_i t$; that is, the evolutions of the angle variables are not dramatically different from that for $\epsilon = 0$. Then condition (5.17) means that the term in Eq. (5.20) with $k = k_0$, $l = l_0$ has a *stationary phase*:

$$\frac{d}{dt}(k_0\theta_1 - l_0\theta_2) = k_0\frac{d\theta_1}{dt} - l_0\frac{d\theta_2}{dt} = 0.$$

We see that the resonant term in Eq. (5.20) is varying very slowly compared to the other terms and thus has much more time (unless V_{k_0,l_0} is accidentally small) to develop large amplitude evolution. The other terms in Eq. (5.20) are rapidly varying, for they produce evolution at much higher frequencies than does the slowly varying resonant term. We don't expect the rapidly varying terms to signficantly affect the average evolution on the longest time scale—that evolution associated with the resonant term. So we assume all the rapidly varying terms average out, and thus we retain only the resonant term in Eq. (5.20).

Let us take the coefficient of the resonant coupling term in Eq. (5.20) to be

$$V_{k_0,l_0}(I_1, I_2) \equiv V_0(I_{10}, I_{20})\exp(i\phi + i\pi/2), \tag{5.21}$$

where we have used the zero-order approximation $I_i \cong I_{i0}$ for the actions, and where ϕ is some constant phase. When we use Eq. (5.21) in Eq. (5.20), and put Eqs. (5.18),

(5.19), and (5.20) into Eqs. (5.16), we obtain the approximate equations of evolution for the action deviations $\Delta I_i \equiv I_i - I_{i0}$:

$$\frac{d(\Delta I_1)}{dt} = \frac{dI_1}{dt} = \epsilon k_0 V_0 \cos\psi \tag{5.22a}$$

$$\frac{d(\Delta I_2)}{dt} = \frac{dI_2}{dt} = -\epsilon l_0 V_0 \cos\psi \tag{5.22b}$$

$$\frac{d\psi}{dt} = k_0\omega_1'\Delta I_1 - l_0\omega_2'\Delta I_2. \tag{5.22c}$$

Here the third dynamical variable is the *resonance phase difference* $\psi \equiv k_0\theta_1 - l_0\theta_2 + \phi$. From Eqs. (5.22a) and (5.22b) it immediately follows that

$$l_0\frac{dI_1}{dt} + k_0\frac{dI_2}{dt} = 0. \tag{5.23}$$

Equation (5.23) first says that in the resonance approximation, there is a new constant of motion, which is

$$l_0 I_1 + k_0 I_2 = \text{constant}.$$

Thus the nonlinear resonance problem has become integrable. Second, Eq. (5.23) states that the changes in I_1 and I_2 are locked together in time, a phenomenon called *autophasing*. We have a vector $\mathbf{I} = (I_1, I_2)$ rotating in action space.

Now let us take the time derivative of Eq. (5.22c). Using Eqs. (5.22a) and (5.22b) the result is

$$\frac{d^2\psi}{dt^2} + \Omega_0^2 \cos\psi = 0, \tag{5.24}$$

where

$$\Omega_0^2 \equiv \epsilon V_0(k_0^2\omega_1' + l_0^2\omega_2'). \tag{5.25}$$

Mathematically we see that Eq. (5.24) is just the second-order equation of motion for the vertical pendulum, where the second-order equation is obtained by combining Eqs. (1.40). The pendulum angle variable is now replaced by ψ, and the pendulum frequency becomes Ω_0. Thus when ϵ is not too large, the pendulum equation determines the slow oscillations in the resonance phase difference ψ. The solution of the pendulum problem (see the end of Section 1.4b and the action-angle variables in Section 3.3b) now describes the evolution on tori that are in the neighborhood of the periodic orbit associated with the resonance condition (5.17). The frequency of this slow evolution is Ω_0.

The introduction of the action deviations ΔI_i and the resonance phase-difference angle ψ can be done more formally by introducing at the outset a canonical transformation to new action-angle variables (Voth, 1986; Rasband, 1990). Physically this constitutes passage into a moving reference frame rotating at a high frequency, such that the predominant frequency in the rotating frame becomes that of the resonance pendulum-like slow evolution. Since approximations are involved, it is not surprising that the literature contains several different canonical transformations that all essentially lead to Eqs. (5.22).

Thus we conclude that the classical dynamics of a simple pendulum describes—at least qualitatively—the nontrivial classical dynamics of *any* perturbed integrable system near a resonance. It is no accident that the chains of resonance islands of regular evolution seen in Poincaré sections for nonintegrable systems (Fig. 5.1) look like repetitions of the one-period invariant curves for the pendulum (Fig. 1.15).

We also conclude that details of the solution of the integrable classical pendulum problem (Sections 1.4b, 3.1b, and 4.3) must give us the basic physics of the regular motion associated with classical nonlinear resonance. However, there is a great difference between the separatrix of the pendulum and the stochastic layer of chaos within the resonance zone of the nonintegrable system.

Even more important for us, details of the solution of the integrable *quantum* pendulum problem (Section 4.3) give us the physics of a corresponding resonant quantum problem. This is the problem of *quantum nonlinear resonance*.

5.6b. Quantum Nonlinear Resonance

The theory of internal quantum nonlinear resonance has numerous potential applications. In molecular chemistry, for instance, there are many models of molecular vibrational mode coupling that exhibit nonlinear resonance, such as two coupled Morse oscillators, a Morse oscillator coupled to a harmonic oscillator, and various other coupled nonlinear oscillators. (For a discussion of internal quantum nonlinear resonance in such systems, see Voth (1986).) In addition, *external* quantum nonlinear resonance plays an important role for atoms, molecules, and doped quantum wells in strong electromagnetic fields; see Sections 8.2 and 8.3.

In our $N = 2$ problem, the intriguing part of the evolution just discussed is connected with the slow phase-difference variable ψ. When the two 1-d systems determined by the Hamiltonians $H_{0i}(I_i)$ in Eq. (5.15) are quantized according to $I_{i0} = (n_i + \frac{1}{2})\hbar$, then the approximate classical constant of motion $l_0 I_1 + k_0 I_2 \approx l_0 I_{10} + k_0 I_{20}$ gives an approximate quantum "constant of the motion" $l_0 n_1 + k_0 n_2$ for a zero-order basis energy eigenstate $\psi^0_{n_1,n_2}$ of our system. Satisfaction of the classical resonance condition Eq. (5.17) at approximately the energy of $\psi^0_{n_1,n_2}$ will imply the existence of near-degeneracies in the manifold of states $\psi^0_{n_1,n_2}$, $\psi^0_{n_1 \pm k_0, n_2 \mp l_0}$, $\psi^0_{n_1 \pm 2k_0, n_2 \mp 2l_0}$, and so forth. Let us call the ranges in I_1 or I_2 over which there occurs classical resonant behavior the "nonlinear resonance widths." If these widths are larger than l_0 or k_0, respectively, then the manifold of states can be strongly coupled, and the near-degeneracy in energy of these states can be removed.

The fast classical evolution at high frequency will produce a ladder of widely spaced quantum energy levels. The slow evolution splits the near-degeneracy of the

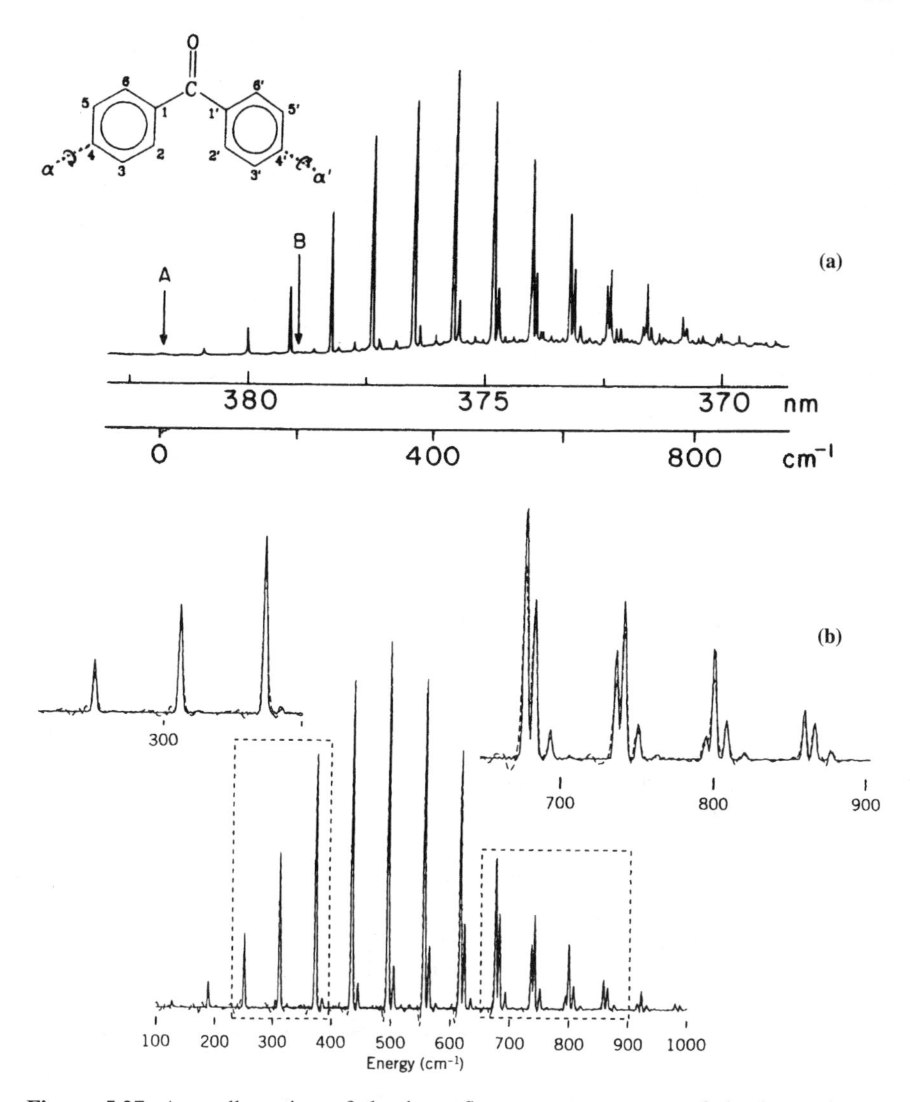

Figure 5.27. A small portion of the laser fluorescence spectrum of the benzophenone molecule. The inset of panel (a) shows the molecule and its benzene-ring torsional rotations about the axes α and α'. Panel (a) also shows an experimental spectrum. The arrows marked A and B indicate, respectively, the onsets of spectral line series for excitation of the symmetric and asymmetric torsional modes of the rings. Reprinted from Holtzclaw and Pratt (1986). Copyright 1986 American Institute of Physics. (b) Model quantum (solid line) and semi-classical (dashed line) numerical results are for the same spectrum. From Sepúlveda and Grossman (1996). Copyright 1996 John Wiley & Sons, Inc. Reprinted by permission of John Wiley & Sons, Inc.

levels on this ladder, changing each level into a distribution of levels. As we noted at the very end of Section 5.6a, the nature of the distributions must be approximately determined by the quantum pendulum problem (Shuryak, 1976).

A clear role for quantum internal nonlinear resonance in a polyatomic molecule has been uncovered in the experimental UV spectrum of the benzophenone molecule (Holtzclaw and Pratt, 1986). The insert in Fig. 5.27a indicates the structure of this molecule. Note that a C=O subgroup is bound to two benzene rings. Figure 5.27a shows fine structure in the fluorescence spectrum of benzophenone, where the spectrum is produced by CW laser excitation of the first vibrational state of the C=O bond. This fine structure is the result of vibrational energy transfer to the two bonds connecting the C=O subgroup to the two rings. Although a total of 66 vibrational degrees of freedom are present in the molecule, only two degrees of freedom significantly influence that portion of the experimental spectrum shown in Fig. 5.27a. Two series of spectral lines are associated with excitation of symmetric and asymmetric torsional modes of the benzene rings, modes that arise from hindered rotations of the rings about the α- and α'-axes; see the inset. As we proceed in energy up the two line series, the coupling between the two torsional modes becomes strong, and the series are increasingly influenced by a 1:1 classical nonlinear resonance. A near-degeneracy of the two series develops and, in conjunction with the strong coupling, produces a spectral transition to subbands. The envelopes of these subbands can approximately be determined by a quantum pendulum approximation for the slow evolution due to the nonlinear resonance. Since time and frequency are inversely related, this slow (long-period) evolution produces the small frequency separations within each subband.

During an investigation of simple models that simultaneously explain the locations and intensities of all spectral lines in Fig. 5.27a, Frederick et al. (1988) found that a nonlinear coupling term $c_2 q_1^2 q_2$ in the coupled-oscillator Hamiltonian was required:

$$H(\mathbf{p}, \mathbf{q}) = \tfrac{1}{2}\omega_1(p_1^2 + q_1^2) + \tfrac{1}{2}\omega_2(p_2^2 + q_2^2) + c_2 q_1^2 q_2 + c_4 q_2^3.$$

Using such a model Hamiltonian and experimental values for the parameters, Sepúlveda and Grossman (1996) obtained some precise and revealing numerical results that go far beyond the quantum nonlinear resonance approximation. A quantum spectrum was obtained by integrating the time-dependent Schrödinger equation using an FFT method (Section B.2e), while a semiclassical spectrum was obtained by pursuing the multiple trajectory approach for semiclassical time propagation (Section 2.1c). The quantum and semiclassical spectra are shown in Fig. 5.27b. The two sets of calculations show very good agreement, indicating that all the dynamical features that go into the splitting of peaks in the spectrum can be associated with the classical system. In particular, the agreement confirms that the energy splitting and line intensities within the subbands are semiclassical.

6

QUANTUM TUNNELING

Quantum tunneling is a flow of probability that is forbidden in a corresponding classical system. We immediately note that this definition precludes tunneling in essentially quantum systems (Section 1.5), for these have no corresponding classical systems. While our definition is made in the time domain, it also includes steady-state situations, such as the familiar partial penetration of a rectangular potential barrier V(x) by a matter wave; see Fig. 6.1a. One characteristic of all tunneling processes is some kind of *barrier penetration* of quantum particles into *classically forbidden regions*. In Fig. 6.1a, a forbidden region labeled II exists in coordinate space and is characterized by V(x) being larger than the energy E of the incident particle in region I.

In this chapter we will extend into phase space the concept of a classically forbidden region, and we'll also consider tunneling problems that involve more than one spatial dimension. We have already learned about various possible classical structures that can exist in phase space, and we'll now see the role these structures can play in tunneling processes. Important interplay can occur between classical structure and quantum penetration of wavefunctions into classically forbidden regions in phase space.

6.1. MATTER-WAVE TRANSMISSION THROUGH 1-d POTENTIAL BARRIER SYSTEMS

6.1a. Steady 1-d Tunneling

We begin by reviewing some salient features of barrier problems in one spatial dimension. Figure 6.1 displays several important situations. In each case, a horizontal line that represents an energy level is superimposed on a potential

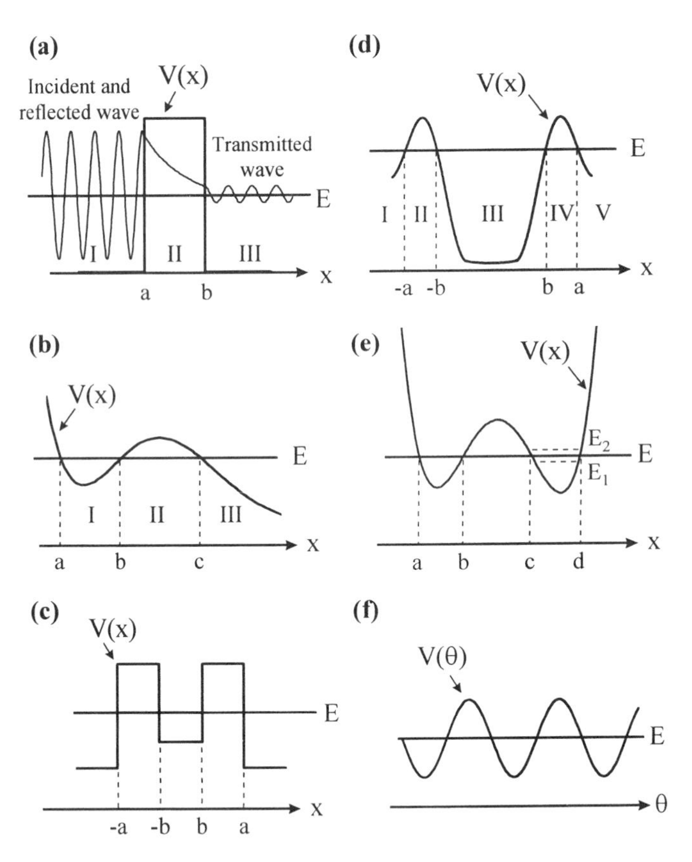

Figure 6.1. Some important one-dimensional potential functions V(x). (a) The single rectangular barrier potential is shown superimposed on the steady-state wavefunction for an incident monochromatic matter wave coming from the left. A portion of this wave penetrates into the barrier region II and is accompanied by a transmitted wave in region III. (b) A repulsive potential with an added potential barrier that produces a potential well. A quantum particle at energy E initially inside the well region I will eventually tunnel through the barrier region II and escape from the well. (c) A symmetric double-barrier potential has a well between the barriers. (d) A more realistic symmetric double-barrier potential has no discontinuities in V(x) or dV/dx. (e) An asymmetric realistic double-well potential is shown, along with the splitting of the degenerate approximate energy level $E \equiv E_0$ of the symmetric well due to the asymmetry and due to tunneling. The split energy levels are E_1 and E_2. (f) A portion of the cosine potential $V(\theta) = V_0 \cos\theta$, in which a quantum particle initially localized in one well will subsequently tunnel through more and more barriers, spreading out to larger and larger values of θ.

V(x). Note that the number of classical turning points—those points a, b, ..., where $E - V(x)$ changes its sign—ranges from two (Fig. 6.1a) up to six or more (Fig. 6.1f).

Some potentials V(x) are simple enough to admit an analytic full quantum solution. In such cases, we first solve for the form of the spatial dependence of the wavefunction in each of the spatial regions that lie in-between neighboring turning points. We then find constants of proportionality, using the quantum requirements that the wavefunction and its first spatial derivative be continuous everywhere. This procedure can be carried out for constant, linear, parabolic, and sine wave portions of V(x). Thus this method can be used to treat tunneling for the potentials of Figs. 6.1a,c,f.

Both the single-barrier problem (Fig. 6.1a) and multiple rectangular barrier problems (Fig. 6.1c) are important for quantum semiconductor devices and therefore nanothechnology. For instance, suppose that an incident charge carrier in Fig. 6.1a has an effective mass m_1 outside the barrier, and a different effective mass m_2 inside the barrier region. (Effective mass was discussed in Section 2.1d.) For a heterostructure made of $GaAs/Al_{0.3}Ga_{0.7}As/GaAs$, the mass values are $m_1 = 0.067m_0$ and $m_2 = 0.092m_0$, where m_0 is the free-electron mass. When $E = \hbar^2k^2/2m_1$ is less than the barrier height V_0, then the steady-flow (time-independent) matter wavefunctions are

$$\psi(x) = \begin{cases} A\exp(ikx) + B\exp(-ikx), & \text{for } x < 0, \\ C\exp(-\gamma x) + D\exp(\gamma x), & \text{for } 0 < x < a, \\ F\exp(ikx), & \text{for } x > a. \end{cases} \tag{6.1}$$

Here $\gamma \equiv [(2m_2/\hbar^2)(V_0 - E)]^{1/2} \equiv [(2m_2/\hbar^2)V_0 - (m_2/m_1)k^2]^{1/2}$. If we follow the full quantum procedure outlined above, the result for the *barrier transmission amplitude* T(E) or T(k) is

$$T(k) \equiv \frac{F}{A} = \frac{2k\gamma\exp(-ika)[2k\gamma\cosh(\gamma a) + i(rk^2 - \gamma^2/r)\sinh(\gamma a)]}{[2k\gamma\cosh(\gamma a)]^2 + [(rk^2 - \gamma^2/r)\sinh(\gamma a)]^2}, \tag{6.2}$$

where $r \equiv m_2/m_1$. Using Eq. (6.2), we can investigate how quantum tunneling in this heterostructure depends on the properties of the semiconductor materials. The analysis can be extended to more complicated heterostructures, such as that shown in Fig. 6.1c. As might be expected, the results are often more unwieldy than Eq. (6.2), especially when many different materials are used. Computer programs, however, do exist for dealing with such complicated problems (Brandt and Dahmen, 1994).

For most potentials V(x), it is easiest to obtain precise full quantum solutions numerically. This is certainly the case for the problems shown in Figs. 6.1b,d,e. An approximate WKB solution (with or without numerical assistance) can help guide the precise numerical work and can help in the interpretation of the numerical results. Beginning with Sections 1.2c and 1.2d, we've seen that the WKB approximation can be acceptably accurate for excited bound energy eigenstates. In a similar fashion, we can explore the usefulness of the WKB approximation in tunneling problems. Connection formulas for a single turning point, similar in spirit to Eq. (1.25), can be developed to splice together WKB wavefunctions within the different spatial regions. To make good connections, the approximate wavefunctions near each turning point of a region—and the WKB wavefunction between these points—must each extend over at least one characteristic spatial length. Thus a barrier width must be at least three characteristic lengths, and the maximum tunneling probability for which the WKB approach can apply is about $(e^{-3})^2 \approx 0.001$. With such restrictions, one-point connection formulas can be used to derive n-point connection formulas, from which the WKB transmission amplitude T and reflection amplitude R can be extracted. For the general case of two complex turning points $x_\pm$, the result is (Kemble, 1937; Fröman and Fröman, 1965; Berry and Mount, 1972)

$$\frac{\exp[-w(x_-, x)]}{\sqrt{p}} + \frac{R\exp[+iw(x_-, x)]}{\sqrt{p}} \leftarrow \psi(x) \rightarrow \frac{T\exp[iw(x_+, x)]}{\sqrt{p}}, \tag{6.3}$$

where

$$T(E) = \frac{\exp[-K(E)/\hbar]\exp[-i\delta(E)]}{\sqrt{1+\exp[-2K(E)/\hbar]}}, \tag{6.4a}$$

$$R(E) = \frac{-i\exp[-i\delta(E)]}{\sqrt{1+\exp[-2K(E)/\hbar]}}. \tag{6.4b}$$

Here $w(x_\pm, x) = \int_{x_\pm}^{x} k(x')\,dx'$, $k \equiv p/\hbar$, and K(E) is the classical action integral under the potential barrier

$$K(E) \equiv \int_{x_-}^{x_+} |k(x)|\,dx. \tag{6.5}$$

This integral over the classically forbidden region is called the *barrier penetration integral* or the *tunneling action*. The *phase shift* $\delta(E)$ is a quantum correction that is important when $E \approx V_0$, that is, when the energy is near the top of a (nonrectangular) barrier. In such a case, K(E) is expanded about $V(x) \approx V_0$ (Miller, 1968):

$$K(E) = -\frac{\pi(E-V_0)}{\hbar\omega_B}\left(1 + \frac{(E-V_0)(\gamma - 5\beta^2/3\alpha)}{16\alpha^2} + \cdots\right). \tag{6.6}$$

Here $\alpha \equiv -d^2V/dx^2|_{V_0} \equiv -V_0''$, $\beta \equiv -V_0'''$, $\gamma \equiv -V_0''''$, and $\omega_B \equiv (|V_0''|/m)^{1/2}$. The first term in Eq. (6.6) comes from the parabolic approximation for V(x) near V_0, which is the approximation used to most simply connect WKB wavefunctions. The quantity $\delta(E)$ in Eqs. 6.4a and 6.4b is obtained from the asymptotic transmitted wavefunction, after this function is written in the form of a constant times $\sin(kx + \delta)$. (See Miller (1968) for application of this WKB approach to the problem defined by Fig. 6.1b.)

6.1b. Wavepacket Tunneling Through Heterostructure Barriers

Now that the steady-state transmission amplitude T is obtained, we will consider the evolution of a (Gaussian) wavepacket incident on a barrier system, using the approach of Section 1.3c. Let us return to our semiconductor barrier problem, where T(k) is given by Eq. (6.2). We take the incident wavepacket to have the form similar to Eq. (1.30):

$$\psi_{inc}(x, t) = \frac{1}{\sqrt{2\pi}} \int_{-\infty}^{\infty} f(k' - k) \exp i\left(k'(x - x_0) - \frac{E(k')t}{\hbar}\right) dk'. \tag{6.7}$$

Here x_0 is the packet location at $t = 0$, and $k = p_0/m$ is the mean value of the initial packet wave vector. The Gaussian distribution $f(k' - k)$ is taken to have a width small enough to make negligible those k' components having values above the barrier, where $\hbar^2 k'^2/2m_1 > V_0$.

We can learn something about the packet's evolution without actually carrying it out. We know that each k' component in Eq. (6.7) will produce a corresponding component in the transmitted packet $\psi_{trans}(x, t)$, with an amplitude $T(k')$ given by Eq. (6.2). Thus

$$\begin{aligned}\psi_{trans}(x, t) &= \frac{1}{\sqrt{2\pi}} \int_{-\infty}^{\infty} T(k') f(k' - k) \exp i\left(k'(x - x_0) - \frac{E(k')t}{\hbar}\right) dk' \\ &\equiv \frac{1}{2\pi} \int_{-\infty}^{\infty} |T(k')| f(k' - k) \exp i\left(k'(x - x_0) + \delta(k') - \frac{E(k')t}{\hbar}\right) dk'. \end{aligned} \tag{6.8}$$

Here $\delta(k')$ is the phase shift contained in Eqs. (6.4). It is reasonable to assume that the major contribution to the integral (6.8) will come from a region where the variation of the phase is slow, that is, where the integrand is not a rapidly oscillating function of k' that would average out close to zero. (This *stationary-phase approximation* will be developed in Section 6.3b.) If the phase is almost a constant function of k', then its derivative is almost zero and we have

$$\frac{\partial}{\partial k'}\left(k'(x - x_0) + \delta(k') - \frac{E(k')t}{\hbar}\right) = 0.$$

For k' close to k, this means

$$t = \frac{m_1}{\hbar k}\left(\frac{\partial \delta(k)}{\partial k} + x - x_0\right).$$

Thus the *tunneling time* (the Wigner phase time) through a barrier of width $x - x_0 \equiv a$ is

$$\delta t = \frac{m_1}{\hbar k}\left(\frac{\partial \delta(k)}{\partial k} + a\right). \tag{6.9}$$

The *tunneling delay time* is defined as the difference between δt and the classical time for the particle to travel the barrier width as if there were no barrier. We see from Eq. (6.9) that the delay time is given by $\hbar(\partial\delta/\partial E) = (m_1/\hbar k)\partial\delta/\partial k$. When we extract the phase $\delta(k)$ out of T(k) given by Eq. (6.2) by writing T(k) in polar form in accordance with Eq. (6.8), and then take the derivative indicated in Eq. (6.9), we obtain an analytic expression for the tunneling time as a function of k and of $r \equiv m_2/m_1$ (Lee, 1993). For $V_0 = 300$ meV, the tunneling time is a decreasing function of increasing k (or of increasing incident particle energy E) up to $E \approx 100$ meV. The classical time of course also decreases with increasing k. Indeed, for a sufficiently narrow barrier width and for the values of m_1 and m_2 given above for a GaAs heterostructure, the tunneling time is just slightly larger than the classical time.

However, as the barrier width is increased to 5 nm and then 7 nm, the tunneling times decrease to roughly five times smaller than the classical time. The tunneling time becomes relatively independent of the barrier width, whereas the classical time is not independent. Note, however, that such results are not valid at energies near V_0, where the incident packet contains above-barrier components.

Having introduced the tunneling time, Eq. (6.9), we hasten to add that other characteristic times associated with tunneling have been defined. At present, the most physical definition of the amount of time a particle spends during its movement through a partial barrier is considered controversial (Xavier and Aguiar, 1997, and reference 1 therein).

We can carry out numerical calculations to investigate the region of incident packet energies $E \sim V_0$. Such calculations are cumbersome if we couple a set of basis states that contains both below-barrier and above-barrier parts. In such situations, a valuable alternative approach is to carry out finite-difference calculations on a discrete space-time grid (Section B.lb.2). An early version of this alternative approach was used by Goldberg et al. (1967). Figure 6.2 shows some results. The top row of packet snapshots at the indicated times is for the case where the mean energy E is $V_0/2$. This is in the region of validity of our derivation of Eq. (6.9). The packet is almost entirely reflected.

More interesting is the bottom row of Fig. 6.2, where $E = V_0$. A large portion of the incident packet is captured in the barrier region. This portion remains trapped for a relatively long period in comparison to the usual transmission time through the barrier (the case of the top row). More snapshots reveal that the captured part of the packet bounces to and fro between the barrier walls, and a small amount of probability escapes at each collision. This is a time-dependent description of the phenomenon called *transmission resonance*.

Another type of transmission resonance occurs for the double-barrier problems defined by Figs. 6.1c and 6.1d. Here the particle can temporarily be trapped within the central well, again leaking out slowly and exponentially in time. The WKB approximation can be used to describe this phenomenon in terms of quasibound states and broadened energy levels within the central well (e.g., Bohm, 1951). The transmission probability $|T|^2$ is given by

$$|T|^2 = \left[1 + \frac{1}{4}\left(4K^2 - \frac{1}{4K^2}\right)^2 \sin^2\frac{1}{2}\left(\pi - \frac{I}{\hbar}\right)\right]^{-1}. \qquad (6.10)$$

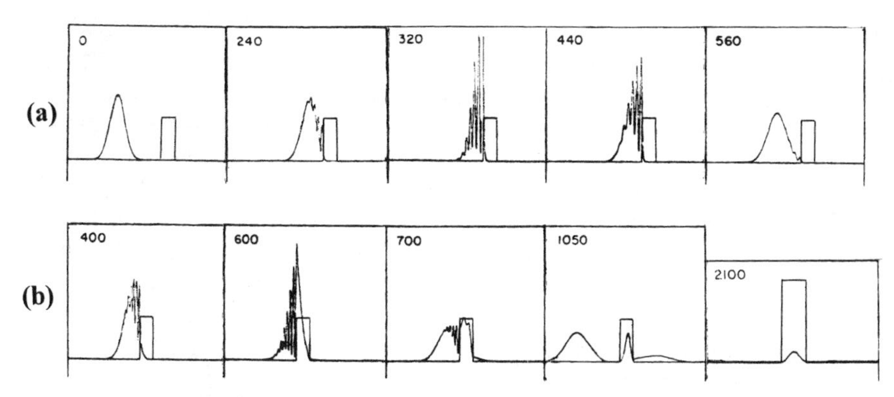

Figure 6.2. Numerical results for the time evolution of a Gaussian matter wavepacket incident on a rectangular potential barrier. In dimensionless units created by $\hbar \equiv 1$, $m \equiv \frac{1}{2}$, the barrier height is chosen to be $V_0 = 2(50\pi)^2$. In these somewhat unusual units, the average velocity of the initial packet is twice the initial momentum p_0. (a) When $p_0 = 50\pi$, the initial mean energy p_0^2 of the packet is one-half the barrier height and there is almost complete reflection. (b) When $p_0 = 70.7\pi$, the initial mean energy is equal to the barrier height and there is significant barrier penetration, matter-wave transmission, and long-time trapping of the particle probability within the barrier region. Reprinted with permission from Goldberg et al. (1967). Copyright 1967 American Institute of Physics and American Association of Physics Teachers.

Here K is the tunneling action—Eq. (6.5)—for one of the barriers, and I is the accumulated action across the central well. The transmission is unity when $I = (n + \frac{1}{2})h$, $n = 0, 1, 2, 3, \ldots$. This condition is the same as the quantization condition Eq. (1.26) for bound states in a true potential well, where the particle cannot escape at all.

6.2. SPLITTINGS IN MOLECULAR SPECTRA AND DYNAMICAL TUNNELING

In this section we consider the effects of tunneling in those quantum systems where particle motion occurs and the potential energy has at least two minima. We take the special case where the potential $V(\mathbf{q})$ is symmetric. Then there will be energy levels in two or more individual potential wells that are (almost) degenerate; these are levels in the different wells that have essentially the same energy values. It is assumed that finite potential barriers (hills or ridges) lie between the wells. For simplicity we consider the 1-d case with $V(-q) = V(q)$. Then we are dealing with a symmetric double-well potential (Fig. 4.3). Figure 6.1e suggests an energy splitting of a degenerate energy level $E \equiv E_0$ into two levels E_1 and E_2, which are separated by $\Delta E = E_2 - E_1$.

Tunneling doublets in molecular vibrational spectra were first observed in a study of the ammonia molecule NH_3 (Hund, 1927). The hydrogen atoms are at the corners of a triangle. The nitrogen atom is outside the plane of the triangle, resulting in a pyramidal structure. The nitrogen atom can be on either side of this hydrogen-atom plane, resulting in two degenerate energy eigenstates that differ only by a reflection through that plane. The coordinate q lies along a straight line perpendicular to the hydrogen-atom plane, with the line passing through the two possible equilibrium points of the nitrogen atom. The potential along this coordinate is similar to that shown in Fig. 4.3, with the top of the barrier lying in the hydrogen-atom plane.

More generally, the tunneling coordinate q is physically different for different quantum systems. It can be a translational or rotational coordinate of a nucleus in a polyatomic molecule, as in NH_3. It can also be a relative coordinate that defines the position of an impurity center in a host solid state environment, or the orientation of an impurity molecule in such an environment (Moerner and Basche, 1993). The coordinate q can also describe different conformational properties of molecular systems, including those of biological interest (Frakenfelder et al., 1988; Peliti, 1992).

6.2a. Theory of Tunneling Splittings

We have just suggested that quantum tunneling can produce a splitting of quantum energy levels that are otherwise degenerate. Let us investigate how this comes about. We begin by rewriting Hamiltonian (4.21) for particle motion in the quartic double-well potential. We introduce dimensionless units slightly different from those for Eq. (4.22); now $q' = (2m\omega^2/E_0)^{1/2}q$, $p' = (mE_0)^{-1/2}p$, and $\gamma^1 = E_0\gamma/4m^2\omega^4 \equiv (64D)^{-1}$. Finally we drop the primes to write

$$H = \tfrac{1}{2}p^2 - \tfrac{1}{4}q^2 + \frac{1}{64D}q^4. \tag{6.11}$$

The quantum parameter D is now the barrier height in units of $E_0 = \hbar\omega_0$, and time is in units of the oscillation period $2\pi/\omega_0$ for motion in either one of the potential wells. For the NH_3 molecule, D approximately equals two.

From what we already know about tunneling through potential barriers and about wavepacket evolution, a packet that starts in the left well will proceed to tunnel through the barrier separating the wells. The existence of tunneling usually implies nonzero particle probability in both wells. Yet our potential is symmetric with respect to reflection about the point $q = 0$ that lies at the top of the barrier. The stationary states (the energy eigenstates) must reflect this symmetry, so we have the symmetric and antisymmetric possibilities—$\phi_S(q)$ and $\phi_A(q)$—shown in Fig. 6.3. In this figure we see superpositions of pairs of wavefunctions where each wavefunction is largely localized in one well or the other. This localization arises because the barrier penetration has been chosen to be small. If $\phi_R(q)$ is a positive wavefunction localized in the right well, then to a good approximation

$$\begin{aligned} \phi_S(q) &= \frac{\phi_R(q) + \phi_R(-q)}{\sqrt{2}}, \\ \phi_A(q) &= \frac{-\phi_R(q) + \phi_R(-q)}{\sqrt{2}}. \end{aligned} \tag{6.12}$$

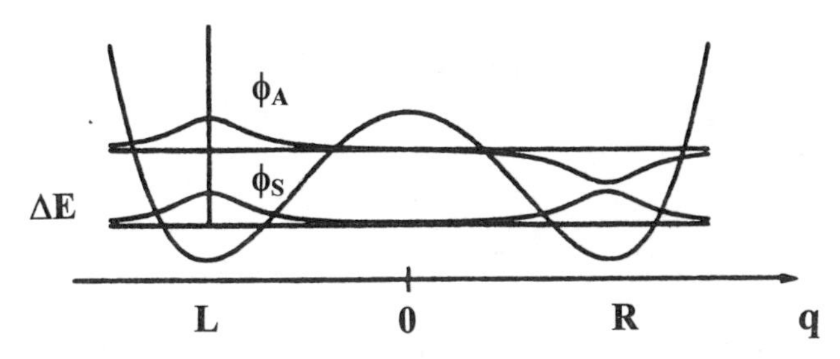

Figure 6.3. Schematic diagram showing the symmetric and antisymmetric energy eigenfunctions ϕ_S and ϕ_A for a quantum particle inside a double-well potential. The energy of the antisymmetric state is above that of the symmetric state by an amount equal to the tunneling splitting ΔE. Adapted from Kilin et al. (1996). Copyright 1996 by the American Physical Society.

These wavefunctions are normalized so that their squares, integrated over both wells, are equal to one.

We can find a good expression for the tunneling splitting ΔE. Following Section 50 of Landau and Lifshitz (1958), we begin by writing Schrödinger's equations for the single-well state $\phi_R(q)$ and the double-well state $\phi_S(q)$:

$$\frac{d^2\phi_R}{dq^2} + \frac{2m}{\hbar^2}[E_R - V(q)]\phi_R = 0, \tag{6.13a}$$

$$\frac{d^2\phi_S}{dq^2} + \frac{2m}{\hbar^2}[E_S - V(q)]\phi_S = 0, \tag{6.13b}$$

We multiply Eq. (6.1 3a) by ϕ_S and Eq. (6.1 3b) by ϕ_R, subtract the two results, and integrate over the right half of the q-coordinate axis; q ranges from the top of the barrier $q = 0$ to $q = \infty$. Note that because of the symmetry, at the top of the barrier we must have $\phi_S(0) = 2^{1/2}\phi_R(0)$, and $d\phi_S/dq|_0 = 0$. In addition, our assumption of little barrier penetration means

$$\int_0^\infty \phi_R(q)\phi_S(q)\,dq \cong \frac{1}{\sqrt{2}}\int_0^\infty \phi_R^2(q)\,dq = \frac{1}{\sqrt{2}}.$$

Gathering the above considerations together produces the result

$$E_S - E_R = -\frac{\hbar^2}{m}\phi_R(0)\left.\frac{d\phi_R}{dq}\right|_{q=0}.$$

When we apply the above procedure again—considering $\phi_A(q)$ and $\phi_L(q) \equiv \phi_R(-q)$—we obtain the same expression for $E_A - E_L$, except for a change in sign. Since the symmetry requires $E_L = E_R$, we have our result

$$\Delta E \equiv E_A - E_S = \frac{2\hbar^2}{m} \phi_R(0) \frac{d\phi_R}{dq}\bigg|_{q=0}. \tag{6.14}$$

The tunneling splitting is proportional to the one-well wavefunction and to its spatial derivative, both evaluated at the top of the barrier.

Using WKB wavefunctions (Sections 1.2c and 6.1a), we can rewrite Eq. (6.14) in terms of the tunneling action K(E); this tunneling action is defined by Eq. (6.5). When properly connected to the WKB wavefunction in the right-hand well, the WKB wavefunction under the right half of the barrier is simply

$$\phi_R^{WKB}(q) = \sqrt{\frac{\omega}{2\pi v_0}} \exp\left(-\frac{1}{\hbar}\int_0^q |p|\, dq'\right),$$

where $v_0 \equiv \{[2(V(0) - E_R)]/m^{1/2}\}$. When we calculate the derivative of this expression for $\phi_R^{WKB}(q)$ and take the upper limits of integrals to be the turning point $q = a$, we obtain the WKB result for the tunneling splitting ΔE:

$$\Delta E = \frac{\omega\hbar}{\pi} \exp\left(-\frac{1}{\hbar}\int_{-a}^{a} |p|\, dq\right) \equiv \frac{\omega\hbar}{\pi} \exp[-K(E)]. \tag{6.15}$$

The splitting depends on both the single-well parameter $\omega(E)$ and the barrier parameter $K(E)$, where $E = E_R = E_L$.

In the case of tunneling in systems with more than one spatial dimension $N \geq 2$, the tunneling splitting formula (6.14) generalizes to Herring's formula (which follows from Green's theorem):

$$E_b - E_a = \frac{\hbar^2}{m} \int_{\Sigma} d\mathbf{S} \cdot (\phi_a \nabla\phi_b^* - \phi_b^* \nabla\phi_a). \tag{6.16}$$

Here ϕ_a, ϕ_b and E_a, E_b are the pairs of exact eigenfunctions and energies. For the symmetrical situation we can take

$$\phi_{a,b}(\mathbf{q}) = \frac{\phi_R(\mathbf{q}) \pm \phi_R(-\mathbf{q})}{\sqrt{2}}. \tag{6.17}$$

The $N - 1$ dimensional surface Σ would then contain $\mathbf{q} = \mathbf{0}$ and should lie in the forbidden region separating one classically allowed region of $\mathbf{q}$-space from another one, such that ϕ_a and ϕ_b should have equal probabilities on each side of Σ (Herring, 1962; Wilkinson, 1986).

6.2b. Dynamical Tunneling

Now that we've shown that the tunneling splitting ΔE exists, it becomes easy to display the effects of tunneling in the time domain. It is simplest to start off with the system localized in one of the wells. Choosing the left well with the wavefunction $\phi_L(q) = \phi_R(-q)$, we can then write the initial state in terms of the eigenstates $\phi_{S,A}(q)$: we invert Eqs. (6.12) to obtain

$$\psi(q,0) = \phi_L(q) = \frac{\phi_S(q) + \phi_A(q)}{\sqrt{2}}.$$

Since the stationary states $\phi_{S,A}(q)$ evolve according to their constant-energy phase factors, we have

$$\psi(q,t) = \exp\left(-i\frac{E_S}{\hbar}t\right)\phi_S(q) + \exp\left(-i\frac{E_A}{\hbar}t\right)\phi_A(q). \tag{6.18}$$

When we reintroduce the one-well states using Eqs. (6.12) and collect terms proportional to each of them, Eq. (6.18) becomes

$$\psi(q,t) = \exp\left(-i\frac{E_S + E_A}{2\hbar}t\right)\left(\phi_L(q)\cos\frac{\delta t}{2} + i\phi_R(q)\sin\frac{\delta t}{2}\right), \tag{6.19}$$

where $\hbar\delta \equiv E_A - E_S$. Thus the wavefunction oscillates back and forth between the wavefunctions for the two single wells. This process is called *dynamical tunneling*. It is an example of two physically coupled quantum states that are started out in a nonequilibrium superposition. Here the two states are $\phi_{L,R}(q)$, and the coupling is due to tunneling. This situation is conceptually very similar to the Rabi oscillations problem of Section 2.2. The difference is that in the Rabi problem, the two states were coupled by an off-diagonal matrix element proportional to the strength of an externally applied field. We note that two-state matrix models of dynamical tunneling can also be based on off-diagonal (coupling) matrix elements having the form $\exp[-K(E)]$ (Anderson et al., 1972).

Of course, we could start off our tunneling system in some other initial state. For example, we might choose the Gaussian wavepacket that is the ground-state wavefunction for a particle in the left well in the harmonic approximation; in this approximation, the last term in Hamiltonian (6.11) is ignored. The initial wavepacket is then $\phi(q,0) = (\omega/\pi)^{1/4}\exp(-\omega^2 q^2/2)$. This case of dynamical tunneling is not a two-level problem, but it has been treated numerically (Bender et al., 1985).

Two periods of dynamical tunneling are shown in Fig. 6.4. In addition to the fundamental oscillations at the *tunneling frequency* δ, we notice other oscillations arising from the multilevel character of the system.

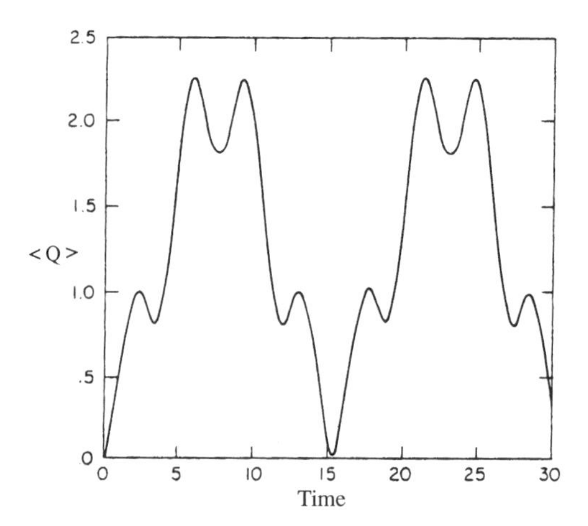

Figure 6.4. Numerical results for quantum dynamical tunneling of a Gaussian wavepacket in the quartic-oscillator double-well potential. The Hamiltonian is taken to be $H = p^2/2 + 4q^2(q - \beta)^2/\beta^2$ with $\beta = 2.5$. At the initial time $t = 0$, the packet $|0\rangle$ is centered at $q = 0$ in the left well. The time dependence of the mean position $\langle 0|q(t)|0\rangle$ of the packet is shown for two periods of oscillation at the tunneling frequency. From Bender et al. (1985). Copyright 1985 by the American Physical Society.

6.3. PATH INTEGRALS AND QUANTUM TUNNELING (ADVANCED TOPIC)

The theory of EBK torus quantization (Section 4.2) provides a satisfactory treatment of regular eigenstates. Consequently, our basic understanding of regular quantum evolution is fairly well-established. This is not the case, however, for problems beyond 1-d, whenever tunneling or classical chaos (or both) play a role.

For $N \geq 2$ it is difficult to extend WKB theory and employ it for Schrödinger's partial differential equation. Huang et al. (1990) describe the situation for multi-dimensional tunneling. In such tunneling problems, two sets of wave fronts must be introduced—the sets of equiphase surfaces and of equiamplitude surfaces. Alternatively, one can work with the rays (paths) defined normal to these surfaces. In this second picture, part of the difficulty is due to the two sets of rays being coupled to one another. There are two situations where this coupling of rays does not occur:

- Normal incidence of the rays upon the turning surfaces or caustics.
- Separable potentials.

Otherwise, the coupling can be avoided by using the complex ray method introduced by Keller (1958b). Keller's method is related to the path integral theory to be discussed in this section.

In the case of chaos, ordinary WKB theory is not at all applicable, so we need to turn elsewhere. Again we need path integral theory, as discussed in Section 7.2. We know that chaos occurs only when $N \geq 2$. Chaos requires just three active dimensions in phase space—a situation that occurs on the energy surface of $N = 2$ autonomous systems, and also in the externally driven 1-d systems of Chapter 8.

Thus for $N \geq 2$, the only practical, reasonably rigorous approach to tunneling and to chaotic systems (other than numerical work) starts with an asymptotic ($\hbar \to 0$) approach that is based on the path integral formulation of quantum mechanics (Section 5.5). This approach gives a zero-order semiclassical theory. As in ordinary WKB theory, it is possible in principle to systematically develop "quantum" corrections to the zero-order theory, with these corrections being higher order in $\hbar$.

In the case of chaos, there is no simple 1-d Hamiltonian example to discuss. But we know that tunneling does occur in 1-d. Because of this possible simplification, we introduce path integral theory here.

6.3a. The Path Integral Approach to Theory for Quantum Evolution

To illustrate the essential elements of what is a very mathematical theory, we consider the 1-d problem of a particle in a potential, following Schulman (1981). The quantum evolution is given by Schrödinger's time-dependent equation (1.1) with $H = T + V = p^2/2\text{m} + V = -\hbar^2/2\text{m}\,(\partial^2/\partial \text{x}^2) + V(\text{x})$.

We introduce the idea of a time propagator K for the evolution of the wavefunction, an operator such that

$$\psi(\text{t}) = K(\text{t}, \text{t}_0)\psi(\text{t}_0). \tag{6.20}$$

The propagation operator K is a Green's function (Morse and Feshbach, 1953) that formally satisfies

$$\left(H - \text{i}\hbar\frac{\partial}{\partial \text{t}}\right)K(\text{t}, \text{t}_0) = -\text{i}\hbar I\delta(\text{t} - \text{t}_0), \tag{6.21}$$

where I is the identity operator. Note that the propagator has a "point source" in time for the solution of Schrödinger's equation—in the same way that the Green's function for an electrostatics problem (where Poisson's differential equation is to be solved) has a point source of charge in space proportional to $\delta(\mathbf{r} - \mathbf{r}_0)$.

In the coordinate-space representation of the operators, we write

$$\text{K}(\text{x}, \text{t}; \text{y}, \text{t}_0) = \langle \text{x}|K(\text{t}, \text{t}_0)|\text{y}\rangle, \tag{6.22}$$

and Eq. (6.21) becomes

$$\left(H_\text{x} - \text{i}\hbar\frac{\partial}{\partial \text{t}}\right)\text{K}(\text{x}, \text{t}; \text{y}, \text{t}_0) = -\text{i}\hbar\delta(\text{x} - \text{y})\delta(\text{t} - \text{t}_0). \tag{6.23}$$

Since our system is to be autonomous—$H \neq H(t)$—the formal solution of the operator equation (6.20) is

$$K(t, t_0) = \Theta(t - t_0) \exp\left(-\frac{iH(t - t_0)}{\hbar}\right). \tag{6.24}$$

Here Θ is the step function. In Eq. (6.24) the exponential function with the operator H in the exponent is defined by a power series expansion. Without loss of generality, we can take $t_0 = 0$ and drop this quantity in the notation. Equation (6.22) then becomes

$$K(x, t; y, 0) \equiv K(x, y, t) = \langle x| \exp\left(-\frac{iHt}{\hbar}\right)|y\rangle. \tag{6.25}$$

Let us call "$it/\hbar$" the quantity λ. Using $e^A = (e^{A/n})^n$ with n an integer, one then needs the Trotter product formula

$$\begin{aligned} K(x, y, t) &= \langle x| \exp\left(-\lambda\frac{T + V}{n}\right) \exp\left(-\lambda\frac{T + V}{n}\right) \cdots \exp\left(-\lambda\frac{T + V}{n}\right)|y\rangle \\ &= \lim_{n\to\infty} \langle x|\left[\exp\left(-\lambda\frac{T}{n}\right) \exp\left(-\lambda\frac{V}{n}\right)\right]^n|y\rangle. \end{aligned} \tag{6.26}$$

For a discussion of the proof of this equation, see Schulman (1981). We insert the identity operators

$$I = \int dx_j \, |x_j\rangle\langle x_j|, \quad j = 1, 2, \ldots, n - 1 \tag{6.27}$$

between each term in Eq. (6.26). Then we note that $V(x)$ is diagonal in coordinate space, so that—for the jth term—

$$\exp\left(-\lambda\frac{V}{n}\right)|x_j\rangle = |x_j\rangle \exp\left(-\lambda\frac{V(x_j)}{n}\right). \tag{6.28}$$

We are left with products of $\langle x_j| \exp(-\lambda T/n)|x_{j+1}\rangle$, which once again we deal with by inserting identity operators. We now use

$$I = \int dp \, |p\rangle\langle p|$$

to obtain

$$\begin{aligned}\langle x_j|\exp\left(-\lambda\frac{T}{n}\right)|x_{j+1}\rangle &= \int_{-\infty}^{\infty} dp\langle x_j|\exp\left(-\lambda\frac{T}{n}\right)|p\rangle\langle p|x_{j+1}\rangle \\ &= \frac{1}{2\pi\hbar}\int_{-\infty}^{\infty}\exp\left(-i\frac{px_j}{\hbar}\right)dp\exp\left(-\lambda\frac{p^2}{2mn}\right) \\ &\quad\times\exp\left(i\frac{px_{j+1}}{\hbar}\right),\end{aligned} \tag{6.29}$$

since $\langle p|x_j\rangle = (2\pi\hbar)^{-1/2}\exp(-ipx_j/\hbar)$. This trick—passing to the momentum representation to deal with the kinetic energy operator—is reminiscent of the similar trick used by Keller to deal with caustics in the EBK theory (Section 4.2).

We next use Gauss's integral

$$\int_{-\infty}^{\infty} dy\exp(-ay^2 + by) = \sqrt{\frac{\pi}{a}}\exp\left(\frac{b^2}{4a}\right),$$

with $y = p$, $a = \lambda/2mn$, and $b = -i(x_j - x_{j+1})/\hbar$, to obtain

$$\langle x_j|\exp\left(-\lambda\frac{T}{n}\right)|x_{j+1}\rangle = \sqrt{\frac{mn}{2\pi\lambda\hbar^2}}\exp\left(-mn\frac{(x_{j+1}-x_j)^2}{2\lambda\hbar^2}\right). \tag{6.30}$$

Inserting Eqs. (6.28), (6.29) and (6.30) into Eq. (6.26) finally gives us our result—Feynman's *path integral expression for the* (exact) *propagator*—

$$\begin{aligned}K(x, y, t) = \lim_{n\to\infty}\prod_{j=1}^{n-1}&\left(\frac{m}{2\pi i\hbar\,\Delta t}\right)^{n/2}\int dx_1\int dx_2\cdots\int dx_{n-1} \\ &\times\exp\left\{\frac{i\Delta t}{\hbar}\left[\frac{m}{2}\left(\frac{x_{j+1}-x_j}{\Delta t}\right)^2 - V(x_j)\right]\right\},\end{aligned} \tag{6.31}$$

where $\Delta t \equiv t/n = \hbar\lambda/in$.

Why do we call Eq. (6.31) a path integral? If we connect the points $y, x_1, \ldots, x_{n-1}, x$ by straight lines, we have a broken-line path from y to x. When we take the limit, there are an infinite number of line segments in the path, which should be a good approximation to a continuous curve (path) from y to x. The integrals over the quantities $x_1, \ldots, x_{n-1}$ in the limit then approximately become an infinite-dimensional integral, summing over all continuous curves (paths) connecting y to x. Note that such a summing would be over the *function space* of all such paths.

When we replace a denumerable number of paths, having a denumerable number of line segments, by a continuous (nondenumerable) number of continuous paths, we seem to have reached a mathematically sticky point. Yet, when we do integrals and other calculations on a computer, we replace sufficiently small segments of continuous quantities by simple differences (straight-line segments) and get results as accurate as we want. It works. It is not the case that such numerical integration can be done for Eq. (6.31). Yet we take the attitude that Eq. (6.31) "works."

In the literature one usually finds the path integral formula (6.31) written in a form that suggests a function-space integral

$$K(\mathbf{q}_1, \mathbf{q}_2, t) = C \int_{\Gamma} d\mathbf{q}(\tau) \exp\left(i \frac{S(\mathbf{q}(\tau))}{\hbar} \right). \tag{6.32}$$

Here C is a normalization constant, adjusted to make K a unitary operator; Γ is one of the complete set of paths connecting coordinate point $\mathbf{q}_1$ to coordinate point $\mathbf{q}_2$; the integral is taken over the function space for Γ; the variable τ prescribes the points on Γ; and S is the action functional of classical mechanics that depends on the classical Lagrangian $L = T - V$ according to

$$S = \int L \, dt.$$

With Eq. (6.32) we have generalized our 1-d result to higher dimensions and have made the connection

$$\sum_{j=0}^{n-1} \Delta t \left[\frac{m}{2} \left(\frac{x_{j+1} - x_j}{\varepsilon} \right)^2 - V(x_j) \approx \int_0^t d\tau \left[\frac{m}{2} \left(\frac{dx}{d\tau} \right)^2 - V(x) \right].$$

6.3b. Stationary Phases for Classical Path Expansions

The number of paths Γ included in the path integral (6.32) is overwhelming. One may need to include complex paths (where space or time is analytically continued into the complex plane); we will need such paths when treating tunneling. Even to reduce the number of paths to a set of classically allowed paths, we will need to settle for an asymptotic ($\hbar \to 0$) treatment. As we see in Section 7.2, the number of classical paths alone can still be challenging when $N = 2$.

Our asymptotic semiclassical theory will require a crucial approximation called the *stationary-phase approximation* (SPA). We'll need to use several versions of the SPA, including an application to the integral (6.32) over the function space of Γ.

Let us look at an ordinary integral on the line. The SPA asserts that if

$$I(\lambda) \equiv \int_a^b dx \, g(x) \exp[i\lambda f(x)]$$

for reasonably well-behaved functions f and g that are independent of λ, and if df/dx has a set of zeros x_j for $a \ll x_i \ll \cdots \ll x_n \ll b$ at which $f''(x) \equiv d^2f/dx^2 \neq 0$, then—when λ is sufficiently large—the saddle-point method gives

$$I(\lambda) \approx \sum_{j=1}^{n} \sqrt{\frac{2\pi i}{\lambda(d^2f/dx^2)|_{x=x_j}}}\, g(x_j) \exp[i\lambda f(x_j)]. \tag{6.33}$$

Generalizations exist for ordinary line integrals in more dimensions (Morse and Feshbach, 1953). In the case of the path integral (6.32), its SPA within function space is

$$K_{WKB}(\mathbf{q}_1, \mathbf{q}_2, t) \approx \sum_{\alpha} \left[\det\left(\frac{i}{2\pi\hbar} \frac{\partial^2 S_{0\alpha}}{\partial q_{1i} \partial q_{2j}} \right) \right]^{1/2} \exp\left(i \frac{S_{0\alpha}}{\hbar} \right) \tag{6.34}$$

(Schulman, 1981, Chapter 14). In Eq. (6.34) the *stationary-phase equation* $\partial S/\partial \mathbf{q}(\tau) = 0$ is assumed to have the solution(s) $\mathbf{q}_\alpha(\tau)$, $\alpha = 1, 2, \ldots$, which are the classical path(s) between the given end points. The quantity $S_{0\alpha}$ is the action functional evaluated along the path $\mathbf{q}_\alpha$, and each $S_{0\alpha}$ satisfies Eq. (3.11). As such, $S_{0\alpha}$ is a function of the end points $\mathbf{q}_1$, $\mathbf{q}_2$ of the path $\mathbf{q}_\alpha$. Thus the matrix of second derivatives of $S_{0\alpha}$ is meaningful. Since the first derivatives of $S_{0\alpha}$ are zero, it is the matrix of second derivatives that determines the linear transformation of a set $\delta\mathbf{q}_1$ of initial deviations into a set $\delta\mathbf{q}_2$ of final deviations associated with the path $\mathbf{q}_\alpha$. We note that Eq. (6.34) applies when $\mathbf{q}_1$ and $\mathbf{q}_2$ lie in the same classically allowed region of coordinate space.

6.3c. Application to 1-d Barrier Penetration

Let us see how Feynman's path integral (6.31) or (6.32) works for the WKB solution of the 1-d single-barrier penetration problem (Section 6.1). Since we are interested in the time-independent (steady-state) tunneling case, we need to transform the propagator K from the time domain to the energy domain. The relationship must be the Laplace transform

$$G(x, y, E + i\varepsilon) = -i \int_0^\infty dt \exp\left[\left(\frac{i}{\hbar} \right) (E + i\varepsilon)t \right] K(x, y, t). \tag{6.35}$$

Since G satisfies

$$\frac{\hbar^2}{2m} \frac{d^2G}{dx^2} + [E - V(x)]G = i\hbar\delta(x - y), \tag{6.36}$$

G is the Green's function for the 1-d time-independent Schrödinger equation (Morse and Feshbach, 1953).

Since x and y now do not lie in the same classically allowed region, Eq. (6.34) cannot be used as the semiclassical propagator for tunneling problems. There is no real classical path (trajectory) Γ connecting x and y at constant energy E. However, our path integral formalism allows complex paths to exist even in classically forbidden regions. For 1-d tunneling problems in particular, we can include paths where t and thus Δt in Eq. (6.31) are complex, having values in the lower half of the complex t plane. When time has a negative imaginary part, the first term in the exponential function in Eq. (6.31) contains a Gaussian function, which guarantees convergence of the integrals.

At this point, we might choose to introduce complex valued paths that will result in complex trajectories, with $\mathbf{q}_\alpha \to x_\alpha$. These trajectories would be complex valued solutions of Newton's equations, related to complex actions that still satisfy the Hamilton–Jacobi equation. In the sense that they don't correspond to real classical orbits, such complex trajectories are "nonphysical" solutions of the stationary-phase equation. These solutions do not live in the classical phase space, but rather in a new phase space that arises from classical phase space due to the complexification of phase-space coordinates. For $N \geq 2$ problems, however, it has recently been shown that these complex trajectories are "physical" in the sense that they produce observable and correct quantum effects (Kus et al., 1993).

For our 1-d tunneling problem, there is another approach—an approach, however, that does not appear easily generalized to $N \geq 2$ (Huang et al., 1990). In this approach we seek real solutions to the classical equations that are related to a stationary-phase approximation in function space. This approximation includes both expanding the action functional in K—Eq. (6.32)—and taking the integral over time—Eq. (6.35)—along an appropriate contour in the complex time plane (McLaughlin, 1972). In the complex time plane one finds a stationary point t_0 that dominates the behavior of the Green's function G, and the SPA gives the final expected result

$$G(x, y, E) = -m[p(x)p(y)]^{-1/2} \exp\left[\frac{i}{\hbar}\int_{x}^{x_L} p(x')\,dx' - \frac{1}{\hbar}\int_{x_L}^{x_R} P(x')\,dx' + \frac{i}{\hbar}\int_{x_R}^{y} p(x')\,dx'\right]. \tag{6.37}$$

Here $p(x) \equiv \{2m[E - V(x)]\}^{1/2}$, $P(x) = \{2m[V(x) - E]\}^{1/2}$, and x_L and x_R are the left and right classical turning points. McLaughlin works out the details for several example potentials V(x), including the linear potential and the quadratic barrier. Of course, McLaughlin's objective was not to find the answers, but to see how the path integral formulation of quantum mechanics worked.

6.4. INTERTORUS TUNNELING (ADVANCED TOPIC)

6.4a. Tunneling in Phase Space

It should be no surprise that the 1-d tunneling we've focused upon so far should have a phase-space picture and a phase-space formulation. Consider the association that exists between the spatial dependence V(q) of the quartic-oscillator double well and the contours of constant energy (the invariant curves or 1-D tori) in the (q, p) phase space—an association shown in Fig. 4.3. Suppose the Huisimi distribution of an initial quantum state with mean energy E is found to be well-localized on a narrow bundle of 1-D tori (periodic orbits) associated with quantum probability only in the left potential well. This initial state is then a regular quantum state. Suppose, as well, there is a final regular quantum state, also with mean energy E localized in the right-hand well, with its bundle of associated 1-D tori. The two bundles of tori associated with the two wells are mirror reflections of one another through the $q = 0$ plane of symmetry. The tunneling between these initial and final regular states involves a passage under the top of the potential barrier along one or more critical (SPA) complexified classical paths.

The picture just described for 1-d tunneling systems generalizes for $N \geq 2$ integrable systems. Now we have N-D tori instead of 1-D tori. The tunneling at constant energy is now along critical paths in a complexified 2N-D phase space, where the tunneling can be considered to start on an initial mean (central) N-D torus within a bundle of tori and to end on a similar N-D torus. If the initial and final sets of tori are symmetry-related, as was the case for the symmetric 1-d quartic-oscillator double well, then there will be classical energy degeneracy of the initial and final mean tori. Ignoring tunneling (Section 6.1), in the symmetrical case there will be corresponding regular (EBK) quantum energy eigenstates with very similar (EBK) energies (Section 4.2). Quantum energy degeneracy is assured. For $N \geq 2$ in the absence of symmetry, however, it is often still possible to achieve degeneracy of EBK energies of regular states that are localized in regions of phase space when the regions are separated by separatrices. This degeneracy can be achieved by adjusting one or more parameters in the Hamiltonian.

We can see $N = 2$ tunneling in action for a two coupled harmonic oscillator case (Fig. 6.5). The Poincaré sections in the center of the figure display the presence of two energy-degenerate classical tori in phase space; the pictures on the left side display the corresponding quantum phase-space probability distributions of the two torus-localized quantum states in the absence of tunneling. For either of the selected quantum states, we count 14 maxima of the wavefunction as we move along the torus guiding the wavefunction. On the right side of the figure, we see phase-space distributions of the two exact energy eigenstates of the full system, where the distributions have been obtained numerically. These states look like linear superpositions of the two tori-localized states. But on closer inspection we see there is no simple superposition in the region of overlap of the tori-localized states. The energy eigenstates must arise from some coupling mechanism. No classical process can play a role here. The coupling mechanism must be quantum mechanical, that is, tunneling. We note that (l) the unexpected nonsuperposition in sectioned phase space is mostly localized over the region where the Poincaré sections of the two tori overlap; and (2) the nonsuper-position involves a modification of three maxima in the quantum phase-space distributions of each of the torus-localized states. Now we will discuss some classical mechanisms that can produce similar nonsuperposition effects.

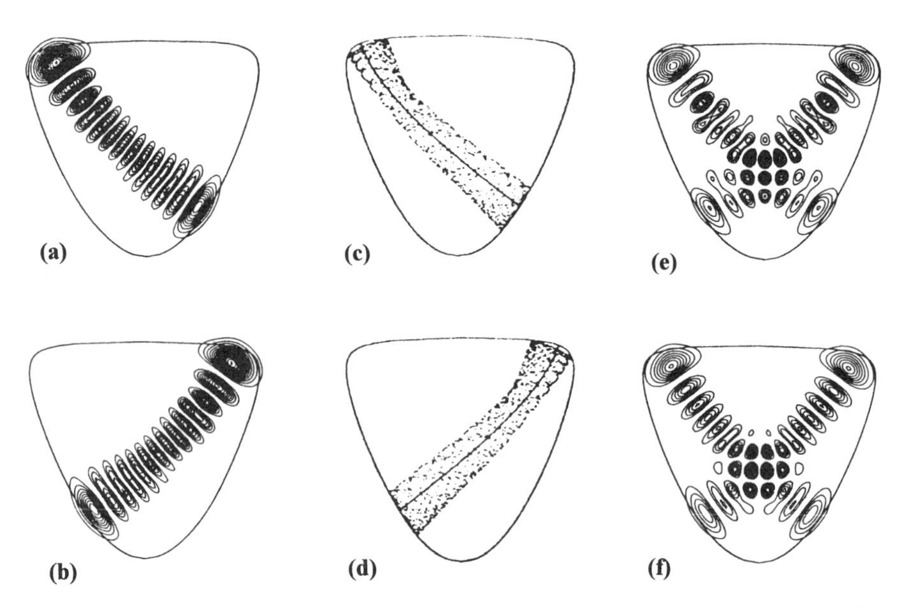

Figure 6.5. Modification of degenerate regular quantum eigenstates due to dynamical tunneling. The stretching of two molecular bonds is modeled by the potential $V(x, y) = \omega_x^2 x^2/2 + \omega_y^2 y^2/2 + \lambda xy^2$, with $\omega_x = 1.0$, $\omega_y = 1.1$, $\lambda = -0.11$, and $m \equiv \hbar \equiv 1$. The variables x and y are the symmetric and asymmetric stretches, respectively. The two energy eigenstates in (x, y) coordinate space are shown in panels (a) and (b), and are well-localized on tori in phase space; see panels (c) and (d). Panels (c) and (d) each show a pair of trajectories, as well as an energy contour at $E = 13.59$; dark lines indicate stable periodic trajectories, while the rest of each panel was generated by a quasiperiodic trajectory. The inclusion of the effects of the dynamical tunneling interaction produces the two energy eigenstates shown in panels (e) and (f). Reprinted with permission from Heller (1995). Copyright 1995 American Chemical Society. Also see Davis and Heller (1981a).

6.4b. Tunneling in Nonintegrable Systems

KAM nonintegrable systems can exhibit tunneling effects qualitatively similar to those of integrable systems—because the separatrix becomes a very thin stochastic layer (Section 5.2b). Path integral theory for this tunneling must involve paths that again connect two regular regions, but now the paths must pass through the stochastic layer. Davis and Heller (1981a) came up with the idea that a classical trapping layer should modify the tunneling process. They dubbed the passage of probability between trapped regular regions "dynamical tunneling," presuming an important role was played by the classical evolution in those regions through which the critical paths pass. Their use of the term "dynamical tunneling," however, is different from the simple two-state time-dependent dynamical tunneling of Section 6.2b.

We will explore the Davis–Heller idea but refrain from using their meaning of the term "dynamical tunneling." Instead, we'll use the specific terms *chaos-mediated tunneling* and *resonance-mediated tunneling*, terms referring to types of classical evolution that do modify what would otherwise be the direct tunneling process of

Sections 6.1 and 6.2. Although direct (two-state) tunneling explains the tunneling splitting of energy-degenerate essentially quantum ground states of quantum systems with symmetry, the two classically mediated tunneling processes can occur in the excited states.

We restrict our study to the case of regular initial and final states—that is, we focus on tunneling from a regular region of phase space through an (at least partially) irregular region into a second, different regular region of phase space. This has been studied numerically for several specific systems (Latka et al., 1994; Tomsovic and Ullmo, 1994; Utermann et al., 1994). Among other possible classically mediated tunneling processes, there is another relatively simple situation of irregular initial and final states that are separated by a regular region (Wilkinson and Hannay, 1987; Ortigoso, 1996).

We should emphasize that tunneling in phase space is different from the more familiar tunneling between two separated regions in coordinate space (Sections 6.1 and 6.2). The EBK-quantized initial-state and final-state tori do not intersect in phase space, but their projections onto coordinate space will usually overlap. There need be no spatial separation of the two wavefunctions that participate in the tunneling process.

Utermann and co-workers offer rather compelling numerical results that support the Davis–Heller suggestion. Their study happened to be on an externally driven system where the Hamiltonian was explicitly time dependent. (Tomsovic and Ullmo, however, have studied two coupled 1-d quartic oscillators.) Since driven systems are discussed in detail in Chapter 8, this is one occasion where we must borrow from a later chapter in order to proceed.

Time is one generalized coordinate for phase space. For time-periodic systems, one usually constructs stroboscopic phase-space portraits, sampling classical trajectories once each period of the driving term in the Hamiltonian. An externally driven 1-d system has a 3-D phase space, which has enough dimensions for the system to be nonintegrable. The system will therefore exhibit stochastic layers, nonlinear resonances, and so forth. Our system will be a periodically driven particle in a 1-d quartic-oscillator double-well potential, where the Hamiltonian is Eq. (6.11) plus an additional periodic driving term $qF\cos(\omega t)$. The interaction of the driver with the quartic oscillator is the electric dipole interaction, and F is the driving electric field strength. Since our problem is time periodic, we can employ Floquet theory; see Section 8.1b. In this theory a quantity called quasienergy plays the role of the usual energy of autonomous systems; see Sections 8.1b and 8.1d. The eigenstates $\psi_\alpha(t)$ of the quantum quasienergy operator are called Floquet states or quasienergy states. The phase-space probability distribution of a Floquet state is periodic along the time axis in generalized phase space.

When the barrier height D in Eq. (6.11) is taken to be eight, only nine pairs of tunneling-doublet *quasienergy* levels exist below the potential barrier. These levels can be ordered by increasing mean energy $\bar{E}_\alpha$ defined by

$$\bar{E}_\alpha \equiv \frac{\omega}{2\pi}\int_0^{2\pi/\omega} dt\langle\psi_\alpha(t)|H(t)|\psi_\alpha(t)\rangle. \tag{6.38}$$

In Fig. 6.6 we can see some stroboscopic phase-space portraits for the driven quartic oscillator, at a driving frequency $\omega = 0.95$ in the dimensionless units of Eq. (6.11). These portraits show the classical evolution for driving strengths F = 0, 0.001, 0.02, and 0.2. At F = 0, we have the integrable unperturbed system of Fig. 4.3. As F is increased, we see two expected phenomena. At F = 0.001, we see a slight blurring of the apparent separatrix, occurring near the unstable fixed point $(x, p_x) = (0, 0)$. More evident are the moon-shaped phase curves within each of the two regions correlated with a potential well. These curves are due to a major classical nonlinear resonance, as can be verified by examining stroboscopic portraits with various constant shifts in the sampling time. The two "moons" rotate clockwise around stable fixed points in the centers of the regular regions, moving at the driving frequency ω_0. At F = 0.02, the nonlinear resonance and the stochastic layer both are major features. Using numerical calculations, we can now find regular Floquet eigenstates whose Huisimi phase-space probability distributions are well-localized in the resonance "moons." These eigenstates rotate with the choice of phase, just like the classical moons do. At F = 0.20, the stochastic layer is huge. What roles do the resonance "moons" and the stochastic layer play in the tunneling?

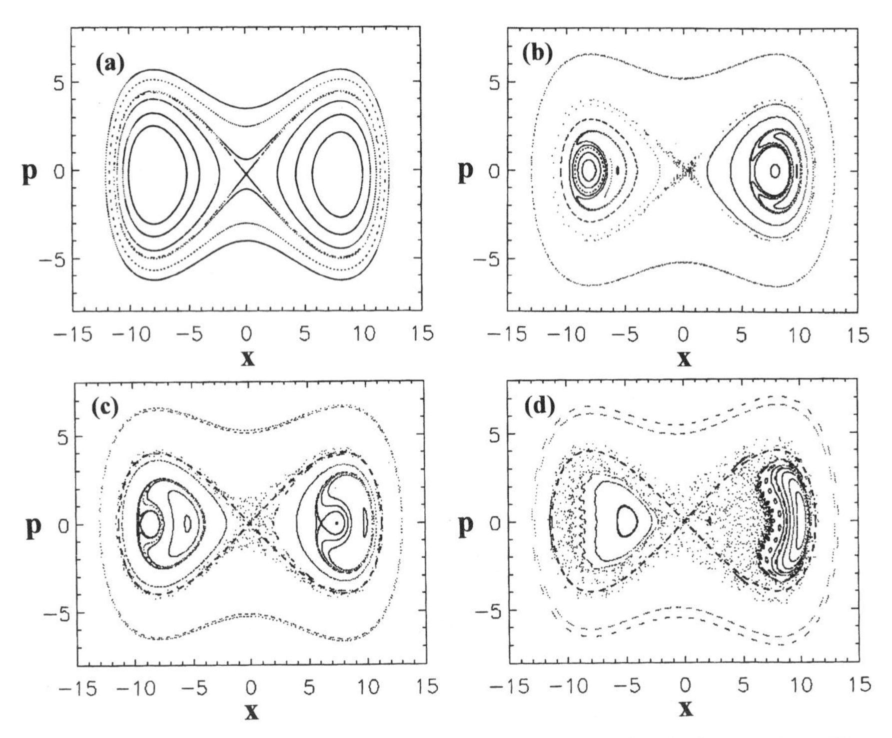

Figure 6.6. Stroboscopic classical phase-space portraits of a particle in the quartic-oscillator double-well potential with Hamiltonian Eq. (6.11), with the particle additionally driven sinusoidally in time by the interaction Hamiltonian $H' = qF\cos(\omega t)$. The parameter values are $D = 8$, $\omega = 0.95$, and (a) F = 0, (b) F = 0.001, (c) F = 0.02, and (d) F = 0.20. From Utermann et al. (1994). Copyright 1994 by the American Physical Society.

6.4c. Chaos-Mediated Tunneling

In order to show how irregular ("chaotic") an initial eigenstate $\psi_\alpha(t)$ becomes as a function of the driving field strength F, Utermann and co-workers defined the eigenstate's mean overlap $\bar{\Gamma}_\alpha$ with the stochastic layer. They used the Huisimi function H_ψ, Eq. (1.37), for the eigenstate:

$$\bar{\Gamma}_\alpha \equiv \frac{\omega}{2\pi}\int_0^{2\pi/\omega} dt \int_{-\infty}^{\infty} dx \int_{-\infty}^{\infty} dp\, H_{\psi_\alpha}(x, p, t)\Gamma(x, p; t). \tag{6.39}$$

Here Γ is called the characteristic function for the stochastic layer. As one numerically integrates Eq. (6.39), Γ is simultaneously generated. One just lets a classical trajectory, started anywhere in the stochastic layer, "tick" boxes in a coarse-grained (discretized) phase space of the desired resolution.

Next, we must numerically compute the quasienergy doublet splittings $\Delta\bar{E}_\alpha$ for different values of F. Then we take the variations of the tunnel splittings $\Delta\bar{E}_\alpha$ with F and compare them to the degree of irregularity $\bar{\Gamma}_\alpha$—see Fig. 6.7—noting the logarithmic scales. As a check of the numerics at low $F \leqslant 0.001$, the dimensionless ground state (doublet number $n = 1$) splitting is 5.20×10^{-18}, which is close to the value $(128D/\pi)^{1/2} \exp(-16D/3) = 5.33 \times 10^{-18}$, the $F = 0$ ground-state splitting. This latter value is obtained using the fully quantum "instanton" complex-time path integral method for ground-state splittings (Grossmann et al., 1991a,b; Auerbach and Kivelson, 1985).

At low F, then, the tunneling—at least for the ground state—is direct tunneling, unmodified by any classical physics. On the other hand, for $F \gtrsim 0.01$, both $\Delta\bar{E}_\alpha$ and $\bar{\Gamma}_\alpha$ increase dramatically with F. A strong correlation exists between the tunnel splittings and the overlaps with the stochastic layer. The steep increases in both functions have similar "threshold behavior" as a function of doublet number. These increases strongly suggest the existence of *chaos-mediated tunneling*. When eigenstate doublets become irregular, their tunnel splittings increase far above the amount ascribed to direct tunneling.

A recent calculation for the annular billiards problem directly confirms that tunneling can be chaos-mediated (Doron and Frischat, 1995). This problem is of the billiards type considered in Section 7.2. The annular billiard consists of the space between two nonconcentric circles of radii R and $a < R$, with one circle completely inside the other. Classically, a particle moves freely in-between specular reflections on the bounding circles. Some trajectories never hit the inner circle and therefore they display the regular motion of circular billiards. Those trajectories that do hit the inner circle produce a region in Poincaré phase-space portraits—a region consisting of a mixture of islands of regular motion and stochastic layers. When parameters are adjusted so there is only one stochastic layer, one can study tunneling between symmetry-related pairs of regular regions where the tunneling goes through the stochastic layer. The dominant tunneling critical paths for path integral theory are found to involve a sequence of three processes: tunneling into the stochastic layer, followed by irregular evolution within that layer, and then tunneling back out.

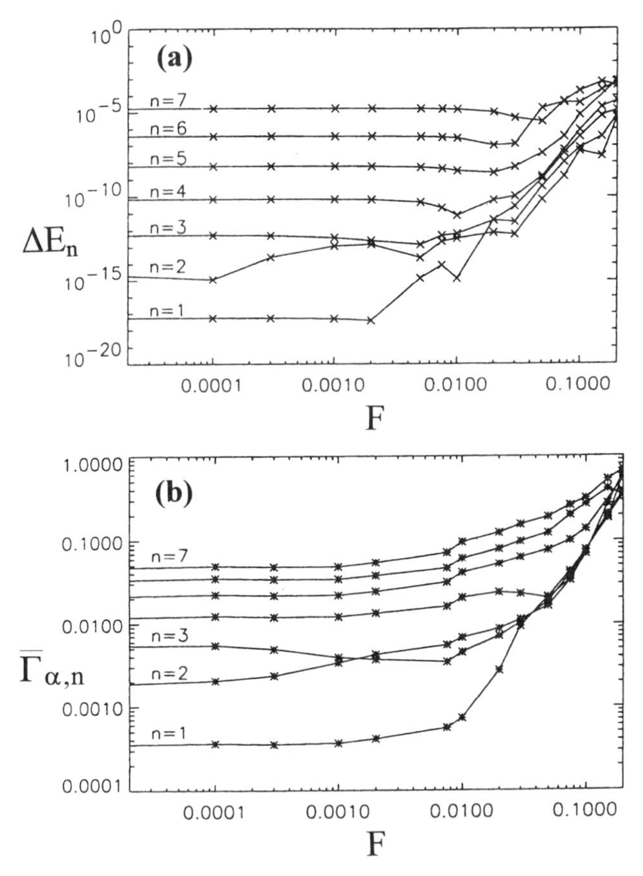

Figure 6.7. Numerical quantum results for the sinusoidally driven particle in the quartic-oscillator double-well potential problem of Fig. 6.6. (a) The dynamical tunneling splittings ΔE_n of the seven lowest tunneling-doublet quasienergy eigenstates $|n\rangle$, as a function of the dimensionless scaled driving electric field F. (b) The overlaps of the states $|n\rangle$ with the F-dependent stochastic layer (Fig. 6.6), as a function of F. These overlaps are defined by Eq. (6.39). From Utermann et al. (1994). Copyright 1994 by the American Physical Society.

6.4d. Resonance-Mediated Tunneling

Figure 6.8 schematically compares Poincaré sections for an integrable system (top panel) with a nonintegrable system (bottom panel). For the integrable case, there is no semiclassical tunneling between tori A and B, since they have the same generating action functions S(x, I) and hence there can be no critical path with a stationary phase. Because the separatrix S in the figure separates torus A (or B) from torus C, semiclassical tunneling does occur from A to C. In contrast, for the nonintegrable case, each torus is separated from every other torus by at least one stochastic layer. So when $\hbar$ is small enough, there will always be tunneling between tori.

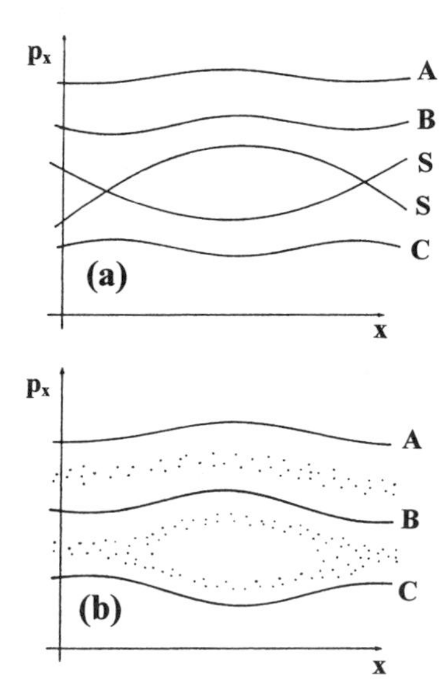

Figure 6.8. A schematic comparison of features in Poincaré sections for (a) an integrable Hamiltonian system and (b) a KAM nonintegrable system. For (a) the tunneling action between tori A and B is infinite, and A can be smoothly deformed into B. However, the tunneling action between A and C is finite, as the separatrix S prevents torus A from being smoothly deformed into torus C. For the KAM system (b), all tori are separated by stochastic layers in phase space, and tunneling actions are finite for any pair of tori. From Wilkinson (1986). Copyright 1986, with permission from Elsevier Science.

In Fig. 6.8b, the situation shown for tori B and C is simply that for *resonance-mediated tunneling*. Theories of doublet splittings due to this process always involve some form of the nonlinear resonance approximation presented in Section 5.6. Sooner or later, the pendulum problem appears in each of these theories.

If we appropriately "preprocess" the full classical problem along the lines of Section 5.6a, then the last theoretical step will involve the exact quantum pendulum solutions (Section 4.3). Alternatively, a path integral approach can lead to critical paths for tunneling where we can use just the tunneling action for the 1-d pendulum potential of Fig. 6.1f. (The potential is extended to an infinite number of sine wave barriers.) This second approach is used in Ozorio de Almeida (1984).

Heller (1995) provides a recent review of the still somewhat mystical, resonance-mediated ("dynamical") tunneling. It is argued that the order (p : q) of the mediating p/q resonance should resemble the quantum number differences between the EBK-quantized initial and final tori. So, for instance, the (2 : 2) classical nonlinear resonance will dominate tunneling between the EBK-quantized 2-D tori having action quantum numbers $(n_1, n_2) = (3, 4)$ and (5, 2).

7

IRREGULAR QUANTUM SYSTEMS

Nonintegrable systems were introduced in Chapter 5, and we found these systems can evolve in an irregular way that is impossible in integrable systems. In classical dynamics, irregular evolution is well-defined by the concept of chaos. For the corresponding quantum dynamics, we were limited in Chapter 5 to an initial look at a few numerical results.

Section 6.3 introduced path integral theory, and we discussed applying the semiclassical version of this theory to 1-d quantum tunneling. Path integral theory is believed applicable to any semiclassical system. It is also possible to develop quantum corrections to this semiclassical theory, corrections that are higher order in $\hbar$. We will now investigate the application of the theory to irregular quantum dynamics, while exploring in some detail two relatively well-studied 2-d autonomous nonintegrable systems.

The first of our two prototype systems is the excited hydrogen atom in a magnetic field. This is a generic semiclassical system that possesses scaling properties. The magnetic field strength can control whether the system is almost integrable, strongly mixed regular and irregular, totally irregular, or spatially open as opposed to having spatially bounded motion only. The system is one of two experimentally accessible systems that can be considered both theoretically and numerically, without any modeling of the Hamiltonian. (The other experimentally accessible system is a driven system—the excited hydrogen atom in a time-oscillating electric field—which is discussed in Section 8.3.)

Our second prototype for studying irregular evolution is a very different class of systems—namely, the billiards problems. (We already mentioned annular billiards in connection with tunneling in Section 6.4c.) Our billiards systems will be spatially closed, meaning that only spatially bounded motion occurs. Depending on the geometry of the boundaries defining the billiards "table," the evolution can be integrable, mixed, or completely irregular. We are especially interested in the completely irregular case.

7.1. ATOMIC ELECTRONS IN STATIC MAGNETIC FIELDS (ADVANCED TOPIC)

7.1a. The Diamagnetic Kepler Problem

To explore our first prototype, we place the hydrogen atom of Section 2.1a into a static magnetic field $\mathbf{B} \equiv B\hat{\mathbf{z}}$. To find the Hamiltonian, we select the Coulomb gauge for the electromagnetic field, $\nabla \cdot \mathbf{A} \equiv 0$, where the vector potential is $\mathbf{A} = \frac{1}{2}\mathbf{B} \times \mathbf{r}$. Next, we recall that the momentum $\mathbf{p}$ in the free-atom Hamiltonian $p^2/2m - e^2/r$ needs to be replaced by $\mathbf{p} + (e/c)\mathbf{A}$. Upon doing this, we obtain the diamagnetic Kepler Hamiltonian in Cartesian coordinates:

$$H = \frac{p^2}{2m} - \frac{e^2}{r} + \omega L_z + \frac{1}{2} m\omega^2 (x^2 + y^2). \tag{7.1}$$

Here the characteristic frequency $\omega \equiv eB/2mc$ is one-half the free-electron cyclotron frequency ω_c, and the component of the electron's angular momentum along the external field direction is $L_z = xp_y - yp_x$. If we pass over to atomic units—and in addition introduce $\gamma \equiv B/B_c$ with $B_c = 2.35 \times 10^5$ tesla—then

$$H = \frac{1}{2}p^2 - \frac{1}{r} + \frac{\gamma}{2} L_z + \frac{\gamma^2}{8}(x^2 + y^2). \tag{7.2}$$

The last term, proportional to B^2, is called the *diamagnetic interaction*. This crucial partial Hamiltonian is responsible for the irregularity arising from the struggle between the Coulomb and magnetic Lorentz forces to control the electron's motion.

We can eliminate the term that is linear in γ—see Eq. (7.2)—by transforming to coordinates in a rotating reference frame; the frequency of rotation is $\boldsymbol{\omega} \equiv \omega\,\hat{\mathbf{z}}$ (Delos et al., 1983a). If we also want to use the axial symmetry of our problem by going over to cylindrical coordinates (ρ, ϕ, z) then Hamiltonian (7.2) becomes

$$\begin{aligned} H &= \frac{1}{2}(p_\rho^2 + p_z^2) + V_{\text{eff}}(\rho, z), \\ V_{\text{eff}} &\equiv \frac{1}{8}\gamma^2\rho^2 + \frac{L_z^2}{2\rho^2} - \frac{1}{\sqrt{\rho^2 + z^2}}. \end{aligned} \tag{7.3}$$

The term $L_z^2/2\rho^2$ is the familiar angular momentum barrier in the effective potential $V_{\text{eff}}(\rho, z)$. When $L_z = 0$, this barrier is absent, and our electron moves in a 2-d potential well; see Fig. 7.1.

At this point, our problem has become qualitatively similar to the Hénon–Heiles 2-d potential well problem (Section 5.3). As in that earlier problem, we find a scaling that results in a scaled Hamiltonian depending on only one parameter. If we scale position, momentum, and energy according to $\mathbf{r}' \equiv \gamma^{2/3}\mathbf{r}$, $\mathbf{p}' \equiv \gamma^{-1/2}\mathbf{p}$, and $H' \equiv \gamma^{-2/3}H$, then

$$H' \equiv \frac{1}{2}[(p'_\rho)^2 + (p'_z)^2] + \left[\frac{1}{8}(\rho'^2) + \frac{(L'_z)^2}{2(\rho')^2} - \frac{1}{\sqrt{(\rho')^2 + (z')^2}}\right]. \tag{7.4}$$

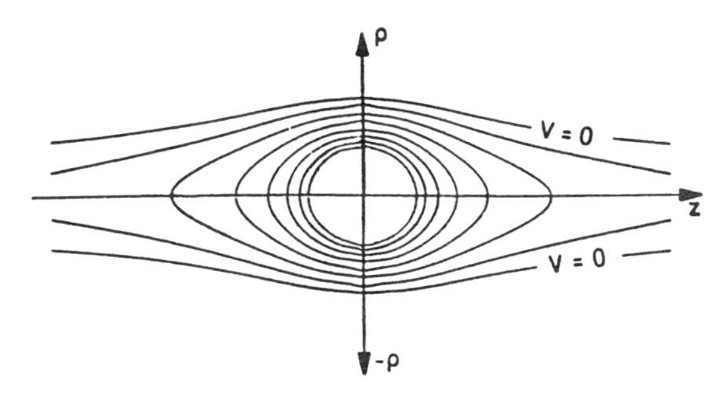

Figure 7.1. Equipotential lines for the motion of an electron in a Coulomb potential and in a uniform static magnetic field $\mathbf{B} = B\hat{\mathbf{z}}$. In cylindrical coordinates (ρ, z) the potential is $V(\rho, z) = \gamma^2\rho^2/8 + L_z^2/2\rho^2 - 1/(\rho^2 + z^2)^{1/2}$, here with $L_z = 0$. The parameter γ is proportional to B. The potential is pictorially extended to negative ρ only to emphasize the existence of a potential well. From Friedrich and Wintgen (1989). Copyright 1989, with permission from Elsevier Science.

The explicit dependence on γ has now disappeared. For the experimentally important case of $L_z = 0$, the scaled classical dynamics depends only on the *scaled energy* $E' \equiv \gamma^{-2/3}E$, not on E and γ separately. By surveying the classical trajectories for the Hamiltonian Eq. (7.4), we learn about all the trajectories for the Hamiltonian Eq. (7.3). We note the scaled action is $S' \equiv \gamma^{1/3}S$.

The scaled potential energy V'_{eff} has a minimum at the point (ρ'_0, z'_0), where $(\rho'_0)^4/4 + \rho'_0 - (L'_z)^2 = 0$. For each value of L'_z, the value of V'_{eff} at that point gives the minimum possible energy value E'_{min}. At $\rho_s = (2L'_z)^{1/2}$, $z_s = \infty$, the quantity V'_{eff} becomes the classical escape energy E'_s, above which the electron can energetically escape from the atomic nucleus.

Figure 7.2 schematically shows characteristics of typical numerically obtained classical trajectories for the diamagnetic Kepler system; these trajectories are for various ranges of L'_z and the normalized shifted scaled energy $f \equiv [E' - [E'_{min}(L'_z)]/[E'_s(L'_z) - E'_{min}(L'_z)]$. For small L'_z (small L_z and/or small B), the trajectories are best described as Kepler ellipses whose orbital parameters vary slowly with time. In this elliptical regime, we can perturbatively calculate trajectories—as well as their associated quantum states—by using the free-atom (Kepler) Hamiltonian as the starting point (Delos et al., 1983a,b; Richards, 1983).

For large L'_z, the magnetic Lorentz force exceeds the Coulomb force, so the electron circles around a magnetic field line and, at the same time, travels slowly back and forth in the z direction. In this helical regime, the atom has the shape of a long cylinder or tube. Between the elliptical and the helical regimes, for f not too small, there is an irregular (chaotic) regime. This regime extends down to $L'_z = 0$ in a narrowing strip near $f = 1$. Finally, below the irregular regime, for $f \leq 0.2$ and $L'_z \sim 1.5$, one sees a transition regime of regular trajectories where the trajectories change character from highly perturbed ellipses to highly perturbed helices.

By transforming to yet another set of coordinates, we can explicitly connect the present problem to 2-d coupled nonlinear oscillator problems, such as the

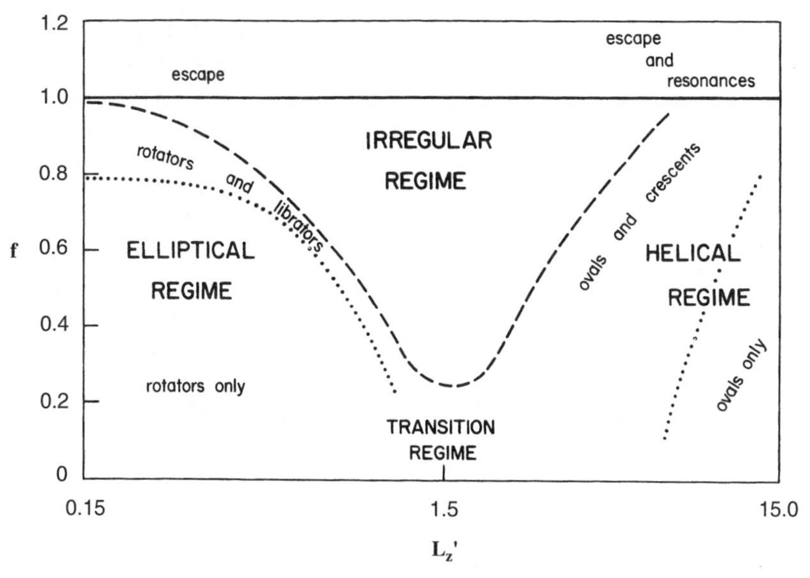

Figure 7.2. Types of electron trajectories most common in the diamagnetic Kepler problem. Regimes of motion are shown in the (f, L_z') parameter space, where f is the dimensionless scaled and shifted electron energy, and L_z' is the dimensionless scaled component of angular momentum along the magnetic field direction; see text. The horizontal axis may be regarded as proportional to $B^{1/3}$. The hydrogen atom can ionize for $f > 1$; see the horizontal line. A wide variety of 3-d trajectories exists; again see text. From Delos et al. (1984). Copyright 1984 by the American Physical Society.

Hénon–Heiles system. The scaled semiparabolic coordinates (ν, μ) are defined by $\nu^2 \equiv \rho' - z'$ and $\mu^2 \equiv \rho' + z'$. The transformation to these coordinates is not canonical. The new momenta must be taken as $p_\nu = d\nu/d\tau$ and $p_\mu = d\mu/d\tau$, where the coordinate-dependent rescaled time τ is determined by $dt \equiv 2\rho'\, d\tau = (\nu^2 + \mu^2)\, d\tau$. Then Hamilton's equations—generated by Hamiltonian (7.4) at a fixed value of the scaled energy E′—are equivalent to new Hamilton's equations for a pseudo-Hamiltonian

$$H(\nu, \mu) = \frac{1}{2}p_\nu^2 + \frac{(L_z')^2}{2\nu^2} + \frac{1}{2}p_\mu^2 + \frac{(L_z')^2}{2\mu^2} - E'(\nu^2 + \mu^2) + \frac{1}{8}\nu^2\mu^2(\nu^2 + \mu^2). \tag{7.5}$$

Here the "pseudoenergy" $H(\nu, \mu) \equiv 2$. We note that different authors define the scaled and semiparabolic coordinates using different numerical factors (Mao and Delos, 1992). For $L_z' = 0$, Eq. (7.5) shows we do have two 1-d harmonic oscillators, coupled by the last term, which arises from the diamagnetic interaction.

Just as in the Hénon–Heiles problem (Section 5.3), when there is particle motion near the bottom of the potential well, there is an approximate third integral Λ of the motion. This integral was found by Solov'ev (1982) and shown by Richards (1983) to be the diamagnetic interaction averaged over a Kepler orbit. In this regime, the magnetic field quantum-mechanically mixes states of different ℓ for each n, called ℓ-mixing, but the n-mixing is negligible. The Kepler ellipse precesses in space, and the third integral Λ describes the mean value of the angle ϕ_p between the Runge–Lenz vector and the line of nodes; for a discussion of these quantities, see Born (1960) or Goldstein (1980). The integral Λ can be viewed as a 1-d effective Hamiltonian having the form of a symmetric double well (Richards, 1983). Thus a separatrix determined by $\Lambda = 0$ lies between regions of ϕ_p motion, where the regions are rotational ($\Lambda > 0$) and librational ($\Lambda < 0$) as in the pendulum. This nearly integrable behavior impacts the elliptical regime at low L_z', see Fig. 7.2. This behavior is also reflected in the quantum eigenstates and energies in the ℓ-mixing regime (Delande and Gay, 1987). The highest energy levels in a single-n diamagnetic multiplet are associated with quantization on tori of rotational motion, while the lowest energy levels are associated with tori of librational (also called vibrational) tori. There are some cross over levels in-between that are associated with separatrix states (Sections 4.3 and 4.4).

For this ℓ-mixing nearly integrable regime, Fig. 7.3a shows a (ρ, p_ρ) Poincaré section for $L_z = 0$. The figure displays the $\Lambda = 0$ unstable fixed point, with the branches of the separatrix passing through this point. The boundary of the region of classical motion is determined by $\Lambda = 4$. This boundary is the trajectory of a stable

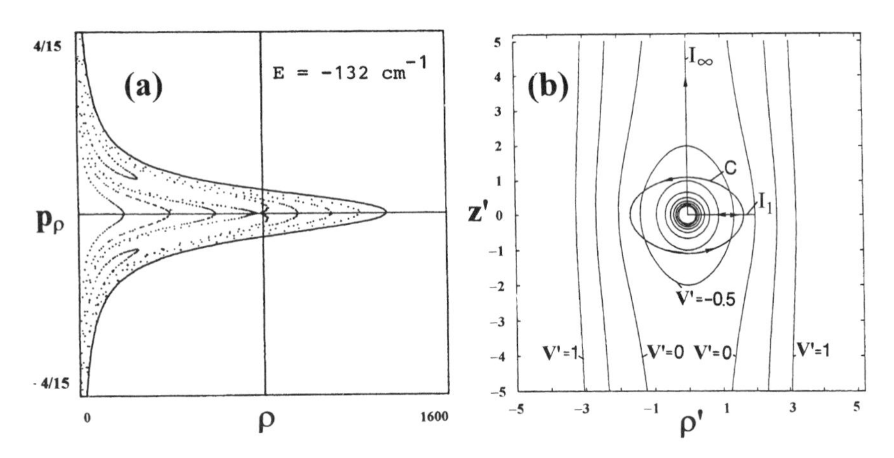

Figure 7.3. Regular electron trajectories in phase space and in the potential well, for the diamagnetic Kepler problem. (a) A (ρ, p_ρ) Poincaré section (atomic units) is shown for the unscaled energy $E = -132\,\text{cm}^{-1}$, $L_z = 0$, and $B = 6$ tesla. The crossing of the vertical and horizontal lines locates an unstable fixed point that lies on the separatrix determined by the Solov'ev integral $\Lambda = 0$. The boundary curve for the allowed motion is determined by $\Lambda = 4$ and is itself a phase-space curve for the periodic orbit I_1, a "rotator" motion in only the ρ spatial coordinate. The separatrix divides phase space into regions of rotational motion $0 < \Lambda \leq 4$ and librational motion $\Lambda < 0$. The vertical boundary at $\rho = 0$ represents the "librator" periodic orbit I_∞, a motion in only the z coordinate. (b) The three periodic electron orbits I_1, I_∞, C at scaled energy $E' = 0$, as seen looking down into the scaled potential energy well $V'(\rho', z')$. From Hasegawa et al. (1989). Copyright 1989 Progress of Theoretical Physics.

periodic orbit I_1 for 1-d purely radial motion in the $z = 0$ plane perpendicular to **B**. This limiting case for "rotational" motion in the effective potential well (pictorially extended to negative ρ for convenience) is shown in Fig. 7.3b. On the other hand, the islands of librational motion shown in the Poincaré section (Fig. 7.3a) contain stable fixed points that correspond to the periodic orbit I_∞ indicated in Fig. 7.3b. This 1-d limiting case for "librational" motion is a Kepler oscillation along the magnetic field direction, where the magnetic Lorentz force is zero.

Figure 7.4 shows a set of Poincaré sections as our $L_z = 0$ system is moved—by increasing the energy E—from the nearly integrable regime to the completely irregular regime. Figure 7.5 shows the regular fraction of the area in the section,

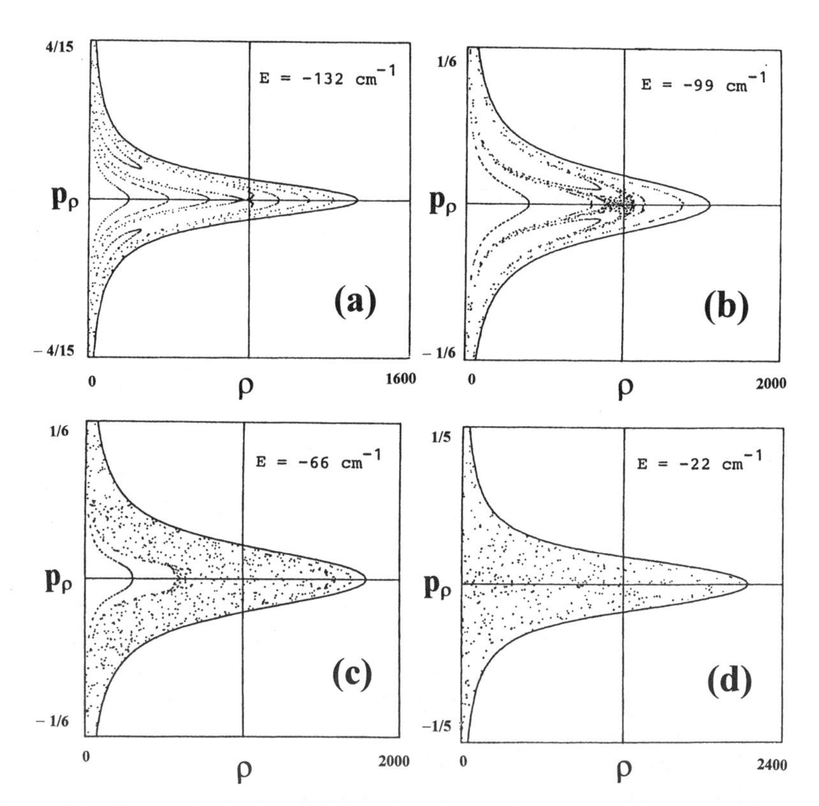

Figure 7.4. Changes in the (ρ, p_ρ) Poincaré section as the energy of the $L_z = 0$ diamagnetic Kepler electron is increased toward the ionization limit $E = 0\,\text{cm}^{-1}$. (a) The nearly regular section that is shown in Fig. 7.3a, at $E = -132\,\text{cm}^{-1}$. (b) The section at $E = -99\,\text{cm}^{-1}$ shows the development of a stochastic layer that is widest near the unstable fixed point, as in Fig. 5.7a. (c) At $E = -66\,\text{cm}^{-1}$, the layer is broad, but a sizable regular region still exists at small (ρ, p_ρ). (d) At $E = -22\,\text{cm}^{-1}$, most of the section exhibits chaotic trajectories. From Hasegawa et al. (1989). Copyright 1989 Progress of Theoretical Physics.

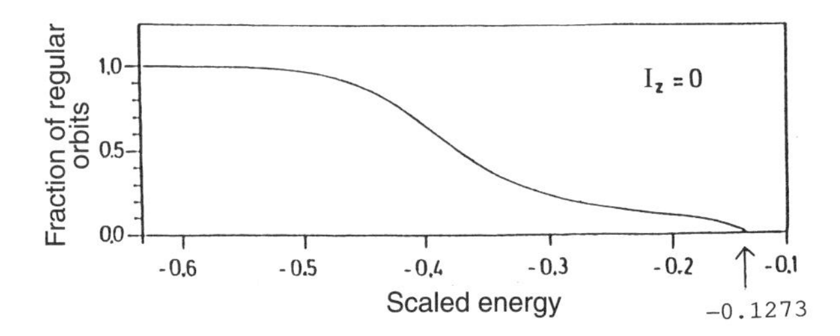

Figure 7.5. Fraction of regular trajectories in Poincaré sections that are similar to those shown in Fig. 7.4, as a function of dimensionless scaled energy E′. At E′ = −0.1273, all tori are believed to be absent. This figure can be compared with similar results for the Hénon–Heiles problem, Fig. 5.13a. From Friedrich and Wintgen (1989). Copyright 1989, with permission from Elsevier Science. Also see Schweizer et al. (1993).

as the scaled energy E′ is increased through the threshold value $E'_c = -0.1275$ for total classical chaos. Compare Fig. 7.5 with Fig. 5.13a for the Hénon–Heiles system.

Quantum mechanical studies of the diamagnetic Kepler problem have been extensive, and much of the work parallels that of the Hénon–Heiles and other 2-d coupled-oscillator systems. We will only make some summary remarks, and then turn to experimental topics related to a discussion of the applicable path integral theory.

Clearly the diamagnetic Kepler problem will have rotational and librational KAM regular quantum eigenstates, analogous to the Q^I and Q^{II} states of the Hénon–Heiles system. The energies of these eigenstates can be obtained by EBK quantization (Delos et al., 1983a). The WKB approximation can also be used; see Richards (1983) for a comparison with precise numerical results.

There will also be irregular quantum eigenstates, analogous to the Hénon–Heiles Type N states. At intermediate values of the scaled energy E′, nonlinear-resonance quantum states coexist with the irregular quantum states and exhibit a clustering analogous to those shown in Fig. 5.27 (Delande and Gay, 1987). Again the projections of true eigenstates onto eigenstates of rotational or librational partial Hamiltonians reveal the transition from primarily regular states in the nearly integrable regions to primarily irregular (Type N) states in the "totally chaotic" regime (Delande and Gay, 1987). Various approaches have been employed for the necessary basis-set expansion numerical calculations (Hasegawa et al., 1989; Ruder et al., 1994). The observed scarring of irregular wavefunctions due to unstable periodic orbits is what we expect from reading Section 5.4.

The diamagnetic Kepler problem has been a testing ground for the complex coordinate rotation technique used to include the continuum (positive energy) states in numerical eigenfunction calculations. In the complex coordinate rotation numerical method—see Section B.2c—the position and momentum of the electron are made complex, according to $\mathbf{r} \to \mathbf{r}e^{i\theta}$ and $\mathbf{p} \to \mathbf{p}e^{-i\theta}$, where the real parameter θ is called the rotation angle. These rotated phase-space variables are substituted into the Hamiltonian (7.2) to obtain a complex Hamiltonian H(θ). The bound spectrum of

H(θ) coincides with the spectrum of H below the ionization threshold. Above this threshold, the complex eigenvalues of H(θ) coincide with the resonances of H that arise from the bound–continuum *and* continuum–continuum close coupling of the selected basis of states; see Section B.2c. This approach is currently the most effective for the continuum. For application of the complex rotation technique to the diamagnetic Kepler problem, see Buchleitner et al. (1994).

Recently Kravchenko et al. (1996) devised a special two-variable series formulation of the diamagnetic Kepler stationary-state problem. The series provides an algorithm for energy eigenstates and eigenvalues that is claimed to be highly accurate for arbitrary values of the magnetic field strength.

7.1b. Spectra of Excited Hydrogen Atoms in Magnetic Fields

One can use a combination of thermal atomic beam and sequential pulsed-laser excitation techniques to resonantly populate excited states of hydrogen atoms in the

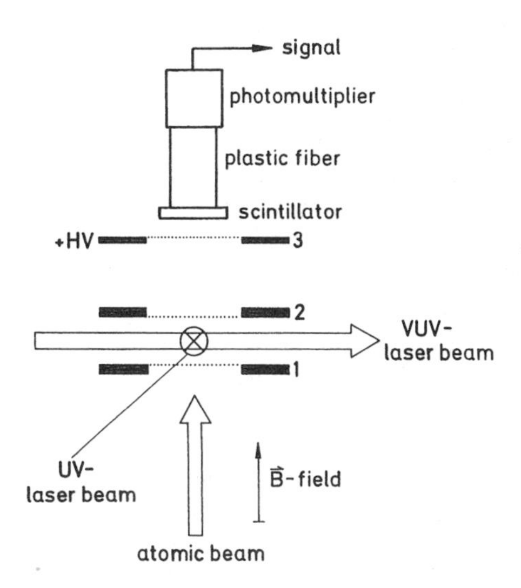

Figure 7.6. Schematic of an apparatus for pulsed-laser electronic excitation spectroscopy of the hydrogen atom in a static magnetic field B. Two laser beams and an atomic beam intersect at the center of a superconducting magnet. The range of operating fields B is $2T \leq B \leq 6T$. The beam intersection point is located between two flat parallel fine-mesh grid electrodes (labeled 1 and 2), which shield electric fields from the excitation region. Laser-excited atoms in the atomic beam pass into the region between electrodes 2 and 3, where a strong electric field ionizes all highly excited states of the atom. The accelerated electrons are converted into an electrical pulse by the scintillator/photomultiplier detection system. From Main et al. (1994). Copyright 1994 by the American Physical Society.

presence of a magnetic field. Holle et al. (1987) and Main et al. (1994) used this approach to experimentally investigate the energy level spectrum for a strong (6 tesla) magnetic field. The polarizations of the two laser beams were adjusted so that only $m = 0$ states were populated. Thus the $L_z = 0$ system was studied. The product atoms drifted out of the laser excitation region into an adjacent electric field region that was located between electrodes 2 and 3 in the schematic of the apparatus shown in Fig. 7.6. The electric field was strong enough to ionize the highly excited product atoms. The electrons so-produced were accelerated into a scintillator/photomultiplier detector and either were individually counted or were collected as a current pulse for each laser pulse.

Experimental photoabsorption spectra have been obtained for various levels of spectroscopic wavelength resolution, with low-resolution results naturally coming first. Figure 7.7b shows a high-resolution spectrum ($\Delta E \sim 0.07\,\mathrm{cm}^{-1}$), which at first appears uninformative. On the other hand, Fig. 7.7a shows a low-resolution spectrum ($\Delta E \sim 0.3\,\mathrm{cm}^{-1}$), and there is a clear oscillatory pattern extending from below to well above the free-atom ionization limit! These data are mostly in the irregular (classically chaotic) regime of energies for the excited atoms produced in the magnetic field. By Fourier transforming the oscillatory pattern, one produces a time spectrum, and we see a large peak at the value of the period of the periodic orbit I_1 displayed in Fig. 7.3b. Here is our first hint of a connection between experimental spectra and path integral theory—through the common thread of a closed orbit of the diamagnetic Kepler system. Because laser excitation of the atom was efficient only when the atomic electron was near the nucleus, only those electron orbits beginning and ending at the nucleus actually affected the experimental spectrum. (This efficiency arises because radiative processes (photon emission or absorption) are accompanied by acceleration of the radiating bound electron.)

The scaling relations producing Eq. (7.4) can be employed to obtain "universal" experimental data. In what is called *constant-scaled-energy spectroscopy*, data are taken only when the experimental magnetic field B and the laser frequency (or, equivalently, the energy of the atom in the field) are adjusted to keep the scaled energy E' fixed (Holle et al., 1988; Main et al., 1994). Such spectra can be Fourier transformed and converted into closed-orbit "recurrence-strength" spectra for different values of the scaled energy. These spectra are scaled-action spectra rather than time spectra.

An example of such experimental spectra is shown in Fig. 7.8a and should be compared with the numerical predictions shown in Fig. 7.8b. In Fig. 7.8a we see a single series of peaks at low scaled energies, such as the value $E' = -0.30$. The peak having the lowest action has the action of the periodic orbit I_∞ that is directed perpendicular to the magnetic field; see Fig. 7.3b. Still at $E' = -0.3$, the next higher action in Fig. 7.8 corresponds to the action of I_1 in Fig. 7.3b; this peak develops into a fan of peaks with increasing E'. We see that the spectral features associated with both periodic orbits extend far into the totally chaotic energy regime, $E' \geq E'_c = -0.1275$. As we might suspect, these peaks are due to the excitation of scarred irregular states (Section 5.4). The fans of peaks arise from bifurcations of the

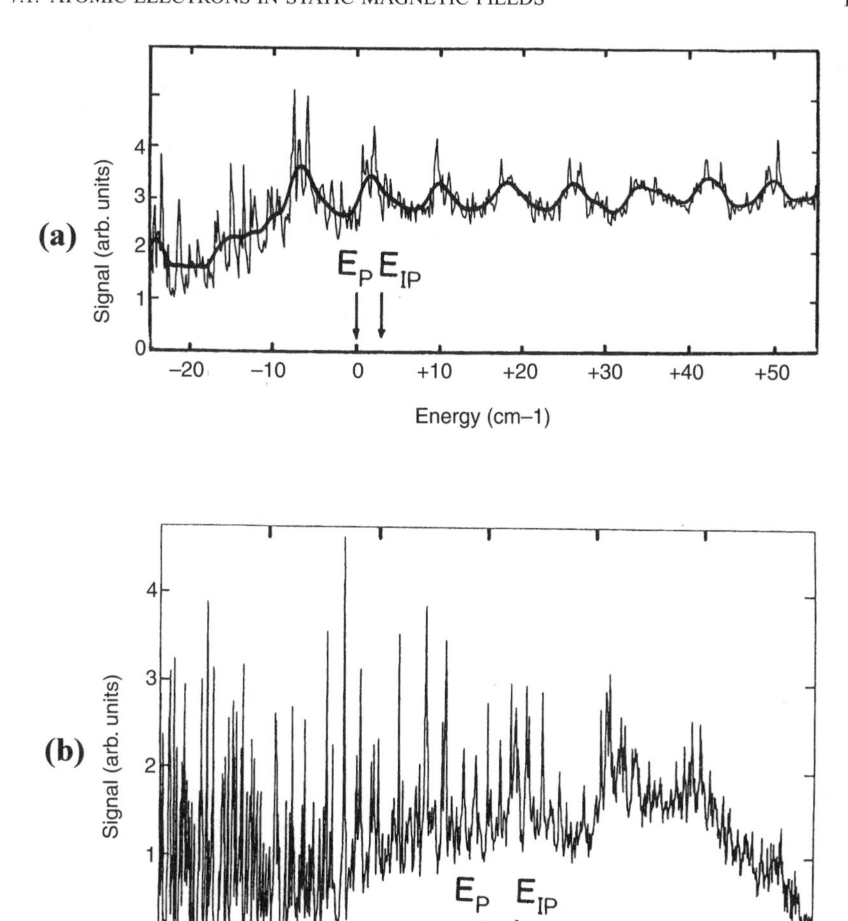

Figure 7.7. Experimental photoexcitation signals obtained from spectroscopic apparatus of the type shown in Fig. 7.6. The range of diamagnetic Kepler energies E includes values both below and above the ionization limit $E = E_p = 0$. The energy E_{IP} is the ionization limit in the absence of the magnetic field. (a) A low-resolution energy scan with resolution $\Delta E = 0.3\,\text{cm}^{-1}$. The magnetic field was 5.96 T, and the second step of the laser photoexcitation was from the $2p_0$ state to high $m = 0$ states. The solid curve is a smoothed version of the data, with the smoothing function FWHM being $2\,\text{cm}^{-1}$. An oscillation is clearly seen that persists above the ionization limit, indicating the existence of an important unstable periodic orbit. From Holle et al. (1986). Copyright 1986 by the American Physical Society. (b) A higher resolution scan—with $\Delta E = 0.07\,\text{cm}^{-1}$—does not so clearly reveal the oscillation seen in panel (a). Panel (b) from Main et al. (1987). Copyright 1987 Springer-Verlag.

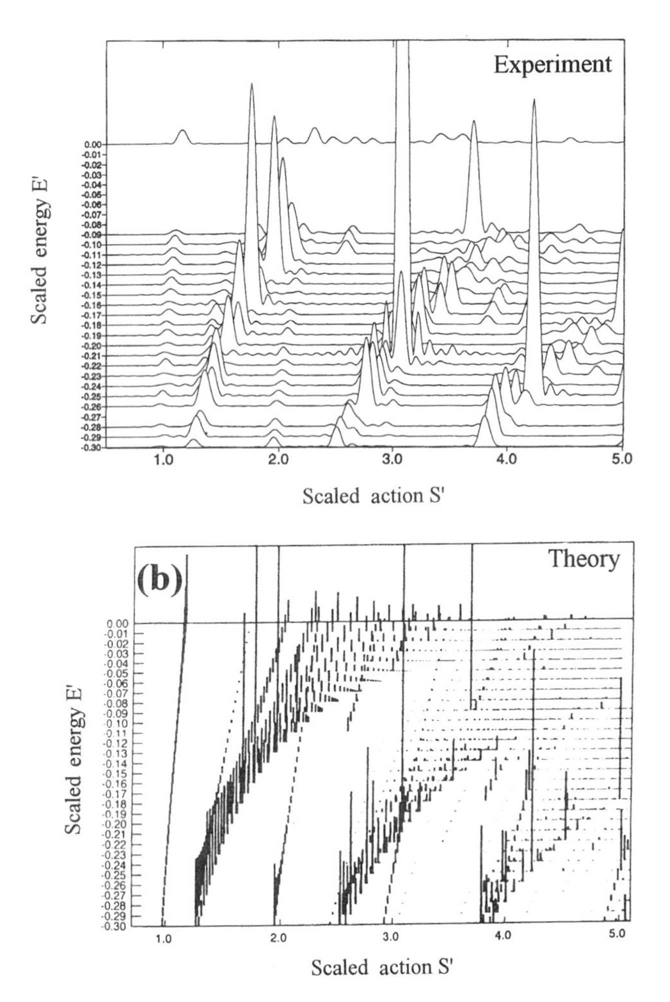

Figure 7.8. Results of high-resolution constant-scaled-energy spectroscopy for the diamagnetic Kepler system. Photoabsorption probability is plotted above the scaled-energy/scaled-action plane (E', S'). Points of maximal probability can be seen that lie on a number of curves (not drawn). Each curve is produced by a periodic orbit in the system, and periodic orbit bifurcations occur where curves split. (a) Experimental data and (b) quantum numerical results. From Main et al. (1994). Copyright 1994 by the American Physical Society.

scarring periodic orbits to produce new periodic orbits. These bifurcations produced the structure in the Lyapunov exponent, shown in Fig. 5.6b. The periodic orbit I_1 sequentially bifurcates to produce new orbits $I_2, I_3, I_4, \ldots$. The first six of the orbits I_n, $n \geq 2$ are shown in cylindrical coordinates in Fig. 7.9. It is remarkable that periodic-orbit bifurcations appear in atomic spectra.

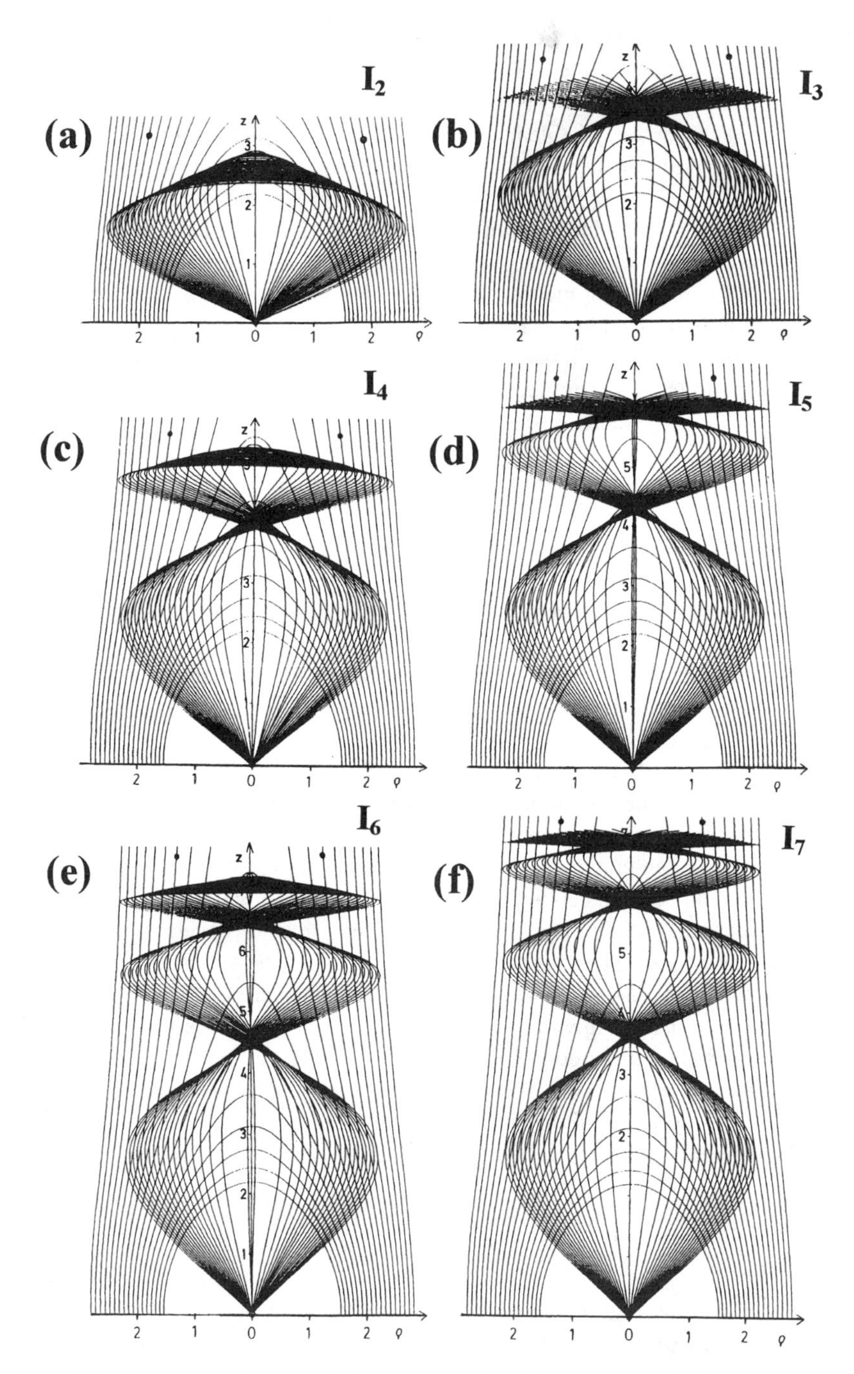

Figure 7.9. Families of diamagnetic Kepler periodic orbits in the (ρ', z') plane. Members of each family have different electron energies E. Panels (a) through (f) show the families of orbits I_n, $n = 2$ to 7, that are produced by bifurcations (Fig. 5.6b) of the orbit I_∞ parallel to the magnetic field (Fig. 7.3b). From Main et al. (1987). Copyright 1987 Springer-Verlag.

7.2. QUANTUM THEORY FOR IRREGULAR SPECTRA AND EIGENSTATES (ADVANCED TOPIC)

The set of all quantum problems with time-independent Hamiltonians can be divided, on the basis of the classical dynamics, into four classes: integrable, near integrable (KAM), strongly mixed regular and irregular, and irregular (classically chaotic). Section 5.5 provided us with an overview of a formal theory for all four types of problems. For the integrable class, from Eq. (4.10), we know that semiclassical energy eigenfunctions have the classical support of a set of nested tori. In this present section we are concerned primarily with eigenfunctions for irregular problems, where classical chaos prevails throughout phase space, except for unstable periodic orbits and their associated stable and unstable manifolds. We've seen much evidence (Section 5.4) that this residual structure in phase space plays an important role for eigenstates, "scarring" their probability distributions.

The eigenstates of integrable systems—Eq. (4.10)—are localized and are supported by a volume of phase space having two or more dimensions. Is there such classical support for the irregular eigenstates? The minimal amount of phase space required would seem to be one periodic orbit, perhaps plus some trajectories in its neighborhood. Identical energies connect the periodic orbit with an energy eigenstate.

We recall our discussions of wavepacket evolution for integrable systems, where the packets are guided by central trajectories and given further support by neighboring trajectories; in particular, see the description of semiclassical propagation for the Coulomb potential (Section 2.1c). In general, the propagation of nonstationary wavefunctions is not believed to require guidance on invariant tori. Nevertheless, from what is known, it is expected that the details of an irregular quantum system's evolution will remain sensitive to the classical dynamics.

We now extend the semiclassical path integral theory introduced in Section 6.3 to classically chaotic quantum systems. A further recent development is also mentioned: the explicit extension of path integral theory to near-integrable (KAM) systems with their classical resonances and stochastic layers (Section 5.1). The remaining case of the strongly mixed quantum system would also seem treatable, but work on this remains to be done.

Some important details of the formal path integral theory are quite complicated and require lengthy elaboration, but we omit these details here. The reader will be referred to various references for additional information.

7.2a. The Semiclassical Propagator

Equations (6.20) and (6.21) still apply for problems involving more than one spatial dimension. From Eq. (6.22) onward, however, we need extensions of the equations in Sections 6.3a and 6.3b. These extensions support very general formulas for the full semiclassical propagator $K(\mathbf{q}', t; \mathbf{q}, t_0 \equiv 0)$, formulas that have the form of Eq. (6.32). For the mathematical development of these extensions, see Schulman (1981) and Bogomolny (1988). In these extensions the wavefunction is taken to propagate from an initial state and is to be supported classically by some time-dependent surface in phase space called a *Lagrangian surface*. The initial state $\langle \mathbf{q}'|\mathbf{q}\rangle = \delta(\mathbf{q}' - \mathbf{q})$ corresponds classically to a uniform distribution in momentum $\mathbf{p}$ along the N-plane $\mathbf{q}' = \mathbf{q}$. This plane then evolves as the Lagrangian surface determined by

$$\mathbf{p} = \frac{\partial S}{\partial \mathbf{q}}(\mathbf{q}', \mathbf{q}, t). \tag{7.6}$$

As time evolves, this Lagrangian surface usually develops caustics, a generalization of the ideas in Section 1.4. Generalizing our discussion of EBK quantization (Section 4.2), the presence of caustics on the Lagrangian surface leads to branching of the surface and to phase jumps $\mu_j'(\pi/2)$ for the wavefunctions corresponding to each of the branches. (The *Morse index* μ_j' for a $\mathbf{q}$-orbit indicates the number of times that the orbit crosses a caustic. This index can be related to the Maslov index μ (Arnold, 1967).)

Away from the caustics, the generalized propagator becomes

$$\langle \mathbf{q}'|K(t)|\mathbf{q}\rangle = (2\pi i\hbar)^{-1/2} \sum_j \left|\det \frac{\partial^2 S_j}{\partial \mathbf{q}'\, \partial \mathbf{q}}\right| \exp\left(\frac{i}{\hbar} S_j(\mathbf{q}', \mathbf{q}, t) - i\mu_j' \frac{\pi}{2}\right). \tag{7.7}$$

Here S_j is a solution of the time-dependent Hamilton–Jacobi equation (3.11), and the indicated sum is over the branches of the Lagrangian surface. Equation (7.7) is an extended version of Eq. (6.34).

Next, we generalize Eq. (6.35) for the Green's function, obtaining

$$G(\mathbf{q}', \mathbf{q}; E) = \frac{1}{i\hbar} \int dt\, K(\mathbf{q}', \mathbf{q}; t) \exp\left(i \frac{Et}{\hbar}\right). \tag{7.8}$$

This quantity satisfies a partial differential equation that generalizes Eq. (6.36) and has a "point source" proportional to $\delta(\mathbf{q}' - \mathbf{q})$. One must now evaluate the integral in Eq. (7.8) using the stationary-phase approximation. (For the complex mathematics here, see Gutzwiller (1971) and Berry and Mount (1972).) The result of the stationary-phase approximation is the sum $G = G_0 + G_{OSC}$, where G_0 is a smoothly varying part arising from very short paths, and where the interesting part G_{OSC} arises from classical paths p connecting $\mathbf{q}'$ and $\mathbf{q}$ at a fixed energy E, irrespective of the time it takes to make the connection. For an N-dimensional coordinate space, we have

$$G_{OSC}(\mathbf{q}', \mathbf{q}; E) = \frac{1}{i\hbar (2\pi i\hbar)^{(N-1)/2}} \sum_p |D_p|^{1/2} \exp\left(i \frac{S_p(E)}{\hbar} - i \frac{\pi \mu_p}{2}\right), \tag{7.9}$$

where

$$D_p \equiv \det \begin{bmatrix} \dfrac{\partial^2 S_p}{\partial \mathbf{q}'\, \partial \mathbf{q}} & \dfrac{\partial^2 S_p}{\partial E\, \partial \mathbf{q}} \\[2ex] \dfrac{\partial^2 S_p}{\partial \mathbf{q}'\, \partial E} & \dfrac{\partial^2 S_p}{\partial E^2} \end{bmatrix}. \tag{7.10}$$

7.2b. Equations for Observable Quantities

By knowing the Green's function (7.8), we can calculate any property of our system. The general theory that uses Green's functions to solve partial differential equations will offer us the necessary connections to the observable properties. (For the mathematics, we refer the reader to standard texts on mathematical physics, such as Morse and Feshbach (1953).) We are interested here in finding the energy-level spectrum ρ_E, also called the *energy density of states*

$$\rho_E = -\frac{1}{\pi}\int \mathrm{Im}G\,(\mathbf{q}', \mathbf{q}; E)\delta(\mathbf{q}' - \mathbf{q})\, d\mathbf{q}'\, d\mathbf{q}, \tag{7.11}$$

or

$$\rho_E = -\frac{1}{\pi}\int d\mathbf{q}\ \mathrm{Im}\ G(\mathbf{q}, \mathbf{q}; E) \equiv -\frac{1}{\pi}\mathrm{Im}(\mathrm{Trace}\ G). \tag{7.12}$$

We can formally show that both Eqs. (7.11) and (7.12) work, if we use an expansion in the complete set of exact eigenfunctions

$$G(\mathbf{q}', \mathbf{q}; E) = \sum_n \frac{\psi_n(\mathbf{q}')\psi_n(\mathbf{q})}{(E + i\varepsilon) - E_n}.$$

We insert this expression into Eq. (7.11) and do the integrals to obtain

$$\rho_E = \sum_n \frac{\varepsilon/\pi}{(E - E_n)^2 + \varepsilon^2} \underset{\varepsilon\to 0}{\rightarrow} \sum_n \delta(E - E_n).$$

If we want to find the expectation values of a relevant quantum operator A that corresponds to some observable, then we calculate the diagonal matrix elements $\langle n|A|n\rangle$ using

$$\begin{aligned}\rho_A &\equiv \sum_n \langle n|A|n\rangle\delta(E - E_n) = -\frac{1}{\pi}\int d\mathbf{q}\ G(\mathbf{q}, \mathbf{q}; E)A(\mathbf{q})\\ &\equiv -\frac{1}{\pi}\mathrm{Im}(\mathrm{Trace}\ GA),\end{aligned} \tag{7.13}$$

assuming that A does not depend on the momentum $\mathbf{p}$. (If A does depend on $\mathbf{p}$, we would need to use a phase-space representation (Eckhardt et al., 1992).)

Lastly, if we want the spatial probability distribution of our quantum system, then we would use

$$|\psi(\mathbf{q})|^2 = \frac{\mathrm{Im}\ G(\mathbf{q}, \mathbf{q}; E)}{\int d\mathbf{q}'\ \mathrm{Im}G(\mathbf{q}', \mathbf{q}'; E)} \equiv \frac{\mathrm{Im}\ G(\mathbf{q}, \mathbf{q}; E)}{\mathrm{Im}(\mathrm{Trace}\ G)}. \tag{7.14}$$

All these quantities—Eqs. (7.12), (7.13), and (7.14)—involve Green's function evaluated at $\mathbf{q}' = \mathbf{q}$, $G(\mathbf{q}, \mathbf{q}; E)$. This special Green's function connects a final point $\mathbf{q}'$ to a source point $\mathbf{q}$, with the two points being the same.

Thus $G(\mathbf{q}, \mathbf{q}; E)$ will be the sum of two terms. There is a singular term representing the wave amplitude that exists at $\mathbf{q}$ because $\mathbf{q}$ is the source point;

that singularity, however, turns out to be in the real part of $G(\mathbf{q}, \mathbf{q}; E)$, and the imaginary part we need is finite. The integral over $\mathbf{q}$ of the imaginary part turns out to be $d\Omega/dE$, where Ω is the volume in phase space bounded by the energy surface $H(\mathbf{p}, \mathbf{q}) = E$. The second term in $G(\mathbf{q}, \mathbf{q}; E)$ contains semiclassical terms (among others) representing waves that propagate outward from $\mathbf{q}$, travel around in position space for awhile, and (for bounded systems) eventually return to pass through the final point, which is at the same place as the source point $\mathbf{q}$. Thus there is a definite possibility for closed classical orbits to appear in a semiclassical approximation for $G(\mathbf{q}, \mathbf{q}; E)$ and in the derived quantities (7.12), (7.13), and (7.14).

7.2c. Calculation of Semiclassical Observable Quantities

We now need to insert the semiclassical Green's function (7.9) into formulas (7.12) and (7.13) and to do the integrals using the stationary-phase approximation (SPA); see Section 6.3b. (We again omit the mathematics, as there are many details.) Recommended is a reading of the original papers by Gutzwiller (1967, 1969, 1970, 1971). A crucial assumption in Gutzwiller's work is that the periodic orbits are isolated from one another, which is true for unstable periodic orbits. The phase will be stationary if the initial and final momenta coincide, which is the condition for the trajectory to be periodic. For every closed path, it is important to separately introduce a coordinate system with q_1 along the path and with $q_2, q_3, \ldots, q_N$ perpendicular to the path. We put off the integral over q_1 until last. Since the SPA uses deviations from trajectories up to second order, only the action S_p along the periodic orbit and a stability matrix M_p around that orbit will appear. The eigenvalues of M_p define the type of fixed point for the periodic orbit in a Poincaré section, see Section 3.1a. For the term arising from a single traversal of a primitive periodic orbit p, Gutzwiller found that

$$\begin{aligned} &\frac{1}{(2\pi i\hbar)^{(N-1)/2}} \int dq_2 \cdots dq_N |D_{S_p}|^{1/2} \exp\left(i\frac{S_p(\mathbf{q})}{\hbar} - i\frac{\mu'_p \pi}{2} \right) \\ &\quad = \frac{1}{|dq_1/dt|} \frac{\exp[(iS_p/\hbar) - (i\mu_p\pi/2)]}{|\det(M_p - 1)|^{1/2}}. \end{aligned} \tag{7.15}$$

The quantities S_p, M_p, and the Maslov index μ_p are independent of the position q_1 along the path. Thus the remaining integral of $dq_1/(dq_1/dt) = dt$ gives the period T_p of the periodic orbit. Realizing that we must sum over multiple traversals j of each orbit, and then putting everything together, we obtain Gutzwiller's *trace formula* for the oscillating term in ρ_E, Eq. (7.12):

$$\rho_{E,OSC} = \operatorname{Im} \frac{1}{\pi\hbar} \sum_p \sum_j \frac{T_p}{|\det(M_p^j - 1)|^{1/2}} \exp\left(i\frac{S_p}{\hbar} - i\frac{\mu_p \pi}{2} \right) j, \tag{7.16}$$

an expression that depends on classical quantities and on $\hbar$.

If an observable $A(\mathbf{q})$ varies little over the important distances in the integrals in Eq. (7.13)—that is, if A varies slowly on the scale of a wavelength—then $A(\mathbf{q}) \approx A(\mathbf{q}_p)$ along the orbit for the region of interest (Eckhardt, 1993). The integral over q_1 then becomes

$$A_p = \int_0^{T_p} dt\, A[\mathbf{q}_p(t), \mathbf{p}_p(t)],$$

and ρ_A is given by Eq. (7.16) with T_p replaced by A_p.

Depending on the type of periodic orbit, the *spectral determinant* in Eq. (7.16) is given by

$$\tfrac{1}{2}[\det(M^j - 1)]^{1/2} = \begin{cases} -i\sinh(j\lambda/2), & \text{unstable orbit,} \\ \sin(j\pi w), & \text{stable orbit,} \end{cases} \tag{7.17}$$

where λ is the Lyapunov exponent and w is the orbit winding number; see the papers by Gutzwiller and also see Tabor (1983). Thus in the unstable orbit case, the amplitudes of the component waves in Eq. (7.16) decrease exponentially with j, whereas these amplitudes oscillate in j for isolated stable orbits. In the unstable case, one can expect the sum (7.16) to converge. This sum yields nearly Lorentzian contributions from each orbit to the level density, where the contributions are centered on energies E_n given by $S_p(E_n) = 2\pi\hbar(n + \mu_p/4)$; see Gutzwiller's papers. Rather than producing a sequence (ladder) of quantum states, a single unstable periodic orbit therefore manifests itself as a modulation of the level density, in accord with observations made in Section 7.1. For isolated stable periodic orbits, an approximation to Eq. (7.16) can be summed analytically to produce delta functions in the density of states at energies E_{nk}, satisfying $S_p(E_{nk}) = 2\pi\hbar[n + (k + \frac{1}{2})w_p + \mu_p/4]$ (Miller, 1975). The quantum number n counts the number of nodes along the periodic orbit, while k gives the degree of excitation of the normal-mode frequency of harmonic perturbations about (transverse to) the orbit. Thus our semiclassical path integral theory predicts ladders of regularly spaced quantum energy levels associated with each stable periodic orbit. This prediction is in accord with numerical results, for instance, for the Hénon–Heiles Q^{I}, Q^{II} states (Section 5.3c).

In the case of a classically integrable system, the periodic orbits are not isolated but are densely distributed in phase space. Gutzwiller's calculations do *not* apply in this situation. Now the periodic orbits are on tori and can be labeled by integers $\mathbf{M} = (jr, js)$, where j is the number of repetitions and the winding number is $w = r/s$. For the contribution $\rho_{\mathbf{M}}$ to the equation obtained by inserting Eq. (7.9) into Eq. (7.11), the action integral is $S_M^0 = 2\pi\mathbf{I}_M \cdot \mathbf{M}$ for all tori of the family of nested tori determined by $\mathbf{M}$. Here $\mathbf{I}_M$ are the action variables for these tori. If we choose combinations of original angle variables such that one new angle variable θ_1 on the torus is constant along the periodic trajectory, then the coordinate integral over $\mathbf{dq}$ can be evaluated to get an analytic expression for ρ_M (Berry and Tabor, 1976, 1977b). In this case of integrable systems, semiclassical path integral theory is formally equivalent to EBK theory, that is, to the torus quantization of Section 4.2. Yet in practice, since $\mathbf{M}$ is specified by a single repetition number j and since j must be summed up to infinity, the sum over all ρ_M is hard to compute convincingly—see Gao and Delos (1994), for example. This difficulty in summation is called the problem of long orbits (Keating and Berry, 1987). Thus for integrable systems, EBK theory is often a more useful approach than periodic orbit theory.

For the near-integrable case with perturbation $\varepsilon H'$, Ozorio de Almeida (1988) has shown that ρ_M should be changed by including a damping factor. This factor originates from a dephasing of the orbit contributions from the family $\mathbf{M}$ and has the form

$$C_M(E) = \frac{1}{2\pi}\int_0^{2\pi} d\theta_1 \; T_M \exp\left(i\frac{\delta S_M(\theta_1)}{\hbar} \right).$$

Here $\delta S_M(\mathbf{q}) = \oint \varepsilon H' \, dt$ is the first-order correction to the action taken on the unperturbed trajectory that starts and ends at $\mathbf{q}$. Recently, Tomsovic et al. (1995) showed that $\delta S_M(\theta_1) \approx \Delta S(\varepsilon) \cos \xi$, where $\Delta S(\varepsilon)$ is the width of the resonance zone associated with the periodic orbit labeled by $\mathbf{M}$ (Section 5.6), and ξ is a new angle variable approximately related to θ_1 by

$$\theta_1 \approx \xi - a(\varepsilon) \sin \xi.$$

The quantity $a(\varepsilon) = (K-1)/(K+1)$, with $K = [-\det(M_U - 1)/\det(M_S - 1)]^{1/2}$. The quantities M_U and M_S are the stability matrices for the unstable and stable periodic orbits associated with the resonance. The integral above for $C_M(E)$ can then be evaluated analytically. For a 2-d quartic-oscillator system, the results of the periodic orbit theory for several resonance contributions to the level density ρ_E agree with precise quantum numerical results to within a few percent or better. Thus it appears that semiclassical path integral theory could now usefully be applied to near-integrable (KAM) systems, if the problem of long orbits can be resolved. (For further developments concerning resonances and semiclassical trace formulas, see Ullmo et al. (1996).)

7.2d. Diamagnetic Kepler Spectra from Closed-Orbit Theory

In applying our semiclassical path integral theory to the diamagnetic Kepler problem (Section 7.1), we first rewrite $\rho_{E,OSC}$, of Eq. (7.16) in terms of the scaled quantities that produced Hamiltonian (7.4). This reformulation is also possible for the expression for the laser light absorption rate or ion signal rate that is measured in experiments; recall the data in Figs. 7.7 and 7.8a. (Details are described in Main et al. (1994) and references therein.) For both these rates an appropriately scaled observable has an oscillating part that can be written in the form

$$\sum_p \sum_j a_p^j(E') \cos[j(\gamma^{-1/3} S'_p(E') - \alpha_p(E')]. \tag{7.18}$$

Here $S'_p(E')$ is a scaled action, $a_p^j(E')$ is an overall amplitude, and $\alpha_p(E')$ is a sum of extra phases. When the scaled energy E' is held fixed, the observables depend on $\gamma^{-1/3}$ only through the linear term in the phase. Thus expression (7.18) is a sum of sinusoidally oscillating functions of $\gamma^{-1/3}$, and its Fourier transform will be a function of S' that exhibits delta function peaks at the values of $jS'_p(E')$. In practice, the sum (7.18) is truncated, either of necessity in the case of numerical evaluations, or because of the finite range and resolution of the experimental data. The Fourier transform will then be smoothed, possessing maxima with finite widths $\Delta S'$.

To begin applying orbit theory, we evaluate a truncated expression for ρ_E once it has been written in the form of Eq. (7.18). Wintgen (1988) carried out this procedure, using two different truncations of the sums over primitive orbits p and orbit repetitions j. First, he included only the two simplest orbits, I_1 and C shown in Fig. 7.3b, with up to one full traversal. Second, he included 13 primitive orbits and all repetitions that give full scaled actions $S' \leq 3$. Figures 7.10a and 7.10b show the results for $\rho_{E'}(\gamma^{-1/3})$, with $E' = -0.2$, compared with appropriately smoothed results obtained from precise basis-set expansion calculations. Over the expected range in $\gamma^{-1/3}$, we see that the smoothed oscillating parts of the energy density of states agree quite well. At low $\gamma^{-1/3}$, the resolution is sharp enough to individually resolve the first nine quantum eigenenergies, as indicated by vertical lines in the figure. The agreement is within a few percent of the local mean level spacing, even down to the ground state! Not bad for just 13 orbits and, at most, three repetitions.

As a check on the applicability of periodic orbit theory, we should be able to extract classical values for the actions S'_p of the periodic orbits, from precise basis-set expansion numerical results for portions of the energy spectrum at fixed E'. A spectrum is Fourier-transformed with respect to $\gamma^{-1/3}$ to obtain the values of S'_p. Wintgen (1988) carried out this procedure and found agreement with classical values of S'_p usually within 1 %, both a little above and a little below the threshold for total chaos, that is, for $E' = -0.2$ and $E' = -0.1$.

What about the experimental photoabsorption signals? The thick curve in Fig. 7.11 shows the dependence of the absorption rate on $2\pi\gamma^{-1/3}$, over a range corresponding to $2.7T \leq B \leq 6.0T$ and for $E' = -0.23$. Such a curve is called a *recurrence spectrum*. The fine curve is the result of closed-orbit theory, calculated by coherently summing the contributions of about 70 closed orbits having scaled actions ≤ 5. There are noticeable differences between the two curves, but the agreement is reasonably good. We recall that the photoabsorption signal is large only for orbits that begin and end at the atomic nucleus, for the same reason as the wavepacket signals discussed in Section 2.1. Thus orbits need only be closed to contribute to the orbit sum. Most of these closed orbits, however, are portions of periodic orbits.

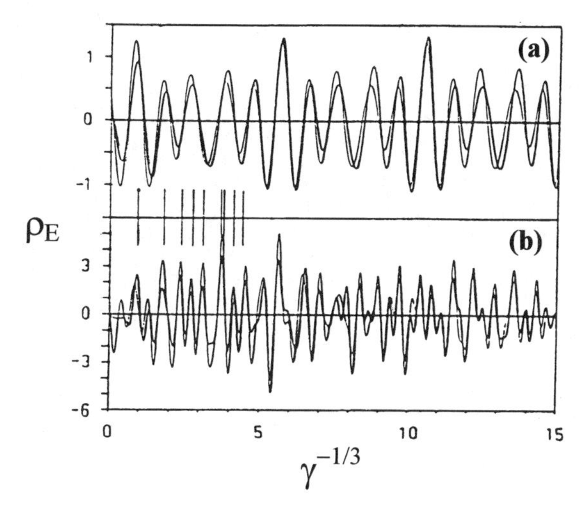

Figure 7.10. Comparison of finite-resolution diamagnetic Kepler energy-density-of-states spectra $\rho_E(\gamma^{-1/3})$ obtained from precise basis-set expansion numerical calculations (thick curves) and from periodic orbit theory (thin curves). The scaled magnetic field γ is introduced in Eq. (7.2). Shown are the appropriately smoothed fluctuating parts of the spectra. (a) Only the two simplest periodic orbits—I_1 and C of Fig. 7.3b—are included in the periodic-orbit theory summation. At low resolution these orbits make the dominant contributions to the smoothed $\rho_E(\gamma^{-1/3})$. (b) At higher resolution (narrower smoothing function FWHM), all 19 contributions from the 13 primitive orbits with scaled actions $S' = \gamma^{1/3}S < 3$ are included. The vertical lines in the center of the figure show the precise spectra for the first few energy eigenstates, which are reproduced by periodic orbit theory. From Wintgen (1988). Copyright 1988 by the American Physical Society. See also Friedrich and Wintgen (1989).

We should remark that high-resolution spectroscopy of the Rydberg lithium atom in strong electric fields has resolved recurrence spectra up to $S' \approx 45$, revealing peaks identified with orbit repetitions j as high as 110 (Courtney et al., 1995).

7.2e. Irregular Wavepacket Evolution Using the Semiclassical Propagator

Recent studies related to regular wavepacket evolution in phase space are quite interesting (Provost and Brumer, 1995; Takatsuka and lnoue, 1997). For instance, they demonstrate that irregular packet evolution provides sufficient information in the time domain for the Fourier transform to accurately determine the phase-space probability density of a *single* scarred irregular quantum energy eigenstate (Provost and Brumer, 1995). To fully appreciate this, we first need to extend path integral theory into a phase-space representation (Schulman, 1981, Chapter 31). Next, we need to know that a phase-space coherent-state representation of the semiclassical propagator K_{SC} properly captures the effects of caustics; in this representation, the singularities at the zeros of the spectral determinant are avoided (Klauder, 1987a).

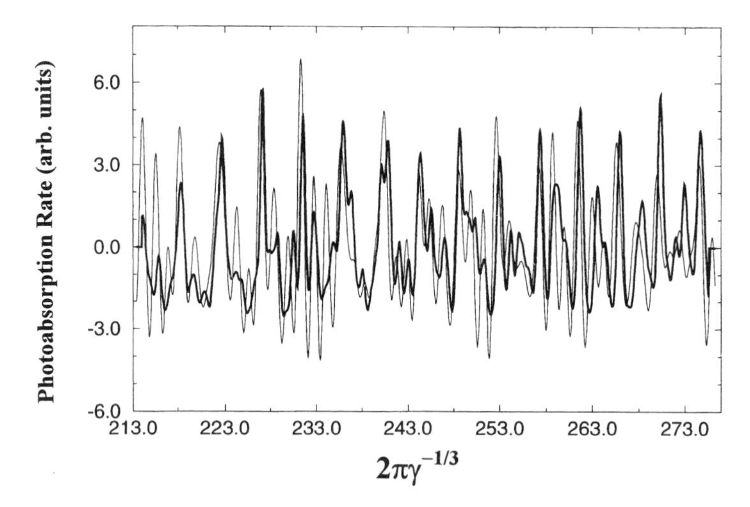

Figure 7.11. Comparison of experimental finite-resolution recurrence spectroscopy results with predictions of periodic orbit theory. The photoabsorption rate at a constant scaled energy E′ is plotted versus the parameter $2\pi\gamma^{-1/3}$, similar to Fig. 7.10. The thick line is experimental, while the thin line is a theoretical spectrum calculated by coherently summing the contributions of about 70 closed orbits whose scaled actions are $S' < 5$. From Main et al. (1994). Copyright 1994 by the American Physical Society.

Then we can follow Provost and Brumer (1995) and let the wavepacket $|\Psi\rangle$ have the coordinate representation $\langle \mathbf{q}|\Psi\rangle$ given by

$$\langle \mathbf{q}|\Psi\rangle = \left(\frac{1}{\pi\sigma^2}\right)^{1/2} \exp\left[-\frac{(\mathbf{q}-\mathbf{q}_0)^2}{2\sigma^2} + \frac{i}{\hbar}\mathbf{p}_0 \cdot \left(\mathbf{q} - \frac{\mathbf{q}_0}{2}\right)\right],$$

with $\sigma = \frac{1}{2}$. We want to evolve $\Psi(\mathbf{q}_i, t) = \langle \Psi|\exp[-iH(-t/\hbar]|\mathbf{q}_i\rangle^*$. Consider the expression

$$\langle \Psi|\exp\left(-i\frac{Ht}{\hbar}\right)|\mathbf{q}_i\rangle = \int d\mathbf{p}_f \langle \Psi|\mathbf{p}_f\rangle\langle \mathbf{p}_f|\exp\left(-i\frac{Ht}{\hbar}\right)|\mathbf{q}_i\rangle. \tag{7.19}$$

To obtain a semiclassical expression for Eq. (7.19), we use in the integral an approximate "nonuniform" semiclassical long-time propagator matrix element in the initial-coordinate to final-momentum representation

$$\langle \mathbf{p}_f|\exp\left(-i\frac{Ht}{\hbar}\right)|\mathbf{q}_i\rangle_{SC} = \frac{1}{(2\pi\hbar)^{N/2}}\sum \frac{1}{|\det(D)|^{1/2}}\exp\left(i\frac{\phi}{\hbar} - i\frac{\nu\pi}{2}\right) \tag{7.20}$$

(Klauder, 1987b; Campolieti and Brumer, 1994). The sum is over all classical trajectories that connect $\mathbf{q}_i$ with $\mathbf{p}_f$ in a time t; the sum is indexed by the initial momenta $\mathbf{p}_i$ such that

$$\mathbf{q}(0) = \mathbf{q}_i, \quad \mathbf{p}(0) = \mathbf{p}_i, \quad \mathbf{p}(t) = \mathbf{p}_f.$$

The matrix D in Eq. (7.20) is one of the four stability submatrices in the phase-space monodromy matrix that satisfies Eq. (3.5),

$$\begin{pmatrix} \delta\mathbf{q}(t) \\ \delta\mathbf{p}(t) \end{pmatrix} = \begin{bmatrix} A & B \\ C & D \end{bmatrix} \begin{pmatrix} \delta\mathbf{q}(0) \\ \delta\mathbf{p}(0) \end{pmatrix}, \tag{7.21}$$

and $\phi(\mathbf{q}_i, \mathbf{p}_i, t)$ is a generator of the classical motion

$$\phi(\mathbf{q}_i, \mathbf{p}_i, t) = -\mathbf{p}(t) \cdot \mathbf{q}(t) + \int_0^t [\mathbf{p}(\tau)\dot{\mathbf{q}}(\tau) - H]\, d\tau.$$

The index $\nu(\mathbf{q}_i, \mathbf{p}_i, t)$ in Eq. (7.20) is given by

$$\nu(\mathbf{q}_i, \mathbf{p}_i, t) = \frac{1}{\pi} \lim_{\varepsilon \to 0} \arg[\det(D + i\varepsilon B)]$$

(Kay, 1994).

Propagator matrix elements like Eq. (7.20) are "two-ended" because the propagator operator couples an initial state in some representation to the final state in some representation. Such matrix elements involve singularities at caustics over which the propagations pass; in the case of Eq. (7.20) the caustics occur when $|\det(D)| = 0$. Fortunately, it is possible to work in an *initial-value representation* (IVR), which is more natural for wavepacket evolution (Miller, 1970; Miller and George, 1972; Levit and Smilansky, 1977; Campolieti and Brumer, 1994).

Initial-value representations possess two very helpful properties: (1) $|\det(D)|^{1/2}$ appears in the numerator rather than the denominator of the terms in the propagator matrix element—the element is therefore "uniform" about the caustics, because there are no divergences; and (2) since initial-value representations are "single ended" rather than "two ended," no trajectory root searches are required for the wavepacket propagation. The construction of propagator matrix elements in various initial-value representations is described in considerable detail in Campolieti and Brumer (1994).

Using the coordinate initial-value representation and inserting expression (7.20) into Eq. (7.19) leads to

$$\langle\Psi| \exp\left(-i\frac{Ht}{\hbar}\right)|\mathbf{q}_i\rangle_{SC} = \frac{1}{(2\pi\hbar)^{N/2}} \int d\mathbf{p}_i \langle\Psi|\mathbf{p}(t)\rangle |\det(D)|^{1/2} \exp\left(i\frac{\phi}{\hbar} - i\frac{\nu\pi}{2}\right) \tag{7.22}$$

(Provost and Brumer, 1995). Note that the determinant indeed is in the numerator, *not* in a denominator. No stationary-phase approximation will be applied to Eq. (7.22). Provost and Brumer used Eq. (7.22) to numerically propagate the wavepacket $\langle \mathbf{q}|\Psi\rangle$ for a particle in a smooth generic 2-d potential, the Hamiltonian being

$$H = \tfrac{1}{2}(p_x^2 + p_y^2) + \tfrac{1}{20}x^2 + (y - \tfrac{1}{2}x^2)^2.$$

For this system at $E = 0.4249$, contours of the 42nd even (in x) energy eigenstate show a large degree of scarring by a symmetric librational unstable periodic orbit; see Fig. 7.12a. The real part of the initial packet is shown in position space in Fig. 7.13; it is chosen large in position space so that it is well-localized in momentum

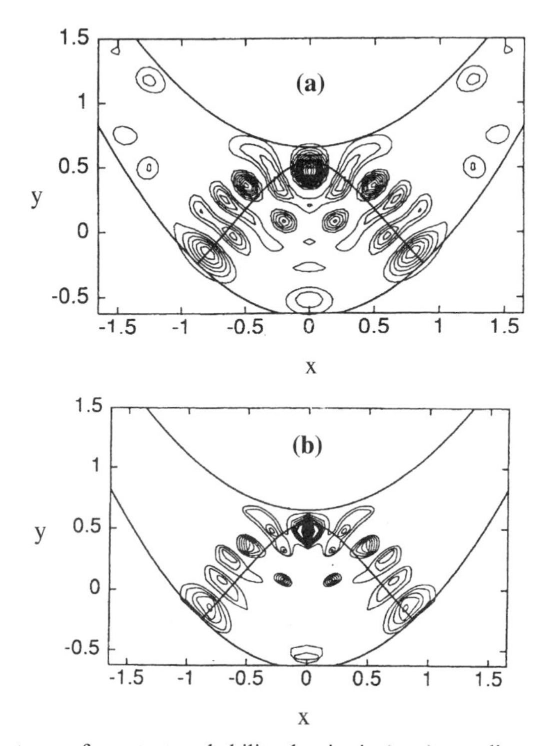

Figure 7.12. Contours of constant probability density in (x, y) coordinate space for quantum states with the Hamiltonian $H = (p_x^2 + p_y^2)/2 + 0.05x^2 + (y - x^2/2)^2$. (a) Shown is the 42nd energy eigenstate that is even in x. The equipotential contour at the energy $E = 0.4249$ of this eigenstate is the pair of U-shaped curves. The eigenstate is strongly scarred by a symmetric librational unstable periodic orbit, also shown. (b) Shown is the probability density distribution produced by semiclassical propagation of the special initial coherent-state (Gaussian) wavepacket that is displayed in Fig. 7.13. The propagation was at $E = 0.4249$ and was taken out to a time equal to 2.5 times the period of the periodic orbit. Most of the structure of the eigenstate in panel (a) is resolved. The agreement with precise quantum wavepacket evolution (not shown) is also very good. From Provost and Brumer (1995). Copyright 1995 by the American Physical Society.

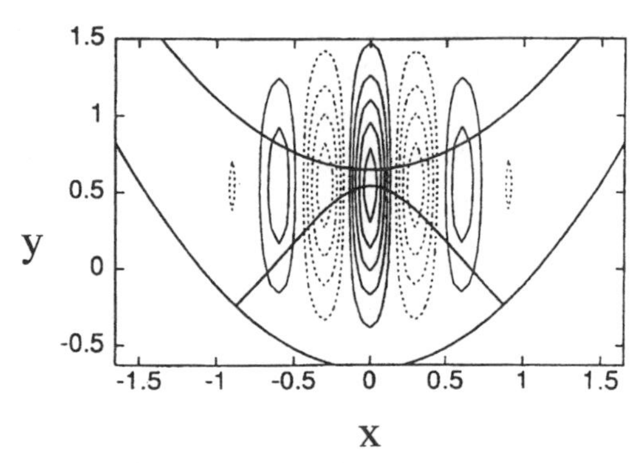

Figure 7.13. The real part of the coherent-state initial wavepacket used in the wavepacket evolution that produced Fig. 7.12b. The solid contours in coordinate space are positive, dashed contours are negative. Since the integral in Eq. (7.22) is over momentum space, this packet is chosen large in coordinate space, as shown; thus it is well-localized in momentum space while "symmetrically centered" on the periodic orbit in coordinate space. From Provost and Brumer (1995). Copyright 1995 by the American Physical Society.

space. The packet center $\mathbf{q}_0, \mathbf{p}_0$ (at $t = 0$) has been placed at the particular point in phase space that corresponds to that starting point in coordinate space on the unstable periodic orbit where the orbit intersects the y-axis. Provost and Brumer (1995) show snapshots of the packet's complicated irregular evolution. When the numerical evolution is stopped after a propagation time of 2.5 to 5 periods τ of the libration, and when the evolution data are Fourier-transformed from the time domain to the energy domain, the result is the state shown in Fig. 7.12b. Most of the true eigenstate's structure has been reproduced. It is important to realize that usual periodic orbit theory in the energy domain, as described in Section 7.2, would not produce the nodal structure seen along the periodic orbit, even when all orbits having periods $\leq 8\tau$ are included. Figure 7.12b demonstrates that semiclassical path integral theory is finding increasing and novel usefulness.

7.3. BALLISTIC ELECTRON TRANSPORT WITHIN WIDE QUANTUM WELLS IN TILTED MAGNETIC FIELDS

7.3a. Biased Semiconducting Heterostructures

Our first example of irregular quantum evolution in solid state nanostructures is quantum charge carriers in a semiconductor quantum well in a magnetic field. We have discussed some general aspects of conducting solid state heterostructures in Section 2.1d. The walls of the quantum well will be determined by edges of potential barriers, barriers through which the charge carriers tunnel into and out of the well. Such tunneling for semiconductor barriers was discussed in Section 6.1b. Now a bias voltage is applied along the spatial direction defining the well and barriers, both to direct the carriers and to control other characteristics of their evolution. We will see that both regular and irregular semiclassical evolutions of the carriers are possible whenever the time is short enough for insignificant carrier energy loss through interactions with the environment. Until now, this *ballistic electron regime* has been achieved by cooling down the solid state system to below 2 K or so.

Semiconductor heterostructures can be constructed to remarkable precision. To start with, heterostructure substrates can be cleaved to atomic-level flatness. In addition, molecular beam epitaxy (MBE) enables the growth of added layers of material—with atomic-level spatial precision and with great control over the material's composition. Thus epitaxial growth of quantum wells yields atomically precise carrier confinement with insignificant levels of impurities. One result is that integrable systems can be achieved where their integrability is not destroyed by irregularities in the carrier-confining potential or by impurity scattering.

The single-barrier heterostructure built by Teissier et al. (1994) is typical of what multilayer MBE nanotechnology techniques can produce. These researchers worked with a GaAs/AlGaAs p-i-n diode structure consisting of eight layers of different materials. The electron emitter structure was three layers of GaAs with increasing levels of n-type doping atoms, the first layer very thick and the next two layers each 25 nm thick. Then came the 10 nm wide $Al_{0.4}Ga_{0.6}As$ barrier, with 5 nm undoped GaAs spacers on each side of it. Next, there was a thick Be-doped GaAs electron collector, and last of all a thick p-type GaAs top contact.

Teissier and co-workers then placed the sample wafer in liquid helium and applied a *forward bias voltage* V across the heterostructure to create the spatial dependence of static electric potential shown in Fig. 7.14. The upper potential is for electrons, the lower potential for positive carriers or holes. An external magnetic field $B \leq 14\,T$ was also applied, parallel to the x-direction and thus perpendicular to the planes of the structure. Under these temperature and field conditions, a thin accumulation layer of electrons, called a *2DEG electron gas*, builds up just to the left of the barrier. The energy of these electrons is fully quantized into discrete Larmour levels, that is, into the levels of a free electron in a magnetic field. Only the ground and first excited levels are populated, as indicated. (In other experiments, significant population of only the ground state has been achieved.)

What ultimately happens to the 2DEG electrons? Two processes can occur that result in the electrons recombining with particles in the collector material. In the first process, the recombination is with holes in the accumulation layer just to the right of the barrier region; see the lower part of Fig. 7.14. The leftmost two long arrows indicate this cross-barrier direct recombination process, which produces accompanying energy-conserving emission of optical photons. In the second process, an electron may tunnel through the barrier (elastically, with no energy change), as indicated by the top two horizontal arrows. These electrons move ballistically in the collector region until they directly recombine with neutral acceptors denoted by A^0 (second pair of vertical arrows), or they lose energy by a quantum interaction with the solid state lattice vibrations. This latter effect is shown on the right-hand side of the figure, where energy loss—accompanied by creation (emission) of one quantum of lattice vibration (a *phonon*)—occurs before recombination ends the ballistic evolution.

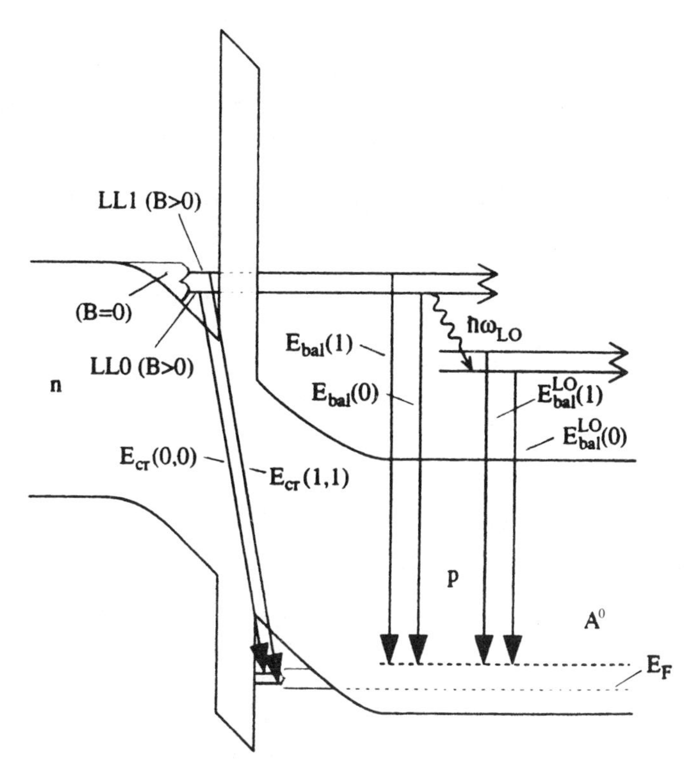

Figure 7.14. Schematic conduction energy band-edge diagram of a single-barrier p-i-n semiconductor tunneling diode. The diode has a sufficiently large forward bias voltage, and there is a large static magnetic field applied parallel to the tunneling direction, horizontally from left to right. The n-type electron emitter material is to the left of the doped diode interface (the spatial region of the barrier), while p-type material is to the right. The upper energy band-edge curve for electrons shows a barrier, while the lower curve is for the positively charged holes. Because of the bias, excess electrons accumulate to the left of the upper curve's barrier, while excess holes accumulate to the right of the lower curve's well. The magnetic field strongly splits the quantum energy levels for the accumulated charge carriers, as indicated. Several transitions observed in electroluminescence optical spectroscopy are also indicated: (i) direct electron–hole recombination lines produced by the energy changes marked $E_{cr}(0, 0)$ and $E_{cr}(1, 1)$; (ii) spectral lines of energies $E_{bal}(0)$ and $E_{bal}(1)$ that are produced after electron tunneling through the barrier and electron ballistic transport in the collector region; and (iii) spectral lines at $E_{bal}^{LO}(0)$ and $E_{bal}^{LO}(1)$ that have energies smaller than those for (ii) because of the emission of a longitudinal phonon of energy $\hbar\omega_{LO}$ before the optical decay. Reprinted from *Superlattices and Microstructures*, **15**, No. 4, R. Teissier, J. W. Cockburn, J. J. Finley, M. S. Skolnick, P. D. Buckle, D. J. Mowbray, R. Grey, G. Hill, and M. A. Pete, "Magneto-optical studies of ballistic electron transport in single barrier heterostructures," pp. 373–376, copyright 1994 by permission of the publisher Academic Press.

By employing sensitive and selective optical spectroscopy techniques, all of the different recombination transitions can be observed with adequate spectral resolution. One can study the variation of the spectrum with changes in the parameters B and V. The results establish three things: (1) that the emitter electron energy distribution over the Landau levels is directly imposed on the ballistic electron spectra; (2) that the tunneling probability can be independent of the initially occupied Landau level; and (3) that energy-randomizing processes (such as electron–electron and electron–hole inelastic scattering) can be negligible. Thus we have the making of an almost ideal microscopic monoenergetic ballistic electron source, followed by electron injection through a barrier into a region that we now will turn into a semiconductor quantum well.

To create the quantum well, we need only add a second potential barrier to Fig. 7.14; see Fig. 7.15. We recall the problem of tunneling through a double-barrier potential (see the end of Section 6.1b). As we adjust the bias voltage V, we change the slope imposed on the spatial dependence of the potential shown in Fig. 7.15. This slope controls the positions of the quantum energy levels for an electron in the well between the barriers. For some values of V, an energy level in the well will equal the energy of an occupied Larmour level in the 2DEG accumulation layer; that is, the energy level will equal the energy of the electrons tunneling through the injection barrier (the first barrier). Then we ideally have a resonant steady-state transmission probability of unity through the entire double-barrier region. For other values of V, Eq. (6.10) gives the relative transmission probability, which is small away from the resonances. We say we have constructed a *resonant tunneling diode* or RTD.

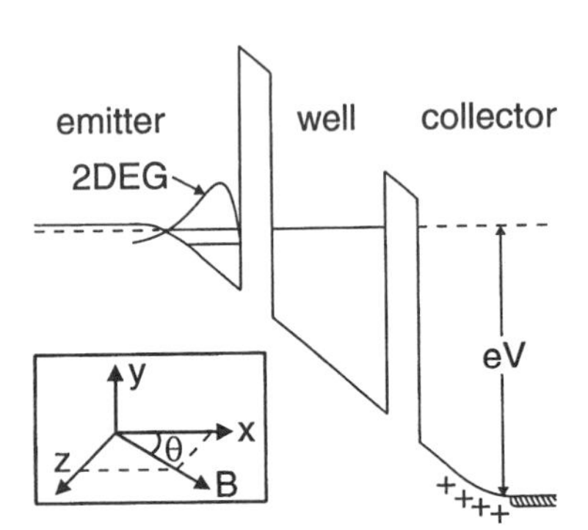

Figure 7.15. Partial schematic band profile for a double-barrier semiconductor resonant tunneling diode (RTD) under forward bias at voltage V and in a magnetic field B that is applied at the field tilt angle θ relative to the electron tunneling direction. Accumulated electrons in the emitter region constitute a 2-d electron gas (2DEG) that occupies quantum energy levels as shown. Tunneling through the left barrier is followed by ballistic electron motion within the well between the barriers and then by tunneling through the right barrier to produce collector current I(V, B, θ). Reprinted from *Superlattices and Microstructures*, **15**, No. 3, A. Fogarty, T. M. Fromhold, L. Eaves, F. W. Sheard, M. Henini, T. J. Foster, P. C. Main, and G. Hill, "Hierarchy of periodic orbits and associated energy level clusters in a quantum well in the regime of classical chaos," pp. 287–291, copyright 1994, by permission of the publisher Academic Press.

Studies with RTDs have a long history as probes of the energy density of states ρ_E in semiconductor quantum wells (Eaves, 1990). A peak in a plot of the diode current I versus bias voltage V must evidence a corresponding peak in ρ_E. Such plots are called *current–voltage characteristics*.

Figure 7.15 also indicates that a magnetic field B is now applied at an angle θ to the tunneling direction (the x-direction); the x-direction is also the direction of the diode internal electric field F. This application of B means an additional force on the electrons—a magnetic Lorentz force that is not along the x-direction. This force thus changes the spatial dimensionality of the system from 1-d to 3-d. For electron motion restricted to be inside the potential well V(x), our Hamiltonian is

$$H = \frac{1}{2m^*}[p_x^2 + (p_y + exB\sin\theta - ezB\cos\theta)^2 + p_z^2] + eF(\tfrac{1}{2}w - x), \qquad (7.23)$$

with $V(x) \equiv \infty$ at the edges of the well. Here $m^* = 0.067m_e$ is the effective electron mass in GaAs (Section 2.1d), w is the width of the well, and $x = 0$ is at the center of the well. Cross terms in x and z are present in Eq. (7.23), as are quadratic terms in these variables. Thus we expect our system to be nonintegrable for nonzero B whenever $\theta \neq 0, \pi/2$; hence the system will be chaotic for some parameter values. We can easily confirm this nonintegrability by numerically evolving Hamilton's equations for Hamiltonian (7.23) and plotting Poincaré sections (Section 1.3). Figure 7.16 shows the transition from regular motion to classically chaotic motion, as revealed in a series of (V_y, V_z) Poincaré sections, with θ being increased from 0° to 24°. Here $B = 11.4$ T, $F = 2.1 \times 10^6$ V/m, $w = 120$ nm, and the electron injection energy is taken to be generated by the bias voltage V across the structure. In fact, an irregular regime is easy to achieve experimentally. For $w = 60$ nm and $B = 10$ T, this regime exists for $V \leq 1.2$ volts when $15° \leq \theta \leq 75°$ (Fogarty et al., 1994). On the other hand, regular regimes exist at small θ or small B, and at large $\theta \approx \pi/2$ or large V.

7.3b. Studies of the Energy Density of States

Figure 7.17 shows RTD current–voltage characteristics, which were obtained using $w = 60$ nm, for two values of the magnetic field tilt angle θ (Fogarty et al., 1994). Actually, the derivative d^2I/dV^2 is plotted versus V. This derivative eliminates several slowly varying changes in the characteristics that are not of interest, characteristics that include the nonoscillating term in ρ_E (Section 7.2b) as well as various experimental effects. The bottom panel in Fig. 7.17 was obtained for $\theta = 90°$, in a regular regime. For $V \lessapprox 0.95$, in this panel we see oscillations with an energy period $\Delta E_p = \hbar/T_p$, where T_p is the period of the periodic ballistic electron motion shown in the left inset. This inset shows a stable "skipping" orbit along the left barrier's interface with the well region. For $V \gtrapprox 0.95$, we see an oscillation with a different energy period having a value equal to that for the periodic traversing orbit shown in the right inset.

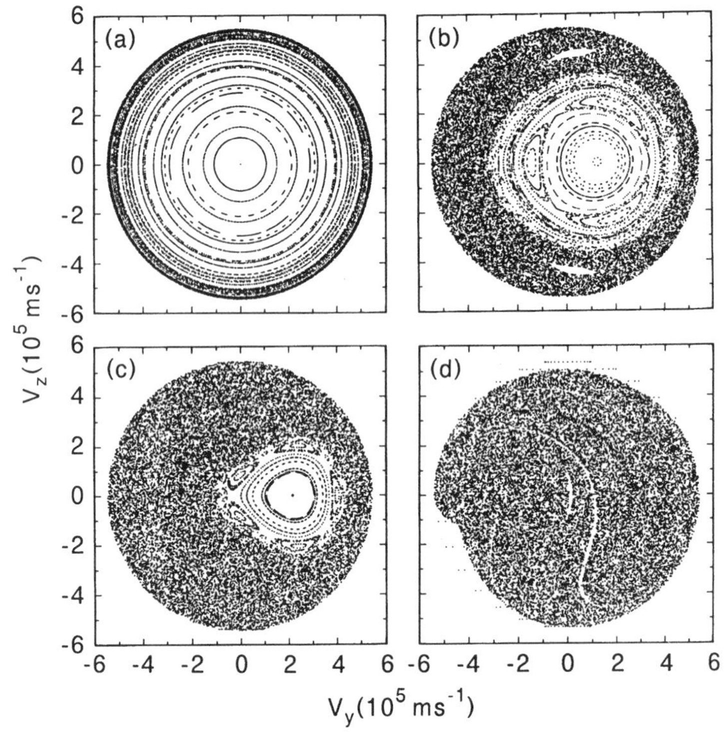

Figure 7.16. Numerical classical (v_y, v_z) Poincaré sections of ballistic electron motion within the barriers of a forward-biased RTD in a tilted magnetic field (see Fig. 7.15). The full phase space is (x, y, z, mv_x, mv_y, mv_z) where **v** is the electron velocity. For B = 11.4 T and a bias electric field F = 2.1 MV/m, 225 starting velocity vectors determined trajectories that were numerically evolved for field tilt angles θ = (a) 0°, (b) 3°, (c) 6°, and (d) 24°. A transition from regular to chaotic classical evolution is seen. The chaos is generated by ballistic electron collisions with the barriers that interrupt at irregular times the otherwise regular motion. From Fromhold et al. (1995). Copyright 1995 by the American Physical Society.

The top panel in Fig. 7.17 displays two very different oscillations that appear when we change θ from 90° to 30°. Poincaré sections show that for this panel the low-V oscillation is in the irregular regime, while the high-V oscillation is in a regular regime. The insets also show the periodic orbits that are believed to be the origins of these oscillations; these insets display projections of the 3-d orbits onto the xy-plane (p_y is constant but nonzero). In accordance with closed-orbit theory (Section 7.2c), the periods of the orbits agree with the dominant peaks in the Fourier transform of the data. Using the experimental approach just described, studies have detected an important role for a number of unstable periodic orbits, with the ballistic electron undergoing multiple bounces in the quantum well before any phonon emission changes the electron's energy.

In order to predict the energy density of states ρ_E and the recurrence spectrum of orbit times, investigators have used quantum basis-set expansion calculations of the

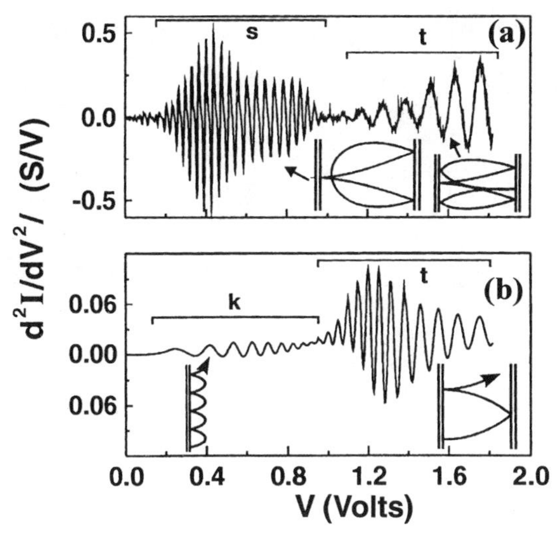

Figure 7.17. Experimental RTD resonant tunneling current signals d^2I/dV^2 plotted versus the forward bias voltage V, for magnetic field B = 10 T and two values of the field tilt angle θ. Oscillations seen in the signals are due to ballistic electron motion between the two barriers of Fig. 7.15. The inset in Fig. 7.15 defines the directions (x, y, z). Tunneling resonances labeled k, s, and t are associated with the periodic electron orbits shown projected on the xy-plane in the insets of the present figure. (a) $\theta = 30^\circ$. (b) $\theta = 90^\circ$. Reprinted from *Superlattices and Microstructures*, **15**, No. 3, A. Fogarty, T. M. Fromhold, L. Eaves, F. W. Sheard, M. Henini, T. J. Foster, P. C. Main, and G. Hill, "Hierarchy of periodic orbits and associated energy level clusters in a quantum well in the regime of classical chaos," pp. 287–291, copyright 1994, by permission of the publisher Academic Press.

energy levels for the irregular regime of w = 120 nm quantum wells in tilted magnetic fields (Fromhold et al., 1995). Figure 7.18a shows the predicted, smoothed density of states for an experimentally observable range of energies. Figure 7.18b shows the recurrence spectrum, which displays peaks associated with periods T_1 to T_4 of the unstable periodic orbits shown in the insets. These results of Fig. 7.18b are in accord with experiments at w = 120 nm.

A different type of experiment gives further evidence of the important role nonlinear classical dynamics plays in the present system. Now we plot the voltages at which the current peaks occur, as a function of B with θ held fixed. If a single stable orbit plays a role, then the peaks would be equally spaced smooth curves; this is essentially the situation shown in Fig. 7.19, at low B for the $\theta = 27^\circ$ data of Müller et al. (1995). At higher B, however, we see regions containing "extra" curves between the low-B curves, as well as regions where all the curves are "jittery" or "disordered." These two new regions can be explained by looking at Poincaré sections, which indicate that the "extra" curves arise from a bifurcation of the periodic orbit, and the "disordered" curves denote a region of classical chaos.

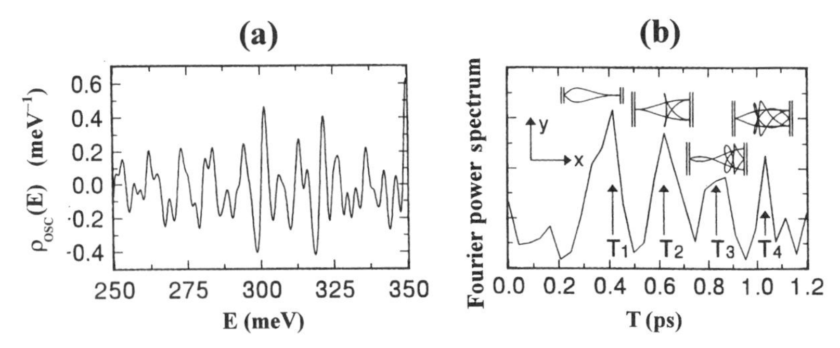

Figure 7.18. Some basis-set-expansion numerical quantum predictions for an RTD in a tilted magnetic field, for the parameters of Fig. 7.17 except for a nearly chaotic intermediate field tilt angle $\theta = 20°$. (a) The oscillating part of the smoothed energy density of states $\rho_{OSC}(E)$ as a function of electron energy E. These results are qualitatively similar to those for the diamagnetic Kepler problem shown in Figs. 7.10 and 7.11. (b) Fourier power spectrum of the curve shown in panel (a), showing four distinct peaks labeled by T_1, T_2, T_3, T_4. In the insets corresponding periodic orbits are shown projected on the xy-plane. From Fromhold et al. (1995). Copyright 1995 by the American Physical Society.

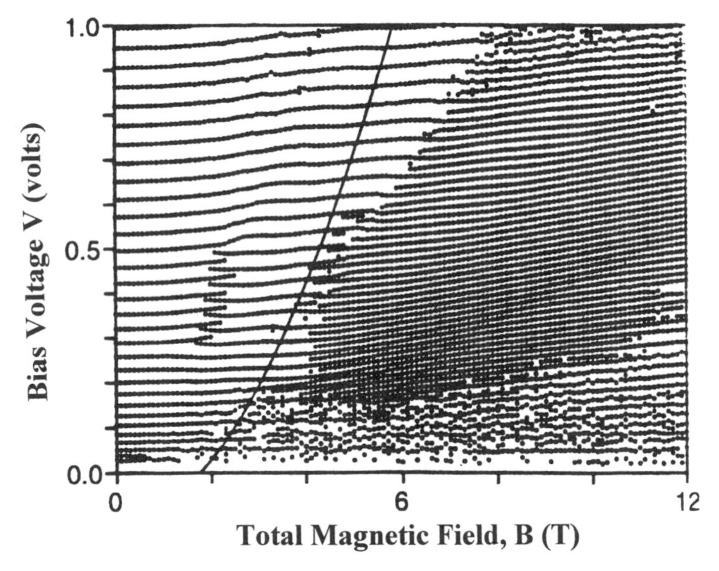

Figure 7.19. Experimental peak positions obtained from RTD tunneling current signals I (V) and plotted in the (V, B) parameter space. The field tilt angle is $\theta = 27°$. The peaks are from curves similar to those shown in Fig. 7.17. The smooth solid curve indicates the location of a periodic orbit bifurcation point obtained from calculations of Poincaré sections. To the right of this curve the effects of the bifurcation are seen in the changed spacing of the experimental peak position curves. Again to the right of this curve, at lower values of V, there are additional randomizing effects that correlate with the presence of classical chaos. From Müller et al. (1995). Copyright 1995 by the American Physical Society.

7.4. BILLIARD SYSTEMS

An oft-studied class of systems is the motion of a particle (a "billiard ball") in two spatial dimensions, where the potential is a constant—except for a set of one or more closed boundary curves, where the potential is infinite. The particle classically moves along straight-line segments and obeys the law of reflection at boundaries. We will see that the geometry and topology of the boundary curves play a pivotal role in determining the dynamics. The dynamics can be integrable for all particle energies E as in elliptical billiards, or the dynamics can be classically chaotic for all E as in stadium billiards. One can also have more generic situations where a parameter controls the dynamics, from integrable through near integrable to strongly mixed and then completely irregular. We will consider deformed circular billiards, where a parameter determines how much the circular boundary is deformed.

From a mathematical point of view, quantum billiards problems appear simple. We have a wave boundary-value problem. To find the energy eigenstates, we must solve the 2-d wave equation for free particle motion, with Dirichlet boundary conditions

$$\frac{\partial^2\Psi}{\partial x^2}+\frac{\partial^2\Psi}{\partial y^2}+\frac{2mE}{\hbar^2}\Psi=0$$

$$\Psi(x,y)=0 \quad \text{on closed curve(s) } C. \tag{7.24}$$

On the other hand, after propagating classical trajectories for a few types of C, we would appreciate that the classical dynamics is often typical of nonintegrable systems. Hence, in spite of the conceptual simplicity of problem (7.24), $\Psi(x, y)$ can contain highly detailed and complex information. We note one special classical property of billiards problems, namely, the point-like disruptions of otherwise free-particle motion. From some choices of the billiard boundary C, this special property may lead to nongeneric behavior (Schulman, 1994).

We note an interesting fact—the wave equation (7.24) also describes transverse magnetic (TM) modes in a cylindrical electromagnetic (EM) cavity that has the billiards boundary curve as the boundary of its transverse cross section. The quantity $2mE/\hbar^2$ is replaced by k_n^2, where $k_n = \omega_n/c$ relates the transverse wave number k_n to the wave mode eigenfrequency ω_n of the cavity. Thus experiments with EM cavities can explore properties of billiards eigenstates and eigenenergies (Gräf et al., 1992; Stein and Stockmann, 1992; Sridhar and Kudrolli, 1994).

7.4a. Elliptical Billiards

For the elliptical billiards problem, the curve C is given by the equation for an ellipse,

$$\frac{x^2}{A}+\frac{y^2}{B}=1.$$

When we have a circle, A = B, and the system is integrable since the rotational symmetry implies conservation of the angular momentum vector. Figure 7.20 shows a typical quasiperiodic trajectory for the circle. The elliptical case, A $\neq$ B, is also integrable because there are two constants of the motion: the energy E and the inner product of the angular momenta around the foci of the ellipse, $\mathbf{J}_1 \cdot \mathbf{J}_2 = (\mathbf{r}_1 \times \mathbf{p}) \cdot (\mathbf{r}_2 \times \mathbf{p})$.

For the elliptic billiards, the Hamilton–Jacobi equation (3.12) separates in elliptical coordinates (ρ, θ) that are related to the coordinates (x, y) according to

$$x = C \cosh \rho \cos \theta,$$
$$y = C \sinh \rho \sin \theta.$$

Here $C \equiv \sqrt{(A - B)}$ is one-half the distance between the foci of the curve *C*. Curves for constant ρ and constant θ are ellipses and hyperbolas, respectively.

Orbits in the 4-D phase space (x, y, p_x, p_y) wind around 2-D tori (Sections 1.4a and 3.3b). The projection of the tori onto position space is generally singular, because the projection mapping is partially many-to-one. Caustics are the curves in position space where the projection is singular; see Figs. 1.14 and 7.21. The caustic is either an ellipse or a hyperbola, depending on whether the trajectory in coordinate space passes outside or inside the foci of *C*. Note the trajectory shares tangential points with its caustic.

To construct Poincaré sections for the elliptical billiards, it's helpful to note that a trajectory is completely determined by the sequence of values of particle outgoing

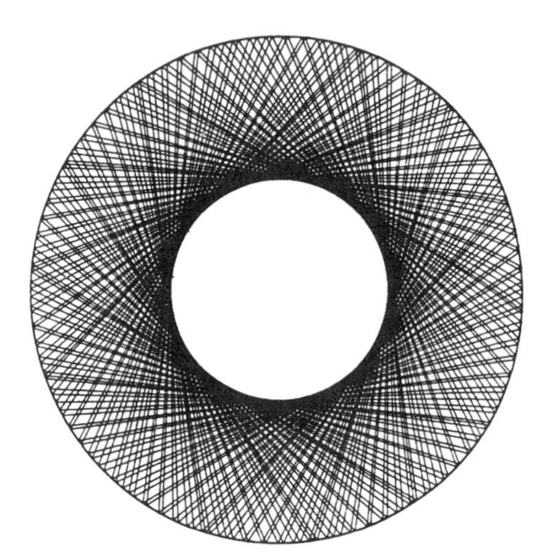

Figure 7.20. A single quasiperiodic particle trajectory within the boundary of the circular billiards problem. Because of conservation of angular momentum, the orbit evolves within an annulus between some minimum radius and the outer radius of the circle. The locus of all inner turning points is an example of a caustic (Section 1.4a). From McDonald and Kaufman (1988). Copyright 1988 by the American Physical Society.

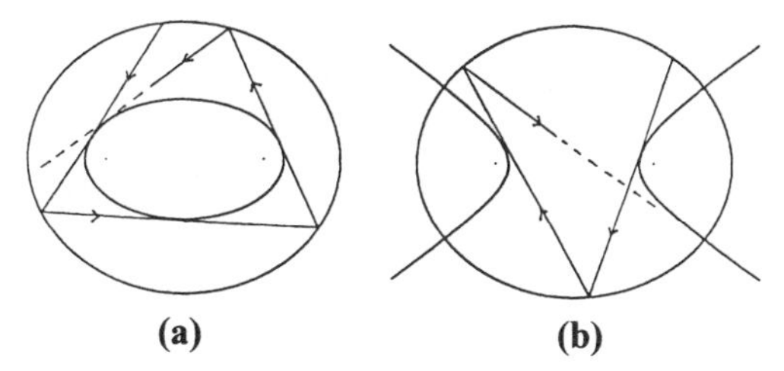

Figure 7.21. Caustics for elliptic billiards determined by $x^2/A + y^2/B = 1$, $A \neq B$. (a) An elliptic caustic lies inside the billiard's boundary. Part of a trajectory is shown, which remains outside the foci of the boundary ellipse. (b) Hyperbolic caustics are associated with trajectories that fall inside the foci. From Crespi et al. (1989).

momentum and particle location for the trajectory's points of impact on the boundary curve *C*. Thus for the 2-D Poincaré section, it is possible to define variables (s, p). Here s is the arc length to the point of impact, measured from some reference point on *C* and normalized to the length of *C*. The normalized tangential momentum p is the sine of the angle between the outgoing direction of the particle and the normal direction to *C*, again at the point of impact. The Poincaré section variables (s, p) are canonical (Berry, 1981, Appendix 1).

Figure 7.22 shows a Poincaré section for A = 1.25, B = 1 using these variables (s, p). Note this section's similarity to the phase-space curves of the pendulum problem (Fig. 1.15b). Given the similarity, it's not surprising that the energy eigenstates of the elliptic billiards are given in terms of Mathieu functions. Figure

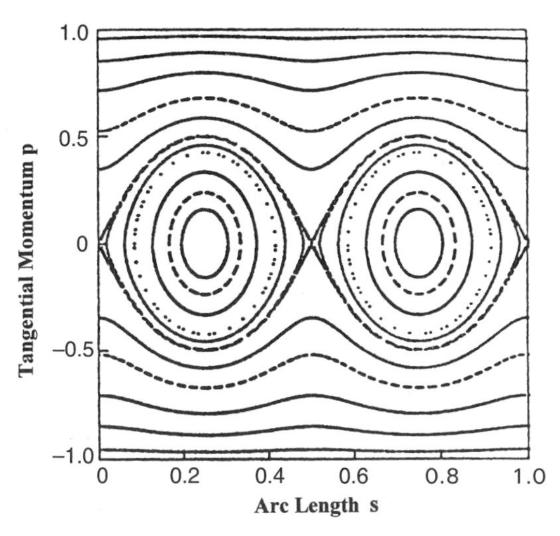

Figure 7.22. An (s, p) Poincaré section of trajectories for elliptical billiards, where the canonical variables (s, p) are defined in the text. The ellipse parameters are A = 1.25, B = 1.00, and the point $(s, p) = (0, 0) \equiv (1, 0)$ corresponds to $(x, y) = (\sqrt{A}, 0)$. The system is integrable, and this Poincaré section is very similar to the phase-space curves for the 1-d vertical pendulum problem (Fig. 1.15b). From Crespi et al. (1989).

7.23 shows examples of eigenfunctions corresponding to q-trajectories having the two kinds of caustics. The (s, p) quantum Poincaré sections for these eigenfunctions are localized on features seen in the classical Poincaré section (Fig. 7.22); see Crespi et al. (1989). The eigenfunction seen in Fig. 7.23a is localized on some of the "rotational" phase curves in Fig. 7.22, while the eigenfunction in Fig. 7.23b is localized on some of the nearly elliptical "librational" phase curves in Fig. 7.22.

7.4b. Deformed Circular Billiards

Let us now consider the nonintegrable system defined by adding a cubic term to the equation that defines the boundary *C*. We set A = B and normalize the spatial variables to $\sqrt{A}$. We then write

$$x^2 + y^2 + \varepsilon x^3 = 1.$$

When $\varepsilon = 0.2$ (Fig. 7.24), the Poincaré section is very different from Fig. 7.22. Figure 7.24 exhibits a stochastic layer covering the region of the separatrix for $\varepsilon = 0$, and filling about 50% of the section's area. There also are islands of regular evolution, both KAM survivors like those labeled "2-cycle" and "wg" in Fig. 7.24, and new islands such as those labeled "3-cycle" in the same figure. From our discussion of nonintegrable systems in Chapter 5, we expect all this to happen.

When quantum Poincaré sections are calculated for the energy eigenstates of the deformed circular billiards, one finds cases of localization on either the stable or unstable portions of the 2-cycle region in Fig. 7.24. We call these eigenfunctions "2s" and "2u" states, respectively. Similarly, there are 3s and 3u states localized on the 3-cycle regions, and wg states localized on the wg regions. (The quantum Poincaré sections are shown in Crespi et al. (1989).) The 2s, 3s, and wg states are examples of regular states as in Fig. 5.15, while the 2u and 3u states are examples of scarred irregular states (Section 5.4).

The probability distributions of typical wg, 2s, and 3s energy eigenstates are shown in Fig. 7.25, along with the caustics for the corresponding sets of classical

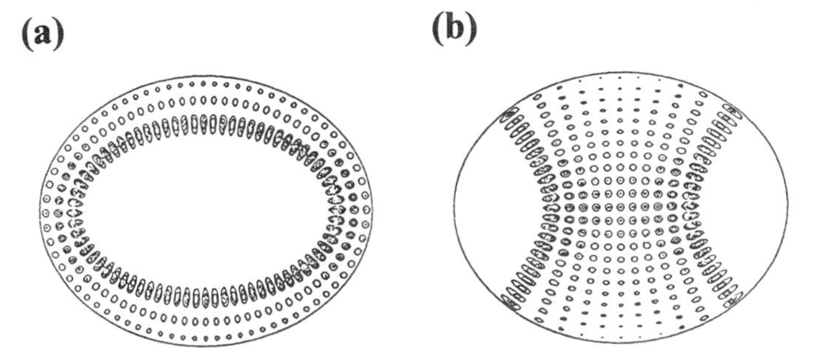

Figure 7.23. Energy eigenfunctions for elliptic billiards, A = 1.25, B = 1.00, at moderate values of the (unitless) wave vector: (a) for an elliptic caustic and k = 46.0971; and (b) for a hyperbolic caustic and k = 46.3316. From Crespi et al. (1989).

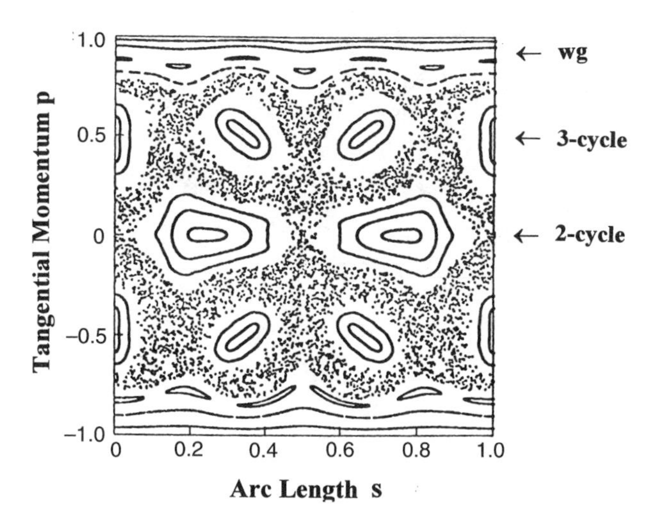

Figure 7.24. An (s, p) Poincaré section for deformed circular billiards, where $x^2 + y^2 + \varepsilon x^3 = 1$ with $\varepsilon = 0.2$. This system is clearly nonintegrable. From Crespi et al. (1993). Copyright 1993 by the American Physical Society.

trajectories. The localization of the eigenstates near the corresponding caustics is quite remarkable. Because all the trajectories in one set are tangent to points on their caustics, there is, near the caustics, positive wave interference of trajectory contributions to the quantum wavefunctions. As the unitless wave vector $k = \sqrt{2mE}/\hbar$ is increased, the spatial resolution of the wavefunctions increases as expected: see panels (b) and (c) in Fig. 7.25. The right-hand side of panel (b) shows that the fine details of a caustic can be present in a quantum wavefunction.

Quantum deformed circular billiards also have irregular eigenstates that do not appear scarred by simple unstable periodic orbits; see Crespi et al. (1989).

Figure 7.26 shows a few energies of the quantum deformed circular billiards, plotted as a function of the deformation parameter ε. Corresponding to the three types of regular eigenstates, the levels are labeled as "wg," as "2s" and "2u," and as "3s" and "3u." We note a large number of apparent level crossings at low deformations, where the Poincaré section is largely regular. We also note a different situation at high deformations, where the section is mostly chaotic. Near the largest value of ε shown, most level crossings are noticeably avoided—that is, the energy levels approach and then repel one another as ε is varied. Thus there seems to be a correlation between two quantities: the irregular fraction of the area in the Poincaré section, and the fraction of energy-level crossings that are *avoided crossings*—with both fractions increasing as ε increases.

This correlation is real—and generic. It can be quantified by considering the distribution function $P(\Delta E)$ of nearest-neighbor energy-level separations ΔE, for various values of the deformation parameter ε. In the almost completely regular

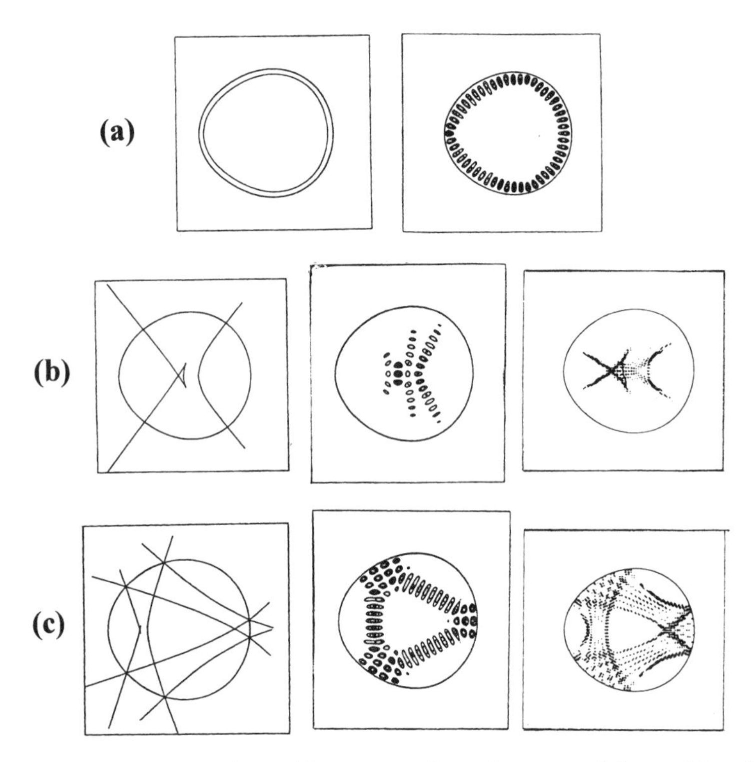

Figure 7.25. Comparisons in position space of caustic curves (left panels) and certain quantum energy eigenfunctions for deformed circular billiards. Unless otherwise specified, the values of the (unitless) wave vector lie within the region $30.0 \leq k \leq 31.3$. (a) A "whispering gallery" (wg) state is localized near a caustic that follows the billiards boundary and lies somewhat inside it. (b) Shown is a 2s "swallow tail" state, named after the shape of its corresponding caustic. The right-hand panel shows this quantum swallow tail at the elevated value $k = 119.0983$. (c) A 3s state is compared with its corresponding caustic. On the right is the wavefunction for $k = 118.9979$. From Crespi et al. (1989).

(KAM) limit, we have a *Poisson level-separation distribution* of the exponential form

$$P_P(\Delta E) = \alpha \exp(-\alpha \, \Delta E), \tag{7.25}$$

where α is a constant. The level splitting ΔE with the highest probability is then zero. In the completely irregular (classically chaotic) limit, we instead have a *Wigner level-separation distribution* of the form

$$P_W = \alpha(\Delta E)^{\alpha} \exp[-\beta(\Delta E)^2], \tag{7.26}$$

where α and β are constants. Now there is zero probability for $\Delta E = 0$: that is, no states are energy degenerate. Asymptotic ($\hbar \to 0$) semiclassical theory predicts Eqs.

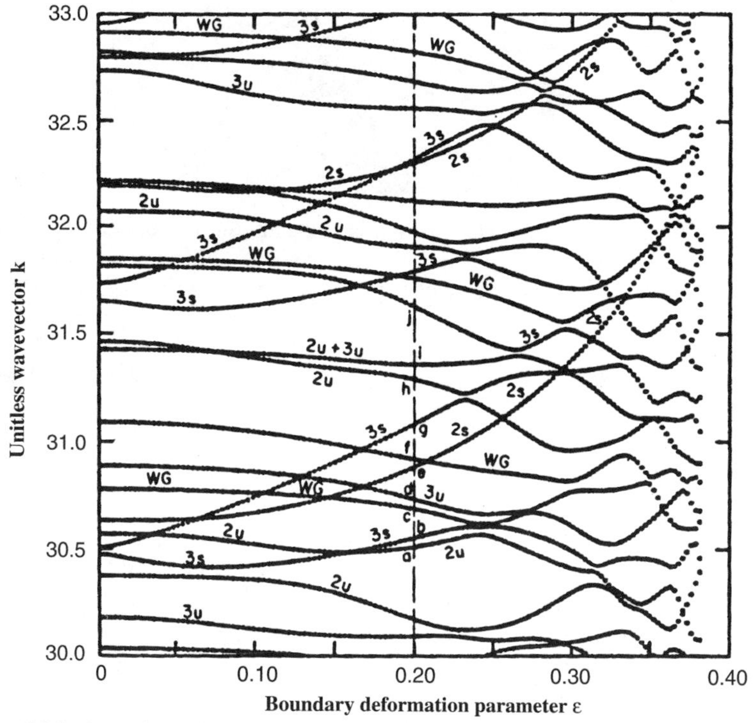

Figure 7.26. A portion of the quantum energy eigenvalue spectrum for deformed circular billiards, as a function of the boundary deformation parameter ε. The ranges of the variables are $30 \leq k \leq 33$ for the vertical axis and $0 \leq \varepsilon \leq \sqrt{\frac{4}{27}}$ for the horizontal axis. The vertical dashed line corresponds to the value $\varepsilon = 0.2$ used for Figs. 7.24 and 7.25. For the present figure, the eigenenergies at values $\varepsilon > 0.35$ are for a strongly chaotic classical phase space; many would-be crossings of eigenenergy curves are avoided in this region of ε values. From Crespi et al. (1989).

(7.25) and (7.26); see Section 9.2d. Numerical results for every semiclassical system discussed in this book have been compared with these equations, with more or less qualitative agreement when approaching the limiting regimes. At a more quantitative level, there often are deviations, indicating correlations arising from the structures in phase space, including the periodic orbits. When phase space is mixed regular and irregular, neither Eq. (7.25) nor (7.26) applies.

Of paramount importance for externally induced quantum evolution is the occurrence of avoided level crossings when a parameter is varied (Section 9.3). We therefore return to the subject of avoided crossings in Section 9.2.

7.4c. Stadium Billiards

For stadium billiards, the boundary C consists of two parallel lines with their ends connected by half-circles; see Fig. 7.27. The figure also shows part of one typical

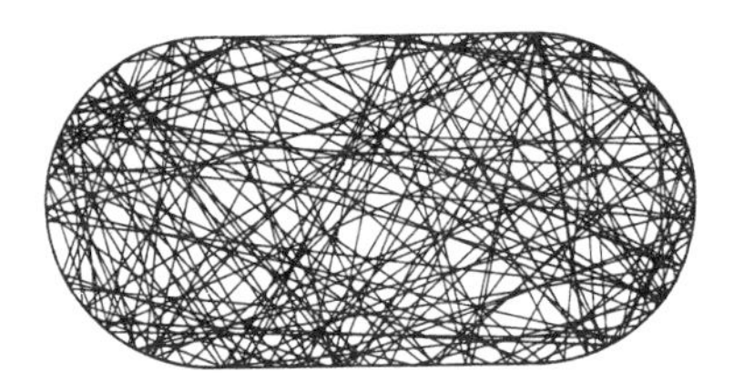

Figure 7.27. A single chaotic trajectory for stadium billiards. Such trajectories are typical for this system. When the length of the horizontal straight section is nonzero, all periodic orbits are unstable. From McDonald and Kaufman (1988). Copyright 1988 by the American Physical Society.

particle trajectory, which has a positive Lyapunov exponent and is chaotic. Periodic orbits exist in the stadium, but they are all unstable, no matter what the particle energy E. For this reason, stadium billiards have been a prototype system for the study of irregular quantum systems, despite the possibility mentioned above that billiards systems may not be generic.

An enormous number of energy eigenstates have been computed and examined for the stadium (McDonald and Kaufman, 1988; Heller et al., 1989). These studies include states corresponding to energy levels up to the 10,000th level of energy excitation. Many states are scarred by identified unstable periodic orbits, while other states may or may not be scarred, the apparently "generic" situation. Figure 7.28 displays a few eigenfunctions.

7.4d. Wavepacket Evolution in the Stadium Billiard

Since the stadium is completely irregular, wavepacket studies have been pursued with considerable interest. If we use the usual method for setting up the initial

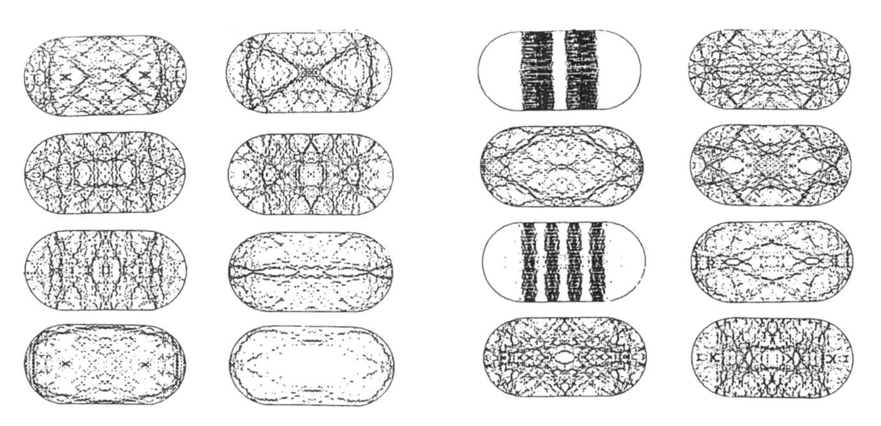

Figure 7.28. A consecutive sequence of energy eigenstates for stadium billiards, going up in energy from about the 8390th eigenstate. The states are in order when read as text. Many of the states are clearly scarred by unstable periodic orbits. From Heller et al. (1989). Copyright 1989 Royal Swedish Academy of Sciences.

packet—the method of diagonalizing a Hamiltonian matrix to get precise eigenstates and energies and then expanding the initial packet in this set of states (Section 1.3c)—we find this approach becomes rather costly when 150 or more states are needed. It's more practical to obtain a direct numerical solution of the time-dependent Schrödinger equation by using grid discretization methods. The evolution of an initially moving Gaussian packet in the stadium is presented in Christoffel and Brumer (1986), where they compare the packet with the appropriate initial ensemble of classical chaotic trajectories. In view of what we saw for the Hénon–Heiles system (Section 5.3d), it is not surprising they found an early classically guided evolution that is accompanied by wavepacket spreading. After an initial time interval, particle probability occurs throughout the stadium and fluctuates quantum mechanically like a bubbling cauldron. We wouldn't expect any early quantum revival from such a situation.

Studies have also followed the evolution of the autocorrelation function $\langle \Phi(0) | \Psi(t) \rangle$, see Eq. (2.14), and have used the semiclassical Green's function in a way similar to that discussed in Section 7.2e. Tomsovic and Heller (1991, 1993) included trajectories in the sum Eq. (2.15) that were homoclinic to the unstable periodic orbit, the orbit that the packet is initially centered on in phase space. All the dynamics, up to ten particle bounces, were included in the periodic-orbit sum. The packet's spatial distribution evolved into a distribution very similar to an irregular energy eigenstate, just as occurred for the 2-d oscillator problem (Section 7.2e).

Very recently, early wavepacket revivals in autocorrelation functions for the stadium billiards system have actually been found numerically (Tomsovic and Lefebvre, 1997). The conditions necessary for such revivals in totally chaotic systems appear to be quite special. The initial wavepackets chosen were stretched ones (Fig. 5.19), spatially localized along a central unstable periodic orbit. Key is the fact that the central unstable periodic orbit is accompanied by stable and unstable manifolds in phase space (Fig. 5.9). We have noted that these manifolds have observable effects on the scarring of irregular quantum eigenstates (Section 5.4 and Figs. 5.25 and 5.26). Returning to Fig. 5.9, it can be shown that the repeated crossing of the stable and unstable manifolds produces a "broken separatrix" in phase space, with each crossing point reflecting the existence of a secondary unstable periodic orbit. The broken separatrix structure encloses a phase-space area in a Poincaré section and, in addition, generates a phase-space area called a "turnstile" that determines the classical flux of trajectories passing in and out of this volume (MacKay et al., 1984). The quantum wavepacket revivals are noticeable when either the flux area and/or the manifold-enclosed phase-space area satisfies apparent quantization conditions. Then a fairly small number of action areas generates an enormous number of correlated periodic-orbit action differences. These correlated differences produce an unexpected constructive interference of amplitudes—associated with a multi-tude of periodic orbits—where the interference is responsible for the early revivals. Thus it appears that wavepacket revivals can occur in irregular quantum systems.

7.5. BALLISTIC ELECTRONS IN CONDUCTING NANOSTRUCTURES

7.5a. Semiconductor Nanostructures

In Figs. 7.29a–c, we see a few nanostructures recently fabricated by electron beam lithography on a multilayered $GaAs/Al_xGa_{1-x}As$ heterostructure of the type mentioned in Section 7.3a. These particular nanostructures—often called quantum dots (Section 2.1d)—are sandwiched in the middle layer of the heterostructure, between electron emitter and collector multilayers, which are not shown. There is the usual heterostructure bias voltage, and current-detection equipment enables the measurement of the heterostructure conductance. This conductance has a relatively small contribution associated with electron flow through the entrance and exit ports of the small nanostructures. These ports are specially coupled to the 2DEG that is located roughly 100 nm below the plane containing the nanostructures.

In cases (a) and (b) in Fig. 7.29, the entrance and exit ports are fixed in size by the lithography and are large enough for simultaneous electron transmission through the individual nanostructures in more than one matter-wave mode. In other words, the

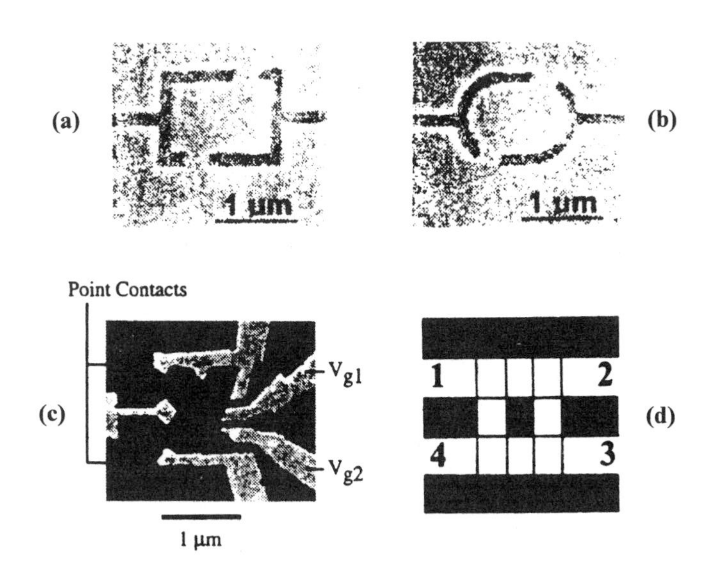

Figure 7.29. Examples of semiconductor nanostructures. (a), (b) Rectangular and stadium billiard structures, respectively, that have been fabricated by electron beam lithography. The entrance and exit ports for ballistic electrons are relatively large, making these devices open systems. From Chang et al. (1994). Copyright 1994 by the American Physical Society. (c) This nanostructure contains a sufficient number of metallic gate electrodes for voltage control of effective port size, from the open to nearly closed regimes. From Folk et al. (1996). Copyright 1996 by the American Physical Society. (d) The design geometry for a four-port ballistic electron waveguide directional coupler. Reprinted from *Superlattices and Microstructures*, **17**, O. Vanbésien and D. Lippens, "Quantum interference electronic switch," pp. 197–200, copyright 1995, by permission of the publisher Academic Press.

de Broglie wavelength of the electron $\lambda_B \lesssim 40$ nm is smaller than the port diameter. Such nanostructures with large ports are said to be *open nanostructures*.

In Fig. 7.29c there are additional metallic gates deposited so that "pincher" gate voltages can be applied to adjust the effective port diameter down to sizes that stop transmission completely. Such ports are called *point contact leads*. When these leads are set to permit some electron transmission in only one mode, the structure is said to be "almost closed." Additionally, in Fig. 7.29c there are other gates—labeled V_{g1} and V_{g2}—whose voltages can be adjusted to deform the shape of the effective interior region of the nanostructure. In other words, these voltages enable the deformation of the shape of the quantum dot. All the gates work because charge depletion layers exist around the gates; the thickness of these layers depends on the gate voltages.

Nanostructure configurations—as determined by lithographically defined boundaries and/or metallic gates—do not possess boundaries having single-atom spatial precision. Because of lithographic imperfections and nonreproducibility that can amount to 50 nm, it is unlikely that such a nanostructure will be an integrable system. Yet many experiments have demonstrated that the nanostructure conductance g depends markedly on the geometry, with the nanostructure area held fixed.

In analogy to microwave directional couplers, Fig. 7.29d shows an idealized four-port matter-wave nanostructure designed to be a waveguide directional coupler. Indeed, present-day nanotechnology includes fabrication of all kinds of matter-wave waveguide components that have two spatial dimensions of the order of the electron de Broglie wavelength. In addition, diodes, transistors, and other active matter-wave nanodevices have been fabricated, including single-electron transistors (Kastner, 1993).

All of these conducting nanosystems can be built small enough for the electron motion to exist locally in the *ballistic regime*. The transport mean free paths and phase-coherence lengths are up to 20,000 nm, while the device major dimensions are only 300 to 2000 nm. To achieve a ballistic regime also requires remarkable purity of materials, as well as operation at very low temperatures. When we have base dilution refrigerator temperatures of 30–70 mK, the electrons inside the nanostructures have temperatures of 70–180 mK. The electron energies are typically 10–25 μeV, with energy spreads of ±30%. Operation typically is near the 100th to 400th excited energy level of the nanostructure, where the mean energy-level splitting Δ is about 20 μeV. Thus ballistic electron energies down to $\Delta/2$ are achievable, meaning that the quantization of energy levels and states can play a decisive role, and that essentially single-mode operation is possible.

7.5b. Open Nanostructures

A great deal of work has focused on open quantum dots, such as the structures shown in Figs. 7.29a,b (Chang et al., 1994; Bird et al., 1995; Clarke et al., 1995). One advantage of these open dots is that the ratio of the sum of entrance port and exit port widths to the dot perimeter—a ratio of about 1 to 6.5—means that most ballistic evolutions within the dot are cut off in time by passage out of the ports (the openings), rather than by the phase-coherence length. Hence very long periodic orbits do *not* play a role.

Often the sample studied contains more than one nominally identical dot, the dots being "wired" to the 2DEG in parallel. Chang et al. (1994) studied 6×8 arrays of dots (actually billiard cavities) with dot spacing of 25 μm, which is larger than the 15 μm phase-coherence length for their samples. One can argue that the total conductance of the array is an average over the lithographically induced fluctuations in dot geometry. Then if we observe structure in the dependence of the conductance g on an adjustable parameter, it would likely be a property of most of the dots, rather than of just one or two abnormal ones.

Many studies of semiconductor nanostructures employ an external static magnetic field **B** that is directed perpendicular to the plane containing the dot. This field is an easily adjusted parameter that changes the electron dynamics, much like the situation of the excited hydrogen atom in a magnetic field (Section 7.1). Figure 7.30a shows the dependence on B of the resistance R(B) for an array of rectangular billiards cavities (quantum dots), which are formed as in Fig. 7.29a. Away from $B = 0$ and at the lower temperatures, we see periodic oscillations that do not depend on the sign of B. These oscillations are reminiscent of the low-resolution oscillations displayed in Figs. 7.7a and 7.18a, where the oscillations arose from the existence of one or more periodic orbits. In these cases, we saw that a Fourier transform of the data produced periods of the orbits. The result of doing the Fourier transform for the data of Fig. 7.30a produces the orbit frequency spectrum shown in Fig. 7.30b. Either of the experimental peaks at 14 and 16 cycles/kG can be explained by the stable periodic orbit shown as the dashed line in the inset of the

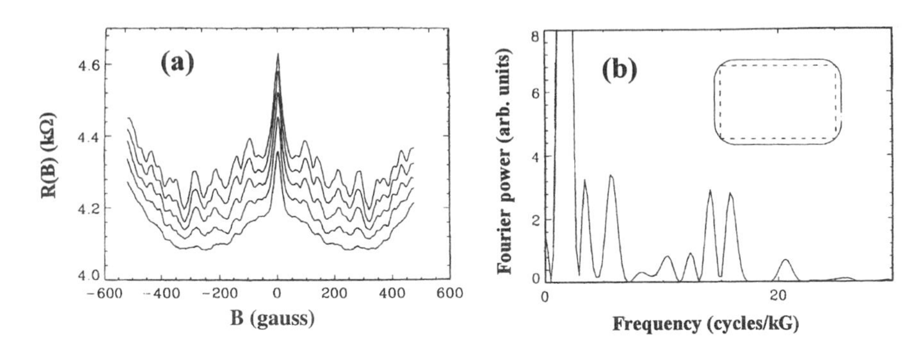

Figure 7.30. Evidence for ballistic electron transport within a semiconductor heterostructure containing 48 of the nominally rectangular billiard nanostructures shown in Fig. 7.29a. (a) The heterostructure resistance is plotted versus the strength and direction of a magnetic field applied perpendicular to the plane of nanostructures. The traces correspond to different heterostructure temperatures, from 50 mK (top) to 1.6 K (bottom). At the lower temperatures, oscillations appear that are qualitatively analogous to the spectra in Fig. 7.18a. (b) As in Fig. 7.18b, the Fourier power spectrum obtained from the oscillations in panel (a) reveals peaks produced by the presence of important periodic orbits. The inset shows a whispering gallery orbit having the properties needed to explain the peaks at 14 and/or 16 cycles/kG. From Chang et al. (1994). Copyright 1994 by the American Physical Society.

figure. As the inset indicates, the rectangular billiard is expected to be modified by a rounding of the corners, thus complicating a precise comparison with numerical predictions.

7.5c. Almost-Closed Nanostructures

Particularly intriguing are recent investigations of almost-closed quantum dots (Chang et al., 1996; Folk et al., 1996). The configuration shown in Fig. 7.29c provides three-parameter flexibility in perturbing this single-billiard system, the parameters being B, V_{g1}, and V_{g2}. The configuration also uses point-contact leads to vary the "bandwidth" of the billiard modes that are coupled to the leads. Folk and co-workers achieved a base temperature of 30 mK, so that $kT_{base} \approx \Delta/2 < \Delta$. In addition, the point-contact leads could be pinched down so that $\hbar/\tau_{escape} \equiv \Gamma \lesssim \Delta/10$. Thus once ballistic electrons were in the billiard cavity, they stayed there long enough for the energy resolution of level spectra to be one-tenth the mean level spacing.

It's important to note that the conditions just described imply that these experiments achieved the condition $\Gamma < \Delta$ by an order of magnitude. This would place the system in the *Coulomb blockade regime*, where there is conductance suppression due to charge quantization. This quantization occurs whenever the energy needed to add a single electron to the dot—$e^2/C \approx 620\,\mu eV$ when $C = 260$ aF—exceeds the available energy eV_{bias} due to the applied bias (Geerligs et al., 1993; Kastner, 1993). When the dot potential is tuned (e.g., by varying V_{g1} and/or V_{g2}) so the number of electrons in the dot can change with a change in the available energy, large conductance peaks are observed for the case $B = 0$, as seen in Fig. 7.31. Each peak marks a change of one in the number of electrons in the dot.

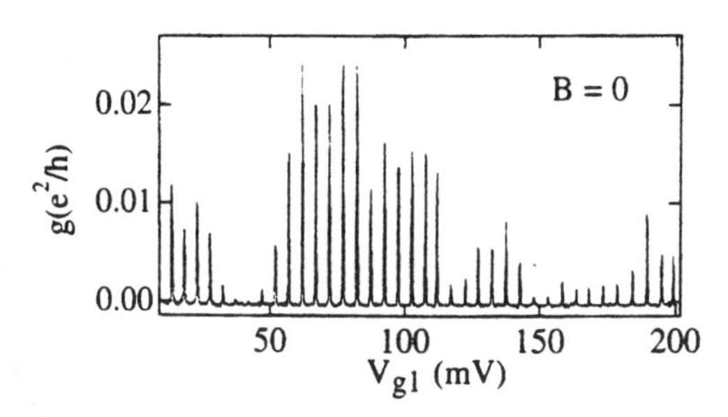

Figure 7.31. Electrical conductance of a semiconductor heterostructure containing the nanostructure of Fig. 7.29c. The nanostructure metallic gate voltages are adjusted for nearly closed ballistic electron ports and for operation in the Coulomb blockade regime; see text. For magnetic field $B = 0$, Coulomb blockade conductance peaks are observed as a function of the nanostructure metallic gate voltage V_{g1} (Fig. 7.29c), which modifies the effective billiards boundary. The distribution of conductance peak heights contains information about relevant ballistic electron periodic orbits within the nanostructure. From Folk et al. (1996). Copyright 1996 by the American Physical Society.

The gate voltage V_{g1} can be swept over 40 peaks before the conductance of the point contacts changes noticeably. Thus we can probe the system's energy-level structure considerably by varying B, V_{g1}, and V_{g2} as we measure values of the peak conductances.

Figure 7.32 shows reproducible variations in the Coulomb blockade peak heights, which are independent of the sign of B to high precision. (Note that the conductance almost drops to zero at several places.) These peak heights, as well as those in Fig. 7.31, depend on the coupling between each point-contact mode and the dot, and this coupling in turn depends on the spatial probability distribution of each dot energy eigenstate and on the location of the point contacts ("leads") along the dot boundary. In the case of lead coupling to one eigenstate, the maximum conductance is

$$g_{\text{peak}} = \frac{e^2}{\hbar} \frac{\pi}{2kT} \frac{\Gamma_1 \Gamma_2}{\Gamma_1 + \Gamma_2}, \tag{7.27}$$

where $\Gamma_{1,2}/\hbar$ are the electron coupling rates to leads 1,2 (Beenakker, 1991). The quantities $\Gamma_{1,2}$ depend on the dot wavefunction near the leads.

If we want to analyze conductance data, such as that from Fig. 7.32, in terms of the scarring of irregular dot eigenstates by periodic orbits (Sections 5.4 and 7.2), then we can adopt a standard procedure that introduces a correlation function $S(E, \lambda)$, where λ is the parameter to be varied and E is the electron's energy in the billiard system. Following the procedure of Heller (1984), let $E_i(\lambda)$ be the eigenenergies, $|E_i(\lambda)\rangle$ be the corresponding eigenstates, and $|\phi\rangle$ be an initial wavepacket with phase-space distribution $\rho_\phi(\mathbf{p}, \mathbf{q})$, and let us define the *overlap*

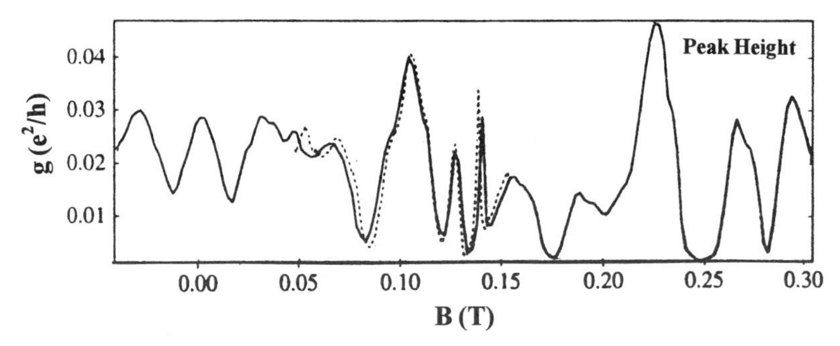

Figure 7.32. The maximum Coulomb blockade peak conductance $g_{\max}$—in V_{g1} scans like the one shown in Fig. 7.31—now plotted versus magnetic field strength B. The superposition of values of $g_{\max}$ for positive B (solid curve) and for negative B (dashed curve) shows the symmetry and repeatability of this experimental spectral data. The data are qualitatively of the same type as that in Fig. 7.30a. From Folk et al. (1996). Copyright 1996 by the American Physical Society.

intensities $p_{\phi i} \equiv |\langle \phi | E_i(\lambda) \rangle|^2$. Then we take $S(E, \lambda)$ to be the Fourier transform of the autocorrelation function for the packet

$$S(E, \lambda) \equiv \frac{1}{2\pi} \int_{-\infty}^{\infty} dt \exp\left(i \frac{Et}{\hbar} \right) \langle \phi | \phi(t, \lambda) \rangle \equiv \sum_i p_{\phi i}(\lambda) \delta[E - E_i(\lambda)], \qquad (7.28)$$

where the $p_{\phi i}(\lambda)$ are coefficients.

Following Tomsovic (1996), let us specialize to the almost-closed stadium billiards quantum dot, and take λ to be the ratio of side length to semicircle diameter. As λ increases, the stadium stretches horizontally, but we scale to preserve the stadium's area. We take $|\phi\rangle$ to be a horizontally stretched wavepacket that is initially centered in the stadium; the packet is then guided by the horizontal-bounce unstable periodic orbit and by neighboring orbits, as in Section 2.1c. Let us choose the packet momentum to make the length of the stadium roughly span 12 de Broglie wavelengths. Figure 7.33 presents a graphical representation of $S(E, \lambda)$ for this situation, shown over small ranges of E and λ. Each small vertical line segment is centered on a numerically calculated energy eigenvalue and a value of λ; the height of each segment is proportional to a calculated value of $p_{\phi i}(\lambda)$. In the figure we see that as λ increases, the energy levels with large overlap intensities characteristically

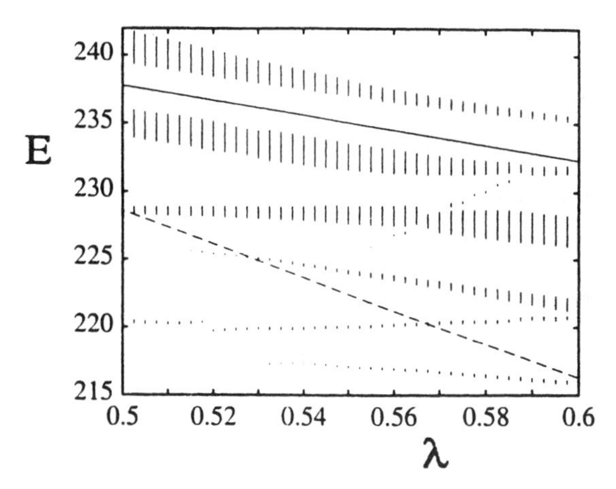

Figure 7.33. A graphical representation of the Fourier transform $S(E, \lambda)$ of the autocorrelation function for the evolution of a ballistic electron wavepacket that is initially oriented and stretched along the horizontal axis of the stadium billiards system. The quantity E is the electron's energy, and λ is a parameter that is varied, such as V_{g1} in Fig. 7.31 or B in Fig. 7.32. In the present figure, λ is the ratio of the side length to semicircle diameter of the stadium. Each small vertical line segment shown in the (E, λ) parameter space is centered on a computed energy eigenvalue and its value of λ; the heights of these line segments are proportional to the overlap intensities of the eigenstates with the initial wavepacket; see text. The overlap intensities are peak heights in the spectrum $S(E, \lambda)$, see Eq. (7.28), which in turn can be calculated from periodic orbit theory. From Tomsovic (1996). Copyright 1996 by the American Physical Society.

decrease. If a state is scarred by the horizontal-bounce periodic orbit and the stadium is smoothly lengthened, then maintaining a constant number of wavelengths across the stadium implies reducing the momentum—in other words, decreasing the energy, as observed in the figure.

From the above, we can infer that signatures of periodic-orbit enhancement of wavefunctions—and thus of dot conductance, Eq. (7.27)—are large values of $p_{\phi i}(\lambda)$, as well as characteristic slopes $\partial E/\partial\lambda$. Thus orbit scarring of some set of energy eigenstates should produce large values of a second correlation function; this function is an average over the N states in an energy band

$$C_{\phi}(\lambda) \equiv \left\langle \tilde{p}_{\phi i}(\lambda) \frac{\partial \tilde{E}_i(\lambda)}{\partial \lambda} \right\rangle_E . \tag{7.29}$$

The tildes indicate that the variables are already zero-centered and rescaled to unitless quantities, with variance one. Tomsovic (1996) showed that, in the absence of scarring, $C_{\phi}(\lambda) = 0 \pm N^{-1/2}$ for *all* choices of $|\phi\rangle$. In the range of the 300th to 400th excited levels of the stadium billiard, $C_{\phi}(\lambda)$ shows large correlations similar to the data in Fig. 7.32. This similarity is expected since g_{peak} is proportional to a quantity similar to $p_{\phi i}(\lambda)$ (Beenakker, 1991). The experimental fluctuations in the locations of the Coulomb blockade peaks are related to the slopes $\partial E/\partial\lambda$.

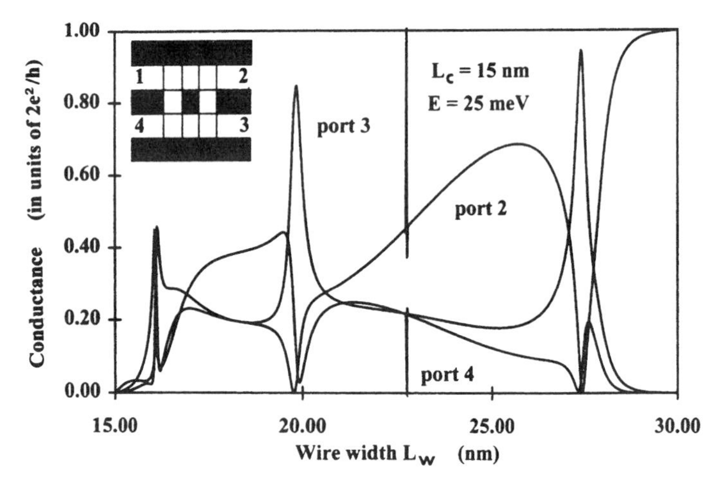

Figure 7.34. Calculated steady-state single-mode ballistic electron matter-wave conductances of the four-port directional coupler shown in Fig. 7.29d, as a function of width L_W of the two parallel quantum wire waveguides connecting ports (1, 2) and (3, 4). Port 1 was taken as the reference port, with forward and backward waves for describing wave injection and reflection. Ports 2, 3, and 4 were assumed to be perfectly matched, with zero reflection. Values of L_W exist where matter-wave interference simultaneously produces zero conductance for port 2 and maximal conductance for port 3. Reprinted from *Superlattices and Microstructures*, **17**, O. Vanbésien and D. Lippens, "Quantum interference electronic switch," pp. 197–200, copyright 1995, by permission of the publisher Academic Press.

7.5d. Waveguide Devices

Among the two-port ballistic electron waveguide nanostructures that have been studied, one finds uniform waveguides, the T-stub, and the double bend (Weisshaar et al., 1991; Wu et al., 1993; Burgnies et al., 1997). Among the four-port nanotechnology devices is a 3 dB directional coupler having 17 dB directivity (Vanbésien and Lippens, 1994). A similar coupler is shown schematically in Fig. 7.29d, a coupler consisting only of rectangular spatial regions. This coupler consists of two parallel quantum wire waveguides connected by two transverse conducting branches. Schrödinger's equation can be solved numerically for this coupler, and the conductances to ports 2, 3, and 4 computed for a ballistic electron matter-wave entering port 1 (Vanbésien and Lippens, 1995).

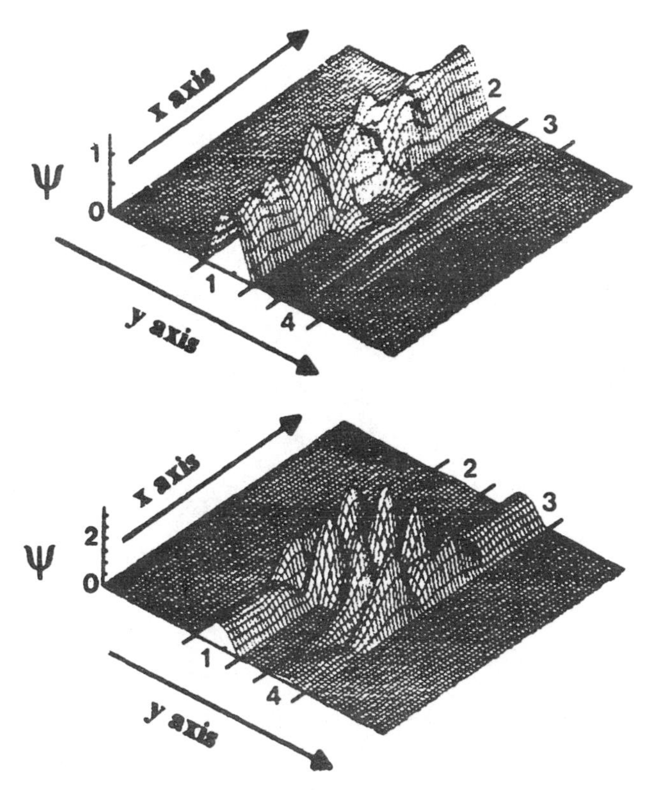

Figure 7.35. Ballistic electron wavefunctions inside the four-port directional coupler. The values of the quantum wire waveguide width are (a) $L_W = 29.8$ nm for straight-through electron transmission through port 2, and (b) $L_W = 27.42$ nm for transmission through port 3. The effective value of L_W can be electrically switched by side metallic-gate voltages on the quantum wires. Reprinted from *Superlattices and Microstructures*, **17**, O. Vanbésien and D. Lippens, "Quantum interference electronic switch," pp. 197–200, copyright 1995, by permission of the publisher Academic Press.

Figure 7.34 shows these conductances as a function of the width L_W of each of the quantum wires, where in practice the width would be controlled by a confinement gate voltage. There is strong quantum wave interference, which varies with L_W. For the indicated transverse branch widths $L_C = 15\,\text{nm}$ and electron energy $E = 25\,\text{meV}$, the conductance through port 3 $G_3 = 94\%$ at $L_W = 27.42\,\text{nm}$, and $G_2 = 99\%$ for L_W near 30 nm. The moduli of the wavefunctions for these two cases are shown in Fig. 7.35. There is little reflection—note the absence of ripples in port 1 for the lower wavefunction. Thus switching L_W between 27.42 nm and 29.8 nm will efficiently switch coherent ballistic electron matter waves between port 2 and port 3. The electron escape time from the device is estimated to be $\sim \frac{1}{2}$ps. The effects of a finite switching-time duration are not considered here, but we discuss switching effects in Chapter 9.

8

SINUSOIDALLY DRIVEN SYSTEMS

8.1. MODULATED QUANTUM STATES, FLOQUET STATES, AND QUASIENERGY

8.1a. Sinusoidally Modulated Quantum States

We know that the transmission of information (signals) by electromagnetic waves involves the imposition of time dependence on one or more of the parameters that characterize the waves. Our bedside radios have AM and FM bands where the carrier wave amplitude A and the carrier frequency ω_0 are modulated, respectively. In another example—that of a sinusoidal phase modulation at the modulation frequency ω—a constant phase factor $\exp(i\phi)$ becomes $\exp[i(\phi + \alpha \cos \omega t)]$. If we measure the frequency spectrum of the wave intensity here, we find sideband peaks at the frequencies $\omega_0 \pm m\omega$, where the *sideband index* m takes on the integer values $m = 0, \pm 1, \pm 2, \ldots$. These peaks arise mathematically from the Fourier decomposition

$$\exp(i\alpha \cos \omega t) = \sum_{m=-\infty}^{\infty} i^m J_m(\alpha) \exp(im\omega t). \tag{8.1}$$

Here J_m is the mth ordinary Bessel function; it determines the amplitude of the sidebands of index $\pm m$. The parameter α depends on the specific properties of the system being modulated and is proportional to the strength of the modulator.

Matter waves can be time-modulated in much the same way as electromagnetic waves can. The quantum system is then interacting with an entity that creates the modulation, called a driver. Rather than being autonomous, the system is therefore a *driven system*. The Hamiltonian of the system now includes the interaction with the driver, and the Hamiltonian becomes time dependent. The quantum system's energy is not constant, and the driven system has no energy eigenstates.

Every autonomous quantum system can be modulated by various types of interactions with external drivers. Such interactions are essential for the everyday usefulness of autonomous quantum systems. There are many types of driven quantum systems of interest. Two of the simplest types are (1) atoms or semiconductor quantum wells placed in time-oscillating electric *fields*—where we have *field modulation*; and (2) semiconductor superlattices or quantum dots with a modulated electric *potential*—where we have *potential modulation*.

Figure 8.1 schematically shows these two types of modulation, for the case of an electron in an infinite square well. In this case the field modulation arises from adding an electric dipole interaction term $eFz\cos\omega t$ to the Hamiltonian (Truscott, 1993). Here z is the position of the electron in the well, and F is the amplitude of the imposed electric field $F\cos\omega t$. In the case of potential modulation, the added term is simply $eV_0\cos\ \omega t$ (Tien and Gordon, 1963). The parameter α in Eq. (8.1) differs for the two cases, with $\alpha \sim \omega^{-2}$ for field modulation, and $\alpha \sim \omega^{-1}$ for potential modulation (Wagner, 1996). (Section 8.2b further considers driven particles in rectangular potential wells.)

Laser spectroscopy experiments have clearly revealed the existence of parameter regimes where the primary effect of the driving is modulation of individual quantum energy eigenstates. Figure 8.2 shows observed microwave-induced photon sidebands in the spectrum of the hydrogen atom, with the sidebands occurring in the highly-excited region of atom energies where the principal quantum numbers are $n = 43,\ 44,\ 45$. The applied driving microwave electric field had a frequency near 7.8 GHz, corresponding to about 1/10 the free-atom local energy-level spacing.

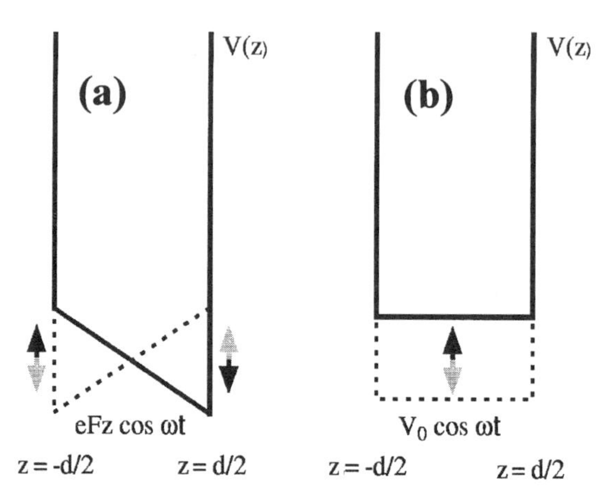

Figure 8.1. Time-dependent effective potentials for a charged particle in an infinite square well. (a) In the case of electric field modulation, the electric dipole interaction of the bound particle with a field $F\cos(\omega t)\,\hat{\mathbf{z}}$ produces the effective potential $V(z) = eFz\cos(\omega t)$. The bottom of the well "seesaws" back and forth at the angular frequency ω. (b) In the case of potential modulation, a spatially uniform potential oscillates as $V_0\cos(\omega t)$. The bottom of the well translates up and down at frequency ω. From Wagner (1996). Copyright 1996 by the American Physical Society.

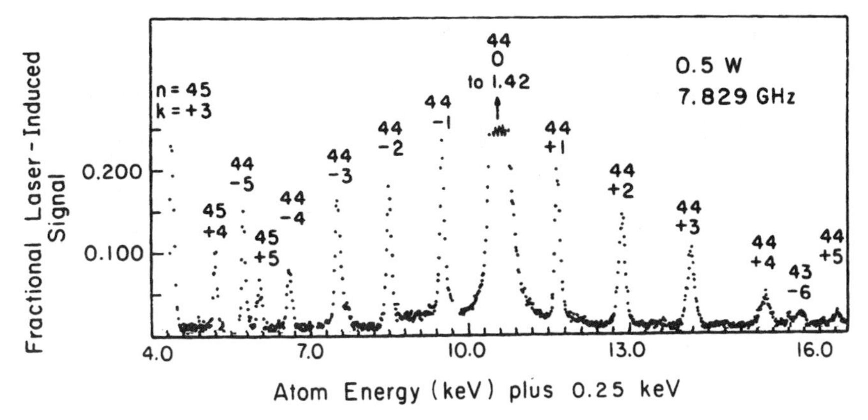

Figure 8.2. Sidebands in the experimental spectrum of highly excited hydrogen atoms that are field-modulated by a monochromatic microwave electric field. While in the microwave field, atoms in a fast atomic hydrogen beam were excited by a monochromatic CW CO_2 midinfrared laser beam, from the $n = 10$ state to energies near that of the $n = 44$ state. The effective frequency of the laser was Doppler-tuned by varying the kinetic energy of the fast atoms. In the absence of the modulation, the $n = 44$, $m = 0$ peak lies at the observed and expected value near the center of the displayed spectrum. Turning on the microwaves produces the $n = 44$ sidebands for $m = \pm 1, \pm 2, \ldots, \pm 5$, as well as some observed sidebands for $n = 43$ and 45. From Bayfield et al. (1981). Copyright 1981 by the American Physical Society.

In this case $\alpha = 3n(n_1 - n_2)F/2\omega$—where n, n_1, n_2 are the parabolic quantum numbers of the atomic energy eigenstate. Locations of the first Bessel function maxima were measured as functions of m, n, and ω (Bayfield et al., 1981). These locations were in good agreement with the form of Eq. (8.1), and they are in accord with a full theory of the laser spectroscopy experiments; see Bersons (1983) and Fig. 8.3.

Microwave photon sidebands have also been observed for the ground and first excited states of electrons in a quantum dot. Figure 8.4 shows a surface-electron-microscopy (SEM) photograph of a sample used by Oosterkamp et al. (1997). As in Section 7.5, the dot was created by voltages on metallic gates, now deposited on a lithography-modified layer within a GaAs/AlGaAs heterostructure. There is a bias voltage between an electron source region and an electron drain region, with the voltage producing the observed dot current. Negative voltages on the outer two pairs of gates created entrance and exit point contacts. The center gates, at potential V_g, confined the electron gas to a $600 \times 300\ \text{nm}^2$ dot. The voltage V_g shifts the quantum energy levels in the dot with respect to the Fermi energy levels of the entrance and exit leads. Thus by adjusting V_g, we tune the system into tunneling resonance (Sections 6.1 and 7.5).

Microwaves are now imposed on the dot via the center gate region, as indicated in Fig. 8.4. Then sidebands will appear on each side of each tunneling resonance peak in the transmitted current. The effect of microwave modulation on the tunneling is

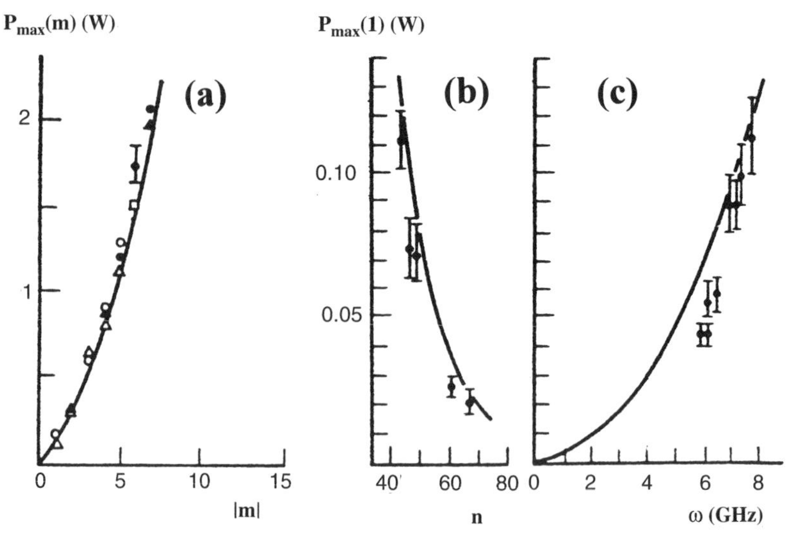

Figure 8.3. Comparison of experimental and theoretically predicted levels P_{max} of microwave waveguide transmitted power that are required to produce maximum excitation of sidebands in the spectrum of field-modulated, highly excited hydrogen atoms. The points are from experimental spectra like the spectrum shown in Fig. 8.2, while the solid curves are theoretical. (a) The dependence on sideband index m, for fixed microwave frequency ω. (b) The dependence on the principal quantum number n of the atom, for fixed ω and m = 1. (c) The dependence on ω, for fixed n and m = 1. Reprinted with permission from Bersons (1984). Copyright 1984 American Institute of Physics.

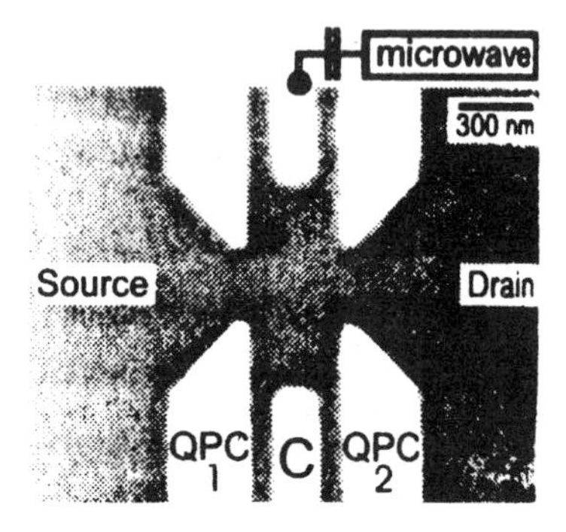

Figure 8.4. A surface-electron-microscopy image of the semiconductor nanostructure used for the study of microwave modulation of the quantum tunneling of carriers out of a quantum dot. The microwaves are introduced at the top central metallic gate electrode. A static potential V_g on metallic gate electrode C shifts the quantum energy levels of the semiconducting dot located in the dark region in the center of the nanostructure, with respect to the Fermi energy levels for electrons coming from the source region and leaving at the drain region. Current will flow when we apply a voltage between the source and the drain regions, and a spectrum of tunneling resonances will be observed in this current as V_g is varied. From Oosterkamp et al. (1997). Copyright 1997 by the American Physical Society.

called *photon-assisted tunneling* (PAT). Figure 8.5 shows some PAT data for dot current plotted above the plane defined by V_g and microwave power. We observe the formation of sidebands about the ground-state tunneling resonance, and we also see the m = 0 central spectral line reduced toward zero as the first zero in $J_0(\alpha)$ is approached.

The inset in Fig. 8.5 shows the current as a function of V_g and $\alpha = eV_\mu/\hbar\omega$. The tunneling rate is in good agreement with a potential-modulation model that assumes the modulation occurs in the barrier regions of the point contacts. The model predicts that the tunneling rate is given by

$$\Gamma_\mu(E, \alpha) = \sum_{m=-\infty}^{\infty} J_m^2(\alpha)\Gamma(E + m\hbar\omega),$$

where $\Gamma(E)$ is the tunneling rate without the microwaves. For both the ground and first excited dot states, the sideband spacings varied as $\hbar\omega$ when the energy-level spacing $\Delta E \equiv E_1 - E_0$ was held fixed. Also, an applied magnetic field B was used to vary E_1 and E_0 with $\hbar\omega$ held fixed. Then each set of sidebands moved together as B

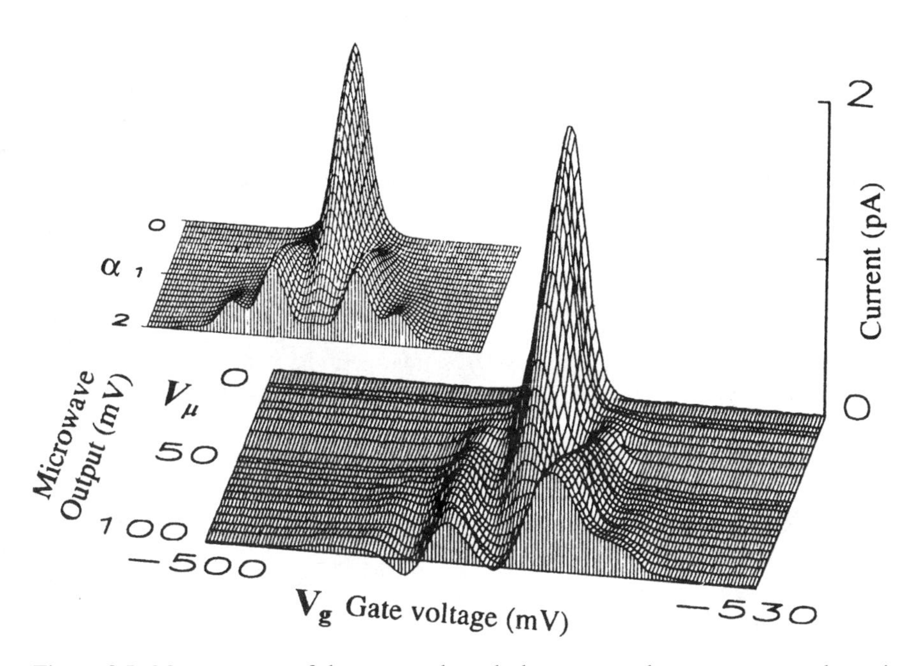

Figure 8.5. Measurements of the current through the quantum dot nanostructure shown in Fig. 8.4, as a function of the gate voltage V_g and of the output voltage V_μ of the microwave power source. The first zero of the m = 0 spectral line is observed near V_μ = 100 mV, and the m = ±1 sidebands are clearly seen. The inset shows the sideband structure expected theoretically, as a function of the microwave voltage parameter $\alpha = eV_\mu/\hbar\omega$ instead of V_μ alone; ω is the microwave frequency. From Oosterkamp et al. (1997). Copyright 1997 by the American Physical Society.

was varied, further supporting the modulation effect of the microwave driving the tunneling electrons (Oosterkamp et al., 1997).

8.1b. Floquet Theory and Quasienergy

When we considerably increase a driving strength parameter (such as F or V_0) beyond the modulated eigenstate regime just described, then energy-changing transitions between the energy eigenstates of the unperturbed (undriven) system become significant. A new picture of the driven quantum system becomes necessary. Note that the Hamiltonian for the driven system is still time periodic. Let the driving period be T. The Hamiltonian formally determines the system's time evolution operator; that is, the Hamiltonian determines the system's propagator—see Eq. (6.24)—which in this chapter we call U. As a result U must be time periodic with period T:

$$U(t+T) = U(t).$$

Thus the evolution repeats itself in time, so that knowing the evolution for a first period means knowing the evolution for all times. The one-period propagator is called the *Floquet propagator*, in recognition of the early work by G. Floquet on the theory of differential equations that have time-periodic coefficients (Floquet, 1883; Jordan and Smith, 1987, pages 245–260). Floquet proved a general theorem that is the basis for the *Floquet theory* of sinusoidally driven systems. A statement of *Floquet's theorem* is as follows. Consider the dynamical system comprised of a set of N ordinary differential equations for the vector $\mathbf{Z}(t)$,

$$\frac{d}{dt}\mathbf{Z}(t) = -i\bar{\bar{W}}(t)\mathbf{Z}(t),$$

where the $N \times N$ coefficient matrix $\bar{\bar{W}}(t)$ is time periodic with period $T \equiv 2\pi/\omega$. Then solutions $\mathbf{Z}_\nu(t)$ labeled by an index ν can be found in the form

$$\mathbf{Z}_\nu(t) = \exp(-i\varepsilon_\nu t)\mathbf{z}_\nu(t),$$

where the vector $\mathbf{z}_\nu(t)$ is time periodic with period T. This latter periodicity means that $\mathbf{z}_\nu(t)$ can be expressed as a Fourier series:

$$\mathbf{z}_\nu(t) = \sum_{m=-\infty}^{\infty} \mathbf{z}_{\nu m} \exp(im\omega t).$$

Thus Floquet's theorem asserts that there exist vector solutions of the dynamical system having the form

$$\mathbf{Z}_\nu(t) = \exp(-i\varepsilon_\nu t) \sum_{m=-\infty}^{\infty} \mathbf{z}_{\nu m} \exp(im\omega t).$$

The quantity ε_ν in general Floquet theory is called a *characteristic exponent*, and there are at most N distinct such exponents within the interval $0 \le \varepsilon_\nu \le \omega$. Each of these N exponents provides a solution $\mathbf{Z}_\nu(t)$ of the dynamical system. Thus the most general solution is a superposition of the form

$$\begin{aligned} \mathbf{Z}_n(t) &= \sum_{\nu=1}^{N} a_{n\nu}\mathbf{Z}_{\nu n}(t) \\ &= \sum_{\nu=1}^{N}\sum_{m=-\infty}^{\infty} a_{n\nu}\exp(-i\varepsilon_\nu t + im\nu t)\mathbf{z}_{\nu m} \end{aligned}$$

in which the constants $a_{n\nu}$ are chosen to ensure given initial conditions.

Returning to our problem of a sinusoidally driven quantum system, the time periodicity has implications for the time-dependent wavefunctions. Solutions can be found that have the form

$$\Psi_\varepsilon(\mathbf{q}, t) = \Phi_\varepsilon(\mathbf{q}, t)\exp\left(-i\frac{\varepsilon t}{\hbar}\right) \tag{8.2}$$

with $\Phi_\varepsilon(\mathbf{q}, t+T) = \Phi_\varepsilon(\mathbf{q}, t)$.

This wavefunction is time periodic, except for a phase factor that has the same form as the time-dependent phase factor for energy eigenstates of autonomous systems. When we insert Eq. (8.2) into the time-dependent Schrödinger equation, we're immediately led to the Schrödinger equation that must be satisfied by Φ_ε

$$[H(t) - \varepsilon]\Phi_\varepsilon = i\hbar\frac{\partial\Phi_\varepsilon}{\partial t}, \tag{8.3}$$

which clearly requires the periodicity of Φ_ε. Wavefunctions of the form (8.2) are called *Floquet states*, and the quantity ε is called the *quasienergy* associated with the Floquet state Φ_ε. As we see below, the role of Floquet states in periodic systems is analogous to that of energy eigenstates in time-independent systems.

We've seen how time-periodic drivers produce sidebands in the undriven system's spectrum. Thus in one approach to strongly driven systems, we deal with the periodic time dependence by Fourier-expanding both the wavefunction and the Hamiltonian operator, and we deal with the spatial dependence by further expanding the wavefunction in a basis $|k\rangle$ of the undriven system (Shirley, 1965). We can then work numerically using the basis of product states $|k\rangle|m\rangle$, where the $|m\rangle$ are the Fourier expansion "states" defined by $\exp(im\omega t) = \langle t|m\rangle$, $\omega \equiv 2\pi/T$. This approach is often a good one for quantum basis-set expansion numerical calculations, although the essential physics of the quantum time evolution is generally not revealed very clearly by the interplay of the coupled basis states $|k\rangle|m\rangle$. In Section 1.2a we remarked about a similar difficulty with the brute force basis-set expansion approach in the case of autonomous systems. So once again, before investigating the driven quantum system, we'll first explore the alternative of developing the classical physics.

8.1c. Extended Phase Space

In dealing with time-periodic systems in a most physical way, the key idea is to include time t itself as a phase-space variable. To have a canonical Hamiltonian formulation, t must have a conjugate momentum that we call p_2. Phase space formally becomes $[\mathbf{q}, \mathbf{p}; t, p_2 \equiv -H(\mathbf{q}, \mathbf{p}, t)]$, and we define a new Hamiltonian by

$$K(\mathbf{q}, \mathbf{p}; t, p_2) \equiv H(\mathbf{q}, \mathbf{p}, t) + p_2 = 0. \tag{8.4}$$

Although it may seem nonsensical to define a Hamiltonian that is identically zero, nevertheless this Hamiltonian K is a function of the extended set of phase-space variables that satisfies an extended set of Hamilton's equations. These equations are simply the usual Hamilton's equations, plus two more for t and p_2. If we use a variable t* to parameterize the system's classical trajectories in the extended phase space, then

$$\frac{dt}{dt^*} = \frac{\partial K}{\partial p_2} = 1, \tag{8.5}$$

or $t = t^* +$ constant. This result explains why the usual Hamilton's equations for $\mathbf{q}$, $\mathbf{p}$ still hold. We also see that

$$\frac{dp_2}{dt^*} = -\frac{\partial K}{\partial t} = -\frac{\partial H}{\partial t} \tag{8.6}$$

then becomes the last Hamilton's equation. Equation (8.6) states that time variations in p_2 are the negative of time variations in H, a fact that justifies identifying p_2 with the energy of the driving subsystem—the subsystem responsible for the variations H(t) of the driven quantum system's "energy." The extension of phase space has enlarged the original driven system to include the driver, thus producing an "autonomous total system."

Now we quantize the new variables t, p_2 in the usual way (Howland, 1980):

$$p_2 \equiv -i\hbar \frac{\partial}{\partial t}, \qquad (t\Psi)(t) \equiv t\Psi(t), \tag{8.7}$$

with $[p_2, t] = i\hbar$. We also have

$$K \equiv H(t) - i\hbar \frac{\partial}{\partial t}, \qquad \text{with} \quad K\Psi = i\hbar \frac{\partial \Psi}{\partial t^*}. \tag{8.8}$$

Thus, in the auxiliary time variable t*, we have an extended time-dependent Schrödinger equation with the formal solution

$$\Psi(t^*) \equiv U(t^*, 0)\Psi(0) \equiv \exp\left(-it^* \frac{K}{\hbar}\right)\Psi(0). \tag{8.9}$$

Note that we've not yet used time periodicity in this extension of phase-space theory.

8.1d. The Quasienergy Operator and Eigenstates

Comparing Eqs. (8.8) with Eq. (8.3) leads us to suspect that as long as we have time periodicity, a connection exists between the operator K and the quasienergy ε. To see this connection we first define the one-period time translation operator $T = \exp(\mathrm{i}\mathrm{T}p_2/\hbar)$, an operator that commutes with K, $[K, T] = 0$. It is then natural to seek simultaneous eigenstates of K and T. Solutions of the eigenvalue equation for T take the precise Floquet form (8.2), without any identification yet being made with ε. However, let us define the eigenvalue equation for K by

$$K[\exp(\mathrm{i}\lambda \mathrm{t})\Phi_\lambda(\mathbf{q}, \mathrm{t}) \equiv \varepsilon[\exp(\mathrm{i}\lambda \mathrm{t})\Phi_\lambda(\mathbf{q}, \mathrm{t})],$$

or

$$\left(H(\mathrm{t}) - \mathrm{i}\hbar\frac{\partial}{\partial \mathrm{t}}\right)[\exp(\mathrm{i}\lambda \mathrm{t})\Phi_\lambda] = \varepsilon[\exp(\mathrm{i}\lambda \mathrm{t})\Phi_\lambda].$$

Expanding the time derivative of the product produces

$$K\Phi_\lambda(\mathbf{q}, \mathrm{t}) = (\varepsilon - \hbar\lambda)\Phi_\lambda(\mathbf{q}, \mathrm{t}).$$

We can remove the λ degeneracy in this equation by restricting K to operating on the subspace of periodic functions with boundary conditions $\Psi(\mathrm{T}) = \Psi(0)$. This creates a modified operator K_T—called the *quasienergy operator*—that satisfies the quasienergy eigenvalue equation

$$K_\mathrm{T}\Psi_\varepsilon(\mathbf{q}, \mathrm{t}) = \varepsilon\Psi_\varepsilon(\mathbf{q}, \mathrm{t}). \tag{8.10}$$

The set of all solutions of Eq. (8.10) is exactly the set of all Floquet states (8.2).

For a simple illustration of Floquet states, let us return to the problem of a two-level atom in a linearly polarized EM electric field $\mathrm{F(t)} = \mathrm{F}\cos\omega \mathrm{t}$ (Section 2.2). There we used expansion (2.17) and, working within the rotating-wave approximation (RWA), we solved Eq. (2.19) for the time-dependent expansion coefficients. On the other hand, in the Floquet picture the expansion coefficients are by definition *independent* of time, and the expansion uses the basis of periodically time-dependent Floquet states along with their quasienergies $\varepsilon_{1,2}$:

$$\Psi(\mathrm{t}) = \mathrm{c}_1\mathrm{u}_1(\mathrm{t})\exp\left(-\mathrm{i}\frac{\varepsilon_1 \mathrm{t}}{\hbar}\right) + \mathrm{c}_2\mathrm{u}_2(\mathrm{t})\exp\left(-\mathrm{i}\frac{\varepsilon_2 \mathrm{t}}{\hbar}\right). \tag{8.11}$$

This is the same form as Eq. (2.17) except for its time dependence of c_1 and c_2. As suggested by Eq. (8.1), one can find u_1 and u_2 in the RWA by using a Fourier decomposition form of Floquet theory (Autler and Townes, 1955). The general procedure for finding the eigenvalues and eigenvectors of a 2×2 Hermitian matrix (Cohen-Tannoudji et al., 1977, Compliment B_{IV}) produces the results (Holthaus and Just, 1994)

$$u_1(t) = \left(\frac{1}{2}\frac{\Omega + \Delta}{\Omega}\right)^{1/2} \exp(i\omega t)\psi_1(\mathbf{r}) + \left(\frac{1}{2}\frac{\Omega - \Delta}{\Omega}\right)^{1/2} \psi_2(\mathbf{r}),$$

$$u_2(t) = -\left(\frac{1}{2}\frac{\Omega - \Delta}{\Omega}\right)^{1/2} \exp(i\omega t)\psi_1(\mathbf{r}) + \left(\frac{1}{2}\frac{\Omega + \Delta}{\Omega}\right)^{1/2} \psi_2(\mathbf{r}),$$

$$\varepsilon_1 = \frac{1}{2}(E_1 + E_2) + \frac{\hbar\omega}{2} - \frac{\hbar\Omega}{2},$$

$$\varepsilon_2 = \frac{1}{2}(E_1 + E_2) + \frac{\hbar\omega}{2} + \frac{\hbar\Omega}{2},$$

where the frequency detuning is $\Delta = (E_2 - E_1)\hbar - \omega = \omega_0 - \omega$ and the Rabi-flopping frequency $\Omega = (\Delta^2 + \omega_R^2)^{1/2}$, as in Section 2.2. Thus apart from a common constant $\frac{1}{2}(E_1 + E_2 + \hbar\omega)$, the eigenvalues of the quasienergy operator K_T are $\pm\hbar\Omega/2$.

The case of exact resonance ($\Delta = 0$) is interesting. Then

$$u_1(t)\exp\left(-i\frac{\varepsilon_1 t}{\hbar}\right) = \frac{1}{\sqrt{2}}\exp\left(+i\frac{\omega_R t}{2}\right)\left[\psi_1(\mathbf{r})\exp\left(-i\frac{E_1 t}{\hbar}\right) + \psi_2(\mathbf{r})\exp\left(-i\frac{E_2 t}{\hbar}\right)\right],$$

$$u_2(t)\exp\left(-i\frac{\varepsilon_2 t}{\hbar}\right) = \frac{1}{\sqrt{2}}\exp\left(-i\frac{\omega_R t}{2}\right)\left[-\psi_1(\mathbf{r})\exp\left(-i\frac{E_1 t}{\hbar}\right) + \psi_2(\mathbf{r})\exp\left(-i\frac{E_2 t}{\hbar}\right)\right].$$

If we choose $\Psi(0) = \psi_2(\mathbf{r})$ at $t = 0$, then we see that on resonance the coefficients in Eq. (8.11) must be $c_1 = -1/\sqrt{2}$ and $c_2 = +1/\sqrt{2}$. From Eq. (2.23) the resonant population inversion time is $T_{inv} = \pi/\omega_R$, which we now see can be written as the condition

$$(\varepsilon_2 - \varepsilon_1)\frac{T_{inv}}{\hbar} = \pi.$$

This is exactly the condition for constructive interference of the two Floquet states at time $t = T_{inv}$. We conclude that Rabi flopping (Section 2.2) must be the same thing as the time-dependent interference of two Floquet states.

We note in passing that Floquet states in the weak-field RWA approximation are often called by another name—*dressed states*. Much is known about these states; see the two-volume treatise by Shore (1990).

As an alternative to solving Eq. (8.10), we can solve the eigenvalue equation

$$U_F\Psi = \exp\left(-i\varepsilon\frac{T}{\hbar}\right)\Psi, \qquad (8.12)$$

where the Floquet operator $U(T, 0) \equiv U_F = \exp(iTK_T/\hbar)$ is obtained by taking $t^* = T$ in Eq. (8.9). As an example of Eq. (8.12), for a free particle in a time-periodic electric field $F(t) = F\sin(\omega t + \phi)$, the Hamiltonian has the form

$$H(t) = p^2 + qF(t).$$

Then, with $\hbar \equiv 1$,

$$U_F = U\left(t = \frac{2\pi}{\omega},\ 0\right) = \exp\left[-i\frac{2\pi}{\omega}\left(\frac{F^2}{2\omega^2} + \frac{F}{\pi\omega}p + p^2\right)\right]$$

(Cycon et al., 1987). From this expression we can immediately pick out K_T.

Now that we have the quasienergy eigenvalue equation (8.10)—which looks much like the time-independent Schrödinger equation for autonomous systems—we can construct a matrix representation that is useful for numerical work. As in Sambe (1973), we consider the composite Hilbert space $R \oplus T$ of square integrable functions $\{g(\mathbf{q}), h(\mathbf{q}), \ldots\}$ on position space with the inner product

$$\langle g(\mathbf{q})|h(\mathbf{q})\rangle \equiv \int_{-\infty}^{\infty} g^* h\, d\mathbf{q}$$

and square integrable functions $\{a(t), b(t), \ldots\}$ on time space with the inner product

$$\langle a(t)|b(t)\rangle \equiv \frac{1}{T}\int_{-T/2}^{T/2} a^* b\, dt.$$

As we know, the time space is spanned by $\exp(im\omega t)$, $m = 0, \pm 1, \pm 2, \ldots$. We can then construct the set of matrix elements of an operator O,

$$\langle mg|O|m'h\rangle = \frac{1}{T}\int_{-\pi/2}^{\pi/2} dt \int_{-\infty}^{\infty} d\mathbf{q}\ \exp(-im\omega t)g^*(\mathbf{q})\ O\ \exp(im'\omega t)h(\mathbf{q}).$$

These can be used to evaluate the mean value of O given by

$$\frac{1}{T}\int_{-T/2}^{T/2} dt\langle\Psi_\varepsilon(\mathbf{q}, t)|O|\Psi_\varepsilon(\mathbf{q}, t)\rangle \equiv \langle\langle\Psi_\varepsilon|O|\Psi_\varepsilon\rangle\rangle. \qquad (8.13)$$

Each quasienergy ε in Eq. (8.10) is multivalued, $\varepsilon_i = \varepsilon_i^1 + m_i\hbar\omega$, $m_i = 0, \pm 1, \pm 2, \ldots$, where the condition $0 \le \varepsilon_i^1 < \hbar\omega$ requires that ε_i^1 have a value in the first Brillouin zone. (The terminology here stems from the fact that the Brillouin theory of crystalline solids—where periodicity is in space and there is a conjugate quasimomentum—is mathematically completely analogous to the present case of periodicity in time with the conjugate quasienergy.) The Floquet states satisfying Eq. (8.10) for given ε_i^1 and different m_i have different phase factors $\exp(+im_i\hbar\omega)$. These phase factors, however, are canceled out by corresponding factors $\exp(-im_i\hbar\omega)$ in Eq. (8.2). Thus the complete wavefunction is the same for all m_i, and it is perfectly acceptable to work only in the first zone.

Alternatively, we can use an extended Brillouin zone in numerical work after an expansion in undriven system position-space basis energy eigenstates $|k\rangle$ is truncated to just N_k states. We then can correlate one-to-one each energy E_k to a quasienergy $\varepsilon = \varepsilon_i^1 + m_i\hbar\omega$ for some specific integer m_i; in other words, i and k are correlated labels. The correlation is done at low amounts of driving strength and away from resonant driving frequencies, just by seeing which ε_i is closest to each E_k for all possible m_i. We'll see results of this procedure within the remaining sections of this chapter.

8.1e. Mean Energy

If we take the operator O in Eq. (8.13) to be the Hamiltonian $H(t)$, we produce a useful quantity $\langle\langle H\rangle\rangle$, which is called the *mean energy* of the Floquet state $\Psi_\varepsilon(\mathbf{q}, t)$. Equations (8.2) and (8.3) imply that

$$\langle\langle H\rangle\rangle = \langle\langle\Psi_\varepsilon|H(t)|\Psi_\varepsilon\rangle\rangle = \varepsilon + \langle\langle\Phi_\varepsilon|i\hbar\frac{\partial}{\partial t}|\Phi_\varepsilon\rangle\rangle.$$

Using the Fourier series expansion

$$\Phi_\varepsilon = \sum_{m=-\infty}^{\infty} B_m(\mathbf{q})\exp(-im\omega t), \qquad \sum_{m=-\infty}^{\infty}\langle B_m|B_m\rangle = 1,$$

we obtain

$$\langle\langle H\rangle\rangle = \varepsilon + \sum_{m=-\infty}^{\infty} m\hbar\omega\langle B_m|B_m\rangle = \sum_{m=-\infty}^{\infty}(\varepsilon + m\hbar\omega)\langle B_m|B_m\rangle. \tag{8.14}$$

Thus the mean energy can be interpreted as the energy accumulated in each sideband of Ψ_ε and averaged with respect to the weights of these sidebands.

8.1f. EBK Quantization for Regular Time-Periodic Evolution

In the above discussion, we found that the extension of phase space to $2(N+1)$ dimensions leads to $(2N+1)$-dimensional quasienergy shells filling the space $(\mathbf{q}, \mathbf{p}, t)$. Classical evolution on these shells is mathematically equivalent to evolution of an autonomous dynamical system (Section 1.1b) having a $2N+1$

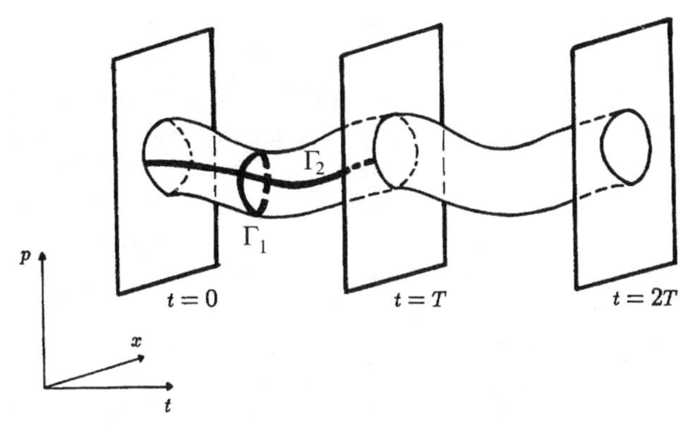

Figure 8.6. A vortex tube in the extended phase space (x, p, t) of a sinusoidally driven 1-d system. The two topologically independent paths Γ_1, Γ_2 on the tube are indicated by heavy curves. The path Γ_1 winds once around the tube, while Γ_2 stretches along the tube such that Γ_2 can be periodically continued. Vortex tubes for sinusoidally driven 1-d systems are analogous to the 2-D tori of 2-d autonomous systems (Fig. 3.1). From Breuer and Holthaus (1991). Copyright 1991 Academic Press, Inc.

phase space. As a result, all the arguments leading up to the EBK theory for regular autonomous evolution (Section 4.2) can be carried through for regular sinusoidally driven evolution (Breuer and Holthaus, 1991; Bensch et al., 1992; Holthaus and Flatté, 1994). The results of such a development are

$$\oint_{\Gamma_i} (\mathbf{p} \cdot d\mathbf{q} - H\, dt) = 2\pi\hbar(n_i + \mu_i), \qquad i = 1, \ldots, N, \tag{8.15}$$

appended by a quantization condition for the quasienergy

$$\varepsilon_{nm} = -\frac{1}{T}\int_{\Gamma_{N+1}} (\mathbf{p} \cdot d\mathbf{q} - H\, dt) + \hbar\omega m, \qquad \omega = \frac{2\pi}{T}. \tag{8.16}$$

In Fig. 8.6 we see the choice of paths for the integrations in the case of $N = 1$, where the evolution is within a 3-D phase space. Since we are considering regular motion, the expected classical evolution mod(T) is on a 2-D torus—which is often called a *vortex tube* when we consider evolution for more than one period T. Note that the path Γ_2 used in evaluating Eq. (8.16) also connects points in (x, p) Poincaré sections taken at two different times separated by the amount T. Taken for trajectories over many periods, such sections are called *stroboscopic Poincaré sections*, regardless of whether the classical evolution in continuous time is regular or chaotic.

8.2. DRIVEN SYSTEMS

Studying a driven system's Floquet states and quasienergies is central to physically understanding the system's time evolution. The time evolution can begin either in a single Floquet state or in some linear combination (such as a packet) of these states. We devote much of the rest of this chapter to what can be learned from Floquet studies, and we consider the results of other numerical calculations for a few specific systems. A number of phenomena we'll uncover are not yet explained by formal quantum theory.

Table 8.1 displays a short list of specific driven 1-d systems of particular interest, both in the past and still today. Their classical Hamiltonians are given in the usual (x, p_x, t) generalized phase-space variables or, when appropriate, in angle-angular momentum (θ, J, t) variables. All these systems are nonintegrable when the time-dependent driving term is included—but they also all are integrable systems (with names easily gleaned from the first column) when the strength (the amplitude) F of the driving term is set equal to zero. The form of the classical Hamiltonian $H_0(I)$ for each undriven system is listed, where I is an appropriate dimensionless scaled-action variable that is generally different for each system. The references listed will either immediately explain the system's action-angle variables or give references for these variables.

Once again we recall that we are concerned with short-time quantum evolution, where radiative decay of excited states can be neglected. In addition, for the last two systems—the driven Morse and Kepler oscillators—we are interested in time intervals smaller than the interval where unit probability occurs for breakup of the undriven system.

We can obtain the quantum energies E_n of the undriven systems by the Bohr–Sommerfeld quantization rule (Section 1.2c), which essentially means we replace the action I in Table 8.1 by nh, where n is an integer quantum number. The dependence of the quantum energy E_n on n varies quite a bit as one goes through the list in Table 8.1. The first three systems are spatially bounded for finite values of energy and have nearest-neighbor quantum energy-level splittings that increase with n. The two driven-pendulum systems are also spatially bounded but are complicated by the undriven system having a separatrix that divides the integrable evolution into different types. The last two systems do not exhibit spatially bounded motion above some finite level of time-averaged energy—thus these two are called spatially open systems. In such spatially open cases, when a finite amount of energy is absorbed from the external driver, the spatial variable x may proceed to infinity. For these systems, note that the energy-level spacing of the undriven system decreases with increasing n, and that there is an energy continuum. Such differences of spatially open systems from the spatially bounded systems without a continuum generate differences in both the classical and quantum driven dynamics. For instance, the spatially open systems classically can have a region of global chaos in phase space, whereas the spatially bounded systems do not. (Spatially open driven systems are discussed later in Section 8.3.)

All the systems listed in Table 8.1 are of physical interest. For instance, to investigate charged-particle dynamics in quantum wells, we would start with the

TABLE 8.1. Some 1-d Model Driven Systems

System	Domain of Variable	Full Hamiltonian	Undriven Hamiltonian (I≡Action)	Recent Reference
Particle in an infinite square well	$-a \leq x \leq a$	$p_x^2/2m + V_{sw}(x) + QFx\cos(\omega t + \phi_0)$	$\pi^2 I^2/8$	Holthaus (1994)
Planar rigid rotor	$0 \leq \theta < 2\pi$	$J^2/2\mathscr{I} + F\cos\theta * \cos(\omega t + \phi_0)$	$I^2/2$	Moiseyev et al. (1994)
Particle in a triangle well	$x \geq 0$	$p_x^2/2m + V_{tw}(x) + QFx\cos(\omega t + \phi_0)$	$(3\pi I)^{3/2}/2$	Holthaus (1995)
Pendulum	$0 \leq \theta < 2\pi$	$J^2/2\mathscr{I} - \alpha\cos\theta + F\cos\theta\cos(\omega t + \phi_0)$	$2\omega_0^2\kappa^2(I) - 1$	Latka et al. (1994)
Angle-modulated pendulum	$0 \leq \theta < 2\pi$	$J^2/2\mathscr{I} - \alpha\cos[\theta + F\cos(\omega t + \phi_0)]$	$2\omega_0^2\kappa^2(I) - 1$	Schlautman and Graham (1995)
Morse oscillator	$x \geq 0$	$p_x^2/2m + D_0\{1 - \exp[-\beta(x - x_e)\}^2$ $+QFD(x)\cos(\omega t + \phi_0)$	$I - I^2/2,\ I \leq 1$	Thachuk and Wardlaw (1995)
Kepler oscillator (SSE)	$x \geq 0$	$p_x^2/2m - e^2/x + eFx\cos(\omega t + \phi_0)$	$-1/2I^2$	Buchleitner and Delande (1995)

$$V_{sw} = \begin{cases} 0, & |x| < a \\ \infty, & |x| \geq a \end{cases} \qquad V_{tw} = \begin{cases} x, & x \geq 0 \\ \infty, & x < 0 \end{cases}$$

$D(x) = x$ *or* $(x + a)\exp[-(x + a)/a]$. Quantity $\kappa(I)$ is obtained by inverting Eq. (4.19).

problems of a particle in an infinite square well and in the triangle well. These problems model quantum wells without and with an applied bias potential difference, respectively. The rigid rotor is a model for the rotational motion of a diatomic molecule, while the Morse oscillator is a model for the vibrational motion of such a molecule. We will see that the driven 1-d Kepler oscillator model is physically relevant for hydrogen atoms in linearly polarized collinear microwave and static electric fields. The Kepler model is also directly applicable to the surface-state electron (SSE) problem; here an electron outside the surface of liquid helium moves perpendicular to the surface, with the electron under the influence of its image charge on the other side of the surface. The usefulness of the two driven pendulum problems is not so evident. However, the phase-modulated pendulum is a reasonable model for a Josephson junction with a time-oscillating externally induced current. The usual driven pendulum is a model for the deflection of an atom beam by a modulated standing light wave.

8.2a. The Driven Planar Rigid Rotor

For the driven planar rigid rotor in Table 8.1, we rewrite the classical Hamiltonian as

$$H(\theta, J, t) = \frac{J^2}{2\mathscr{I}} - \frac{F}{2}\cos(\theta + \omega t) - \frac{F}{2}\cos(\theta - \omega t). \tag{8.17}$$

We see there are two nonlinear-resonance driving terms (Section 5.6). Rather than the periodic motion in a second spatial dimension present in the internally driven resonance case, we now have an external driver providing the second frequency. As noted before, time can be considered as just another coordinate in the generalized phase space; thus it's not surprising that the theory of nonlinear resonance (Section 5.6) carries over to our set of sinusoidally driven 1-d oscillators in Table 8.1. This carryover includes the usefulness of classical and quantum pendulum approximations made in the rotating reference frame.

When one of the resonance terms in Hamiltonian (8.17) is dropped, the system becomes integrable. For a few paragraphs, let us set the last term equal to zero. We quantize the angular momentum according to Eq. (4.16), define the new angle variable $\phi = \theta + \omega t$, and transform the wavefunction $\Psi(\theta, t)$ into the new wavefunction

$$\Phi(\phi, t) = \exp\left[i\left(\ell_1\phi - \frac{\hbar}{2\mathscr{I}}\ell_1 t\right)\right]\Psi(\phi, t),$$

where $\hbar \ell_1 / \mathcal{I} = -\omega$ gives a quantum number ℓ_1 associated with the classical nonlinear-resonance condition. The problem of finding the Floquet states from Eq. (8.10) then reduces to the solution of the Mathieu equation

$$\frac{\hbar^2}{2\mathcal{I}} \frac{d^2\Phi_\varepsilon(\phi)}{d\phi^2} + \left(\varepsilon - \frac{F}{2}\cos\phi\right)\Phi_\varepsilon(\phi) = 0, \qquad \Phi_\varepsilon(\phi + 2\pi) = \Phi_\varepsilon(\phi).$$

The Fourier series expansions of the periodic Mathieu functions $\Phi_\varepsilon(\phi)$ can be evaluated numerically to obtain numerical Floquet states $|n_\varepsilon\rangle$. Since the unperturbed (free-rotor) states $|j\rangle \equiv (1/\sqrt{2})\exp(ij\phi)$ provide the basis, expansion coefficients A_j of these Floquet states can also be obtained numerically; Fig. 8.7 shows some results. Here the Floquet states are labeled by their Floquet-state quantum number n_ε, which is determined by ordering the states in increasing mean energy, Eq. (8.14).

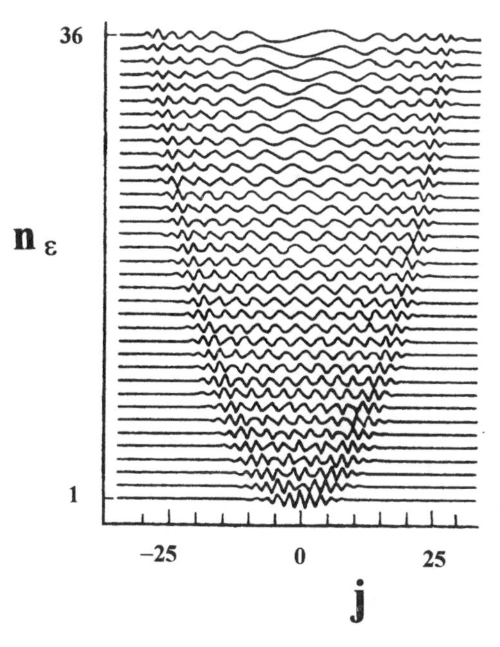

Figure 8.7. Probability amplitudes A_j of numerically obtained Floquet states $|n_\varepsilon\rangle$ for an isolated nonlinear resonance in the quantum sinusoidally driven rigid rotor, in terms of the basis set of free rotor states $|j\rangle$. The Floquet-state quantum number n_ε labels each distribution over the free-rotor quantum number j. The distributions are displaced vertically from one another to aid the eye, and they are ordered by increasing mean energy of the Floquet state. Lines connect the points, again to guide the eye. The dimensionless parameters chosen are $F = 0.4$ and $\hbar^2 = 10^{-3}$. The quasienergies ε for these regular Floquet states of an integrable system lie within the wells of the effective potential $V(\phi) = (F/2)\cos\phi$ and increase with the quasienergy quantum number n_ε. Reprinted with permission from Berman et al. (1987). Copyright 1987 American Institute of Physics.

These states are located below the top of the cos ϕ potential well (Fig. 6.1f) that is formed in ϕ-space due to the last two terms in Hamiltonian (8.17).

It is often useful to construct measures of the extension of Floquet states in position space, in momentum space, or in phase space. Figure 8.7 is a quantum distribution in the free-rotor action (momentum) space. The behavior shown is characteristic of Floquet states that are associated with an (undestroyed) classical nonlinear resonance. We can define the *momentum-space width* ℓ_w of these Floquet states simply as the mean-square width

$$\ell_w \equiv 2\left(\sum_{j=1}^{\infty}(j - \langle j \rangle)^2 |A_j|^2\right)^{1/2}, \qquad \langle j \rangle \equiv \sum_{j=1}^{\infty} j|A_j|^2, \tag{8.18}$$

where the A_j are the amplitudes plotted in Fig. 8.7. Figure 8.8 shows the results. The widths of the Floquet states increase with increasing Floquet-state quantum number n_ε, until the separatrix value of n_ε is approached where the mean energy is close to the top of the cosine well.

Let us now return to our original problem, the nonintegrable driven rotor with the full classical Hamiltonian (8.17). As the stroboscopic section of Fig. 8.9 reveals, when $F = 1$ and $\omega = 1$, then the widths of the two individual nonlinear-resonance islands produced by Hamiltonian (8.17) have grown to strongly overlap, resulting in their almost complete destruction. (The second resonance is not shown in Fig. 8.9; just reflect the figure using $J \to -J$.) These resonance islands remain as only small regular regions within a large but bounded region of chaos. There also is an infinitely large outer regular region, where the motion is that of a KAM system—in this case, the weakly perturbed (hindered) rotor.

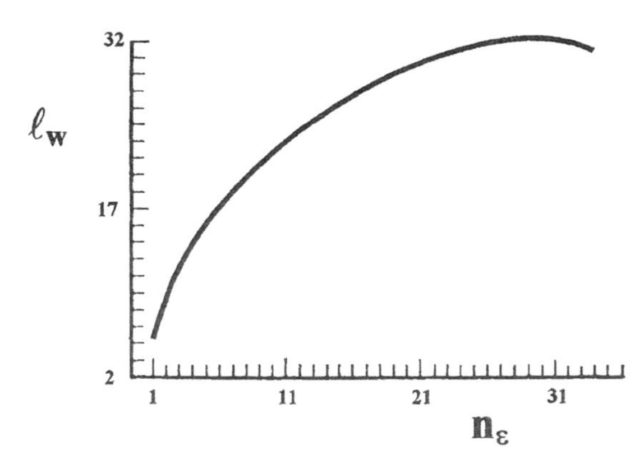

Figure 8.8. The momentum-space width ℓ_w, Eq. (8.18), of the Floquet states shown in Fig. 8.7, as a function of the Floquet-state quantum number n_ε. The widths of the eigenfunctions increase as the separatrix in phase space (or the well tops in position space) is approached, reaching a value characteristic of Floquet states lying in the region of the separatrix. Reprinted with permission from Berman et al. (1987). Copyright 1987 American Institute of Physics.

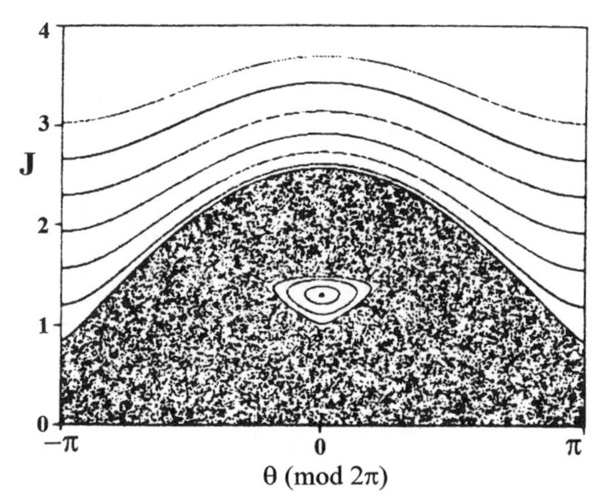

Figure 8.9. A (J, θ) stroboscopic Poincaré section for the nonintegrable full Hamiltonian, Eq. (8.17), of a driven rigid rotor. The dimensionless classical parameters are taken to be $F = 1.0$ and $\omega = 1$. Only the upper region, $J \geq 0$, is shown because of symmetry. There are fundamental nonlinear resonances with winding numbers ω and $-\omega$ that produce islands of regular "librational" motion in the upper and lower J half-planes, respectively. All points shown in the stochastic layer result from a single chaotic classical trajectory. The layer is bounded by regions of regular "rotational" motion, where approximately $J(\theta) = \pm[F(c - \sin^2 \theta/2)]^{1/2}$ and c is a continuous parameter that labels different KAM phase curves. From Moiseyev et al. (1994). Copyright 1994 Springer-Verlag.

For future comparison with the properties of the Floquet states, let us now consider the classical time evolution of an initial ensemble of trajectories. We choose these trajectories to approximately correspond to the ground rotational state of a diatomic molecule. The initial ensemble is 8000 points with angular momentum $J = 0$, the points equally distributed in the angle variable θ over the interval $(0, 2\pi)$. We evolve these trajectories by numerically integrating Hamilton's equations and take *stroboscopic snapshots* of the ensemble points in the (θ, J) phase subspace, one snapshot after each period of the driving frequency $T = 2\pi/\omega$. A few of the snapshots are shown in Fig. 8.10, along with the probability distributions for J obtained by simply projecting the phase-space points onto the J-axis. We see that a complicated evolution of "whorls" develops over a short time interval of about 12 periods. The whorls then begin to fill the chaotic region of phase space and ultimately produce an asymptotic (long-time) classical probability distribution in J that characterizes the chaotic region alone. We now obtain this probability distribution from the quantum Floquet states.

We repeat the steps that were used above to obtain Figs. 8.7 and 8.8—except that now our system is nonintegrable, the Mathieu equation is not really relevant, and we carry out instead a basis-set expansion quantum numerical calculation. For the situation shown in Fig. 8.9, a basis set of 200 free-rotor states has been found sufficient (Moiseyev et al., 1994). The scaled Planck's constant $\hbar'$ was taken to be

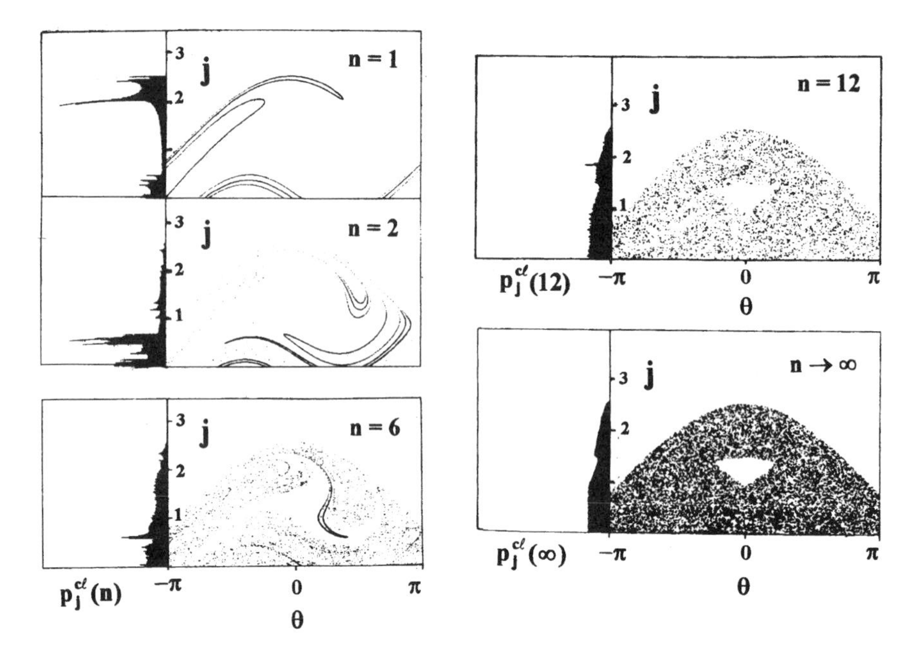

Figure 8.10. Time evolution of an ensemble of 8000 trajectories for the driven rigid rotor. The initial conditions for the ensemble were $J = 0$ and a uniform distribution in the angle variable θ over the interval $0 \leq \theta \leq 2\pi$. The variable plotted along the vertical axes is $j \equiv J/\hbar'$ with $\hbar' = 0.02$. This classical calculation mimics a diatomic molecule that, initially in its ground rotational state, becomes rotationally excited due to an interaction with a CW laser electric field. The location of the ensemble in (j, θ)-space is shown stroboscopically, at times $t = nT$, $n = 1, 2, 3, \ldots$, where $T = 2\pi/\omega$. At the left of each panel, the classical excitation probability distribution $\rho_j^{c\ell}(n)$ for j, summed over θ, is shown as a histogram. At $n = 12$, $\rho_j^{c\ell}(n)$ is already close to the long-time distribution for $n \to \infty$. From Moiseyev et al. (1994). Copyright 1994 Springer-Verlag.

0.02, small enough to nicely show the quantum–classical correspondence, yet large enough for tractable numerical work. As expected from Fig. 8.9, some Floquet states were found distributed exclusively over one of the three classical phase-space regions (the regular resonance, chaotic, and KAM regions). An *extended Floquet state* is distributed over a wide range of free-rotor states, and Fig. 8.11 shows a typical distribution for the driven rotor system. There are 92 Floquet states having such extended distributions, a fact we can display after calculating mean energies of all the Floquet states. Since $\langle\langle\Psi_k| - i\hbar(\partial/\partial t)|\Psi_k\rangle\rangle = 0$ for a Floquet state Ψ_k, $k \equiv n_\varepsilon$, we need only calculate the expectation value $\langle\Psi_k|H_0|\Psi_k\rangle \equiv \langle H_0\rangle_k$ to obtain the mean energy of the state. The distribution of $\langle H_0\rangle_k$ is shown in Fig. 8.12. The 96 states covering the resonance zone (regular part plus chaotic part) all have mean energies near 0.75, while the hindered-rotor KAM states have mean energies lying on a smooth curve with values above one.

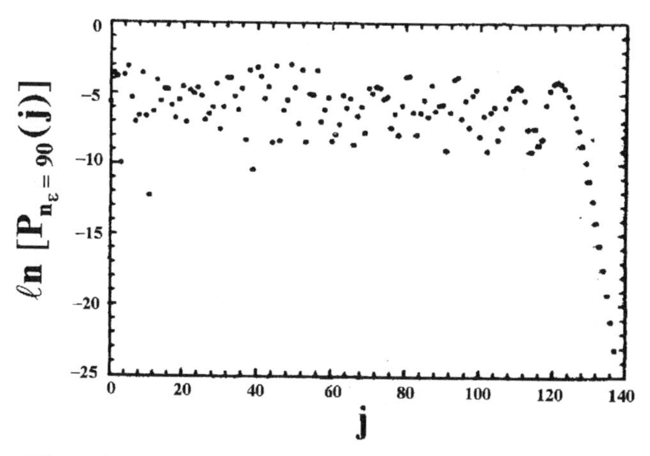

Figure 8.11. The probability distribution P of a typical extended Floquet state, $|n_\varepsilon = 90\rangle$, of the driven rigid rotor, obtained by an expansion in the basis of free-rotor states. This Floquet state is supported classically by the region of the stochastic layer shown in Fig. 8.9. Note the logarithmic vertical scale. From Moiseyev et al. (1994). Copyright 1994 Springer-Verlag.

The regular resonance Floquet states can be distinguished from the irregular resonance Floquet states by a more sensitive procedure. For each Floquet state Ψ_k, the Shannon entropy defined by

$$S_k \equiv -\sum_j A_{k,j} \ln A_{k,j} \tag{8.19}$$

is calculated from the free-rotor state amplitudes $A_{k,j}$ and then is plotted versus $\langle H_0 \rangle_k$; see Fig. 8.13. The four states $k \equiv n_\varepsilon = 89, 87, 82, 93$ that are localized on the

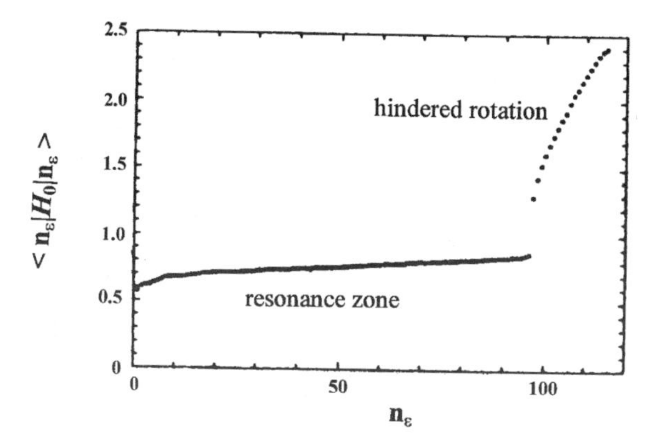

Figure 8.12. Expectation values of the Hamiltonian H_0 (the free-rotor Hamiltonian), $\langle n_\varepsilon | H_0 | n_\varepsilon \rangle$, where $\{|n_\varepsilon\rangle\}$ are the Floquet states of the driven rotor. The nonlinear-resonance states, $n_\varepsilon \leq 96$, are clearly distinguished from the regular hindered-rotor states, $n_\varepsilon \geq 97$. From Moiseyev et al. (1994). Copyright 1994 Springer-Verlag.

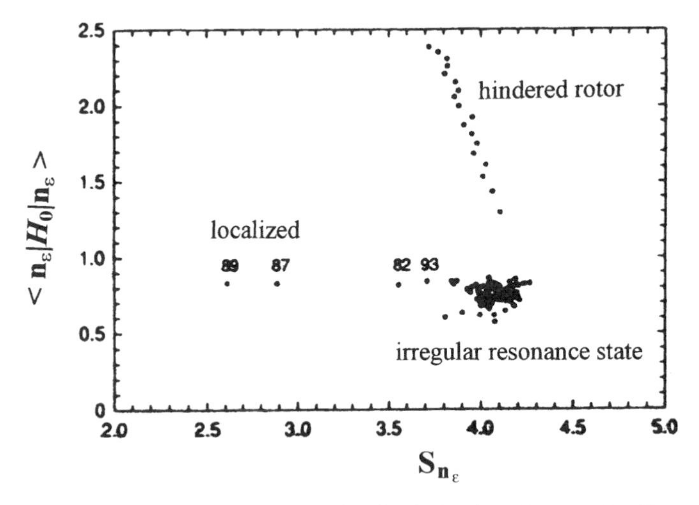

Figure 8.13. The expectation values shown in Fig. 8.12, plotted versus the Shannon entropy Eq. (8.19). Now not only are the hindered rotor states identified, but also two pairs of localized (regular) nonlinear-resonance states are distinguished from the extended nonlinear-resonance states. From Moiseyev et al. (1994). Copyright 1994 Springer-Verlag.

regular resonance island lie along a nearly horizontal straight line, well separated from the cluster of points arising from the irregular states and from a rapidly rising line produced by the KAM hindered-rotor states.

Now we can calculate an overall probability distribution arising from just the 92 irregular Floquet states, by simply summing up their free-rotor-state probability

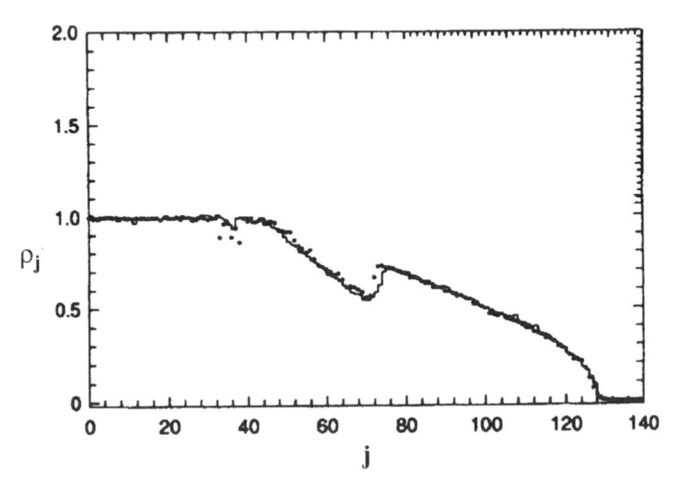

Figure 8.14. Comparison of the $n \to \infty$ classical probability distribution, shown at the bottom right in Fig. 8.10, with a quantum mechanical distribution that is a sum over the 92 extended Floquet states; see the text. There is a striking quantum–classical correspondence. From Moiseyev et al. (1994). Copyright 1994 Springer-Verlag.

distributions. The result is shown in Fig. 8.14, along with the classical probability distribution shown in the last panel of Fig. 8.10. The agreement is rather impressive. The irregular Floquet states accurately contain detailed information about the character of the chaotic region in phase space.

8.2b. The Driven Particle in an Infinite Square Potential Well

Let us now turn to the first problem listed in Table 8.1. For the case of a semiconductor nanostructure, the quantity m is the effective mass of the particle of charge Q, and the quantity a is the half-width of the square well. For the choice of phase constant $\phi_0 = \pi/2$, the field oscillates as $\sin(\omega t)$. Next, we define dimensionless space and time variables by $q \equiv x/a$ and $\tau \equiv \omega t$. The dimensionless driving amplitude is $\kappa \equiv F/m\omega^2 a$, the dimensionless momentum is $p = p_x/m\omega a$, and the dimensionless particle kinetic energy is $K = K_x/m\omega^2 a^2$. The Hamiltonian then becomes

$$H(q, p, \tau) = \tfrac{1}{2}p^2 + V_{SW}(q) + \kappa q \sin\tau,$$
$$V_{SW}(q) = \begin{cases} 0, & |q| < 1 \\ \infty, & |q| \geq 1. \end{cases} \tag{8.20}$$

The segments of the particle's motion within the well are segments of free-particle motion and hence are integrable. They satisfy

$$q(\tau) = \kappa[\sin\tau - \sin\tau_0] + [-\kappa\cos\tau_0 + p_0](\tau - \tau_0) + q_0. \tag{8.21}$$

One way to start the evolution of a classical trajectory is to use a root-finding algorithm (see Section A.4) to obtain the smallest $\tau \equiv \tau_1$, such that the particle hits a wall of the potential well at $q = \pm 1$. Starting at (q_0, p_0, τ_0), Eq. (8.21) then gives $q(\tau)$, $\tau_0 \leq \tau \leq \tau_1$. We then reverse the momentum, $p_1 = -p_0$, reset time, $\tau_0 \to \tau_1$, and reset the coordinate q_0 as appropriate. Repeatedly using the rootfinder and resetting the variables, we can use this procedure to find the parameters for as many segments of the trajectory as we desire.

We can then construct a (q, p) stroboscopic Poincaré section by sampling the set of segments at $\tau = 0 \pmod{2\pi}$. Figure 8.15 shows sections for two values of κ: 0.0267 and 0.0800. For the larger of these two driving amplitudes, the stochastic layer at small $|p|$ has merged with a second stochastic layer surrounding the regular resonance islands, as was the case in Fig. 8.9. However, the size of the present regular resonance island remains much larger. Thus we expect that regular classical evolution will play a greater role in the quantum evolution contained in the Floquet states than the role displayed in Fig. 8.13. For our discussion of the available quantum results, we will keep the dimensionless Planck's constant $\hbar' = \hbar/ma^2\omega$ for the present system at a value of either 0.020 or 0.022. These values are essentially the same as the value 0.020 used above for the driven rotor.

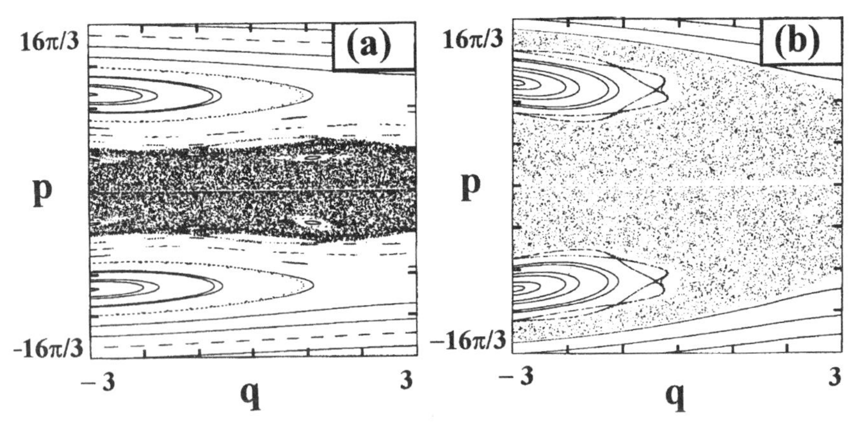

Figure 8.15. Two (q, p) stroboscopic Poincaré sections for a sinusoidally driven particle in a symmetric infinite square-well potential. This problem is a model for a charge-carrying particle in a semiconductor quantum well exposed to a CW laser electric field. The parameters are chosen to be $\phi_0 = 0$, $m = 1$, $a = 3$, $\omega = 5$, and (a) $F = 2$, (b) $F = 6$. From Shin and Lee (1994). Copyright 1994 by the American Physical Society.

Figure 8.16 shows part of a numerically calculated set of the quasienergies, plotted as a function of the driving amplitude κ. The quasienergies displayed originate at $\kappa = 0$ from the lowest 30 energy eigenvalues of the undriven system, as determined by the procedure (see the end of Section 8.1d) for making such connections between Floquet states and undriven system states.

We note that when the driving amplitude κ is very low but nonzero, the regions of phase space occupied by nonlinear resonances are too small to satisfy semiclassical quantization conditions; the Floquet states will then closely resemble the eigenfunctions of the undriven system. With increasing κ, however, the main nonlinear resonances will grow and be able to support more and more semiclassical Floquet states.

A subset of the quasienergy curves in Fig. 8.16 is labeled by an integer $k = 0, 1, 2, \ldots$. A calculation of approximate quasienergies based on the quantum pendulum approximation (Sections 5.6b and 4.3) gives excellent agreement with this subset of curves shown in Fig. 8.16, except at high values of $\kappa \gtrsim 0.05$; see further Holthaus (1995). Thus k is the Mathieu index (Section 4.3). The $k = 0$ Floquet state is that regular resonance state most tightly localized in the centers of the regular resonance islands in Fig. 8.15. In Fig. 8.16, the $k = 1, 2, \ldots, 14$ quasienergy curves all rise with increasing κ to a maximum value and then fall at sufficiently large κ—at least in the pendulum approximation. At the lower values of κ that still give quasienergy values above the maximum, these Floquet states are localized about central nonlinear-resonance vortex tubes. As we increase the Mathieu index k, these tubes lie increasingly outside the $k = 0$ central vortex tube. Let us now look at some of the evidence for these alleged properties.

On the left-hand side of Fig. 8.17, we see that the (q, p) Huisimi distribution for the Floquet state labeled by $k = 0$ varies with κ. In the top panel the value of κ is

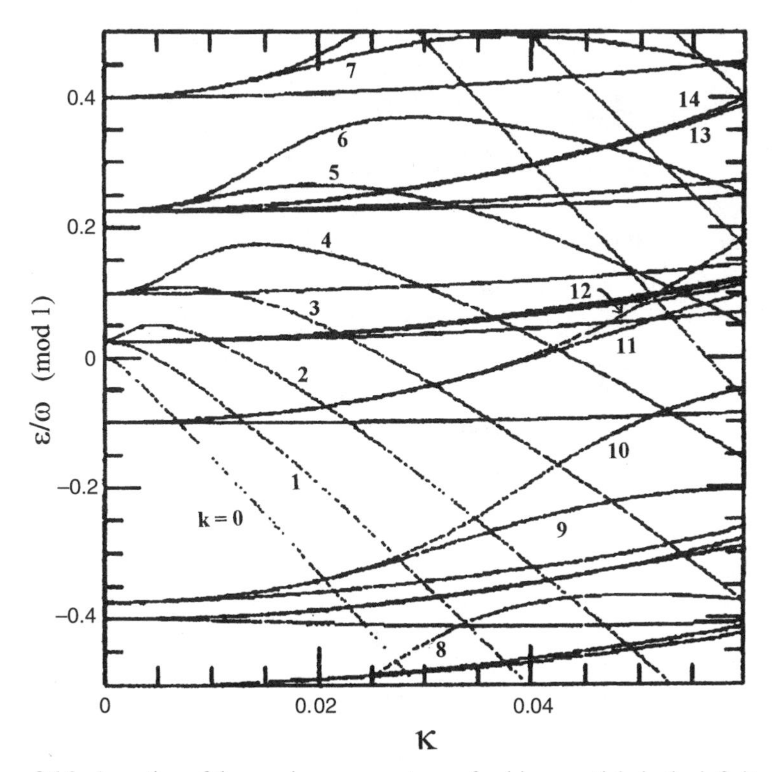

Figure 8.16. A portion of the quasienergy spectrum of a driven particle in the infinite square well—with $\omega = 5\pi^2$, $m = 1$, $a = 3$, and $\hbar = 1$—as a function of the dimensionless driving amplitude κ. The displayed normalized quasienergies ε/ω originate at $\kappa = 0$ from the lowest 30 energy eigenvalues (mod 1) of the undriven system. Some of the quasienergies are labeled by the Mathieu index k, as for the quantum pendulum problem (Fig. 4.1). From Holthaus (1995). Copyright 1995, with permission from Elsevier Science.

very small; it is increased as we go down the panels, rising to values near those on the right-hand side of the quasienergy plots in Fig. 8.16. For the smallest value of $\kappa = 0.0005$, the $k = 0$ Floquet state is almost identical to that for the corresponding state of the undriven system (which is not shown). As κ is increased past the maximum in the quasienergy (the maximum is not discernible in Fig. 8.16), the Huisimi probability distribution increasingly moves over to the regular resonance island region. For values of $\kappa \gtrsim 0.01$, the probability is entirely localized in this island region. That is, the $k = 0$ Floquet state has become completely localized on vortex tubes, and it remains so up to the highest values of κ included in Fig. 8.16. This state has become a regular nonlinear-resonance state.

What we see on the left-hand side of Fig. 8.17 is a fascinating phenomenon. As the driving amplitude κ is increased from zero, the initial probability density of the Floquet state smoothly evolves into a state entirely "trapped" in the regular resonance island region of Fig. 8.15. We thus conclude that when the quantum

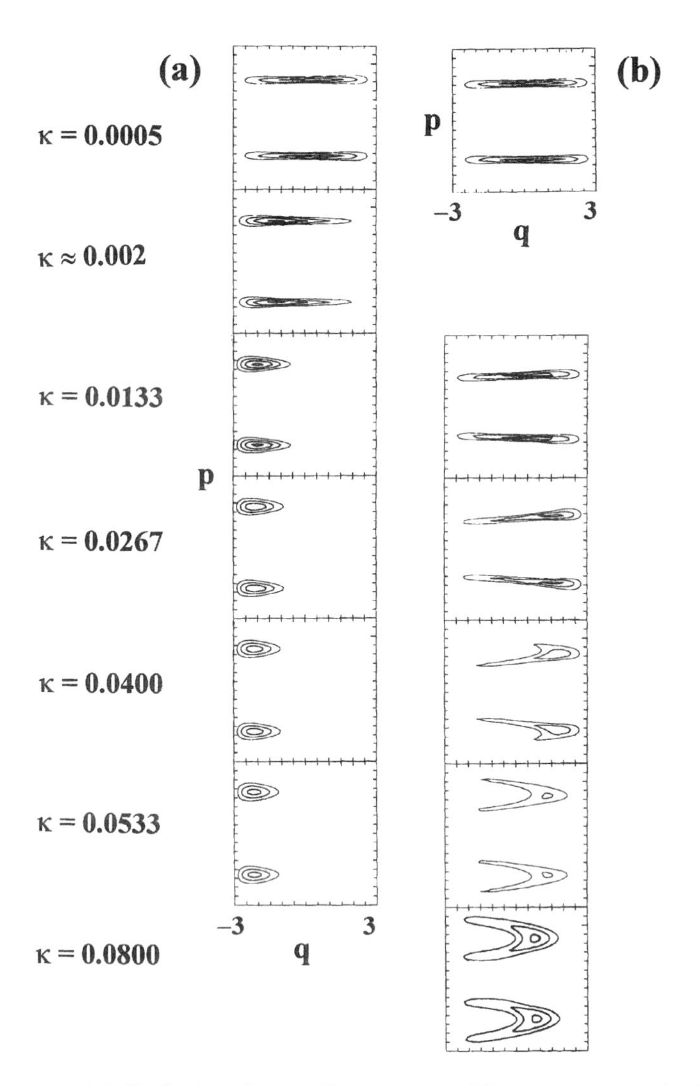

Figure 8.17. Huisimi distributions for two Floquet states of the same system as in Fig. 8.15, for various values of dimensionless driving amplitude κ. Again $\hbar \equiv 1$. The dimensionless particle momentum p is plotted versus the dimensionless particle position q within the infinite square well. For each panel, the range of values for p begins at $-16\pi/3$ at the bottom and ends at $+16\pi/3$ at the top (not indicated). Similarly, $-3 \leq q \leq +3$, as shown. (a) As we proceed from top to bottom, a Floquet state with a low value of the Mathieu index k changes from a spatially symmetric undriven state at $\kappa = 0$ into a regular Floquet state that is localized at negative q on the large fundamental nonlinear-resonance island shown in Fig. 8.15. (b) A Floquet state with a particular intermediate value of k changes from an undriven state into an extended state at $\kappa = 0.0533$. This extended state is supported by the chaotic region at positive q in Fig. 8.5, where there is the fundamental unstable fixed point. From Shin and Lee (1994). Copyright 1994 by the American Physical Society.

system remains in this single Floquet state as κ increases, the quantum evolution is actually a *quantum transport process*. At the chosen stroboscopic sampling times, one-half of the probability density in phase space is transported from a region with $q > 0$ into the resonance island region located at negative q. In coordinate space the particle is localized near the left wall of the potential well. We also conclude that in the problem of the driven particle in an infinite well, the characteristic behavior of the quasienergy of an island-trapped Floquet state is that its quasienergy decreases rapidly with increasing κ; see Fig. 8.16.

Since the existence of the regular resonance island is a classical phenomenon, we might ask whether the quantum resonance-island trapping process just noted has an analog in the corresponding classical system, as the classical system is again slowly varied by increasing κ from zero. Indeed, there is a corresponding classical island-trapping transport process, discussed in Section 9.1b.

Let us return to Fig. 8.17 and look at the set of panels on the right-hand side. Now we view the change in Huisimi distributions for the $k = 8$ Floquet state, which, according to Fig. 8.16, has a quasienergy that rises with increasing κ until a maximum is reached near the large value $\kappa = 0.05$. The evolution of this state with increasing κ is very different from the evolution of the $k = 0$ state shown in the left-hand panels. The transport of probability density becomes large only at relatively large κ, and that transport is toward the opposite wall (again for the chosen stroboscopic sampling times). At $\kappa = 0.04$, there is appreciable localization of the probability density near the unstable fixed point that must exist on the right-hand side of Figs. 8.15a,b (recall the phase-space picture for the pendulum, Fig. 1.15). This type of localization is simply that of states scarred by unstable periodic orbits (Section 5.4 and much of Chapter 7). We note that in Fig. 8.17 the localization of the $k = 8$ Floquet state is never as complete as for the $k = 0$ state. As κ is increased beyond the inflection point of the quasienergy function $\varepsilon(\kappa)$, there is increasing movement of probability density away from the unstable fixed points into portions of the region of classical chaos. Thus the density of scarring by the unstable periodic orbit decreases at high values of κ.

In Fig. 8.18 we see the (q, t) probability distributions of the $k = 0$ state and of the $k = 4$ state, which can be obtained from these states' distributions in the full (q, p, t) phase space. The distributions shown are obtained for values of κ where the $k = 0$ state is localized on vortex tubes and where the $k = 4$ state is strongly scarred. We clearly see the influence of the stable and unstable periodic orbits, which evolve in time back and forth across the potential well, out of phase. Note that unlike the $k = 0$ case, in the $k = 4$ case there is probability density outside the apparent region of influence of the unstable periodic orbit. This difference in (q, t) distributions is quite consistent with the (q, p) distributions of Fig. 8.17.

At $\kappa = 0.058 \approx 0.06$, the expectation value $\langle \Psi_k | H_0 | \Psi_k \rangle \equiv \langle H_0 \rangle_k$ has been calculated for $k = 0$ to 16; the results are displayed in Holthaus (1995). A transition in the k dependence is found to occur for k values that are associated with the location of the maximally scarred state. This transition is similar to that shown in Fig. 4.2.

Shin and Lee (1994) have displayed Huisimi distributions for 15 Floquet states centered about the resonance $k = 0$ state, for the parameter value $\kappa = 0.080$. At this

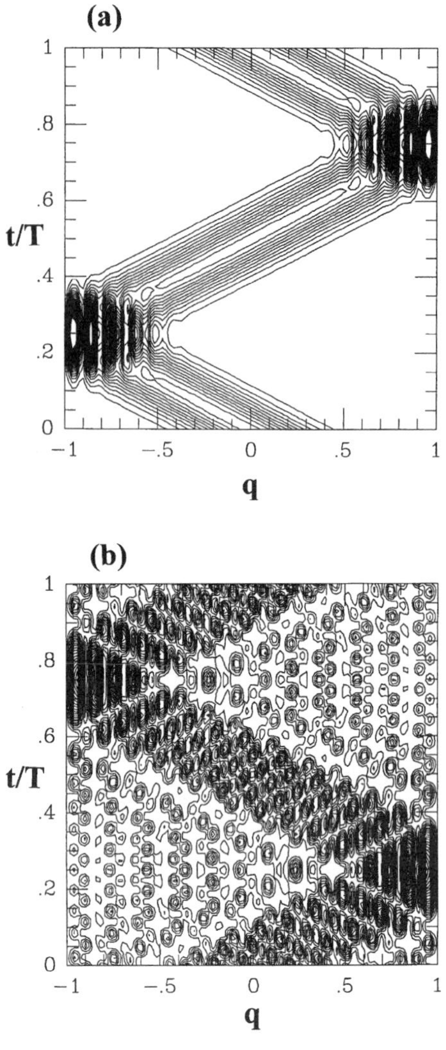

Figure 8.18. Contour plots of (q, t) probability densities for Floquet states having the parameters of Fig. 8.16. (a) The resonance-induced ground state (Mathieu index $k = 0$) becomes well-localized about the fundamental stable periodic orbit when $\kappa = 0.05778$. This corresponds to the situation in Fig. 8.17a for $\kappa = 0.0533$. (b) The $k = 12$ Floquet state at $\kappa = 0.05778$, which is extended over all (q, t) but is additionally scarred by the fundamental unstable periodic orbit. This state corresponds to the situation in Fig. 8.17b for $\kappa = 0.0533$. From Holthaus (1995). Copyright 1995, with permission from Elsevier Science.

high value of κ, only two of the Floquet states are localized on the usual regular resonance vortex tubes, and two others are highly scarred. The remainder of the states are not as easily characterized, as is the case for some other irregular quantum states already discussed (Sections 7.4b and 7.4c).

8.3. DRIVEN SPATIALLY OPEN SYSTEMS (ADVANCED TOPIC)

8.3a. About Experiments and Numerical Calculations

To complete our overview of sinusoidally driven systems, we must turn to the last two items in Table 8.1, namely, the driven Morse and driven Kepler systems. These are the best-studied driven spatially open systems. One of their key properties is that absorption of a finite amount of energy from the driver suffices to completely break apart the undriven system. We can see this in Fig. 8.19, where the excited energy levels of the hydrogen atom are compared with the vibrational energy levels of the ground electronic states for the I_2^+, H_2, and HF diatomic molecules. The ionization limit of the atom and the dissociation energies of the molecules have all been set

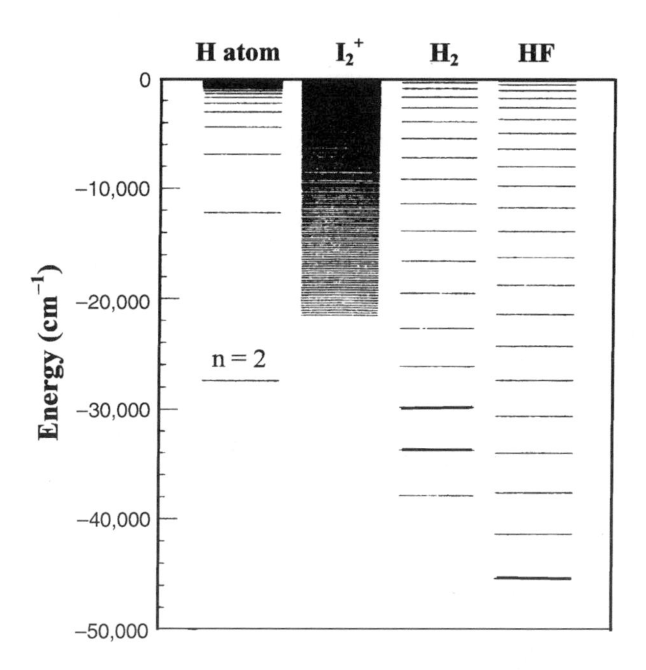

Figure 8.19. A comparison of the energy levels of the hydrogen atom H(n), $n \geq 2$, with the energy levels of Morse oscillators whose parameters correspond to the I_2^+, H_2, and HF diatomic molecules. The anharmonicities of the lower levels of H_2 and HF are quite different from that for H(n). The spectra of I_2^+ and H(n) each include over 100 highly excited states. From Tanner and Maricq (1989). Copyright 1989 by the American Physical Society.

equal to zero. The number of bound-state energy levels indicated in the figure is infinite for the atom but finite for the molecules. All the systems, of course, have an infinity of additional continuum levels lying above the energy $E = 0$.

We now address a new issue—quantum evolution to final states of positive energy, that is, quantum evolution into the continuum. What is the classical evolution like, and how does it impact the quantum evolution? These questions are most clearly addressed in certain parameter regimes. In Fig. 8.19, the lowest levels in each of the energy-level ladders are ΔE =2–5 eV below $E = 0$, where ΔE corresponds to the combined energy of 20–50 laser photons for a middle infrared (IR) laser. To absorb such a number of photons in a single process, we would need to achieve very high laser electric fields. One would need to employ a pulsed CO_2 laser (in the picosecond range of pulse times) and focus its output beam tightly in space.

Although they have come close, studies to date have not yet achieved intense IR laser-pulse switching times of one field oscillation period or less. Thus present experiments use pulses where the switching can play an important role; see Chapter 9. However, results from Section 9.3a indicate that if the switching time is long (with many optical periods), then we could achieve an adiabatic evolution of single Floquet states. This evolution could include states exhibiting changing spatial features as a function of laser field strength (Fig. 8.17). (Such adiabatic evolution is not always the case, as we will see in Sections 9.1b, 9.1c, and 9.2b.) For initial quantum states prepared in the laboratory, adiabatic evolution simplifies quantum calculations that propagate the particular linear combination of Floquet states. We would have to include the effects of the continuum ($E > 0$) states, perhaps by finding appropriate Floquet-state decay rates Γ_α and by making the quasienergies complex with imaginary parts $i\Gamma_\alpha$. Such quantum computations are challenging, but we'll see there are ways to proceed with the numerical work.

Returning to Fig. 8.19, we notice the hydrogen atom (H atom) and I_2^+ molecule are special in that many energy levels lie within 1000 cm^{-1} of the continuum; that is, they lie in the 100 meV energy band below $E = 0$. If the driving electric field oscillates in the microwave (MW) to far infrared (FIR) regions, we can achieve that interesting regime where the photon energy is comparable to, or less than, the energy spacings between the initial quantum eigenstate and its immediate neighbors. Then the undriven system requires the net absorption of many photons of the driving electromagnetic field in order to break the system apart; the impact of the nonintegrable classical physics on the quantum evolution is quite apparent in this regime. However, since neither MW nor FIR experiments have yet achieved pulse switching times less than one driving-field oscillation period, the above remarks about adiabatic and nonadiabatic switching are relevant once again.

One advantage of studying highly excited states in MW or FIR electric fields is that driving-field strength requirements are greatly reduced. For the case of the hydrogen atom, we see this in the scaling with principal quantum number n (Table 2.1); this scaling is n^{-4} for the electric field strength. While studies on low excited and ground states are also interesting, we will focus on the highly excited state regime, as considerable MW laboratory and numerical data are available.

The Morse and Kepler problems defined in Table 8.1 are models in one spatial dimension. Given that real atoms and molecules reside in three spatial dimensions, there is the question of what relevance, if any, the 1-d models have for experiments. The answer depends in part on how controlled the experiments are.

For the case of a linearly polarized driving electric field, the field direction defines one spatial direction, which we call x. The challenge is adjusting the atom or molecule to minimize the importance of the remaining degrees of freedom. We can accomplish this by selectively exciting the initial quantum state in the presence of an additional static electric field. If the molecule has a permanent electric dipole moment (like HF, but not the symmetric molecule H_2) or if the atom is hydrogen, which acquires a permanent moment when in a static field, then the selective excitation can be to that quantum state having a maximal dipole expectation value in the x-direction. Thus if we apply both the static field and the oscillatory driving field along the same x-direction, we can hope to maintain the initial maximal alignment of quantum states. This does occur and can be understood both classically and quantum mechanically.

Such experiments, called *nearly 1-d experiments*, have been implemented in the case of the hydrogen atom (Bardsley et al., 1986; Bayfield and Sokol, 1988). Although beams of field-oriented molecules have been produced, the rest of the agenda remains undone for molecules. Insufficiently addressed are the effects of the combined rotation and vibration of the molecule while in the oscillating field—where these effects in essence produce a full 3-d problem with a 7-D phase space.

There is more to the relevance of 1-d models for experiment. When the driving field is present only for short times, say, 100 oscillation periods, then a particularly high driving amplitude is required for nonzero ionization or dissociation probability. Large, rapid changes will arise in the atom or molecule's evolution from such driving, and these changes are largely confined to the component of the motion along the driving field's direction. Analysis of the classical physics for the driven 3-d hydrogen atom problem predicts such changes (Meerson et al., 1982). In nearly 1-d experiments, experimental state analysis of the atoms after exposure to the microwave pulse confirms these predictions (Bayfield, 1987). Quantum mechanically, we can consider the ionization probabilities of the 3-d Floquet states shown in Fig. 8.20 from Buchleitner and Delande (1997). The Floquet state in the center panel of the figure is both rapidly ionizing and strongly "stretched" along the vertical driving-field direction. Thus some experiments with 3-d atoms or with molecules can also be expected to exhibit signs of 1-d model behavior, a fact that is well-established for the hydrogen atom problem.

Figure 8.21 shows experimental data for the scaled ionization threshold field strength $F_0(10\%)$ for microwave ionization of highly excited hydrogen atoms. The data are plotted versus microwave oscillation frequency in units of the Kepler orbit frequency of the initial state, $\omega_0 = n_0^3\omega$, with n_0 the initial value of the atom's principal quantum number. The threshold field is defined as that for 10% ionization probability and is scaled to the characteristic Coulomb field n_0^{-4} in atomic units. In the figure, data taken using the maximal-aligned initial state and a static electric field

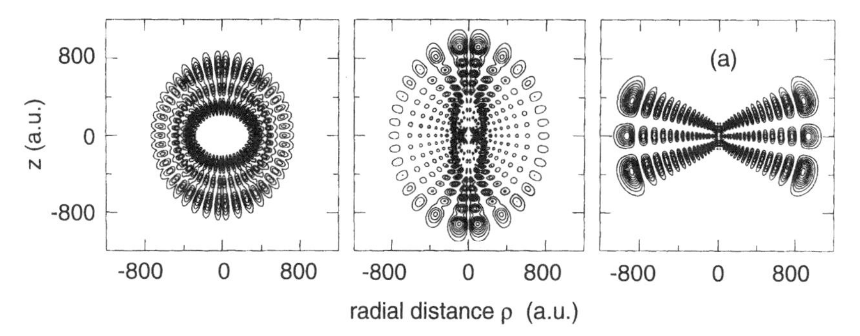

Figure 8.20. Contour plots of the electron spatial probability density in cylindrical coordinates (ρ, z), for three different Floquet states of the $L_z = 0$ 3-d hydrogen atom in a linearly polarized, sinusoidally oscillating electric field directed along the z-axis. These Floquet states originate from the n = 23 manifold of undriven hydrogen atom states and are shown for the scaled driving frequency $\omega_0 = n^3\omega = 1.304$ and for the scaled driving-field strength $F_0 = n^4F = 0.03$. From left to right, the states are localized near a circular orbit, near the electric field axis, and near the plane perpendicular to the field. Only the middle state ionizes at $F_0 = 0.03$. The quantum calculations of quasienergies, ionization rates, and Floquet states used complex-rotation grid numerical techniques (Section B.2c). From Buchleitner and Delande (1997). Copyright 1997 by the American Physical Society.

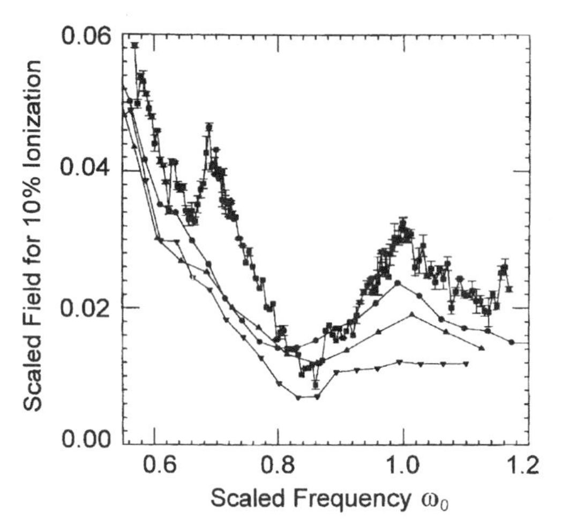

Figure 8.21. Experimental values for the threshold-scaled microwave electric field strength for pulsed ionization of highly excited hydrogen atoms, as a function of scaled microwave frequency ω_0. The data showing the most structure are for nearly one-dimensional atoms, where the Floquet states are like that shown in the center panel of Fig. 8.20. The other three sets of data are for ensembles of atoms where only the principal quantum number is initially well-defined (Van Leeuwen et al., 1985; Galvez et al., 1988; Bellerman et al., 1996). (The data points for these sets of data taken at zero static field have been adjusted for this comparison with the nearly 1-d data taken with a static field.) An increase in the threshold field indicates increased stability against ionization. The maximum observed near the fundamental scaled frequency $\omega_0 = 1$ arises from the existence of the fundamental (1 : 1) regular nonlinear-resonance region in phase space. From Bayfield et al. (1998).

are compared with three other sets of data obtained using a statistical ensemble of initial states and no static field.

All the data show a maximum in a region centered near $\omega_0 = 1$. The primary 1 : 1 nonlinear resonance exists here, with the microwave and Kepler frequencies about equal. We can explain the basic nature of this region between $\omega_0 = 0.85$ and 1.15 in terms of the adiabatic evolution of Floquet states that support the resonance-island trapping process (Fig. 8.17a). Highly trapped states are harder to ionize than other nearby states, requiring more MW electric field strength. (To produce the 10% ionization probability, the trapping process is accompanied by a small amount of coupling to the continuum.) A maximum in $F_0(10\%)$ arises close to $\omega_0 = 1$ because the evolution then maximally involves the most stable $k = 0$ Floquet state—that state most localized on vortex tubes that are closest to the stable periodic orbit. These tubes require the highest fields for tube destabilization. We note that purely classical model calculations also come close to explaining the $\omega_0 \approx 1$ region, which therefore is called a *region of classical stabilization against ionization*.

In Fig. 8.21, as we lower ω_0 from 0.85 to 0.68, the ionization threshold rises. Classical calculations show this rise closely follows the threshold field for probability flow from the initially KAM state into the stochastic band of the 1 : 1 resonance zone (Bayfield et al., 1998); the classical flow comes after a temporary partial localization of the trajectory ensemble near the unstable periodic orbit. Thus in this region of lower microwave frequencies, some experimentally populated Floquet states behave as shown on the right-hand side of Fig. 8.17. Near the peak of the pulse, these Floquet states are maximally scarred by the unstable periodic orbit. Since scarring is the result of constructive quantum wave interference (in spite of the presence of chaos), we call this frequency region a *region of quantum stabilization against ionization*. Note that this region is delineated in Fig. 8.21 from a yet different region below $\omega_0 = 0.68$.

8.3b. The Floquet States of the Driven Morse Oscillator

In Fig. 8.22 we see a stroboscopic section of the classical evolution of a driven Morse oscillator; the oscillator is driven at a frequency $\omega = 0.016015$ a.u. (atomic units), with a constant amplitude $F = 0.05$ a.u. These parameters correspond to a laser photon energy of 0.436 eV and a laser intensity of $8.77 \times 10^{13}\ \mathrm{W/cm^2}$. The parameters for the Morse model of the diatomic molecule were taken to be the parameters for the local O—H stretching mode of a H_2O molecule (Henkel and Holthaus, 1992).

Looking again at Fig. 8.22, a region of regular resonance evolution is centered near $\alpha(x - x_0) = 0.7$, $p = 0$. Shown around the center are closed curves that surround areas of action S, areas that are determined by the quantization condition $S = 2\pi\hbar(k + \frac{1}{2})$, $k = 0, 1, 2, 3$. The two innermost curves clearly are sections of vortex tubes of the type that classically support regular resonance-island Floquet states; see the top two quantum–classical comparisons in Fig. 8.23.

The bottom of Fig. 8.23 shows a quantum–classical comparison of an irregular Floquet state lying on a stochastic layer in phase space and scarred by an unstable periodic orbit. Note that the outer boundary of the irregular Floquet state follows the

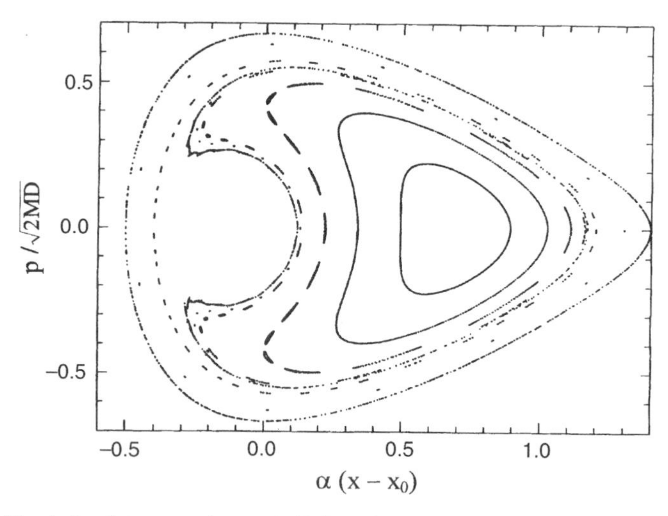

Figure 8.22. A (q, p)-type stroboscopic Poincaré section for the sinusoidally driven Morse oscillator model for the vibrational motion of a diatomic molecule. The fundamental stable fixed point is near (0.62, 0) and the unstable one near (−0.30, 0). The driving frequency is $\omega = 0.0160$ a.u. and the driving-field strength is $F = 0.05$ a.u. From Henkel and Holthaus (1992). Copyright 1992 by the American Physical Society.

outer boundary of the stochastic layer rather nicely. The unstable orbit is the mate of the stable periodic orbit lying at the center of the regular resonance nest of vortex tubes shown in the top comparison. The top and bottom comparisons in Fig. 8.23 can be compared with the $k = 0$ and $k = 4$ comparison in Fig. 8.18 for the driven particle in the square well. A similar comparison for corresponding states of the driven Kepler system can be found in Henkel and Holthaus (1992).

Thus the general characteristics of Floquet-state probability distributions are now fairly evident for states that do not ionize or dissociate—that is, for states that have no contributions from undriven states with energies in the continuum, and that have no classical support from a global chaotic region of phase space.

Let us now consider the I_2^+ molecule (Fig. 8.19) in a far infrared laser field with frequency 60 cm^{-1}. The Morse parameters chosen are discussed in Tanner and Maricq (1989). This quantum Morse oscillator has 179 bound states. In Fig. 8.24 we see the squared amplitudes $|\langle n|\alpha\rangle|^2$ of the projections of the Floquet states $|\alpha\rangle$ on the undriven Morse states $|n\rangle$. The lower panel is for a driving laser field F that is about 7.2 times larger than the laser field in the upper panel. The upper panel shows the existence of KAM Floquet states for low α and n; these KAM states are almost the same as single Morse energy eigenstates. At medium values of α and n in the upper panel and at low values in the lower panel, we see Floquet states composed of local bundles of Morse states, Morse states that are similar to those in Fig. 8.23. Note, however, that in Fig. 8.24 there are states at high α that spread out to very high values of n, and therefore extend out spatially to large values of x. If the continuum were included in the calculation, these states at high α would be extended Floquet

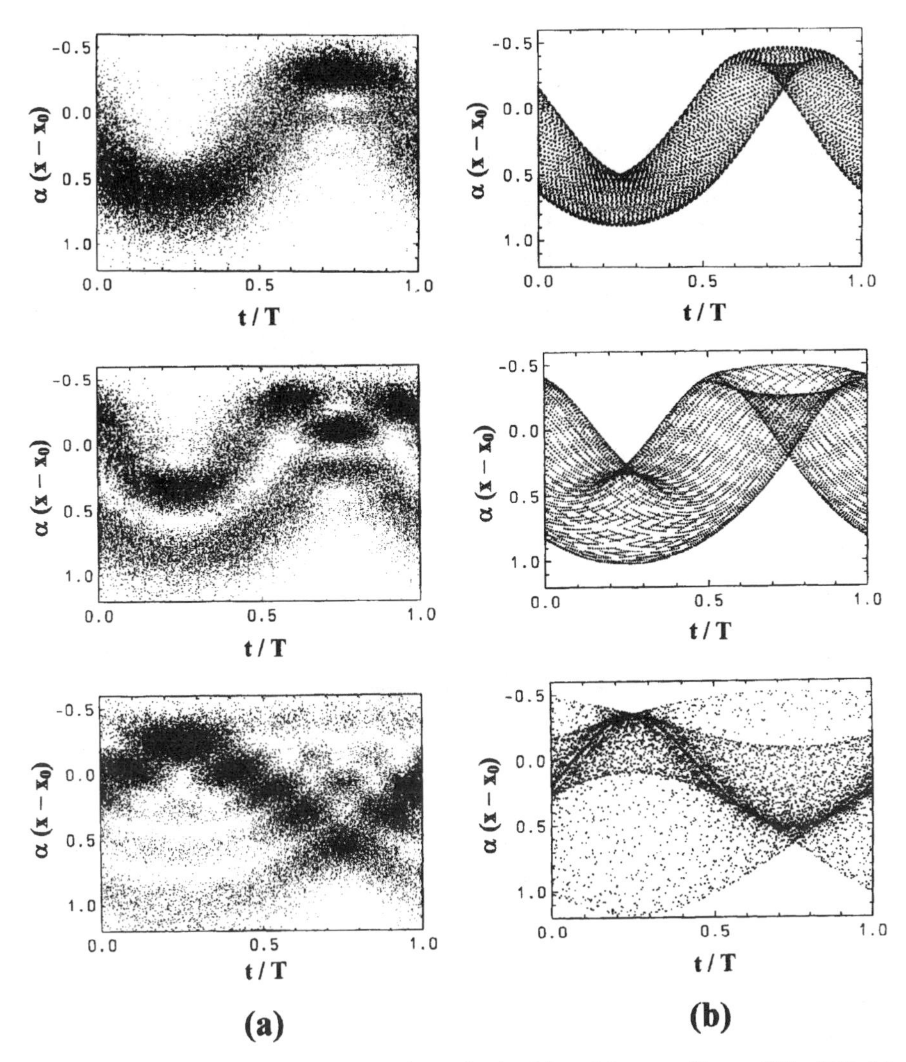

Figure 8.23. Quantum–classical comparisons for the driven Morse oscillator of Fig. 8.22. (a) Floquet-state probability densities (left panels) and (b) classical trajectories, after projecting the trajectories in 3-D phase space onto the (q, t) subspace. The top two pairs of panels show the k = 0 and k = 1 regular Floquet states, which are localized in phase space on sets of nested vortex tubes. The bottom pair of panels shows an irregular Floquet state that extends over a stochastic layer and is scarred by the fundamental unstable periodic orbit, similar to the situation shown in Fig. 8.18b. From Henkel and Holthaus (1992). Copyright 1992 by the American Physical Society.

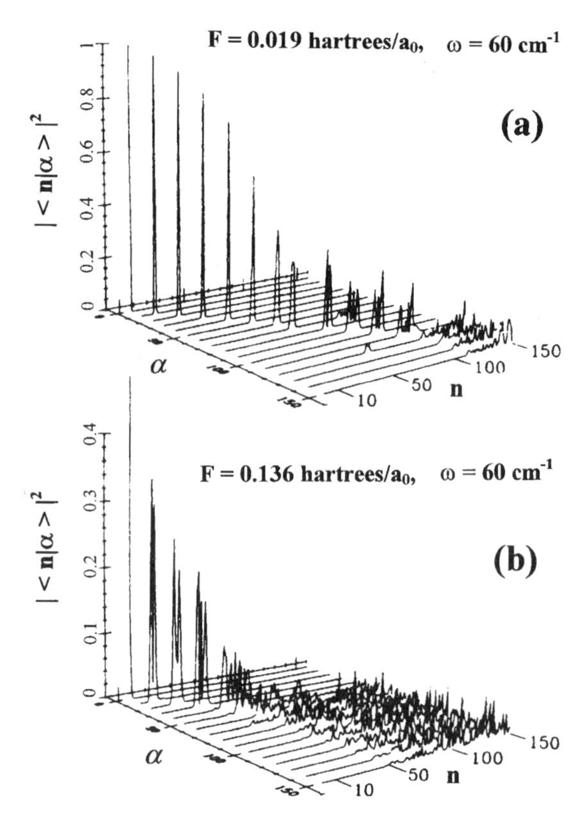

Figure 8.24. Squares of projections $\langle n|\alpha\rangle$ of driven Morse oscillator Floquet states $|\alpha\rangle$ onto undriven Morse oscillator states $|n\rangle$, for a far infrared driving laser with frequency $\omega = 60\ \mathrm{cm}^{-1}$ and driving fields (a) 0.019 hartrees/a_0 and (b) 0.136 hartrees/a_0. The Morse parameters modeled the spectrum of I_2^+ (Fig. 8.19). Three types of Floquet states can be discerned: KAM states dominated by a single Morse state $|n\rangle$; less-localized states comprised of a local bundle of Morse states; and extended (delocalized) states. From Tanner and Maricq (1989). Copyright 1989 by the American Physical Society.

states that dissociate to some degree. The division of Floquet states into the three types—KAM, localized bundle, and extended—also exists for the driven Kepler problem (Bardsley et al., 1986; Blumel and Smilansky, 1989). Driven spatially open systems typically display these three types of Floquet states.

8.3c. Dissociation and Ionization Probabilities

Accurate ab initio numerical basis-set expansion calculations of dissociation probabilities of molecules (and of ionization probabilities of atoms) require a rigorous and convergent numerical procedure that includes contributions of the continuum states of the undriven system. Typically such calculations require a good deal of supercomputer CPU time. The issue of convergence to precise results can be a delicate and/or tricky business, best left to the real experts. Several approaches to the continuum problem have been pursued.

One approach is to replace the continuum by a carefully selected quasicontinuum of closely spaced discrete states. In this method one assumes that within a discretization interval the amplitudes of all continuum states add up to the nondecaying amplitude of the discrete state representing that interval. This approach has been tried for the driven Morse oscillator (Graham and Höhnerbach, 1992) and for the driven Kepler oscillator (Susskind and Jensen, 1988).

A second approach uses basis-set expansion quantum calculations that include matrix elements for the bound-state coupling to the continuum, but that omit continuum–continuum coupling as well as back coupling to the bound states. This effectively means working with a damped set of bound and discretized continuum states, which then have complex energies. For the driven Morse problem, this approach can give results in good agreement with those of the first approach (Graham and Höhnerbach, 1992).

Using another method to effectively damp discrete states, we begin by complexifying the spatial variable x at the outset. This complex coordinate method—also called the *complex coordinate rotation method*—makes the replacements $x \rightarrow xe^{i\theta}$ and $p \rightarrow pe^{-i\theta}$, where θ is a real parameter called the rotation angle. The rotated variables are inserted into the Hamiltonian function $H(x, p, t)$ to form a complex rotated Hamiltonian $H(\theta)$ whose complex eigenvalues are the resonances of the initial Hamiltonian H. Actually, this procedure is generally applicable to both autonomous and driven systems and has been used in work on the driven Kepler problem (Buchleitner et al., 1994). The procedure is discussed further in Section B.2c.

Still yet another approach is splitting the Hamiltonian into two parts in a nonphysical way—with the "unperturbed" Hamiltonian selected to have no continuum! This procedure can be used for the driven Kepler problem where the selected complete set of discrete states is the so-called Sturmian states; see Section B.2b. Calculations have been carried out for the driven Kepler problem (Casati et al., 1987). More recent calculations involving hydrogen atoms use both the complex rotation method and an expansion in Sturmian basis states.

In contrast, important numerical methods exist for solving Schrödinger's time-dependent partial differential equations, methods that do not resort to any of the above basis-state expansions. Often called *FFT grid methods*, these approaches readily handle the continuum evolution. We define the wavefunction Ψ on a spatial grid (see Section B.1a), use an explicit differencing scheme for the time propagation of the wavefunction, and deal with the kinetic energy term in the Hamiltonian propagator by fast Fourier transformation to and from momentum space during each time step (see Section B.1b.2). We must choose the spatial-grid spacing small enough (and the grid in momentum space large enough) for all the interesting details of Ψ to be adequately sampled. Timesteps must be consistent with the higher energies that appear; this means that if the range of momenta increases with time, the range of energies goes up, and the timestep must be decreased accordingly. Given these caveats, we can simply deal with the effects of the continuum states of the undriven system by always using a large enough spatial grid. All significant amounts of wavefunction amplitude are then far from the edges of that grid. If we work in an appropriate interaction representation to eliminate free-particle wavepacket spreading, we can greatly enhance the chance of actually achieving this avoidance of the grid edges.

8.3d. Laser Dissociation of a Morse Oscillator

Let us now return to the driven Morse model of I_2^+ in a laser field, where some Floquet-state distributions are shown in Fig. 8.24. Using the FFT grid method to numerically evolve the wavefunction, Tanner and Maricq (1989) calculated the time evolution of an initial Morse state $|n_0\rangle$. In Fig. 8.25 we see the results for the evolution of the dissociation probability for a range of initial vibrational quantum numbers n_0 that corresponds to 100–200 photons absorbed. Similar numerical results for the dissociation probability have been obtained using Morse parameters for the NO molecule (Ting, 1994). Above a threshold field intensity, there is an

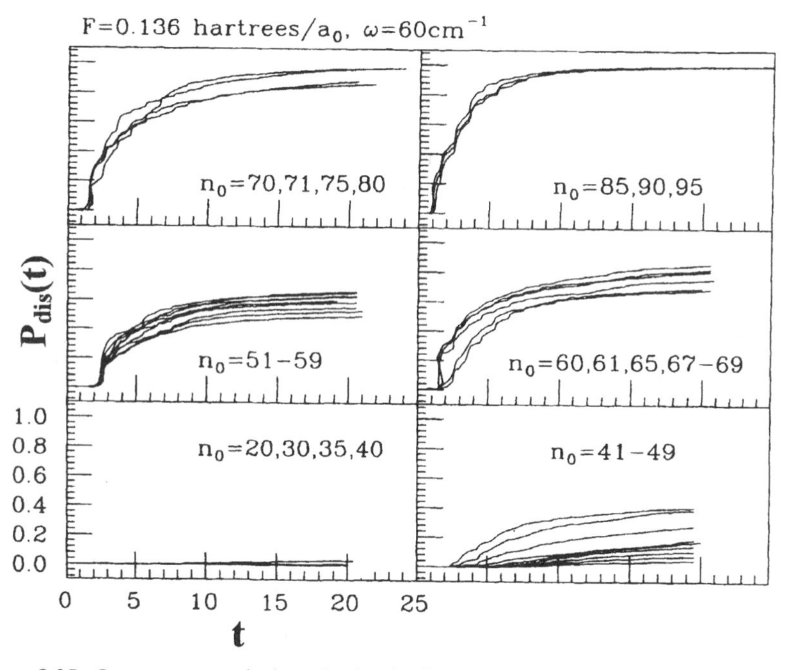

Figure 8.25. Quantum numerical results for the time evolution of the dissociation probability for the driven Morse oscillator of Fig. 8.24b. The dissociation is the consequence of extended initial Floquet states. The panels show dissociation curves for a wide range of initial Morse states, obtained using the FFT grid method for numerical quantum time evolution (Section B.1b.2). From Tanner and Maricq (1989). Copyright 1989 by the American Physical Society.

initial rapid flow of oscillator probability into the continuum. The dissociation rate subsequently slows down, however, and the remaining bound portion of the wavefunction is localized to a relatively narrow range of Morse eigenstates (not shown).

The initial Morse state evolves as a set of Floquet states. Let us see then whether the information in Fig. 8.24 can explain the long-time values of the dissociation probabilities in Fig. 8.25. We sum squared probability amplitudes (such as those in Fig. 8.24), but only over the third type of Floquet state—the extended states. The result $P_B(n_0) = \sum_{\text{extended}} |\langle n_0|\alpha\rangle|^2$ is the probability that the initial Morse state "belongs" to the subspace of extended Floquet states.

In Fig. 8.26 we compare this probability obtained from Floquet theory with the dissociation probability obtained from the numerical time evolutions. The agreement is remarkable. The probability of whether or not the Morse oscillator eventually dissociates is primarily determined by the projection of the initial state onto the extended states. For this driven Morse system and the chosen set of parameters, we conclude from Fig. 8.26 that the extended Floquet states would, if accurately calculated, contain portions of the continuum.

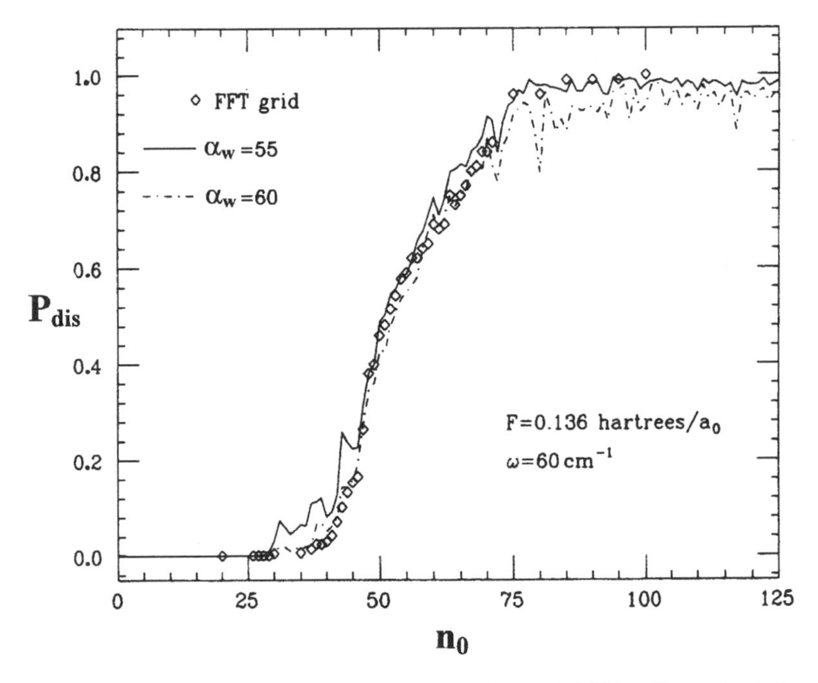

Figure 8.26. A comparison of the long-time dissociation probabilities (data points) shown in Fig. 8.25 with the sum of squared projections of the initial Morse state $|n_0\rangle$ on the set of extended Floquet states shown in Fig. 8.24b. Two choices of the definition of the onset α_w of the extended states are considered and give the solid and dashed curves. The degree of dissociation is clearly associated with the fractional initial population of extended Floquet states. From Tanner and Maricq (1989). Copyright 1989 by the American Physical Society.

8.3e. Spatial Growth of Driven Kepler Wavepackets

The evolution of properly localized initial wavepackets is very helpful for understanding quantum evolution, as we've already seen for a number of systems (Sections 1.3c, 2.1c, 2.4c, 5.3d, 6.1b, 7.2e, and 7.4d). Let us see this for a driven system, choosing the 1-d hydrogen atom in an oscillating electric field as the example. Following Latka et al. (1993), we choose three separate minimum uncertainty wavepackets P_1, P_2, P_3 to have centers in (x, p_x)-space at $p_0 = 0$, but with the differing values $x_{1,0} = 8000$ a.u., $x_{2,0} = 10{,}000$ a.u., and $x_{3,0} = 12{,}000$ a.u. Those initial electron distances from the atomic nucleus are indicated in Fig. 8.27 by the vertical lines superimposed on a stroboscopic section for $n_0 = 66$, $\omega_0 = 1$, and $F_0 = 0.03$. A value of position equal to 20,000 a.u. corresponds to the maximum of an undriven atom Huisimi distribution for $n = 100$ (recall Fig. 1.10 and Table 2.1). Thus 20,000 a.u. locates the outer turning point of the Kepler orbit for $n = 100$. As the initial packets are chosen to have spatial widths of about $\sigma_x = 700$ a.u., we see from Fig. 8.27 that packet P_1 is primarily embedded in the 1 : 1 regular resonance zone, yet far off the primary island's center. Packet P_2 just barely overlaps this zone, whereas packet P_3 is totally outside it in the chaotic portion of phase space. The chaos exists in a wide stochastic layer. As there is no static electric field, this layer at $F_0 = 0.03$ is not strongly linked to the region of global chaos and ionization.

For the quantum evolution of the driven Kepler wavepackets, Latka and coworkers carried out standard basis-set expansion calculations in the undriven atom

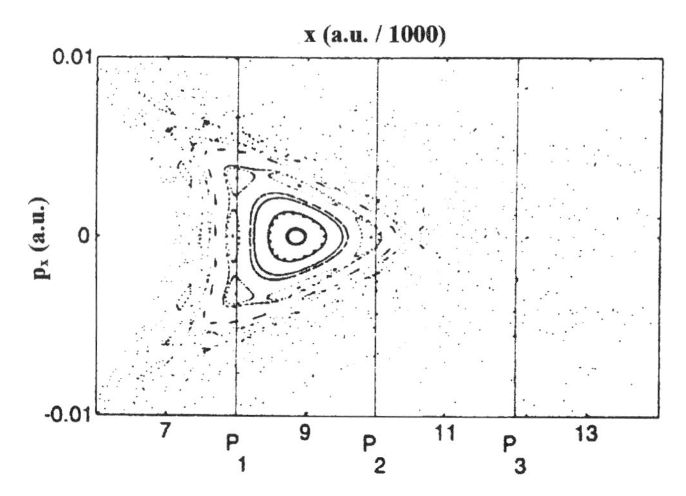

Figure 8.27. An (x, p_x) stroboscopic Poincaré section for the sinusoidally driven 1-d hydrogen atom, with superimposed vertical lines indicating locations in the position x of three selected minimum-uncertainty initial wavepackets that differ only in their locations in phase space. Packets P_1, P_2, and P_3 are all centered at $p_x = 0$, and at $x_{1,0} = 8000$, $x_{2,0} = 10{,}000$, and $x_{3,0} = 12{,}000$ a.u., respectively. The spatial widths, full-width at 20% of the maximum, are taken to be about 700 a.u. The scaled system parameters are $\omega_0 = 1$ and $F_0 = 0.03$. From Latka et al. (1993). Copyright 1993 by the American Physical Society.

basis, omitting the continuum. Figure 8.28 shows the time evolutions of the autocorrelation function $C_i = |\langle\psi_i(t)|\psi_i(0)\rangle|^2$ (also called the *survival probability*) of the three packets, for the first 150 driving periods. The regular packet P_1 exhibits Rabi oscillations (Section 2.2), now with some amplitude modulation. These oscillations occur because the initial packet P_1 decomposes into just two Floquet states with a joint 90% probability, along with one other state with the remaining 10% of the probability. Snapshots of the Huisimi distributions show that the oscillations in $C_1(t)$ arise from a small coherent displacement of the packet from its initial position, not from wavepacket spreading. The second moment (variance) of packet P_1's position evolves as shown in the lowest curve in Fig. 8.29. This evolution confirms that the regular packet P_1 grows very slowly in size.

Time evolution is more complicated for packet P_2, which starts at the edge of the resonance zone. Figure 8.28b shows a rapid initial collapse in the survival probability for the early time interval up to 30 periods, followed by signs of a revival starting near 100 periods of the evolution. This pattern is reminiscent of packet collapse and revivals for the undriven 2-d hydrogen atom problem (Section 2.1b) and of the nonintegrable driven two-level atom (Sections 2.4c and 2.4d). Packet P_2's growth in the position variance is more rapid than for P_1; see Fig. 8.29.

This growth is still more rapid for the "chaotic" (irregular) packet P_3, which—after an initial jump—grows linearly with time for the first 60 periods. This growth exemplifies an *apparent quantum transient diffusion process* in position space, a "ballistic" filling up of the stochastic layer. The decomposition of packet P_3 into Floquet states shows major contributions from Floquet states delocalized over the

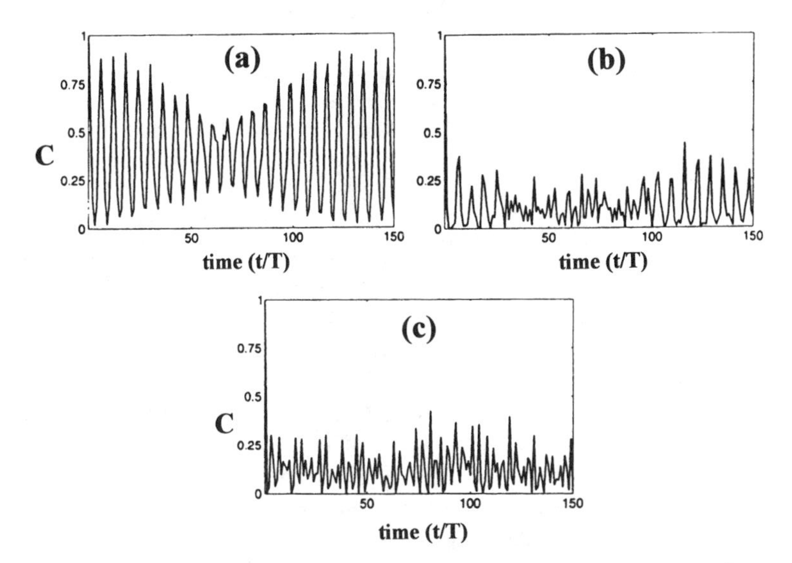

Figure 8.28. The evolutions of the autocorrelation functions $C(t) \equiv |\langle\psi(t)|\psi(0)\rangle|^2$ of the three wavepackets indicated in Fig. 8.27. These evolutions of (a) the regular packet P_1, (b) the intermediate packet P_2, and (c) the irregular packet P_3 are discussed in the text. From Latka et al. (1993). Copyright 1993 by the American Physical Society.

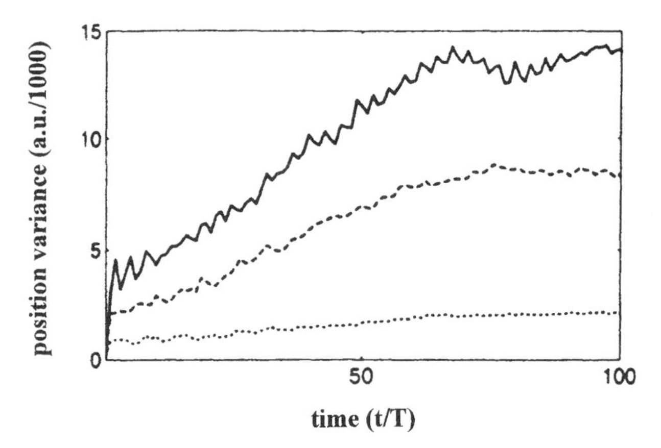

Figure 8.29. Time evolution of the position variance of the three packets of Figs. 8.27 and 8.28. The regular packet P_1 spreads out very little (dotted line), while the other packets show an abrupt growth followed by a period of further growth that is approximately linear in time. The solid curve is for the irregular packet. From Latka et al. (1993). Copyright 1993 by the American Physical Society.

layer. One of these delocalized Floquet states is also responsible for much of the growth of packet P_2 seen in Fig. 8.29. The apparent diffusion is not true diffusion, of course, since the quantum evolution is time reversible—while true classical diffusion is an irreversible random process.

8.3f. Quantum Transient Diffusion and Dynamical Quantum Localization

For driven, spatially open systems it is intriguing to see what happens when the initial state is located in the region of global chaos. To explore this experimentally, it is a good idea to use the diamagnetic Kepler system (Section 7.1) with the magnetic field parameter γ high enough so that the system is above the threshold for "total" classical chaos. We then add a microwave electric driving field so the system can absorb energy. Figure 8.30 shows, for one choice of parameters, the predicted classical growth in energy with time. The second moment (variance of the energy) again increases linearly with time. We expect this linearity for the evolution of a large ensemble of chaotic trajectories in phase space, since chaos is a sort of deterministic randomness. If the quantum initial state evolves similarly under the action of the Schrödinger equation, at least for some initial time period, then we again speak of *quantum transient diffusion*. This diffusion would be illusory because the quantum evolution is in fact reversible, as already noted.

If growth of some particular physical quantity appears diffusive for both a quantum system and its classical counterpart, then at any moment in time the two probability distributions of the quantity should be similar, at least when the distributions are smoothed to reduce quantum interference effects. For some choices of parameters ($\omega_0 \lesssim 1$ and sufficiently high F_0) and for the driving fully switched-on

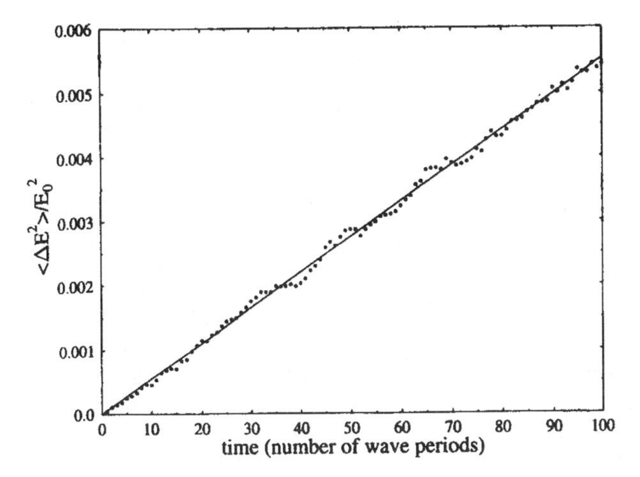

Figure 8.30. Numerical results for the square of the average growth $\Delta E(t)$ in mean energy of 100 classical electron trajectories for the 3-d hydrogen atom in collinear static-magnetic and microwave electric fields B_0 and F_0. The parameters are initial energy $E_0 = -1/(2n_0^2)$, $\omega_0 = 0.05$ and $F_0 = 0.003$, and the magnetic field alone places the diamagnetic Kepler system above the threshold for global chaos (see Figs. 7.4 and 7.5). When the microwave field is turned on at time $t = 0$, the preexisting chaos immediately leads to classical diffusive energy excitation that is characterized by $(\Delta E)^2$ being proportional to elapsed time. From Benvenuto et al. (1997). Copyright 1997 by the American Physical Society.

at the initial time $t = 0$, the driven Kepler system exhibits similar diffusive early probability distributions in the classical action variable and the quantum principal quantum number. Figure 8.31 shows results for an average elapsed time of 60 periods, with the averaging over ± 20 periods. The quantum distributions for initial energy eigenstates with different quantum numbers n_0, n_0' are found to scale when $n_0 P(n_0'/n_0)$ values are plotted versus n_0'/n_0—at least for n_0, n_0' between 20 and 100. These results were obtained by iterating the standard basis-set expansion equations using a large undriven 1-d hydrogen atom bound-state basis set.

One shouldn't conclude such behavior is typical for all parameter values that place the initial state above the threshold for global chaos. In particular, for ω_0 above about 1.5, the situation in Fig. 8.32 applies for a range of driving-field strengths F_0 that lie just above the chaos threshold value. In the left panel we see that the quantum variance in quantum number follows classical expectations only for a finite time—in this case, only five field periods. After this diffusive transient period in time, the growth in the quantum variance stops, or grows only slowly. The result is that quantum and classical probability distributions increasingly differ for times past the transient period; see Fig. 8.32b. The quantum distribution appears almost frozen (except for time fluctuations in the sharp jagged structure), with an average dependence on the quantum number that is exponentially decaying on the side open to the continuum.

Why does the quantum evolution stop while the classical chaotic diffusion continues? The quantum mechanism cannot be tunneling since tunneling occurs

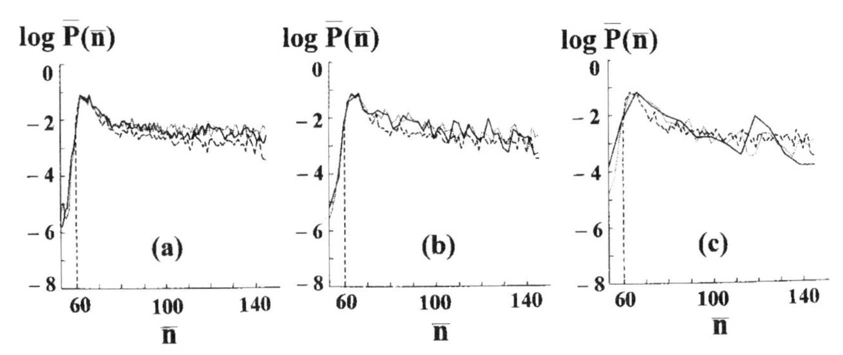

Figure 8.31. Calculated time-averaged quantum and classical final-state probability distributions for the 1-d hydrogen atom with initial quantum number n_0 in a microwave electric field having the low scaled frequency $\omega_0 = 0.7$ and $F_0 = 0.04$. The rescaled projections $\overline{P}(\overline{n}) = (n_0/66)P(n)$ onto the hydrogen atom basis are shown as a function of the rescaled quantum number $\overline{n} = (66/n_0)n$. The dashed lines in the panels are classical results, while the other curves are quantum results for various values of n_0: (a) $n_0 = 100$ and 66; (b) $n_0 = 45$ and 30; and (c) $n_0 = 20$ and 10 (solid curve). To average out rapid fluctuations in $\overline{P}(\overline{n})$, a time average was made over the 40 microwave periods within the range of times $40T < t < 80T$. Above the boundary for chaos near $\overline{n} = 60$, the quantum results qualitatively agree with the classical results over the entire range of n_0. This and the observed scaling indicate a role in the quantum time evolution of a classical diffusion process of the type shown in Fig. 8.30. From Casati et al. (1987). Copyright 1987, with permission from Elsevier Science.

even during the period of quantum transient diffusion. The mechanism appears to be wavepacket spreading accompanied by a strongly destructive wave interference that is characteristic of the classically chaotic regime! In Fig. 8.29 we've already noted that wavepacket spreading can be more rapid in the irregular (chaotic) regime than in the regime of regular classical motion. But now the chaos is global rather than confined to a stochastic layer. The effect shown in Fig. 8.32 is called *dynamical quantum localization*. A similar but nondynamical effect exists for quantum particles that are present in spatial lattices with randomized properties at the lattice sites. This quantum localization is in space rather than in time and is called *Anderson localization*. Dynamical quantum localization and Anderson localization are built on very similar mathematical foundations (Casati et al., Grempel et al., and Shepelyansky in Casati and Chirikov, 1995).

The two signatures of dynamical quantum localization—quantum transient diffusion, followed by stable localized probability distributions exhibiting exponential decay—have been seen in a number of other systems that differ from the driven Kepler system. These other systems include the momentum distribution of an atom moving in a phase-modulated standing light wave (Moore et al., 1994), and the center-of-mass motion of a single ion confined in a electromagnetic trap and driven by an optical field. In the latter case, dynamical localization can occur simultaneously in both the position and momentum spaces (El Ghafar et al., 1997).

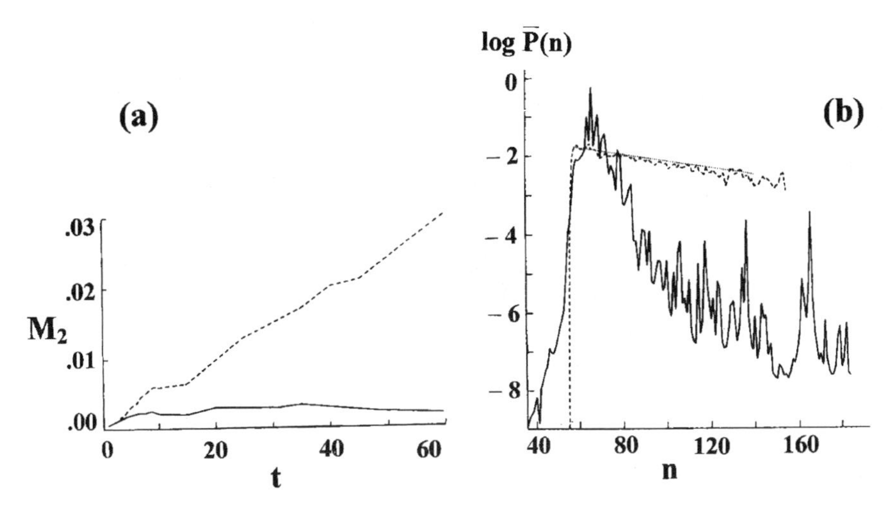

Figure 8.32. Numerical results showing dynamical quantum localization of the electron in the 1-d hydrogen atom in a high-frequency microwave electric field, with $\omega_0 = 2.5$, $F_0 = 0.04$, and $n_0 = 66$. (a) Time evolution of the second moment M_2 of the probability distribution P(n) in the hydrogen atom basis, $M_2 \equiv \langle (n - \langle n \rangle)^2 \rangle / n_0^2$. The classical growth (dashed curve) is linear in time (diffusive), while the quantum growth (solid line) stops after about five microwave periods T. (b) A logarithmic plot of P(n) time-averaged over the interval $560T < t < 600T$. The envelope of the quantum distribution (solid curve) has been stable for a long time, while the classical distribution has continued to evolve out to large values in n. From Casati et al. (1987). Copyright 1987, with permission from Elsevier Science.

Both quantum and classical irregular evolutions can be quite complicated if we start a driven quantum system in a region of global classical chaos that contains or closely borders regular regions of phase space. The chaotic classical trajectories can linger near the regular regions for various finite times. They can leave and maybe return, and so forth. Quantum effects complicate this evolution even further. Sorting this out is still underway.

9

EXTERNALLY INDUCED TRANSPORT

9.1. INDUCED CLASSICAL TRANSPORT (ADVANCED TOPIC)

Many interesting and important phenomena occur when we modify a quantum system's evolution by externally adjusting, in a time-dependent way, either the system or its interaction with the environment. We studied one of the simplest cases in Chapter 8, where the adjustment was sinusoidally oscillating in time. Because of time periodicity we were able to find the Floquet states with their quasienergies, with the Floquet states playing the roles of energy eigenstates of autonomous systems. We might think of adding more driving frequencies—an addition that requires a generalization of Floquet theory that is presently being developed. Random (stochastic) driving is another situation that is being explored theoretically, and utilized in the laboratory as well. More generally, we can employ an arbitrary time dependence of one or more parameters of an unperturbed system's Hamiltonian, and/or we can employ an arbitrary time dependence of the system's interaction with its environment.

We will need, for instance, a complete understanding of externally induced quantum evolution that transports probability from one region to another in phase space. Understanding this evolution helps us answer questions asked in Section 9.3c: How can we externally adjust a quantum system in a given initial state (perhaps in extended phase space) so that we achieve a given final state at a particular later point in time? And how well can we control the path of the evolution between the initial and final times?

We are a long way from answering these challenging questions. But we do know about a sizable number of externally controlled processes for the quantum transport of probability in phase space. Our understanding of these processes is often greatly abetted by studies of corresponding classical transport—a relationship recognized early in the twentieth century (Ehrenfest, 1917, 1923).

9.1a. Adiabatic Invariants and the Averaging Method

We begin with the relatively simple situation where only one of a system's parameters is varied slowly relative to the characteristic fixed-parameter times of all the motions within the system. Let us consider the problem of a simple harmonic oscillator whose frequency is being slowly varied in time, $\omega = \omega(t)$. This is an example of a frequency "chirped" system, with the chirping being "slow." We need a small *slowness parameter* ε that determines a time scale for the effects of the chirping. Given ε exists, we construct the ε-scaled dimensionless time $\tau \equiv \varepsilon t$. We consider just the range of early times where $\omega(\tau)$ changes significantly only when τ becomes of order unity, corresponding to an actual time of order $1/\varepsilon$. By this we mean that if $d\omega/dt = \varepsilon(d\omega/d\tau) \equiv \varepsilon\omega'(\tau)$, then $|d\omega/d\tau| \lessdot \omega$ for $\tau \lessdot 1$. We define the rate parameter $R(\tau) \equiv \omega'(\tau)/2\omega(\tau)$. A change in $\omega(\tau)$ is called *adiabatic* if the change $\Delta\omega$ in $\omega(t)$ is small compared to ω, over one period of the oscillator at the initial frequency $\omega = 2\pi/T$. That is, $\Delta\omega \approx T|d\omega/dt| \ll \omega$, which means that the dimensionless *classical adiabaticity parameter* defined as $A_{cl}(\tau) \equiv \varepsilon R(\tau)/\omega(\tau)$ will be small compared to unity.

The Hamiltonian of our frequency-chirped oscillator is written as

$$H(p, q, \tau) = \frac{p^2}{2m} + \frac{1}{2}m\omega^2(\tau)q^2.$$

It is important to first consider the case $\tau =$ constant—where $H(p, q, \tau = \text{const.})$ is called the *frozen Hamiltonian*—and to transform this Hamiltonian to action-angle variables (I, θ) of the fixed-frequency (unperturbed) oscillator. The result is $H = I\omega(\tau)$. As we shall see, the term "unperturbed" takes on real meaning when we use these variables: the equations of motion take a form where we can implement perturbation theory expansions in powers of the slowness parameter ε. If we now let τ vary, then we need to derive a transformation $(q, p, \tau) \to (\theta, I, \tau)$ via an F_1-type generating function, as in Section 3.3. However, we must extend Section 3.3 to include time-dependent transformations (Percival and Richards, 1982). Using $F_1(q, \theta, \tau) = \frac{1}{2}m\omega(\tau)q^2 \cot\theta$ and $q^2 = (2I/m\omega)\sin^2\theta$, the time-dependent Hamiltonian becomes

$$H'(I, \theta, \tau) = H(I, \theta, \tau) + \frac{\partial F_1(q, \theta, \tau)}{\partial t} = I\omega(\tau) + \varepsilon R(\tau) I \sin 2\theta$$

and the extended Hamilton's equations are

$$\begin{aligned} \frac{d\theta}{dt} &= \frac{\partial H'}{\partial I} = \frac{\partial}{\partial I}\left(H + \frac{\partial F_1}{\partial t}\right) = \omega(\tau) + \varepsilon R(\tau)\sin 2\theta, \\ \frac{dI}{dt} &= -\frac{\partial}{\partial \theta}\left(H + \frac{\partial F_1}{\partial t}\right) = -2\varepsilon R(\tau)\cos 2\theta. \end{aligned} \tag{9.1}$$

Note that dI/dt is periodic in θ with zero mean value. We'll use this fact to advantage by taking an appropriate θ-averaging of the fast motion of the unperturbed oscillator, thus leaving only the oscillator's time-dependent response to the slow changes in the frequency.

Now we make the perturbation expansion in ε:

$$\begin{aligned} \theta(t) &= \theta^0(t) + \varepsilon\theta^1(t) + O(\varepsilon^2), \\ I(t) &= I^0(t) + \varepsilon I^1(t) = O(\varepsilon^2). \end{aligned} \tag{9.2}$$

If we insert Eqs. (9.2) into Eqs. (9.1), equate powers of ε, and integrate the zero-order and first-order terms in turn (consistently keeping only lowest order contributions), we find (Percival and Richards, 1982)

$$\begin{aligned} I(t) &= I_0[1 - A_{cl}(\tau)\sin 2\theta^0(t) + A_{cl}(0)\sin 2\theta_0] + O(\varepsilon^2), \\ \theta(t) &= \theta^2(t) - A_{cl}(\tau)\cos 2\theta^0(t) + A_{cl}(0)\cos 2\theta_0 + O(\varepsilon^2), \end{aligned}$$

where $\theta^0(t) = \int_0^t \omega(\tau)\, dt + \theta_0$. Since $\theta^0(t)$ is rapidly varying (increasing by π in about a time interval π/ω), then for small $A_{cl}(\tau)$, $\theta(t)$ must also be fast. However, $I(t)$ is slowly varying and almost equal to the initial value because, by our construction above, $A(\tau) \ll 1$ for times $t \leqslant 1/\varepsilon$. We say that the action is (almost) an *adiabatic invariant*, as it is constant to order ε for times $t \leq 1/\varepsilon$. Ehrenfest (1917) pointed out the importance of such classical adiabatic invariants for quantum mechanics.

An action is an adiabatic invariant in any Hamiltonian system where the system has one fast degree of freedom that can be described by an angle variable when τ is constant. There are several ways to prove this. *Canonical adiabatic perturbation theory* is based on finding an F_1-type canonical transformation $(I, \theta, \mathbf{p}, \mathbf{q}) \rightarrow (I', \theta', \mathbf{p}', \mathbf{q}')$ that generates a new Hamiltonian $H' = H + \varepsilon G + O(\varepsilon^2)$, which is independent of θ' to first order in ε (Lichtenberg and Lieberman, 1992, Section 2.3b). These authors show that

$$I' = I + \frac{\varepsilon}{\omega}[G - \langle G \rangle_\theta] \tag{9.3}$$

is the adiabatic invariant, where $\langle G \rangle_\theta$ is an average over the fast θ-motion and ω is the frequency of that motion.

Another proof is more general in that the dynamical system need not even be Hamiltonian. This proof applies the *method of averaging* (Bogoliubov and Mitropolsky, 1961; Lichtenberg, 1969), which we now apply to our Hamiltonian case. We define a vector $\mathbf{y} \equiv (\mathbf{p}, \mathbf{q}, I)$ and consider the set of differential equations

$$\begin{aligned} \frac{d\mathbf{y}}{dt} &= \varepsilon \mathbf{f}(\mathbf{y}, \theta), \\ \frac{d\theta}{dt} &= \omega(\mathbf{y}) + \varepsilon g(\mathbf{y}, \theta). \end{aligned} \tag{9.4}$$

We expand $\mathbf{y}$ and θ in powers of the slowness parameter ε. As before, to zero order in ε, only the phase θ varies, and it varies at the constant rate ω so that $\theta = \omega t$. Also to zero order, $\mathbf{y} = \mathbf{y}_0 = \text{constant}$. Then we change from $\mathbf{y}$ to a new variable $\mathbf{z}$ defined according to

$$\begin{aligned}\mathbf{y} &\equiv \mathbf{z} + \int_0^t \frac{d\mathbf{y}}{dt}\,dt - \int_0^t \left\langle \frac{d\mathbf{y}}{dt} \right\rangle_\theta dt \\ &= \mathbf{z} + \varepsilon \int_0^t \mathbf{f}(\mathbf{y}, \theta)\,dt - \varepsilon \int_0^t \langle \mathbf{f}(\mathbf{y}) \rangle_\theta\,dt,\end{aligned} \tag{9.5}$$

where $\langle \mathbf{f}(\mathbf{y}) \rangle_\theta$ is independent of θ because it is an average over θ. Inserting the expansions of $\mathbf{y}$ and θ into Eq. (9.5), we can again solve each order separately by iteration. We keep terms in Eq. (9.5) through first order in ε, differentiate the result with respect to time, and use the first of Eqs. (9.4) to obtain

$$\frac{d\mathbf{z}}{dt} = \varepsilon \mathbf{f}(\mathbf{y}_0) + \mathcal{O}(\varepsilon^2).$$

This is an equation of a reduced dynamical system in which the fast changes of the phase θ are averaged out; there is no dependence on θ. Since I is the conjugate variable to θ, we again see that the action must be an (approximate) adiabatic invariant.

9.1b. Nonlinear Resonance Trapping During Frequency Chirping

The theory we've just seen assures us there exists a regime of low chirping rates at which the evolution is essentially adiabatic for some early period of time. Unfortunately, because we don't actually know what the slowness parameter is, the theory doesn't provide a result for the actual change in the adiabatic invariant when the Hamiltonian varies at a given finite rate. However, there is no problem with integrating Hamilton's equations numerically, with the explicit time dependence of the frequency chirping included. This has been done for the driven Morse and Kepler systems of Section 8.3. Let's see what happens.

First consider the driven Morse oscillator, with linear frequency down-chirping given by $\omega(t) = \omega(0)(1 - t/2T_p)$. Figure 9.1 shows the evolution of the action I(t) of a few trajectories, for the initial scaled frequency $\omega(0) = 1.1$. What happens depends on the initial values I(0), $\theta(0)$ of the action and angle variables; see the values given in the figure caption. For three separate values of I(0), we see the evolution for two values of $\theta(0)$ that were selected to display different behavior. The difference in each case is whether or not we start outside the 1 : 1 nonlinear resonance zone. If we start outside the zone, then the action does not grow upward to produce molecular dissociation. There is such upward growth in the action, however, if we start either inside the regular resonance island (upper two cases in the figure) or close to the island (the bottom case). We can overlay (I, θ) stroboscopic snapshots of the

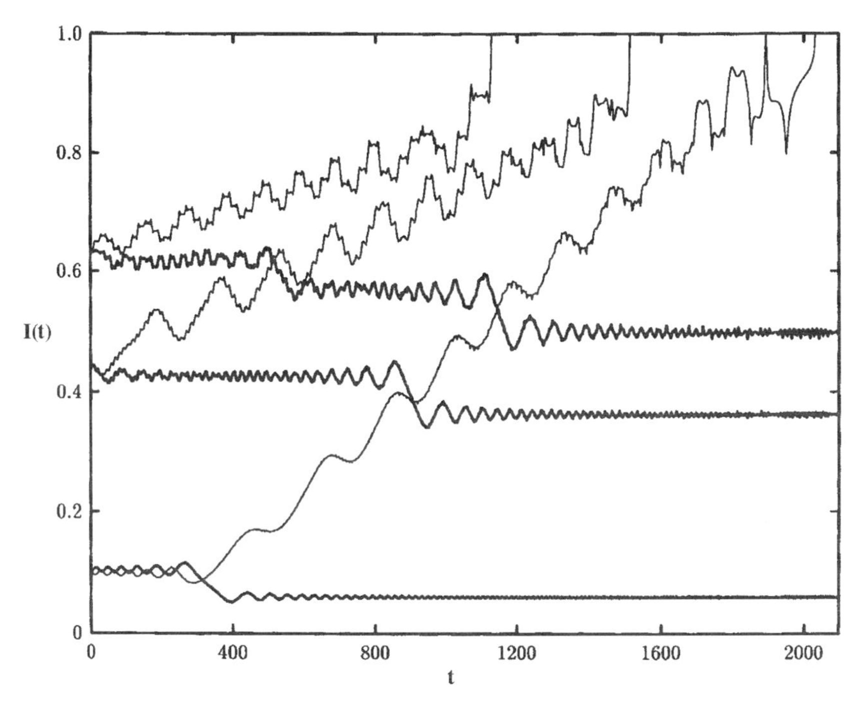

Figure 9.1. Time evolutions of the action I(t) of some classical trajectories of the frequency-chirped sinusoidally driven Morse oscillator model for the vibrational motion of a diatomic molecule. The downward chirp is linear in time, $\omega(t) = \omega(0)[1 - t/2T_p]$, with $\omega_0 = 1.1$ and $T_p = 2000$ driving periods. The Hamiltonian is $H = H_{Morse} - Fx\cos[\omega(t)t]$, $H_{Morse} = p^2/2 + (e^{-2x} - 2e^{-x})$ in dimensionless units, and F is taken to be 0.003. Depending on the initial action-angle variables $\{I(0), \theta(0)\}$ for the trajectory, the early evolution can reflect or not reflect growth in the mean action that leads to subsequent dissociation. During the chirp a trajectory can switch from nondissociating to dissociating. The initial conditions for the bound trajectories are $\{I(0), \theta(0)\} = (0.1, \pi)$, $(0.45, 1.6\pi)$, $(0.6333, \pi)$, while the dissociating trajectories are for $(0.1, 1.8\pi)$, $(0.45, 1.8\pi)$, and $(0.6333, 0.8\pi)$. From Liu et al. (1995). Copyright 1995 by the American Physical Society.

trajectories (Gray, 1983) onto long-time stroboscopic Poincaré sections made at fixed values of ω taken equal to the instantaneous values (Cohen and Meerson, 1993). The instantaneous (I, θ) locations of these trajectories that ultimately dissociate are found to lie within the 1 : 1 resonance zone. Thus we have *nonlinear resonance trapping* of classical trajectories during frequency chirping. As the frequency is slowly varied, the resonance is moving in phase space and the trajectories are moving with it.

In classical physics, nonlinear resonance trapping is also called *autoresonance*—because when the system's parameters are varied adiabatically, the resonantly driven nonlinear oscillator and the driving oscillations are persistently phase-locked. There is a sizable list of applications (Friedland, 1997), including the resonant excitation of soliton waves in nonlinear dispersive optical media (Aranson et al., 1992).

Assuming that the action I is an adiabatic invariant, one can explain the average slope of the growth of I(t) for the dissociating trajectories in Fig. 9.1. The argument uses the classical pendulum approximation (Section 5.6a), applied to the case of driven oscillators (Liu et al., 1995; Yuan and Liu, 1998). The adiabatic invariant is then given by Eq. (4.19). No doubt we are seeing adiabatic evolution during frequency chirping—until the action reaches high values where nonadiabatic transport leads to actual dissociation.

As we would now expect, resonance trapping during frequency chirping also occurs for the 1-d driven Kepler problem. Figure 9.2 shows the precise numerical evolution of an ensemble of trajectories initially trapped in the 1 : 1 resonance zone, with the driving frequency varying as $\omega(t) = 1/(1 + \alpha t^2)$ with $\alpha = 2.5 \times 10^{-5}$. The numerical results for the evolution of both $N(t) \equiv I(t)/I(0)$ and $\theta(t)$ follow adiabatic predictions derived in the pendulum approximation, up to time $t \approx 400$. (See Meerson and Friedland (1990) for the theoretical details.) After this time the evolution becomes nonadiabatic, leading to ionization of the 1-d hydrogen atom. This ionization occurs in spite of the fact that the driving-field strength is below the fixed-frequency ionization threshold value, for all instantaneous values of frequency during the frequency chirp.

Returning to Fig. 9.1, note the dissociating trajectory starting at $I(0) = 0.1$. During the early part of the chirp, the action is almost constant. Near time $t =$ 250, however, the trajectory quickly becomes dissociating via the resonance-trapping adiabatic transport mechanism. This change seems analogous to the field

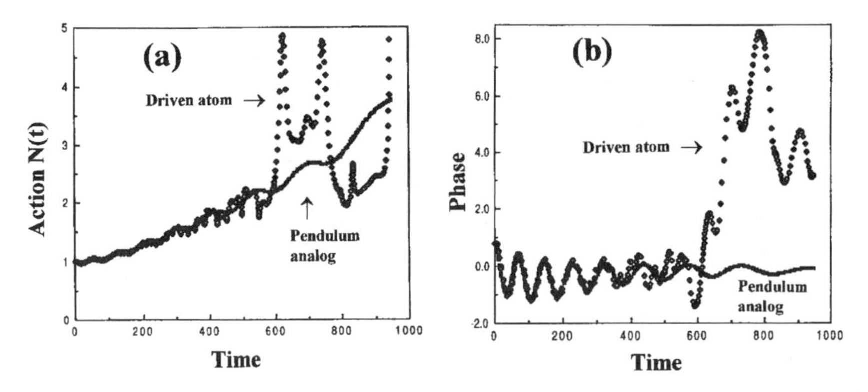

Figure 9.2. Nonlinear-resonance trapping and phase-locking in the frequency-chirped sinusoidally driven 1-d Kepler system (the hydrogen atom). The downward chirp is $\omega(t) = \omega_0/(1 + \alpha t^2)$, $\omega_0 = 1$, $\alpha = 2.5 \times 10^{-5}$. (a) As for the initially dissociating trajectories in Fig. 9.1, the normalized action $N(t) = I(t)/I(0)$ of an initially trapped ensemble of electron trajectories grows with time t until about $t = 600$, where the evolution becomes nonadiabatic because of a growing stochastic layer in phase space. The smoothly oscillating curve for the integrable approximate pendulum system stays adiabatic. (b) The phase follows the oscillations of the phase-locked approximate pendulum system (Section 5.6a), again until $t = 600$. From Meerson and Friedland (1990). Copyright 1990 by the American Physical Society.

dependence of the Floquet state shown on the left side of Fig. 8.17, except that here the frequency is varying. What caused the change in the classical behavior? It must be that the early evolution—which involves transport in a time-evolving region of KAM (slightly perturbed) undriven oscillator motion—has changed into evolution within the time-evolving resonance zone. Thus near time $t = 250$, the trajectory must have crossed the thin stochastic layer between the KAM and resonance regular regions of phase space. In an integrable system the layer would be replaced by a separatrix, and the crossing called *separatrix crossing*. Let's look a bit further into these important classical nonadiabatic transitions.

9.1c. Passage Through Separatrices and Stochastic Layers

We know that the period of evolution on a separatrix trajectory is infinite, which means the frequency $\omega(\tau)$ is zero on the separatrix. Hence the condition for applying adiabatic theory—$A_{cl}(\tau) \ll 1$—is not satisfied in the vicinity of a separatrix. No matter how small the slowness parameter ε, $A_{cl}(\tau)$ is arbitrarily large in some neighborhood of the separatrix.

For a simple example of trajectory passage through a separatrix, we consider an autonomous 1-d system, such as the particle in a quartic double well (Fig. 4.3). The "figure eight" separatrix divides phase space into three identifiable regions: the two lobes a and b inside the separatrix, and the region c outside the separatrix. The location of the separatrix in phase space depends on the parameters of the system, such as the height of the barrier within the double well. As we vary this height, the separatrix sweeps through phase space, growing or shrinking as the height is raised or lowered. At some initial time we could consider a trajectory point located in phase space in any one of the regions a, b, c. We could then ask what can happen as time goes on, with the trajectory evolving and the separatrix also evolving with the time variation of the barrier height. Figures 9.3a,b,c show the basic different possibilities for the transport: (1) from one lobe to the other; (2) from one lobe to region c outside the separatrix; or (3) from region c to inside one of the lobes.

Let the phase-space areas of lobes a and b be $A_a(\tau)$ and $A_b(\tau)$. Let us consider evolution initially along a periodic trajectory, which has the action I_i. We assume that trajectory evolution is adiabatic outside the neighborhood of the separatrix. If this neighborhood could be ignored, then there may come an ε-scaled time $\tau_x = \varepsilon t_x$, where the initial value I_i of the adiabatic invariant (the action) becomes equal to the sum $A_a(\tau_x) + A_b(\tau_x) \equiv A_c(\tau_x)$. At this *pseudocrossing time* τ_x, we see that $I_i(\tau_x) - A_c(\tau_x)$ would change its sign. Since I_i is the phase-space area inside the trajectory, the trajectory subsequently must cross the separatrix.

However, we cannot ignore the special dynamics in the neighborhood of the separatrix. We know that the value of the stationary action of the frozen (unvaried) system jumps as we move through a separatrix (Section 1.4b). It's not surprising then that the theory of separatrix crossing by a trajectory finds, to lowest order in ε, that the final value of the adiabatic invariant is the trajectory action I_f for the new region that is entered after the crossing. The action jumps by the amount $\Delta I_{if} \equiv I_f(\tau_x) - I_i(\tau_x)$. But to next order in ε, we must consider the near-separatrix dynamics,

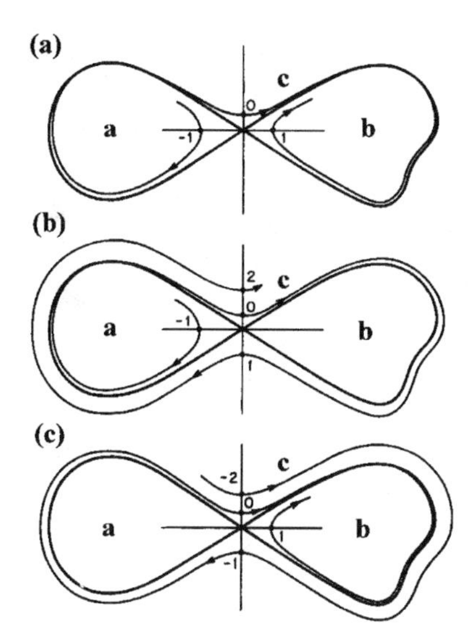

Figure 9.3. Schematic illustrations of the possible types of evolution of a classical trajectory near a figure-of-eight separatrix, as a parameter in the Hamiltonian is slowly varied. (a) A double crossing of the separatrix, for a particle beginning in lobe a and ending in lobe b. Single crossings are also shown, for a particle that (b) begins in a lobe and (c) ends in a lobe. Actually, the motion loops of the trajectory stay roughly constant in size while the separatrix-enclosed areas change with the parameter variation. From Cary et al. (1986). Copyright 1986 by the American Physical Society.

develop a special theory for it, and appropriately "join" this with the theory for the adiabatic evolution. We have seen this idea work before; it is at the heart of WKB theory with its connection formulas (Section 1.2c). In the present separatrix crossing problem the method of matched asymptotic expansions in ε is used, as described in Farrelly (1986). The separatrix crossing theory for 1-d systems has been developed in Cary et al. (1986) and in Cary and Skodje (1988). An approximate Hamiltonian is found for the near-separatrix dynamics, where this Hamiltonian is of course valid only near the separatrix. For this separatrix region, the approximation is made that the system's energy is close to that of the separatrix. The order-by-order expansion matching process leads to a dominant correction of order $\varepsilon \ln \varepsilon$ to ΔI_{if}. This correction depends on a new parameter that is related to the initial phase θ_0 and depends in a somewhat complicated way on three more parameters describing the near-separatrix dynamics.

Thus the change in action ΔI_{if} depends on the initial phase θ_0. This dependence is important for problems where the separatrix is crossed by a trajectory more than once due to the chosen character of the parameter time dependence. We can take, for instance, the problem of the barrier in the double well. If we slowly raise the top of the barrier past the mean adiabatic energy of a trajectory and then bring the barrier back down again, we have a time sequence of two separatrix crossings, with an adiabatic evolution of some sort between them. The first crossing sets an initial phase for this adiabatic evolution between crossings; this evolution accumulates a phase change $\Delta\phi$ that affects the final outcome. Hence correlations persist between the two separatrix crossings, and corrections to zero-order theory that account for $\Delta\phi$ must be included (Cary and Skodje, 1988).

In the top panel of Fig. 9.4 we see the numerically obtained precise evolution for the position variable q(t) of a trajectory for the frequency-swept quartic oscillator with Hamiltonian

$$H = \tfrac{1}{2}p^2 - \tfrac{1}{2}\omega^2(t)q^2 + bq^4,$$
$$\omega^2(t) = \pm 2a\left(1 - \varepsilon t + \frac{\sin(2\pi\varepsilon t)}{2\pi}\right), \qquad \varepsilon = \frac{1}{50}, \qquad t = 0\text{–}50.$$

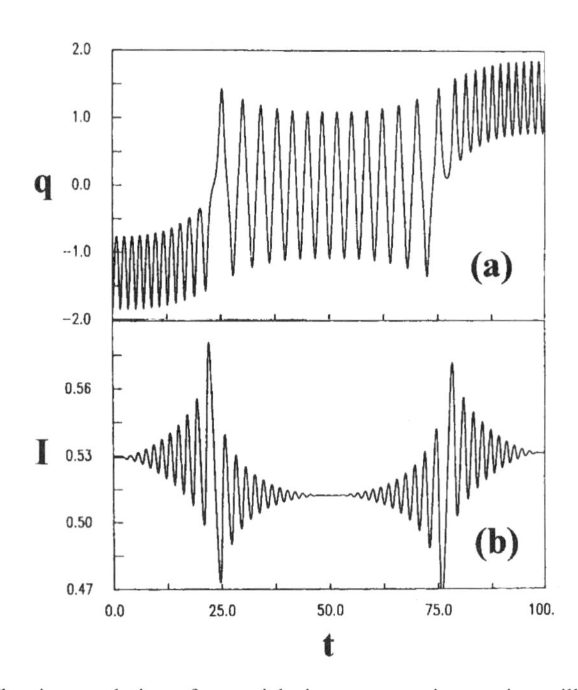

Figure 9.4. The time evolution of a particle in a symmetric quartic-oscillator double-well potential, with an inverted-barrier frequency $\omega(t)$ that first is decreasing during the time interval $0 \leq t \leq 50$; the frequency sweep is then reversed and is therefore increasing for $50 \leq t \leq 100$. A (lobe a $\rightarrow$ region c) separatrix crossing occurs near $t = 23$, and then a (region c $\rightarrow$ lobe b) crossing occurs near $t = 77$ (Fig. 9.3a). (a) The evolution q(t) of the position of the particle begins as oscillations in the left potential well, changes to oscillations above the potential barrier after the first separatrix crossing, and becomes oscillations in the right potential well after the second crossing. This is confirmed by the mean values of q(t) near $t = 0$, $t = 50$, and $t = 100$. (b) The corresponding time evolution I(t) of the action variable shows the changes in action associated with each separatrix crossing. The changes are equal and opposite since the quartic potential is symmetric. From Cary and Skodje (1989). Copyright 1989, with permission from Elsevier Science.

The positive sign is taken when the frequency $\omega(t)$ is decreasing during $0 \leq t \leq t_1$, and the negative sign is taken when $\omega(t)$ is increasing, $t_1 \leq t \leq t_2$. The quantity $\omega(t)$ is called the *inverted barrier frequency* because it is the frequency of a particle in the potential $+\omega^2 q^2/2$.

In the top panel of Fig. 9.4, we note changes in the behavior of q(t) that occur near the two separatrix crossings at $t_x = 23$ and 77. Referring to Fig. 9.3, we have case (a). The trajectory moves from lobe a to region c during the first separatrix crossing, and from region c to lobe b during the second crossing. We have detrapping from lobe a and then retrapping in lobe b. Note this is confirmed by the range in values of the excursions in q(t) during each of the time intervals spent in regions a, c, and then b. As expected, the frequency of oscillations for region c is lower than the frequencies for a and for b. Although the system begins to adjust to an impending crossing quite a few oscillation periods before t_x, there is a rapid switchover near t_x in the character of the motion. The switchover takes less than an average oscillation period. This switchover is also seen in the evolution of the instantaneous action function I(t); see the lower panel in Fig. 9.4. This continuous action function takes on asymptotic values that are simply the expected invariant action when the trajectory is in lobes a and b, but one-half that action when the trajectory is in region c. As expected for an adiabatic invariant, the mean value of I(t)—averaged over a local oscillation period—is quite constant, except very near the crossing times t_x.

To make our study of separatrix crossing transport more relevant to quantum mechanics, we must consider ensembles of trajectories rather than just one trajectory. These ensembles might initially correspond to a quantum wavepacket or to an energy eigenstate (Section 1.3b). Let us return again to the driven Morse oscillator (Section 8.3). We consider an initial energy eigenstate so that the trajectory ensemble at $t = 0$ has action $I = \text{constant} \equiv I_0$ and has a uniform distribution in angle θ over the interval $[0, 2\pi)$. We let the driving field be $F_{max} \sin^2(\pi t/T_p) \cos \omega t$, $0 \leq t \leq T_p$. Thus we now consider a *pulsed driving field* with the pulse envelope $\sin^2(\pi t/T_p)$ and a full pulse length T_p.

Following Dietz et al. (1992) we numerically start off the ensemble as if it were the quantum ground state of vibrational bond stretching of an O—H molecular bond, which means $I_0 = 0.0226$ in the scaled units of the Morse oscillator. We choose the driving frequency $\omega = 0.8868$, which will place the 1 : 1 resonance zone near the initial ensemble. We take $F_{max} = 0.0273$, which is below the threshold for dissociation but large enough for the resonance zone to be quite large at the peak of the pulse; see Fig. 9.5. Finally we take T_p equal to 105 driving periods $T = 2\pi/\omega$; this puts the peak of the pulse at 52.5T. We numerically integrate Hamilton's equations; see Section A.2.

In Fig. 9.6 we see stroboscopic snapshots of the ensemble evolution in (I, θ) space. The time periods increase from (a) to (d). Initially the ensemble is on a horizontal line at $I_0 = 0.02266$. At $t = 30T$ the ensemble is near the finish of a nearly adiabatic evolution and is still distributed like a weakly perturbed oscillator (KAM) state that spans the full range of θ. The system is still in region "a" of phase space.

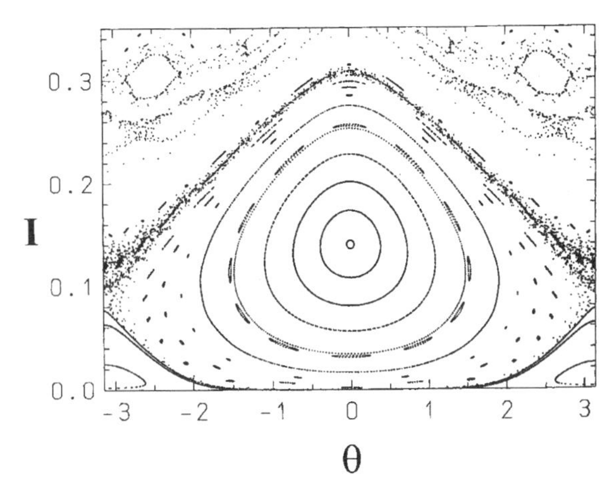

Figure 9.5. An action-angle (I, θ) stroboscopic Poincaré section for a sinusoidally driven Morse oscillator with dimensionless driving parameters $\omega = 0.8868$ and $F = 0.02712$. There is a large fundamental nonlinear-resonance island bounded by a thin "separatrix" stochastic layer. The value F of the driving-field strength is the same as the maximum field value F_{max} for the field-pulsed system of Fig. 9.6. From Dietz et al. (1992). Copyright 1992 by the American Physical Society.

However, at $t = 50T$, almost all of the ensemble is distributed inside the 1 : 1 regular resonance island. Now we are in a different region "c" in phase space. Something like a separatrix crossing has occurred for a certain range of field strengths—although we know we really have a stochastic layer. The ensemble has been spreading out within the resonance island, moving along a path that is near closed phase-space curves lying on particular vortex tubes. The ensemble is rotating at the slow frequency given by the pendulum approximation (Section 5.6a). This second nearly adiabatic evolution continues for a while as the pulse amplitude reaches its peak and then drops. Again the field strength for the "separatrix crossing" is reached, after which part of the ensemble returns to region "a" and the rest to a new region "b."

At the end of the pulse the ensemble is observed to be split into two narrow bands of action values, as we see in panel (d) of Fig. 9.6. The center of one band is near the initial action, the center of the other band is an equal distance in action on the opposite side of what was the center of the 1 : 1 resonance zone. This particular double-peaked distribution is a distinguishing feature produced by the double "separatrix-crossing" transport process within phase space.

Nearly 1-d experiments on hydrogen atoms in microwaves have shown the double-peaked action distributions of this double separatrix-crossing transport process (Bayfield and Sokol, 1988; Bayfield et al., 1995). All the driven Kepler oscillator parameters were very similar to those given just above for the driven Morse oscillator. The pulse envelope was $\sin(\pi t/T_p)$ with $T_p \approx 130T$. The results provide strong evidence for a quantum analog to stochastic-layer crossing. The

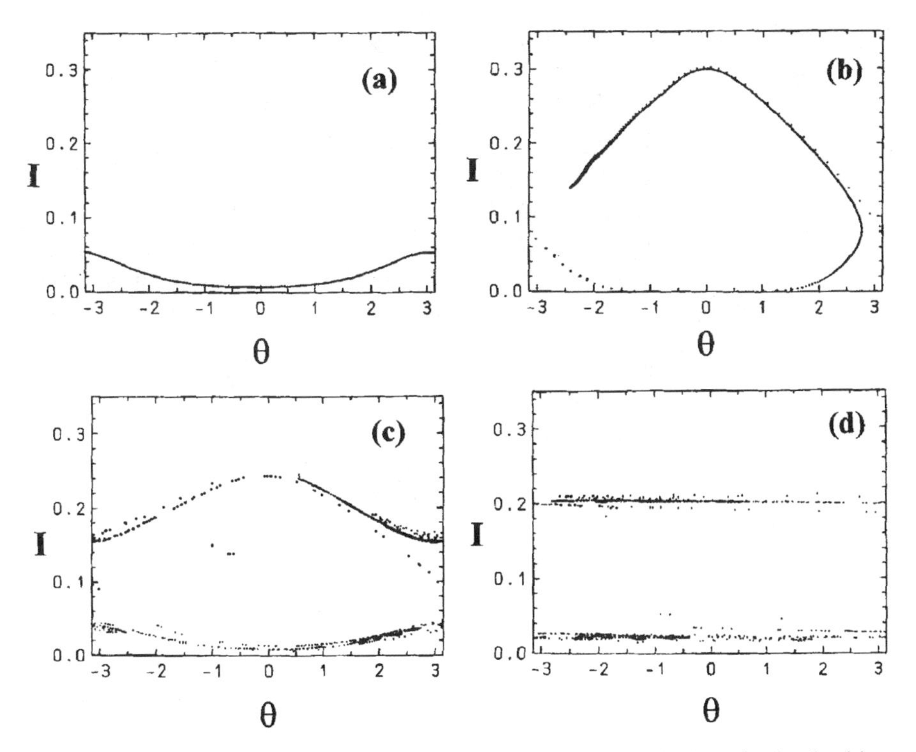

Figure 9.6. (I, θ) snapshots of the time evolution of an ensemble of trajectories for the driven Morse oscillator of Fig. 9.5, during the electric field pulse $F_0(t) = F_{max} \sin^2(\pi t/T_p)$ with $T_p = 105$ driving periods T. The initial state of the ensemble is $I(0) = 0.02264$ with a uniform distribution in θ. (a) The ensemble at $t = 30T$, before the first separatrix crossing that will occur near the unstable fixed point at $\theta = 0$ or 2π. (b) By $t = 50T$, separatrix crossing of the ensemble has almost completely occurred. (c) By $t = 80T$, the second separatrix crossing has occurred during the fall of the pulse. Each trajectory lies near either an upper or lower KAM curve, depending on the trajectory's phase at its second separatrix-crossing point. (d) The ensemble at the end of the pulse, with the double-peak final distribution in the action variable characteristic of the double separatrix-crossing transport process in phase space. From Dietz et al. (1992). Copyright 1992 by the American Physical Society.

quantum transport process must be something more than adiabatic evolution of a single Floquet state that evolves from the initial quantum energy eigenstate prepared in the experiments. The final system parameters are the same as the initial ones—but the final state is double peaked in principal quantum number (action), while the initial state is single peaked. Quantum transitions between Floquet states must have occurred. Something nonadiabatic must have happened quantum mechanically. In the next two sections we investigate some aspects of quantum transitions arising from time-changing parameters.

9.2. AVOIDED ENERGY-LEVEL CROSSINGS

9.2a. Adiabatic and Diabatic States and Their Couplings

The energy or quasienergy levels of a quantum system depend on a number of parameters present in the Hamiltonian. It is natural to investigate the behavior of the levels as a parameter is changed from one *fixed* value to another. We saw examples of this in Figs. 7.26 and 8.16. A new question immediately arises: What will happen if the parameter is actually changing in time, either explicitly in the case of a driven system, or implicitly via coupling to another degree of freedom in the case of an autonomous system?

The diatomic molecule provides a good system for investigating the case of autonomous systems. Such a molecule is comprised of two relatively heavy atomic nuclei and one or more light electrons. Because of the difference in masses, the electronic and nuclear motions occur on different time scales. It then makes considerable sense to view the slowly varying nuclear coordinates as parameters that determine the faster electronic motion.

In a laboratory (fixed) frame of reference, let the nuclear separation vector $\mathbf{R} = \mathbf{R}_a - \mathbf{R}_b$ be the set of relative coordinates for the nuclei a, b; and let $\mathbf{r} = \{\mathbf{r}_i\}$ be the set of electronic coordinates. The kinetic energy of the relative nuclear motion is then $T^{nuc} = (\mathbf{p}^{nuc})^2/2M$ with $M = m_a + m_b$. The full Hamiltonian for the molecule has the form

$$H = T^{nuc} + H' = T^{nuc} + [T^{el} + V(\mathbf{r}, \mathbf{R})]. \tag{9.6}$$

Here T^{el} is the kinetic energy of the electrons, and $V(\mathbf{r}, \mathbf{R})$ is the sum of the Coulomb potential energies of interaction for all the electrons and nuclei, including the nuclear Coulomb repulsion.

For a fixed nuclear configuration, the Hamiltonian becomes simply H', with $\mathbf{R}$ being a parameter. Suppose we have solved the time-independent Schrödinger equation for the Hamiltonian H', obtaining for each value of $\mathbf{R}$ a set of electronic energy eigenfunctions $\psi_k(\mathbf{r}, \mathbf{R})$ that form an orthonormal complete basis set with respect to the electronic variables $\mathbf{r}$, $\langle j|k\rangle \equiv \int \psi_j^* \psi_k \, d\mathbf{r} = \delta_{jk}$. Then we can expand a wavefunction $\Psi(\mathbf{r}, \mathbf{R})$ for the full Hamiltonian in the form

$$\Psi(\mathbf{r}, \mathbf{R}) = \sum_k \chi_k(\mathbf{R}) \psi_k(\mathbf{r}, \mathbf{R}). \tag{9.7}$$

The functions $\chi_k(\mathbf{R})$ depend only on $\mathbf{R}$, because T^{nuc} is the only difference between H and H'. Let us consider the set of identities

$$\int \psi_j^*(\mathbf{r}, \mathbf{R})(H - E)\Psi(\mathbf{r}, \mathbf{R}) \, d\mathbf{r} = 0, \qquad j = 1, 2, \ldots. \tag{9.8}$$

Inserting Eq. (9.7) into Eqs. (9.8) produces an infinite set of coupled equations that can be represented by a matrix equation. The matrices $\bar{\bar{U}}$ and $\bar{\bar{\mathbf{P}}}$ appear, with the matrix elements

$$\mathrm{U_{jk}}(\mathbf{R}) = \langle \mathrm{j}|\mathrm{H}'|\mathrm{k}\rangle,$$
$$\mathbf{P}_{\mathrm{jk}}(\mathbf{R}) = \langle \mathrm{j}|\mathbf{p}^{\mathrm{nuc}}|\mathrm{k}\rangle.$$

It's not surprising that only these two matrices are needed to completely describe the matrix equation (Smith, 1969).

We proceed with the formal theory of the rotating and vibrating molecule by transforming the electronic coordinates from the laboratory frame of reference to another frame of reference rotating with the nuclear separation vector $\mathbf{R}$ (Kolos, 1970). We then say we have transformed our molecular problem into a *molecular representation*. The two angular coordinates Θ, Φ describing the orientation of $\mathbf{R}$ do not completely separate out, and thus there is a coupling of the electronic states $\psi_k(\mathbf{r}, \mathbf{R})$, where the coupling arises from the rotational motion of the molecule. Nevertheless, in the rotating frame, the important matrices $\bar{\bar{U}}$, $\bar{\bar{P}}_R$, and $\bar{\bar{P}}_\Theta$ are all independent of molecular orientation, that is, independent of Θ and Φ. Therefore we can define a unique *adiabatic molecular representation* by the requirement that the potential energy matrix $\bar{\bar{U}}^{ad}(R)$ be diagonal for all $R = |\mathbf{R}|$. All the coupling between states in this representation will clearly be due to $\bar{\bar{\mathbf{P}}}^{ad}(R)$. We call the set of exact energy eigenstates in this representation the *adiabatic states*.

Similarly, we can define the *standard diabatic molecular representation* by the requirement that $\bar{\bar{P}}_R^{sd}(R)$ be diagonal for all R, where $P_R \equiv \hbar/i\ (\partial/\partial R)$. Of course, if we diagonalize this operator, which produces vibrational motion, then we can't expect operators for the electronic motion to remain diagonal. So $\bar{\bar{U}}^{sd}(R)$ will have radial off-diagonal matrix elements $U_{jk}^{sd}(R)$. Said another way, the unitary transformation from the adiabatic representation to the standard diabatic representation will make $\bar{\bar{U}}^{sd}$ nondiagonal. Since $[P_R, U]$ vanishes at $R = \infty$, we can choose a boundary condition such that the diabatic and adiabatic representations are identical at $R = \infty$. Then the molecule dissociates to the same states in both representations, and we normally take the asymptotic states to be products of unique atomic states.

To see the above concepts in action, let us consider a nonrotating molecule, with only two states being of interest. The Hamiltonian matrix in the diabatic representation can be written

$$\bar{\bar{H}}^{d} = \frac{1}{2m}\begin{bmatrix} -\partial^2/\partial R^2 & 0 \\ 0 & -\partial^2/\partial R^2 \end{bmatrix} + \begin{bmatrix} U_{11}^d(R) & U_{12}^d(R) \\ U_{12}^d(R) & U_{22}^d(R) \end{bmatrix}. \tag{9.9}$$

Using the notation of Smith (1969), the 2×2 unitary matrix that diagonalizes $\bar{\bar{U}}^{d}(R)$ must have the form

$$\begin{bmatrix} \cos\alpha(R) & \sin\alpha(R) \\ -\sin\alpha(R) & \cos\alpha(R) \end{bmatrix}, \qquad \tan 2\alpha(R) = \frac{2U_{12}^d(R)}{U_{22}^d(R) - U_{11}^d(R)}. \tag{9.10}$$

Using Eq. (9.10) to transform $\bar{\bar{H}}^{d}$—which is given by Eq. (9.9)—into the adiabatic representation, we obtain the adiabatic potential energy matrix

$$\bar{\bar{U}}^{ad}(R) = \begin{bmatrix} 1 & 0 \\ 0 & 1 \end{bmatrix} U_{Av}(R) + \begin{bmatrix} 1 & 0 \\ 0 & -1 \end{bmatrix} u(R), \tag{9.11}$$

where

$$2U_{Av}(R) = [U_{11}^{d}(R) + U_{22}^{d}(R)],$$
$$4u^{2}(R) = [U_{11}^{d}(R) - U_{22}^{d}(R)]^{2} + [U_{12}^{d}(R)]^{2}.$$

The adiabatic radial momentum matrix (Smith, 1969) is

$$\bar{\bar{P}}_{R}^{ad}(R) = \begin{bmatrix} 0 & -i \\ i & 0 \end{bmatrix} \hbar \frac{\partial\alpha(R)}{\partial R}; \qquad \hbar \frac{\partial\alpha(R)}{\partial R} \equiv p_{R}^{ad}(R),$$

$$p_{R}^{ad}(R) = \frac{1}{4u^{2}(R)} \left(U_{12}^{d}(R) \frac{\partial}{\partial R} [U_{11}^{d}(R) - U_{22}^{d}(R)] - [U_{11}^{d}(R) - U_{22}^{d}(R)] \frac{\partial U_{12}^{d}(R)}{\partial R} \right). \tag{9.12}$$

It often happens that for some particular value R_x of R, the difference of the diabatic potential energies vanishes:

$$U_{11}^{d}(R_x) - U_{22}^{d}(R_x) = 0. \tag{9.13}$$

We then say there is an *energy-level crossing* at R_x in the diabatic representation. We can define a length scale ΔR_x that characterizes the *effective width of the crossing region* by considering the parameter entering into Eq. (9.12):

$$\left. \frac{\partial\alpha(R)}{\partial R} \right|_{R_x} = \frac{1}{2U_{12}^{d}(R_x)} \left. \frac{\partial[U_{11}^{d}(R) - U_{22}^{d}(R)]}{\partial R} \right|_{R_x} \equiv \frac{1}{\Delta R_x}. \tag{9.14}$$

Figure 9.7 qualitatively shows some typical aspects of a two-state level-crossing system. In the lower panel we see the diabatic energies and their coupling, while the upper panel shows the corresponding adiabatic energies and their radial momentum coupling.

It follows from Eqs. (9.11) and (9.13) that the difference in adiabatic potential (electronic) energies has a minimum at R_x equal to $U_{12}^{d}(R_x)$. We say that the adiabatic energy levels have an *avoided crossing* near R_x. We also note that since u(R) has a minimum at R_x, the quantity $p_{R}^{ad}(R)$ given by Eq. (9.12) will usually be large in the avoided crossing region. This large $p_{R}^{ad}(R)$ can create difficulties for basis-set expansion numerical calculations made in the adiabatic representation.

As indicated in Fig. 9.7b, the coupling in the diabatic representation is usually smaller. This is confirmed in the case of the first four $^{1}\Sigma^{+}$ states of the BH molecule;

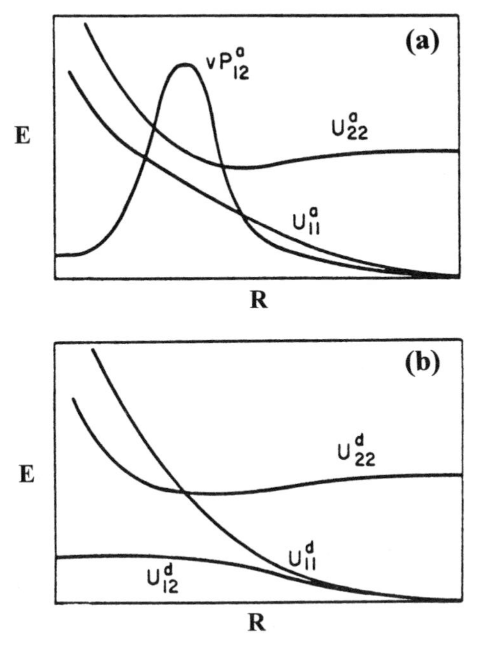

Figure 9.7. Qualitative local behavior of two electronic potential energies $U_{11}(R)$, $U_{22}(R)$ and their coupling, for an atom–atom (collision) system having a quantum energy-level crossing. (a) The adiabatic potential energies exhibit an avoided crossing region at values of R where the radial momentum coupling Eq. (9.12) is large. (b) The diabatic potential energies cross where the diabatic potential coupling U_{12}^{d} is significant. From Smith (1969). Copyright 1969 by the American Physical Society.

the calculations of the energies and coupling in the two representations are discussed in Cimiraglia (1992). Figure 9.8a shows the adiabatic energies. We see several avoided level crossings for the three excited states. These crossings correspond to large peaks in the relevant nonadiabatic coupling; the radial momentum coupling matrix elements are shown in Figs. 9.8b,c. In Fig. 9.9a we see that the avoided level crossings have become true level crossings in the diabatic representation. Figure 9.9b shows the much smaller off-diagonal matrix elements $U_{jk}^{d}(R)$ in the diabatic representation.

In a general way, useful diabatic states are a basis set defined by the set's minimizing property for momentum or kinetic energy couplings. As we might expect, the "character" of such diabatic states does not vary with R, meaning that their Huisimi and spatial probability distributions do not change qualitatively as a parameter is varied through a diabatic level-crossing region. As a consequence of this constant character, the diabatic energies cross.

On the other hand, within the crossing region, the adiabatic states obtained using transformation (9.10) are strong mixtures of the diabatic states—indeed there are 50% of each for $R = R_x$. As the system passes through the crossing region, each

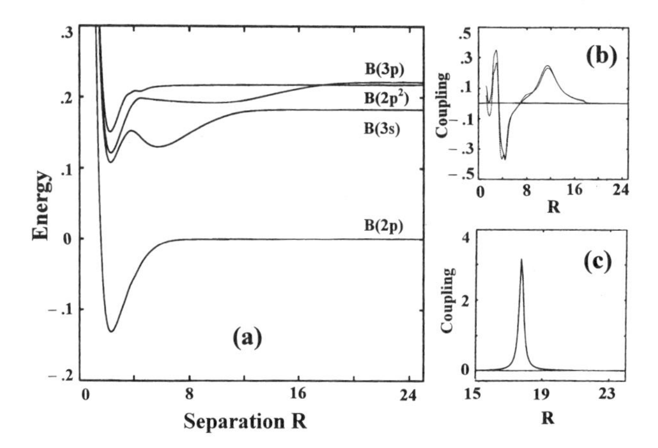

Figure 9.8. Adiabatic potential energy functions for the first four $^1\Sigma^+$ electronic states of the BH molecule, with the nuclear separation variable R in atomic units. (a) The potential energies are labeled by quantum numbers of the valence electron(s) of the boron atom. An avoided crossing occurs for the 3p and $2p^2$ states near R = 18 a.u.; their coupling is shown in panel (c) and it peaks near the crossing, as expected. The $2p^2$ and 3s states have avoided crossings near R = 12 a.u. and 3 a.u.; the relevant coupling is shown in panel (b), where the heavier line is the precise result of a finite-difference numerical calculation. From Cimiraglia (1992). Copyright 1992 Plenum Publishing.

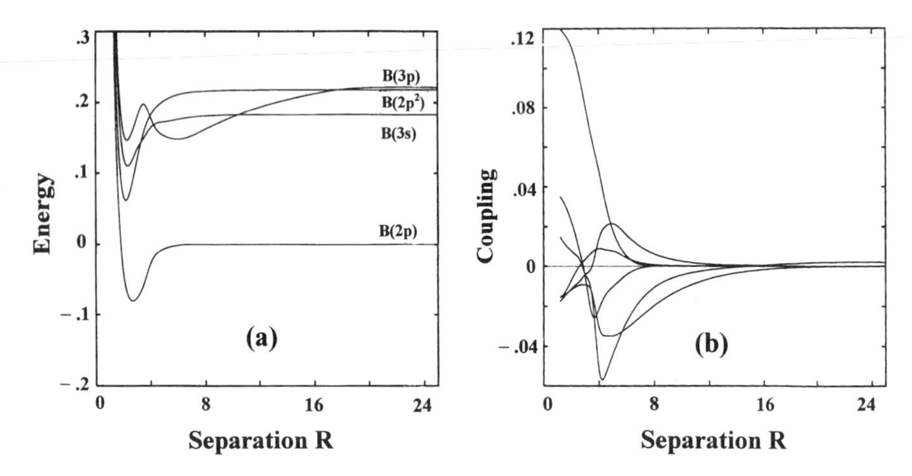

Figure 9.9. For the same states of BH as in Fig. 9.8, (a) the diabatic potential energy functions and (b) the diabatic potential couplings $U^d_{jk}(R)$. These couplings tend to be smaller and smoother than those in Figs. 9.8b,c; this behavior facilitates basis-set expansion calculations of time evolution of the molecule. From Cimiraglia (1992). Copyright 1992 Plenum Publishing.

adiabatic state first has the character of one diabatic state, then switches to another character of a different diabatic state. Thus a good (useful) diabatic representation provides an easy interpretation of the problem under study, disentangling the changes in character that otherwise occur near avoided crossings.

There is a set of analytical models for two-state problems, and some of these models can be used to parameterize some level-crossing systems. Special diabatic energy and nonadiabatic coupling functions are chosen such that the diabatic wavefunctions become hypergeometric functions (Bayfield et al., 1973; Nikitin and Umanskii, 1984). After using Eq. (9.10) and proceeding away from the level-crossing region, asymptotic formulas for the hypergeometric functions yield simple formulas for the amplitudes and phases of the adiabatic states. Among these hypergeometric models is the well-known Landau–Zener (LZ) model; see Zhu and Nakamura (1995) for a semiclassical treatment.

The LZ model is defined by very simple relationships involving the potential energy matrix elements of the diabatic Hamiltonian (9.9),

$$\begin{aligned} U^{d}_{22}(R) - U^{d}_{11}(R) &= \gamma(R - R_x), \\ U^{d}_{12}(R)F = U^{d}_{21}(R) &= V_0, \end{aligned} \tag{9.15}$$

where γ and V_0 are constants. Away from R_x, Eqs. (9.15) describe dependences on R that rarely are physical. So the LZ model will make sense only when the width $\Delta R_x = 2V_0/\gamma$ of Eq. (9.14) is sufficiently small, that is, the level-crossing region is well-localized. For the LZ model the adiabatic potential energies (9.11) become

$$\begin{aligned} \bar{U}^{ad}_{\pm}(R) &= \tfrac{1}{2}[U^{d}_{11}(R) + U^{d}_{22}(R)] \pm \tfrac{1}{2}[(U^{d}_{22}(R) - U^{d}_{11}(R))^2 + 4(U^{d}_{12}(R))^2]^{1/2} \\ &\equiv U^{d}_{ave} \pm \frac{1}{2}[\gamma^2(R - R_x)^2 + 4V_0^2]^{1/2}. \end{aligned} \tag{9.16}$$

We can use these results to make estimates of quantum transport probabilities; see Section 9.3.

Sometimes it is useful to treat the internuclear motion of a diatomic molecule classically, while treating the lighter electrons quantum mechanically. One assumes there is an effective potential V_{eff} in which the nuclei move. The nuclei follow a classical trajectory $\mathbf{R} = \mathbf{R}(t)$, which can be obtained from V_{eff} by integrating Hamilton's equations analytically or numerically. Let us consider the electronic time-dependent Schrödinger equation

$$H'(\mathbf{r}, \mathbf{R})\Psi(\mathbf{r}, t) = i\hbar \frac{\partial\psi(\mathbf{r}, t)}{\partial t}, \tag{9.17}$$

where H' now depends on t through $\mathbf{R}(t)$ (Tully and Preston, 1971). We now expand Ψ in terms of the adiabatic electronic basis functions $\psi_k(\mathbf{r}, \mathbf{R})$ and their adiabatic energies $E_k(\mathbf{R})$:

$$\Psi(\mathbf{r}, t) = \sum_k a_k(t)\psi_k(\mathbf{r}, \mathbf{R}) \exp\left(-\frac{i}{\hbar}\int^t E_k(\mathbf{R})\, dt\right). \tag{9.18}$$

Inserting Eq. (9.18) into Eq. (9.17) leads to the coupled equations

$$i\hbar \frac{da_k}{dt} = \sum_j a_j(t)\left\langle \psi_k \middle| -i\hbar \frac{\partial}{\partial t} \middle| \psi_j \right\rangle \exp\left(-\frac{1}{\hbar}\int^t (E_j - E_k)\, dt\right), \qquad k = 0, 1, \ldots,$$

where the chain rule gives the further replacement

$$\left\langle \psi_k \middle| -i\hbar \frac{\partial}{\partial t} \middle| \psi_j \right\rangle = -i\hbar \frac{d\mathbf{R}}{dt}\left\langle \psi_k \middle| \frac{\partial \psi_j}{\partial \mathbf{R}} \right\rangle.$$

Thus the coupling of the adiabatic states is due to the nuclear relative velocity vector multiplied by the matrix elements $\langle \psi_k | \partial\psi_j/\partial\mathbf{R}\rangle$; for 2-d motion and polar coordinates, the product becomes

$$\frac{dR}{dt}\left\langle \psi_k \middle| \frac{\partial}{\partial R} \middle| \psi_j \right\rangle + \frac{d\theta}{dt}\left\langle \psi_k \middle| \frac{\partial}{\partial \theta} \middle| \psi_j \right\rangle, \tag{9.19}$$

a sum of radial and rotational coupling terms.

The approach we have discussed for the diatomic molecule is also useful for pulsed driven systems, where the adiabatic basis states are taken to be the Floquet states, and the time dependence $F_0(t)$ of the pulse envelope plays the role of $\mathbf{R}(t)$ according to $\partial/\partial t = [dF_0(t)/dt]\partial/\partial F_0$. We discuss this topic in Section 9.3.

9.2b. Wavepacket Evolution in a Level-Crossing System

Broeckhove et al. (1992) reported an interesting example of the impact of an electronic level crossing on vibrational molecular quantum evolution. They studied the nonrotating N_2 molecule. The potential energies $U^d_{11}(R)$ and $U^d_{22}(R)$ of two known diabatic excited electronic states $|1\rangle$ and $|2\rangle$ were modeled by Morse potentials, using appropriate parameter values. The two states chosen had the same symmetry such that $U^d_{12}(R)$ would be nonzero and the adiabatic energies would possess an avoided crossing. The coupling $U^d_{12}(R)$ was taken to be a broad Gaussian function centered at the crossing point R_x, a slowly varying coupling function that properly goes to zero at $R = \infty$. In Fig. 9.10a we see the diabatic electronic potential energy curves and their quantum vibrational energy-level ladders. Compared to the potential well for state $|2\rangle$, the potential well for state $|1\rangle$ is centered at a smaller value of R. The available parameters of the two Morse potential wells permit the independent adjustment of each well width, as well as each

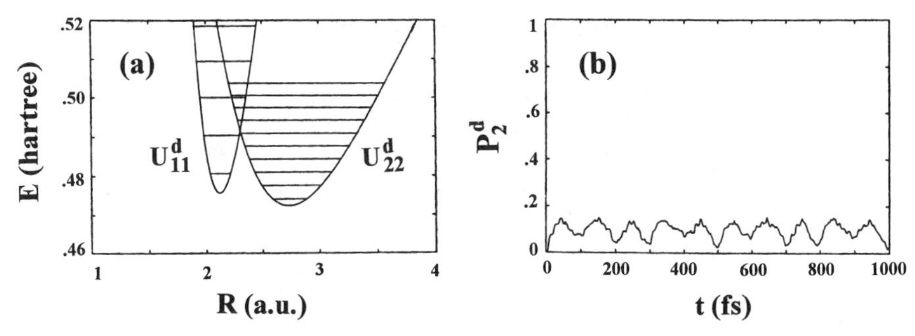

Figure 9.10. Vibrational wavepacket dynamics within the nonadiabatically coupled b′, c′ $^1\Sigma_u^+$ electronic states of the N_2 molecule. (a) The diabatic electronic potential energy functions $U_{11}^d(R)$, $U_{22}^d(R)$ are modeled by fits to Morse potentials. Note the level crossing near $R_x = 2.3$ a.u., and the near energy-degeneracy of the second vibrational state for $U_{11}^d(R)$ and the sixth vibrational state for $U_{22}^d(R)$. The vibrational ground-state eigenfunction of the X $^1\Sigma_g^+$ electronic state (not shown) is approximated by a Gaussian wavefunction. A laser pulse excites this ground state via a vertical (Franck–Condon) transition, creating a Gaussian wavepacket centered near $R = 2.07$ a.u. on the repulsive part of the inner diabatic potential well $U_{11}^d(R)$. The packet acquires kinetic energy sliding down the inner side of the inner well. (b) Numerical results for the time evolution of the occupation probability of the partial wavepacket within the outer potential well $U_{22}^d(R)$. Part of the packet crosses over to the outer well, reflects from the outer wall of this well, and after a certain recurrence time, the whole two-state vibrational process repeats itself. From Broeckhove et al. (1992). Copyright 1992 Plenum Publishing.

vertical and horizontal location of the wells in Fig. 9.10a. Thus one could separately vary the potential-energy level-crossing location and the relative positions of the quantum levels in the two wells.

We now consider an initial vibrational state wavepacket. This packet is created by laser excitation onto the repulsive part of the inner potential well 1 and is due to a vertical (Franck–Condon) transition out of the N_2 ground electronic state. An appropriate fast laser pulse would create this packet, in a way similar to the laser atomic wavepacket production technique (Section 2.1). Here the ground-state wavefunction, and hence the initial wavepacket, was fitted by a Gaussian spatial distribution. In order to calculate the evolution of an initial vibrational wavepacket in the coupled two-well system, Broeckhove and co-workers used the FFT grid method and the split operator technique; see Section B.2.

Let us choose the energy of the initial packet to be somewhere between the bottom of well 1 and the location of the potential energy curve crossing. While the packet slides further down into well 1, it also moves outward. When the packet enters the crossing region, the coupling $H_{12}^d(R)$ becomes effective and wavepacket amplitude is transferred to the outer well 2. Since no dissociation can take place at the selected initial energy, this transmitted part of the wavepacket reflects from the wall of well 2. A process of particle bouncing in both wells—combined with partial quantum transmission through the crossing region—develops for some time, and so we expect time recurrences in the packet evolution to occur, analogous to those described in Section 2.1. The recurrences are seen in Fig. 9.10b in the plot of the probability $P_2^d(t)$ for occupying the second diabatic state $|2\rangle$. We see an overall recurrence period of roughly 100 fs, as well as other oscillations with a period ~11.5 fs, and still faster fluctuations. The behavior of the corresponding adiabatic probability $P_2^{ad}(t)$ is qualitatively the same.

By changing an additive constant in the energy $U_{22}^d(R)$, we now shift well 2 vertically with respect to well 1. The maximum probability $P_2^{d,max}$ of the recurrence oscillations is plotted against the additive constant in Fig. 9.11. These results were calculated keeping only low-frequency components of the wavepacket, as in the rotating-wave approximation (Section 2.2). Figure 9.11 shows resonances in $P_2^{d,max}$ that are approximately equidistant, with spacing equal to that of the energy-level ladder for well 2. If we investigate the situation (Fig. 9.12a) we find that—at the resonances—each level in well 1 is approximately equal to a level of well 2; the separation of levels in well 2 is about one-third the separation of well 1.

Figure 9.12b shows the time evolution $P_2^d(t)$ when the system is tuned into one of the resonances. The behavior is very different from the off-resonance case of Fig. 9.10b. The value of $P_2^{d,max}$ is indeed large, but a longer time scale has also appeared. We conclude that for large probability transfer between two wells, it is important to have a near matching of quantum energy levels within the wells.

After varying other parameters in the two Morse potentials and varying the coupling $U_{12}^d(R)$, one can conclude that each resonance is due to a pair of levels—

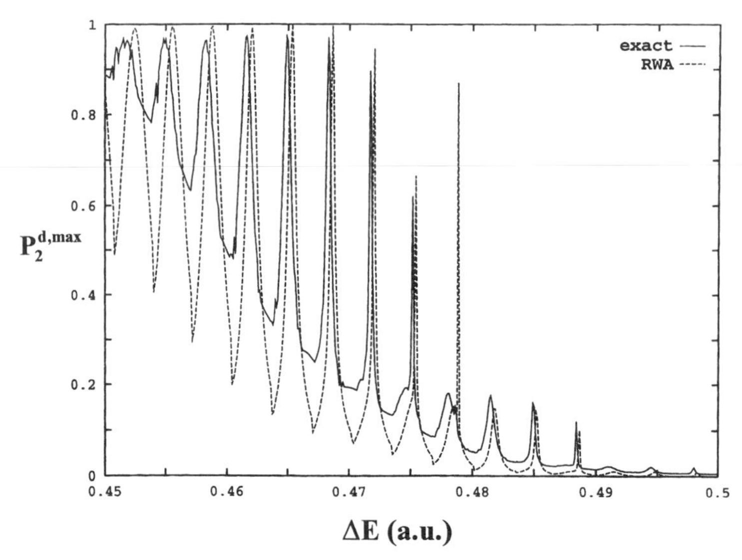

Figure 9.11. The maximum probability of oscillations like those shown in Fig. 9.10b, as a function of a (vertical) energy shift ΔE of potential well $U_{22}^d(R)$ relative to well $U_{11}^d(R)$. A series of resonance peaks are seen, with the peak spacing corresponding to the vibrational frequency of the outer well. This indicates that the probability transfer strongly depends on the relative position of the vibrational energy levels of the two diabatic Morse potential wells. From Broeckhove et al. (1992). Copyright 1992 Plenum Publishing.

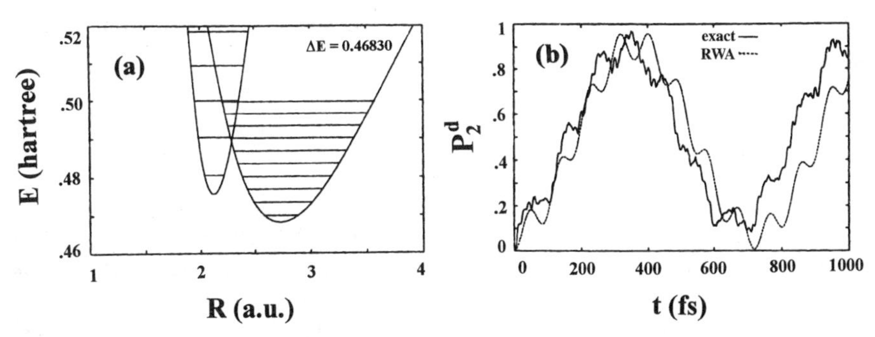

Figure 9.12. Time evolution of the occupation probability of the outer well of N_2 (Fig. 9.10) except for an energy shift $\Delta E = 0.4683$ a.u., where a resonant maximum appears (Fig. 9.11). (a) There now are near-degeneracies for two pairs of vibrational levels in the two wells. (b) The time evolution now has a large oscillatory component with a period much longer than that for the oscillations in Fig. 9.10b. From Broeckhove et al. (1992). Copyright 1992 Plenum Publishing.

one for each well—with turning points for the classical motion nearly coinciding with the crossing point R_x.

9.2c. Sets of Avoided Level Crossings (Advanced Topic)

We have seen that multilevel quantum systems display numerous avoided crossings in the energy (or quasienergy) spectrum as a parameter in the Hamiltonian is varied; see Figs. 7.26 and 8.16. As we follow along one of the diabatic energy (or quasienergy) curves, we see successive avoided crossings of the adiabatic curves. It is natural to ask whether the members of some subset of the avoided crossings share a common property. This could occur when a subset of the diabatic states associated with the curves that cross our selected diabatic curve share a common property—a property that is not possessed by our selected diabatic state. For instance, the property might be the spatial delocalization of irregular states; the selected state would be a localized (regular) state, while the interacting diabatic states would be delocalized. In such a situation, we say there is a persisting *set of avoided level crossings*.

Persisting sets of avoided crossings can be generated by considering certain primitive Hamiltonian models. Consider the case where there is a linear dependence on a parameter τ, so that $H(\tau) = H_0 + \tau V$, and suppose that the matrix representation of $H(\tau)$ happens to have the form

$$\bar{\bar{H}}_1(\tau) = \begin{bmatrix} p\tau & u_1 & u_2 & u_3 & \dots & u_N \\ u_1 & y_1 & 0 & 0 & \dots & 0 \\ u_2 & 0 & y_2 & 0 & \dots & 0 \\ u_3 & 0 & 0 & y_3 & & \vdots \\ \vdots & \vdots & \vdots & & & \vdots \\ u_N & 0 & 0 & \dots & \dots & y_N \end{bmatrix},$$

with $y_n < y_{n+1}$. The parameter τ is sometimes called the "fictitious time." The eigenvalues of $\bar{H}_1(\tau)$ are the roots x_i of the equation

$$p\tau = x - \sum_{n=1}^{N} \frac{u_n^2}{x - y_n}$$

(Bixon and Jortner, 1968; Demkov and Osherov, 1968). The energy spectrum contains N horizontal levels in an $E(\tau)$ plot; these levels are crossed by one extra level, whose slope $dE_{N+1}(\tau)/d\tau$ is given by the value of the parameter p. Each and every crossing is avoided. The "repulsion" between the levels is determined by the coupling parameters $\{u_i\}$. For equal spacing between the horizontal levels, $y_n = \alpha n$, the succession of avoided crossings mathematically is analogous to a particle-like wave entity called a soliton. As Remoissenet (1996) explains, "A *solitary wave* is a spatially localized wave that propagates along one spatial dimension only, with undeformed shape. A *soliton* is a large-amplitude coherent pulse—the exact solution of a wave equation whose shape and speed are not altered by a collision with other solitary waves." The soliton propagates with the velocity p through a spatial lattice of period α, if the horizontal levels in the $E(\tau)$ plot are indefinitely extended in their number N, thereby extending the range of x to $-\infty < x < \infty$.

If instead of the above, we take the Hamiltonian matrix to be

$$\hat{H}_2(\tau) = \begin{bmatrix} q\tau + b & w & v & v & v & \cdots \\ w & p\tau & u & u & u & \cdots \\ v & u & a & 0 & 0 & \cdots \\ v & u & 0 & 2a & 0 & \cdots \\ v & u & 0 & 0 & 3a & \cdots \\ \cdot & \cdot & \cdot & \cdot & \cdot & \cdots \\ \cdot & \cdot & \cdot & \cdot & \cdot & \cdots \\ \cdot & \cdot & \cdot & \cdot & \cdot & \cdots \end{bmatrix}, \tag{9.20}$$

then one obtains the "two-soliton" energy spectrum shown in Fig. 9.13; see further Gaspard et al. (1989). The solitons are associated with the two diabatic energy levels that intersect the horizontal diabatic levels; although not explicitly shown, all these levels are easily drawn by eye.

Now we mention a second example of sets of crossings. When the nonintegrable quantum driven Morse and Kepler oscillators (Section 8.3b) have low driving fields F, many but not all of the Floquet states have either KAM character or nonlinear-resonance character, and quasienergy levels for states of different character exhibit almost true crossings. The resonance character occurs for various values of the winding number of central periodic orbits (Section 5.6a). However, when F is increased past a value F_b, a central periodic orbit bifurcates to form a new periodic orbit, and a new resonance zone appears with new character. Quantum mechanically, a new sequence of associated avoided crossings appears in the quasienergy spectrum $\varepsilon_k(F)$; the sequence is located in the range of the field strengths $F > F_b$ (Perotti et al.,

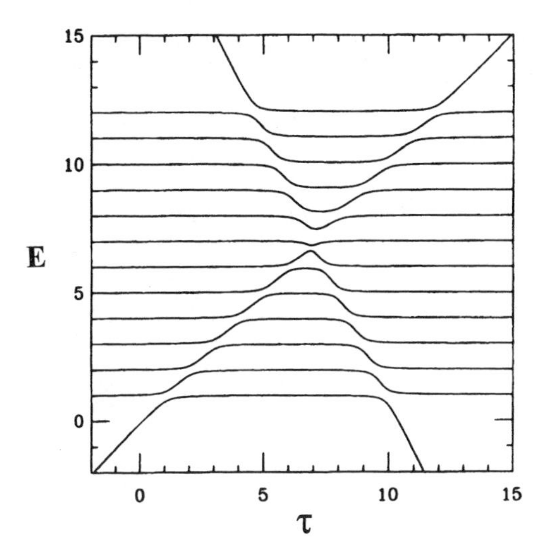

Figure 9.13. Energy eigenvalues of the "two-soliton" Hamiltonian matrix—Eq. (9.20)—as a function of the "fictitious time" variable τ. The parameters in the 14×14 matrix are $p = 1$, $q = -2$, $u = 0.2$, $v = 0.3$, $w = 0.4$, $a = 1$, and $b = 21$. Two sets of avoided energy-level crossings lie along straight lines that themselves intersect. Each set can be considered as a single entity. From Gaspard et al. (1989). Copyright 1989 by the American Physical Society.

1998). Locations of these avoided crossings in the (ε, F) plane can be estimated using an extension of the quantum nonlinear-resonance pendulum approximation (Section 5.6b).

Another quantum dynamical system of present interest is a collection of mesoscopic quantum spins exposed to a time-periodic (AC) spatially varying external magnetic field (Van Hemmen et al., 1997). Avoided quasienergy crossings, as a function of varying the parameters, have been identified as a mechanism for reduced tunneling probabilities at a fixed time. Discrete sets of special parameter values that reduce the probabilities have also been found. However, a connection between the avoided crossings and the special parameter values is unknown at present.

9.2d. Energy-Level Statistics

In nonintegrable quantum systems, almost true crossings are observed to completely give way to avoided crossings when a system is moved from an almost totally regular regime to a totally irregular (classically chaotic) regime (Fig. 7.26).

By studying the statistics of the energy levels, we can quantify the degree to which avoided crossings are present for any selected parameter value. Let us order the energies $E_1 < E_2 < E_3 < \cdots$ and consider their nearest-neighbor spacings $s_i \equiv E_{i+1} - E_i$. We assume the probability distribution of spacings is approximately given by a smooth function $P(s)$, and we consider the part of the energy spectrum near an energy E_0. Then P(s) ds is equal to the product of a probability P_1 (that the energy is between E and E + dE if $E < E_0$) times the probability P_2 (that E is less than E_0). Let us define $P_1 = \mu(s)\,ds$. We note that $P_2 = \int_s^\infty P(s')\,ds'$. Putting these probability factors together, we have

$$P(s)\,ds = \mu(s)\,ds \int_s^\infty P(s')\,ds,$$

which has the solution

$$P(s) = c\mu(s)\exp\left(-\int_{s_0}^{s} \mu(s')\,ds'\right),$$

where we can determine the constants c and s_0 by the conditions

$$\int_0^\infty P(s)\,ds = 1, \qquad \int_0^\infty sP(s)\,ds = 1.$$

Let us now consider two particular cases for the function $\mu(s)$. First, if the energy levels are uncorrelated, then μ will be constant. This leads to a Poisson level-separation distribution $P_P(s) = \exp(-s)$. In this case the most probable spacing is zero, and there are many true crossings. Second, if there is correlation such that the probability P_1 increases linearly with s, as Eugene Wigner surmised, then we obtain the Wigner level-separation distribution $P_W(s) = (\pi/2)s\exp[(-\pi/4)s^2]$. In this case there are no true crossings, and the most probable spacing is nonzero.

For generic classically integrable quantum systems, there is a semiclassical proof that P(s) is indeed a Poisson distribution (Berry and Tabor, 1977a). To make this result accurate for a quantum system, the distribution had best not include energy levels of essentially quantum eigenstates; such eigenstates normally include the ground state and some low excited states.

Besides P(s), there are a number of other statistical measures of the energy levels. One is the so-called $\bar{\Delta}_3$-statistic, defined as the mean-square fluctuations of the integrated level density $N(E) = \int^E P(E')\,dE'$ around the best-fitted straight line over the interval L:

$$\bar{\Delta}_3(L) \equiv \left\langle \frac{1}{L}\min_{A,B}\left(\int_{E_0}^{E_0+L} [N(E') - AE' - B]^2\,dE'\right)\right\rangle.$$

This statistic is sensitive to long-range correlations. For integrable systems, the $\bar{\Delta}_3$-statistic increases linearly with L.

In the case of irregular (classically totally chaotic) systems, there is a theory of spectral fluctuations based on the statistical properties of infinite random matrices. Unlike the Hamiltonian matrices that produced "solitons" earlier in this section, the random matrices have matrix elements that are statistically independent. Also, the matrices are assumed to all have the same symmetry properties. Thus, if our system is time-reversal invariant, then the matrices must be invariant under orthogonal transformations, and they are real symmetric matrices. If there is no time-reversal symmetry, then the matrices are invariant under all unitary transformations, and therefore they are complex Hermitian matrices. In either case, the distribution of matrix elements is Gaussian (Porter, 1965). For this reason, the two sets of random matrices are called, respectively, the Gaussian orthogonal ensemble of matrices (GOE) and the Gaussian unitary ensemble (GUE).

Recently—using semiclassical theory—it has been mathematically shown that irregular quantum systems have the same statistics as random matrices (Andreev et al., 1996). In the case of GOE statistics, the nearest-neighbor level-spacing distribution is $P_W(s)$, as given above. For irregular systems the GOE $\bar{\Delta}_3(L)$

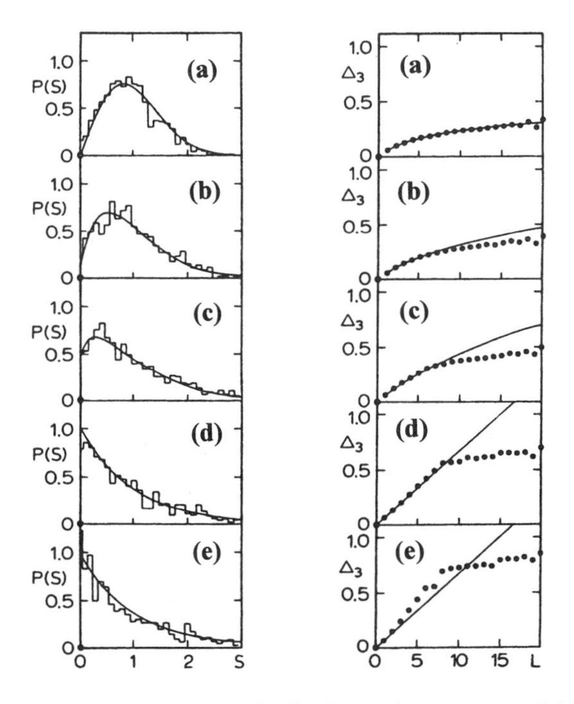

Figure 9.14. Numerical results for the distribution P(s) of nearest-neighbor energy-level spacings s (left panels), and for the corresponding $\bar{\Delta}_3(L)$-statistic (right panels). The dots and histograms represent results for the dimensionless 2-d coupled anharmonic oscillator Hamiltonian $H = p_1^2/2 + p_2^2/2 + 100(1.56q_1^2 - 0.61q_1^4 + 0.32q_1^6) + 100(0.69q_2^2 - 0.12q_2^4 + 0.03q_2^6) + V(-1.00(q_1 - q_2)^2 + 0.25(q_1 - q_2)^4 + 0.08(q_1 - q_2)^6)$. The system is quantized with $\hbar^2 = 0.2$. The values for the coupling strength parameter V are 100, 40, 30, 10, 0 for the panel pairs (a) through (e), top to bottom. For $V = 100$, there is almost complete classical chaos in the energy region occupied by the first 400 energy levels. For $V = 0$, the problem is separable and therefore integrable. The solid curves are predictions of a model based on the properties of certain sets of matrices; see the text. From Seligman et al. (1984). Copyright 1984 by the American Physical Society.

distribution has been derived from the trace formula of periodic orbit theory (Section 7.2c and Berry, 1985). Of course there will be quantum corrections to these semiclassical results.

These ideas have been investigated—numerically and sometimes experimentally—for many of the irregular systems we've considered. Random matrix theory is roughly followed, but evidence exists for other spectral correlations. One example is those systems that display dynamical localization (Section 8.3f and Dittrich and Smilansky, 1995). The new correlations appear to arise from the fact that dynamical localization stems from quantum interference, whereas usual correlations in the matrix theory reflect a localization within physical spatial boundaries for the classical motion.

As is the case for integrable systems (Feingold, 1985), one must do excellent numerical calculations of many energy levels to find evidence for quantum corrections to the semiclassical theory. Even at high energies, all the calculated energies must be accurate to a small fraction of the local mean level spacing. Such calculations have been used to explore the transition from regular to irregular evolution in the nonintegrable system of two particles in 1-d potential wells, with a coupling depending on the particle separation (Seligman et al., 1984). Figure 9.14 shows changes in P(s) and $\bar{\Delta}_3(L)$ as the chaos fraction in the Poincaré section varies from zero to greater than 95%. In panels (b) and (c), the solid curves were obtained by a matrix model that was constructed to interpolate between the Poisson and GOE limits in the energy-level statistics. The agreement is reasonably good.

9.3. CONTROLLED QUANTUM TRANSPORT

9.3a. Adiabatic Parameter Switching of Floquet States

This section addresses the time evolution of sinusoidally driven quantum systems that can arise when one or more of the parameters of the driving interaction varies relatively slowly in time. The quantum regime of adiabatic evolution will be explored, with the help of results from Chapter 8 and Section 9.2. Whenever the Floquet states reveal the influence of classical trajectories (Sections 8.2 and 8.3), the classical dynamics (Section 9.1) may play a role.

To introduce the idea of adiabatic evolution of driven quantum systems, we return to the Floquet-state treatment of the two-level problem that began with Eq. (8.11). In that discussion the parameters of the driving interaction were constant. Now we let the magnitude of the driving electric field vary slowly with the time dependence of a pulse: the field starts at zero, rises to a maximum, and falls back to zero. We briefly postpone addressing how slowly this rise and fall must occur for the evolution to be quantum mechanically adiabatic. We saw that on resonance ($\Delta = 0$), the condition for first population inversion during Rabi flopping was $(\varepsilon_2 - \varepsilon_1)T_p/\hbar = \pi$. In the Floquet picture this expression describes the first time $t = T_p$ at which the superposition of Floquet states (8.11) undergoes completely constructive wave interference. For a slowly pulsed system, the field amplitude F(t) can be "frozen" at any particular time t to define instantaneous quasienergies and instantaneous Floquet states. The response of the driven quantum system to the envelope of the pulse is adiabatic; that is, the moduli of the amplitudes of the two instantaneous Floquet states remain constant, while their phases accumulate as determined by the time-integrated quasienergies. For our pulse, the condition for constructive interference and population inversion must reflect the final accumulations of the phases, thereby becoming

$$\frac{1}{\hbar}\int_0^{T_p} dt\,[\varepsilon_2(t) - \varepsilon_1(t)] = \pi. \tag{9.21}$$

For the resonant two-level system, the RWA quasienergies (Sections 2.2 and 8.1d) immediately give us $\varepsilon_2(t) - \varepsilon_1(t) = eDF_0(t)$. Equation (9.21) then becomes

$$\frac{eD}{\hbar}\int_0^{T_p} dt\,F_0(t) = \pi. \tag{9.22}$$

This states the *area theorem for population inversion* in a two-level system: the area under the pulse envelope must be equal to $\hbar\pi/eD$. The shape of the envelope is not important as long as the evolution is adiabatic.

Equation (9.21) does not depend on the use of the RWA. At high field strengths there can be a big difference between the evolution of two interfering exact Floquet states and two interfering RWA approximate eigenstates. The Floquet states can be linear combinations of many energy eigenstates. As an example of this, let us return to the problem of the driven Morse oscillator (Section 8.3b), use the parameters for the HF molecule (Walker and Preston, 1977), and use the pulse envelope $F_0(t) = F_{max}\sin^2(\pi t/T_p)$. For the driving frequency $\omega = 0.016489$ a.u., there is initial resonance between the ground ($n = 0$) and sixth ($n = 5$) vibrational energy eigenstates. For $T_p = 100(2\pi/\omega)$ and $F_{max} = 0.0431$ a.u., Eq. (9.21) is satisfied. Figure 9.15 shows the results of a precise, direct numerical integration of the time-dependent Schrödinger equation. The bottom panel shows the projections of the wavefunction onto the two Floquet states that correlate with $n = 0$ and $n = 5$ energy eigenstates at zero field strength. We see that indeed only two Floquet states are involved and that full adiabatic population inversion $n = 0 \rightarrow n = 5$ is achieved. The top panel shows the evolution of the projections onto the Morse oscillator energy eigenstates; many eigenstates are involved and the two-state character of the evolution is not at all apparent. This situation has been called that of a generalized π-pulse (Holthaus and Just, 1994).

We note in passing that the two-state quantum transport evident in Fig. 9.15b is supported classically by the separatrix-crossing transport process of Section 9.1 (Holthaus and Just, 1994).

Now let us consider an important issue: the conditions for adiabatic quantum transport. In Section 9.1 some effort was required to find the conditions for adiabatic classical transport. Here we begin by finding a quantum condition using the RWA

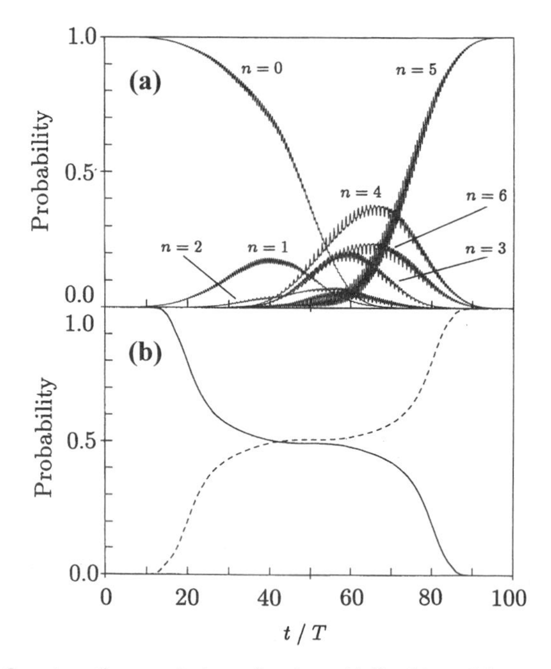

Figure 9.15a,b. Quantum time evolution of a sinusoidally driven Morse oscillator, with the driving-field strength pulsed according to $F_0(t) = F_{max} \sin^2(\pi t/T_p)$. The driving parameters are $\omega = 0.016489$ a.u., $F_{max} = 0.0431$ a.u., and $T_p = 100$ driving periods (0.922 ps). (a) The evolution of the occupation probabilities of the unperturbed Morse oscillator energy eigenstates $|n\rangle$. The pulse somehow produces 100% population inversion from the ground state $|0\rangle$ to the excited state $|5\rangle$, with at least seven Morse states being involved in the quantum transport process. (b) Occupation probabilities of the instantaneous Floquet states, for the same process as in (a). Only two Floquet states account for all of the dynamics. From Holthaus and Just (1994). Copyright 1994 by the American Physical Society.

two-level Hamiltonian. The wavefunction is the linear combination of (RWA) Floquet states, Eq. (8.11). Because of the field-pulsing, these states now are time-dependent linear combinations of initial energy eigenstates ψ_1, ψ_2 that must be determined by a state-mixing parameter β that is now slowly varying. Letting $w_{1,2}(t) \equiv u_{1,2}(t) \exp[-\varepsilon_{1,2}(t)t]$,

$$\begin{pmatrix} w_1(t) \\ w_2(t) \end{pmatrix} = \begin{bmatrix} \cos\beta(t) & -\sin\beta(t) \\ \sin\beta(t) & \cos\beta(t) \end{bmatrix} \begin{pmatrix} \psi_1 \\ \psi_2 \end{pmatrix}. \tag{9.23}$$

We calculate the time derivatives

$$\frac{dw_1}{dt} = -[\psi_1 \sin\beta(t) + \psi_2 \cos\beta(t)] \frac{d}{dt}\beta(t) = -w_2 \frac{d\beta}{dt},$$
$$\frac{dw_2}{dt} = w_1 \frac{d\beta}{dt},$$

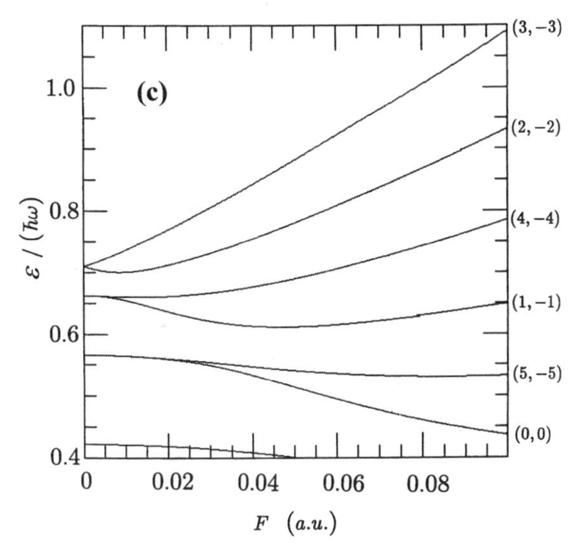

Figure 9.15c. Quantum time evolution of a sinusoidally driven Morse oscillator, with the driving-field strength pulsed according to $F_0(t) = F_{max}\sin^2(\pi t/T_p)$. The driving parameters are $\omega = 0.016489$ a.u., $F_{max} = 0.0431$ a.u., and $T_p = 100$ driving periods (0.922 ps). (c) Some quasienergy curves for the system as a function of field F. The dynamics is completely determined by the temporary removal of the degeneracy of the quasienergies labeled $(n, -N) = (0,\ 0)$ and $(5, -5)$, where n is the Morse quantum number and N is the photon sideband number. The transport process is a strong-field (nonperturbative) $\Delta N = 5$ multiphoton absorption process. From Holthaus and Just (1994). Copyright 1994 by the American Physical Society.

and obtain the time derivative of the full wavefunction (8.11):

$$\frac{\partial \Psi(t)}{\partial t} = \left(\frac{dc_2(t)}{dt} - \frac{d\beta}{dt}c_1(t)\right)w_2(t) + \left(\frac{dc_1(t)}{dt} + \frac{d\beta}{dt}c_2(t)\right)w_1(t).$$

Within the RWA we use the fact that $H\Psi(t) = \varepsilon_2(t)c_2(t)w_2(t) + \varepsilon_1(t)c_1(t)w_1(t)$, to within a multiplicative time-independent phase factor. We take $\varepsilon_{1,2}(t) = \mp\Omega(t)/2 = [\Delta^2 + \omega_R^2 t)]^{1/2}/2$. From the time-dependent Schrödinger equation for $\Psi(t)$, we then obtain the coupled equations for the coefficients $c_{1,2}$:

$$\frac{d}{dt}\begin{pmatrix} c_1(t) \\ c_2(t) \end{pmatrix} = -\frac{i}{2}\begin{bmatrix} \Omega(t) & -2i\dfrac{d\beta}{dt} \\ 2i\dfrac{d\beta}{dt} & -\Omega(t) \end{bmatrix}\begin{pmatrix} c_1(t) \\ c_2(t) \end{pmatrix}. \tag{9.24}$$

Were it not for the off-diagonal matrix elements $\pm 2i\ (d\beta/dt)$, this equation would provide uncoupled equations for the coefficients $c_{1,2}(t)$,

$$c_{1,2}(t) \cong \exp\left[\mp\frac{i}{2}\int_0^t dt'\ \Omega(t')\right]c_{1,2}(0),$$

the expected amplitudes for adiabatic wavefunctions that have accumulating phases. Thus in the RWA two-level problem, the condition for adiabatic evolution is just the condition necessary for ignoring the off-diagonal (coupling) matrix elements in comparison to the diagonal matrix elements:

$$\frac{d\beta}{dt} \ll \Omega(t) = [\Delta^2 + \omega_R^2(t)]^{1/2}.$$

Noting from Section 8.1d that

$$\cos\beta(t) = \left(\frac{1}{2}\frac{\Omega(t)+\Delta}{\Omega(t)}\right)^{1/2},$$

we consider $d[\cos\beta(t)]/dt$ and carry out some algebra to obtain

$$\left|\Delta\frac{d\omega_R}{dt}\right| \ll \Omega^3(t).$$

Allowing the detuning Δ to also be slowly time varying leads to the generalization

$$\left|\Delta(t)\frac{d\omega_R}{dt} - \omega_R(t)\frac{d\Delta}{dt}\right| \ll [\Delta^2(t) + \omega_R^2(t)]^{3/2}. \tag{9.25}$$

Rather than relying on inequalities such as expression (9.25), one would like to have quantitative analytical ways of determining the extent of nonadiabatic behavior. To obtain simple analytical expressions we must resort to useful models that do not, however, always describe reality quantitatively. As an example, let us consider the Landau–Zener (LZ) model—Eqs. (9.15)—for a two-state system exhibiting a localized avoided energy-level crossing. In our model we ignore the kinetic energy matrix in Eq. (9.9), so that $\bar{\bar{H}} = \bar{\bar{U}}$; the time dependence comes only through a dependence $R = R(t)$ that is assumed known. Then not only can the diabatic Hamiltonian be diagonalized to obtain the adiabatic energies (9.16), but the adiabatic eigenstates can also be obtained analytically. The transformation from diabatic to adiabatic states must have the mathematical form of Eq. (9.23). In the case of the atom–atom separation distance R for the diatomic molecule problem (Section 9.2a), $\tan(2\beta) = 2H_{12}(R)/[H_{11}(R) - H_{22}(R)]$. In our LZ model, we consider the constant-velocity approximation $R(t) = vt$. Expanding the unknown solution of the time-dependent Schrödinger equation in terms of the diabatic states, $\Psi = b_1(t)\psi_1^d + b_2(t)\psi_2^d$, and introducing $a_i(t) = b_i(t)\exp[-i/\hbar \int^t H_{ii}^d(t')\,dt']$, one can find the second-order equation

$$\frac{d^2a_2}{dt^2} + \left(\frac{i}{\hbar}(H_{11}^d - H_{22}^d) - \frac{1}{H_{12}^d}\frac{dH_{12}^d}{dt}\right)\frac{da_2}{dt} + \left(\frac{H_{12}^d}{\hbar}\right)^2 = 0.$$

The solution of this equation can be written in terms of Weber functions, a type of hypergeometric function. If $a_2(t = -\infty) \equiv 0$ and $|a_1(t = -\infty)| \equiv 1$, then the large-time asymptototic expansions of the Weber functions determine the transition probability from the initial diabatic state to the other diabatic state:

$$P^d_{1\to 2}(t = +\infty) = |a_2(t = +\infty)|^2 = 1 - \exp(-2\pi\delta^R_{LZ}).$$

Here the Landau–Zener (LZ) parameter for the atom–atom problem is

$$\delta^R_{LZ} = \frac{[H^d_{12}(R_x)]^2}{\hbar v|(d/dR)[H^d_{11}(R) - H^d_{22}(R]|_{R=R_x}} = \frac{V_0^2}{\hbar v\gamma}, \tag{9.26}$$

with R_x being the location of the crossing of the diabatic energies. The quantities V_0 and γ were introduced in Eqs. (9.15). Since $P^d_{1\to 2}$ is also the probability for remaining in the initial adiabatic state, the probability for a nonadiabatic transition is just $1 - P^d_{1\to 2} = \exp(-2\pi\delta^R_{LZ})$; this probability is almost zero when either the velocity v is sufficiently small or the separation $2V_0$ of the adiabatic energies at R_x is sufficiently large. We see that strong nonadiabatic behavior occurs when $\delta^R_{LZ} \sim 1$ and the evolution is diabatic when $\delta^R_{LZ} \gg 1$.

For the related LZ problem of a localized avoided crossing of two quasienergies $\varepsilon_\pm$ as functions of an electromagnetic field strength F_0, we choose coordinates in an $\varepsilon(F_0)$ plot so that $\varepsilon_\pm = \pm\frac{1}{2}\delta\varepsilon(1/2\delta F_0)[(2\delta F_0)^2 + F_0^2]^{1/2}$. Here $2\delta F_0$ is the *field* splitting at the avoided crossing, and $\delta\varepsilon$ is not an energy gap but rather the quasienergy separation at the crossing field F_x in an $\varepsilon(F_0)$ plot in one Brillouin zone, as in Fig. 8.16. The LZ parameter for the field-switched two-level problem then becomes

$$\delta^{F_0}_{LZ} = \frac{(\delta F_0)^2}{|(dF_0/dt)(\delta F_0/\delta\varepsilon)|_{F_x}} \text{ a.u.},$$

similar in form to Eq. (9.26). Here dF_0/dt is the rate that the field is changing during passage through the crossing region of field strengths.

If the frequency ω of the electromagnetic field is varied in time, rather than $F_0 = F_0(t)$, then

$$\delta^{\omega}_{LZ} = \frac{1}{2}\frac{(\delta\omega)^2}{|(d\omega/dt)(\delta\omega/\delta\varepsilon)|_{\omega_x}} \text{ a.u.},$$

where $\delta\omega$ is the *frequency* splitting at the crossing point ω_x in an $\varepsilon(\omega)$ plot. The factor of $\frac{1}{2}$ arises from the fact that the adiabatic change of the period $T = 2\pi/\omega$ directly affects the Floquet states through the periodicity in their definition (Breuer and Holthaus, 1989).

In a multistate Floquet problem (more than two Floquet states), the situation is more complicated. Again within the RWA, Messiah (1962) obtained a perturbative bound on the probabilities $P_{0\to i}(t)$ for finding the system in the excited (RWA) Floquet state $|\varepsilon_i\rangle$, given unit initial probability in the ground Floquet state $|\varepsilon_0\rangle$:

$$P_{0i}(t) \leq \left| \frac{\langle \varepsilon_0 | dH^{eff}/dt | \varepsilon_i \rangle}{(\varepsilon_0 - \varepsilon_i)^2} \right|^2 . \tag{9.27}$$

The effective RWA Hamiltonian matrix is $\bar{\bar{H}}^{eff} = \bar{\bar{\Delta}} + \bar{\bar{\omega}}_R$, where $\bar{\bar{\Delta}}$ is a diagonal matrix of detunings and $\bar{\bar{\omega}}_R$ is an off-diagonal matrix of electric dipole couplings (Peterson and Cantrell, 1985). These workers showed that when the detunings are time independent, Eq. (9.27) leads to

$$\left| \frac{F_0(t)}{dF_0/dt} \right|^2 \gg \sum_{i\neq 0} \left| \frac{\langle \varepsilon_0 | \Delta | \varepsilon_i \rangle}{(\varepsilon_0 - \varepsilon_i)^2} \right|^2 .$$

Letting $T(t) \equiv F_0(t)/(dF_0/dt)$ we have the *quantum adiabatic criterion*

$$T(t) \gg \tau \equiv \left[\left| \sum_{i\neq 0} \frac{\langle \varepsilon_0 | \Delta | \varepsilon_i \rangle}{(\varepsilon_0 - \varepsilon_i)^2} \right|^2 \right]^{1/2} . \tag{9.28}$$

A similar equation may be obtained if the frequency is varying, $\bar{\bar{\Delta}} = \bar{\bar{\Delta}}(t)$.

Peterson and Cantrell have numerically evaluated Eq. (9.28) for a model system possessing six RWA Floquet states. The quasienergies ε_i are shown for one particular laser frequency ω in Fig. 9.16a. Figure 9.16b shows the results for τ in nanoseconds as a function of frequency and field strength. The peaks in τ usually indicated conditions for achievable adiabatic population inversion, when the field switch-on was taken to have the form $F_0(t) = F_m/[1 + \exp(-t/T_p)]$, $-70 \leq t < 70$ ns.

The Floquet states in Fig. 9.16a correspond to an undriven system having a nondegenerate ground state and an almost-degenerate band of five excited states. At low field strengths, the five excited states are taken to have equal level splittings; the ground-state quasienergy mod(ω) is the fourth curve up from the bottom of the figure. For a particular nearly resonant frequency and a particular field strength, the time evolution of the total population in the excited states is shown in Fig. 9.17, for three different values of the switch-on time parameter T_p. The situations of adiabatic, almost adiabatic, and sudden switching are nicely demonstrated.

We end this subsection noting that rigorous quantum adiabatic theorems exist for some classes of problems (Kato, 1950; Avron et al., 1987).

9.3b. Adiabatic Probability Transport Using Frequency-Chirped Laser Pulses

Now we turn to the conceptually interesting problem of a two-state quantum system exposed to a linearly polarized laser electric field, where the field strength F is pulsed in time while its frequency ω is also time dependent or "chirped." After discussing

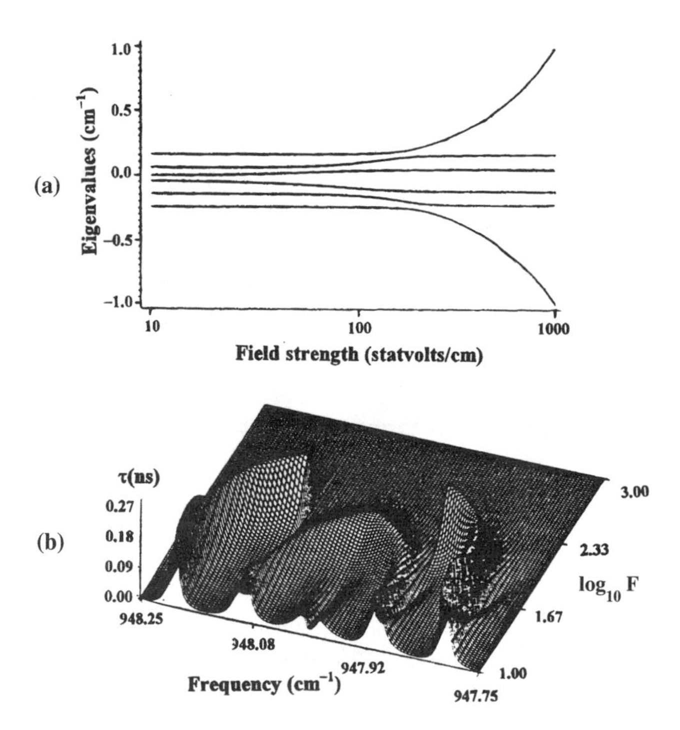

Figure 9.16. The quantum time scale for adiabatic evolution, for a model sinusoidally driven multilevel system consisting of a ground state and a band of five excited states. (a) The quasienergies mod(ω) of the system are shown as a function of driving-field strength F_0 for the particular driving frequency $\omega = 947.96\ \mathrm{cm}^{-1}$. Note the various avoided crossings. (b) The adiabatic time scale τ of Eq. (9.28), plotted in nanoseconds over a portion of the (ω, $\log_{10} F_0$) parameter space. From Peterson and Cantrell (1985). Copyright 1985 by the American Physical Society.

this system, we will look at simultaneously pulsed and chirped three-level systems, and at their important application to selective bond-breaking of molecules.

In our discussion of Rabi flopping (Section 2.2), the phase choices were chosen so that

$$\hbar\dot{\zeta}_{1,2} = \mathrm{E}_{1,2} + \tfrac{1}{2}\hbar(\omega_0 \mp \omega).$$

In the present section, the extension from two-level to three-level systems is more conveniently done using slightly different phase choices (Shore, 1990), adding $(\omega_0 - \omega)t/2$ to $\zeta_1(t)$ and $\zeta_2(t)$. In the presence of simultaneous pulsing and chirping, the RWA Hamiltonian matrix then takes the asymmetric form

$$\bar{\bar{\mathrm{H}}}(t) = \frac{\hbar}{2}\begin{bmatrix} 0 & \omega_R(t) \\ \omega_R^*(t) & 2\Delta(t) \end{bmatrix}, \tag{9.29}$$

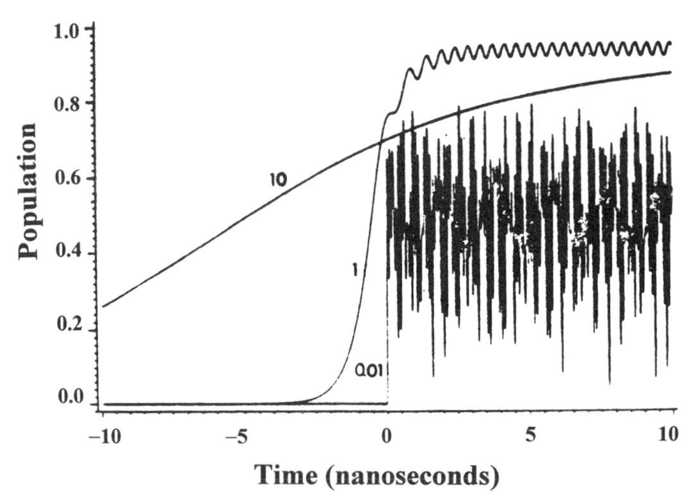

Figure 9.17. The time evolution of the multilevel system of Fig. 9.16, in response to the field switch-on function $F_0(t) = F_{max}/[1 + \exp(-t/T_p)]$, $-70 \leq t \leq 70$ ns. Shown is the net occupation probability in the band of five excited states, given an initially populated ground state. The parameters are $\omega = 947.82\ \text{cm}^{-1}$ and $F_{max} = 150$ statvolts/cm, locating a point on a high ridge with $\tau \approx 0.25$ in Fig. 9.16b. For $T_p = 10$ ns $\gg \tau$, the quantum evolution is adiabatic and results in population inversion. For $T_p = 1$ ns $\approx 4\tau$, the evolution is still close to adiabatic, with small Rabi oscillations being present. For $T_p = 0.01$ ns $\ll \tau$, the switch-on is sudden and there is no inversion. From Peterson and Cantrell (1985). Copyright 1985 by the American Physical Society.

where again $\omega_R(t)$ is the Rabi frequency $DF_0(t)/\hbar$ and $\Delta(t)$ is the frequency detuning $(\omega_0 - \omega)$.

For linear chirping $\Delta(t) = 2bt$, Melinger et al. (1994) numerically calculated (weak-field) quasienergies for the RWA problem defined by Hamiltonian (9.29). For the case of the frequency sweep rate $b = 0.18\ \text{cm}^{-1}/\text{ps}$, laser pulse intensity $I(t) = \exp[-4(\ell n2)t^2/T_p^2]$, laser pulse FWHM $T_p = 20$ ps, and peak Rabi frequency $\omega_R^{max} = 2\pi \times 10\ \text{cm}^{-1}$, the results are the quasienergies shown in Fig. 9.18a. Because of our phase choice, the diabatic ground-state quasienergy is represented by a horizontal line with quasienergy $\varepsilon_1 = 0$, and the excited-state diabatic quasienergy reflects the temporal change of the laser frequency to become a straight line of negative slope. The peak of the pulse occurs at 50 ps. Figure 9.18a suggests that combined pulsing and chirping may result in adiabatic evolution from one state to the other, with unit efficiency. If we start in the ground state, this adiabatic evolution is confirmed by numerical results shown in Fig. 9.18b for the field dependence of the excitation probability. The probability approaches unity with increasing instantaneous Rabi frequency, and Eq. (9.25)—the condition for adiabaticity—becomes more and more satisfied. The probability then remains unity. For comparison we also see the excitation probability for a bandwidth-equivalent unchirped pulse with $T_p = 2$ ps, which produces Rabi flopping.

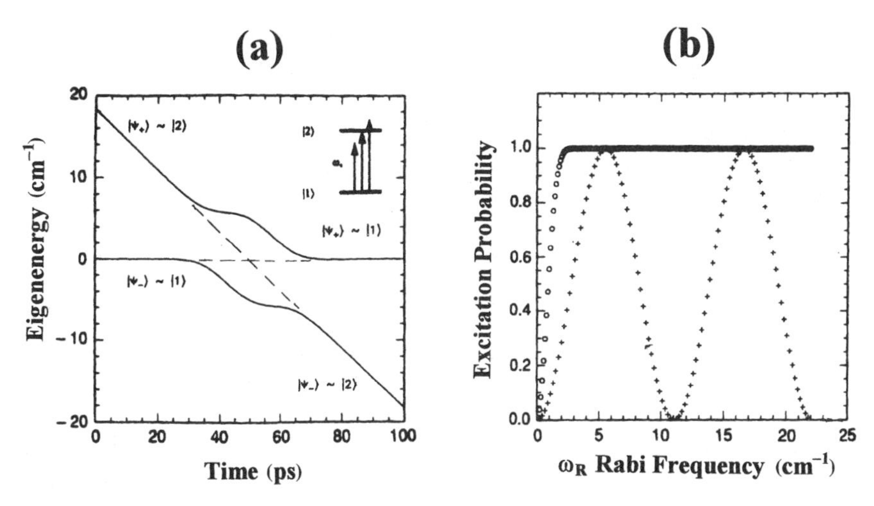

Figure 9.18. The time evolution of a frequency-chirped sinusoidally driven two-level quantum system described by the Hamiltonian matrix Eq. (9.29). (a) The evolution of the primary (rotating-wave approximation) quasienergies shows a combination of the effects of a 20 ps Gaussian pulse in the field strength and an upward linear frequency chirp at the rate $b = 0.18\ cm^{-1}/ps$ (see the inset). The pulse produces a bulge in the quasienergy separation near the field maximum F_{max}, which occurs at $t = 50$ ps. The chirp produces a crossing of the diabatic quasienergy levels; see dashed lines. (b) The occupation probability of the upper level $|2\rangle$, assuming the initial state is the ground state $|1\rangle$, as a function of F_{max} or the instantaneous Rabi frequency $\omega_R = DF_0/\hbar$. For $\omega_R \gtrsim 2\ cm^{-1}$, the curve of circles indicates adiabatic evolution on the lower quasienergy curve (solid line) shown in (a). The curve of crosses is for a 2 ps unchirped pulse, which has the same bandwidth as the 20 ps chirped pulse; now simple Rabi flopping is observed. Reprinted with permission from Melinger et al. (1994). Copyright 1994 American Institute of Physics.

It is worthwhile to note that an oft-used geometrical representation of the general time-dependent two-state problem exists. This representation is usually termed the (Bloch) *state-vector model* (Feynman et al., 1957; Shore, 1990, Section 8.5). There is a conservation of probability condition for the two complex amplitudes $c_1(t)$, $c_2(t)$, given by $c_1c_1^* + c_2c_2^* = 1$. In addition to this, the evolution of the system is entirely determined by three real quantities:

$$\begin{aligned} r_1 &\equiv c_1c_2^* + c_2c_1^*, \\ r_2 &\equiv i(c_1c_2^* - c_2c_1^*), \\ r_3 &\equiv c_1c_1^* - c_2c_2^*. \end{aligned}$$

Here the state vector $\mathbf{r} \equiv (r_1, r_2, r_3)$ evolves in its space according to an equation having the same form as that for a magnetic moment classically precessing in 3-d in the presence of a magnetic field:

$$\frac{d\mathbf{r}}{dt} = \boldsymbol{\omega} \times \mathbf{r}.$$

For the case of a $\Delta m = 0$ electric dipole interaction $V_{12} = -D_{12}F(t)$, the vector $\boldsymbol{\omega}$ has the three components $\omega_1 = (V_{12} + V_{21})/\hbar = -2D_{12}F(t)/\hbar$; $\omega_2 = -i(V_{12} - V_{21})/\hbar = 0$; and $\omega_3 = \omega_0 \equiv (E_2 - E_1)/\hbar$. If the amplitude of the field is slowly pulsed while its frequency is adiabatically chirped from above resonance to below resonance, the time dependence causes $\boldsymbol{\omega}$ to rotate through an angle of 180 degrees, and $\mathbf{r}$ precesses around $\boldsymbol{\omega}$ as it follows the rotation of $\boldsymbol{\omega}$; see Fig. 9.19. During the chirp, $\boldsymbol{\omega}$ makes the angle $\theta(t)$ with the downward vertical axis, and this angle is given by

$$\theta(t) = \left(\frac{\pi}{2}\right) + \tan^{-1}\left(\frac{\Delta(t)}{\omega_R(t)}\right).$$

There is an approach for efficiently dissociating a diatomic molecule and perhaps selectively breaking a bond in a larger molecule. The method is to use sequential frequency chirping through adjacent resonances to move probability up the vibrational ladder of energy levels (Chelkowski et al., 1990; Chelkowski and Bandrauk, 1993). They used the driven Morse oscillator as a model. The anharmonicity of the energy levels permits separation of the time intervals used for each adiabatic population inversion. However, introduction of a realistic dipole moment function D(R) results in a bottleneck in the process at the higher vibrational levels, where D(R) becomes small and the two-level transitions cease to be adiabatic at non-

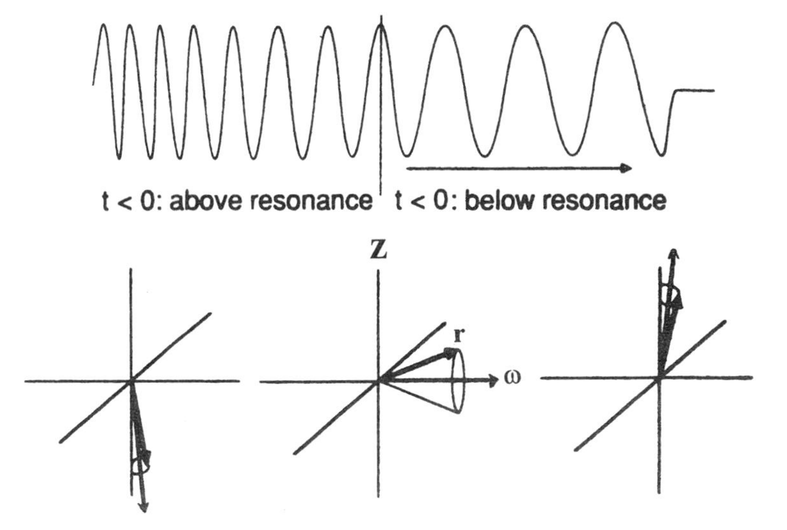

Figure 9.19. A schematic of quantum adiabatic population inversion in the frequency-chirped sinusoidally driven two-level system, using the state-vector model; see text. As the frequency is chirped downward through resonance (top), the state vector precesses about the field-interaction vector $\boldsymbol{\omega}$, while $\boldsymbol{\omega}$ moves from the $-Z$ to the $+Z$ direction to achieve the inversion. Reprinted with permission from Melinger at al. (1991). Copyright 1991 American Institute of Physics.

ionizing field strengths. Existing schemes to avoid this problem are based on a sequence of three-level transitions, each temporarily involving the same electronically excited intermediate molecular state. So let us turn to three-level systems.

Figure 9.20 shows the three types of three-level systems—the ladder system, the "Λ" system, and the "V" system. These are discussed in some detail in Chapter 13 of Shore (1990). When the energy ordering of the levels is $E_1 < E_2 < E_3$, then the 3 × 3 RWA Hamiltonian matrix can be written

$$\bar{\bar{H}} = \frac{1}{2}\begin{bmatrix} 2\Delta_1 - i\Gamma_1 & \omega_{R1}^* & 0 \\ \omega_{R1} & 2\Delta_2 - i\Gamma_2 & \omega_{R2}^* \\ 0 & \omega_{R2} & 2\Delta_3 - i\Gamma_3 \end{bmatrix}, \qquad (9.30)$$

where we have included possible phenomenological radiative decay terms $-i\Gamma_n$ that complexify the energies and where the detunings Δ_n are defined in Fig. 9.20. The two laser frequencies are ω_1 and ω_2, and the Rabi frequencies ω_{R1}, ω_{R2} are proportional to the coupling matrix elements V_{12}, V_{23}, respectively. The RWA is valid only if exponentials with frequencies ω_1, ω_2, and $\omega_1 \pm \omega_2$ oscillate many times during the excitation process, and only if all the Δ_n and ω_{Rn} are smaller in magnitude

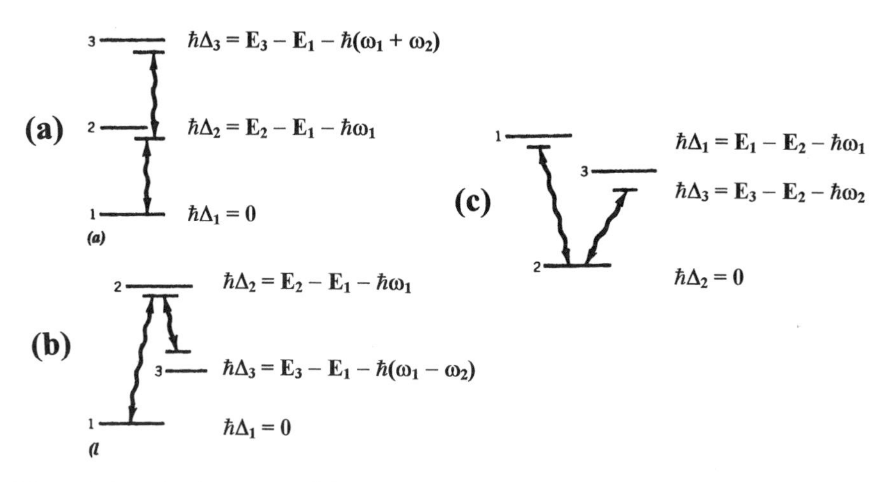

Figure 9.20. The three types of three-level systems, as determined by the nonzero coupling matrix elements between the levels and the energy-ordering of the levels. The level energies are E_1, E_2, and E_3; the photon energies for transitions are $\hbar\omega_1$ and $\hbar\omega_2$; and the detunings from resonance are defined as Δ_1, Δ_2, and Δ_3. (a) The ladder system admits two steps of photon absorption that each increase the energy. For the indicated labeling of the levels, the Hamiltonian matrix is Eq. (9.30). (b) The Λ system admits energy transfer from a lowest energy level 1 to a higher level 3 that is not coupled back to level 1. (c) The V system is an inverted Λ system. From Shore (1990). Copyright 1990 John Wiley & Sons, Inc. Reprinted by permission of John Wiley & Sons, Inc.

than all these frequencies. The phase choice taken for Fig. 9.20 also means that, for the ladder system, the wavefunction expansion is

$$\Psi^{\rm ladder}(t) = \exp\left(-i\frac{E_1 t}{\hbar}\right)\{c_1(t)\psi_1 + c_2(t)\psi_2 \exp(-i\omega_1 t) + c_3(t)\psi_3 \exp[i(\omega_1 + \omega_2)t]\}.$$

Efficient population transfer in the three-state ladder system has been demonstrated experimentally by Broers et al. (1992) for the 5s, 5p, 5d energy levels of the Rb atom; see Fig. 9.21a. A single chirped laser pump pulse induced transitions at both wavelengths λ_{12} and λ_{23} to pump up the system from the 5s ground state; after that, the population of just the 5d upper state was probed by a relatively long and weak ionizing laser pulse. The bandwidth-limited pump-pulse duration was 310 fs, and an entire pump-probe experimental event was completed within a time interval three orders of magnitude less than the state radiative lifetimes $\hbar/\Gamma_n > 10$ ns. A special pulse shaper produced a pump pulse with an electric field in the frequency domain given by $F(\omega) = |F(\omega)| \exp[i\alpha(\omega - \omega_0)^2]$, where α is the chirp and has units ps^2, and $|F(\omega)|$ is a rectangular-shaped function of ω. A frequency-increasing chirp ("red to blue") corresponds to $\alpha > 0$, a "blue to red" chirp, $\alpha < 0$.

Figure 9.21b shows the ionization signal as a function of the chirp α, for a pump-pulse bandwidth $\Delta\lambda = 5.8$ nm, energy fluence $F = 500\,\mu J/cm^2$, and mean

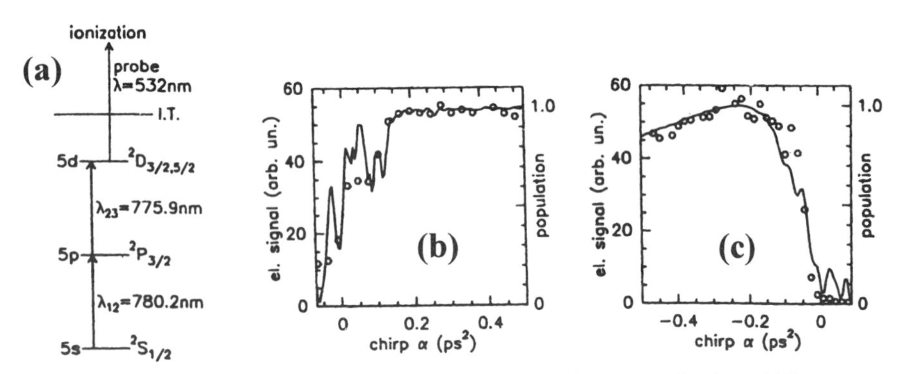

Figure 9.21. Population transfer within a three-state ladder system in the rubidium atom, achieved by a single frequency-chirped pump laser pulse. (a) The energy levels and energy eigenstates of the ladder are indicated along with the resonant laser wavelengths λ_{12} and λ_{23}. The population of the 5d upper level was determined by time-delayed laser photoionization at $\lambda = 532$ nm. (b) The circles denote the measured photoelectron signal (left scale) and achieved upper-level population (right scale) as a function of the frequency chirp rate parameter α. The central wavelength $\lambda_c = 777.5$ nm and bandwidth $\Delta\lambda = 5.8$ nm of the laser pulse were chosen so that both λ_{12} and λ_{23} were within the spectrum of the pulse. The pulse energy fluence was $500\,\mu J/cm^2$. For $\alpha > 0.2$ ps^2, an upward frequency chirp produces full population transfer into the 5d upper level. The results of basis-set expansion calculations are shown as the solid curve. (c) The same as in (b), except that $\lambda_c = 780.8$ nm. Now full population transfer occurs for a downward chirp with $\alpha = -0.2$ ps. From Broers et al. (1992). Copyright 1992 by the American Physical Society.

wavelength $\lambda_0 = 777.5$ nm. Note carefully that $\Delta\lambda$ is large enough for *both* λ_{12} and λ_{23} to be within the bandwidth of the pulse. Unity occupation probability of the upper (5d) state was essentially achieved when the chirp was $\alpha \gtrsim 0.2$ ps^2. Similarly, Fig. 9.21c shows equally high population inversion when $\alpha = -0.2$ ps^2 and when $\lambda_0 = 780.8$ nm is the mean wavelength centered above the wavelengths of both atomic transitions. The solid curves in Figs. 9.21b,c are numerical results based on Hamiltonian (9.30). The agreement is quite good.

The experimental and numerical results can be explained by adiabatic evolution of initially populated RWA Floquet states; see Fig. 9.22a. Note that quasienergy ε_2 is independent of the wavelength chirp $\lambda(t)$ because one takes the phase choice so that Δ_2 in Eq. (9.30) is zero (rather than $\Delta_1 = 0$ as indicated in Fig. 9.20). Thus the quasienergies $\varepsilon_1, \varepsilon_2$, and ε_3 are for RWA Floquet states $|5s, +\omega\rangle$, $|5p\rangle$, $|5d, -\omega\rangle$, respectively. There are three avoided crossings seen in Fig. 9.22a. For the case of the red-to-blue chirp that produces the unity inversion of Fig. 9.21b, the adiabatic passage from the 5s state 1 to the 5d state 3 is mediated by two avoided crossings, as indicated by the path marked "A" in Fig. 9.22a. On the other hand, the "counter-intuitive" case of Fig. 9.21c involves adiabatic passage again from state 1 to state 3, mediated by the third avoided crossing, as indicated by the path marked "B" in Fig. 9.22a. This case is particularly interesting, as little transient population of the 5p intermediate state 2 occurs during the evolution; compare Fig. 9.22c to Fig. 9.22b. In other applications of this case of chirped-pulse adiabatic inversion in ladder systems, radiative decay of the intermediate state would be relatively unimportant. Such decay is also an important issue in population transport studies in the promising Λ three-level system now to be discussed.

Experiments demonstrating complete population transfer in the Λ system were reported by Gaubatz et al. (1990). Double-laser two-photon resonant transitions were used, so that $\Delta_3 = 0$ in Fig. 9.20b and in Eq. (9.30). Such processes are called *stimulated Raman scattering* processes, resonant if $\Delta_2 = 0$ but otherwise nonresonant. The nonresonant case can achieve little transient population of the upper level 2 at energy E_2. The experiments utilized a time delay between the pump laser pulse with frequency ω_1 and the so-called Stokes laser pulse at frequency ω_2 that for $\Delta_2 = 0$ would be resonant with the second or downward transition between level 2 and level 3. It is necessary of course for the two laser pulses to partially overlap in time. The experiments were carried out in the counterintuitive manner of the Stokes laser pulse coming first, which introduces avoided crossing of the quasienergies that are then modified by the pump pulse, which initiates the flow of probability. The experiments are understood in terms of the RWA quasienergies and Floquet states (Band and Magnes, 1994).

If an experimenter would want to dissociate a molecule via sequential Λ three-level population inversions of some kind, he/she would wish to avoid using different pairs of different lasers for each step. Fortunately, this complication is not necessary. We choose the level 2 for each three-level inversion to be the ground vibrational level of an excited *electronic* state. Then we need only one unchirped nonresonant pulsed Stokes laser, say, an Nd:YAG laser with wavelength 1064 nm, plus a chirped pump laser with wavelength centered during the middle of the chirp so that $\Delta_3 = 0$ for a Λ transition in the middle of the sequence of inversions. Stokes-pump time

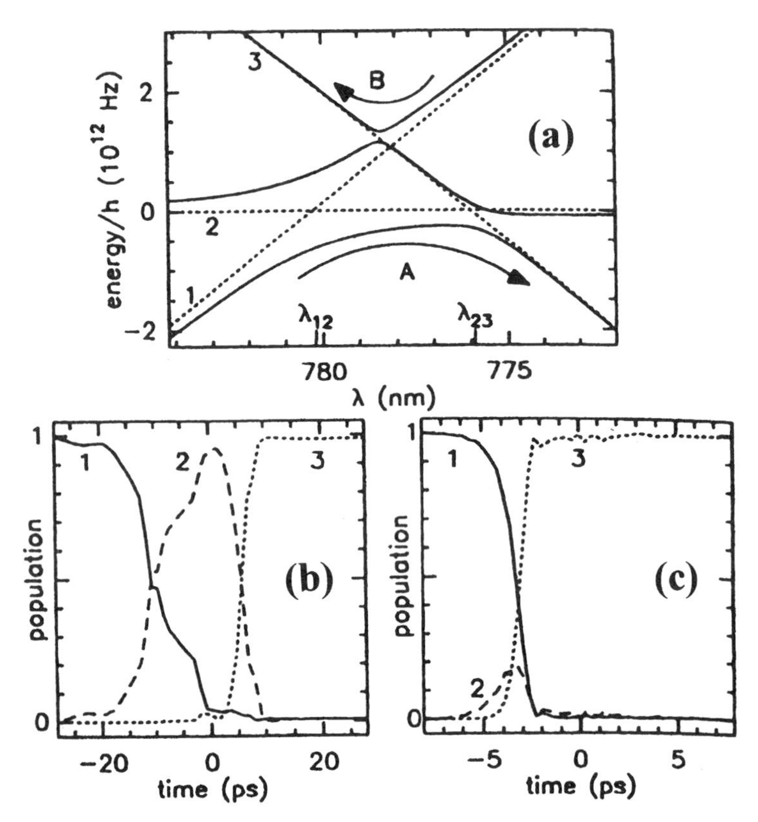

Figure 9.22. Features of the quantum adiabatic transport mechanism responsible for the experimental results shown in Fig. 9.21. (a) The primary (RWA) quasienergies for the three Floquet states, as a function of the instantaneous mean laser wavelength (or of the time). The curves labeled 1, 2, and 3 correspond to the RWA states $|5s, +\omega\rangle$, $|5p\rangle$, and $|5d, -\omega\rangle$, respectively. The path A is followed adiabatically to produce the $\alpha > 0.2$ ps^2 population inversion shown in Fig. 9.21b, while path B similarly explains Fig. 9.21c. (b) Calculated instantaneous populations of the free-atom states for the evolution along path A. The population of the 5p intermediate state 2 is about unity during the middle of the pulse. (c) The same as panel (b), except now the path is B. The transient population in state 2 is greatly suppressed, reducing possible population loss due to this state's spontaneous radiative decay. From Broers et al. (1992). Copyright 1992 by the American Physical Society.

delay is needed. Numerical calculations demonstrating the efficacy of this approach have been carried out by Chelkowski and Gibson (1995). They modeled the ground $X'\Sigma_g^+$ and excited $B\Sigma_g^+$ states of H_2 using Morse potentials with appropriate parameters, and took the pump laser to produce a nonresonant two-photon transition. The total electric field was chosen to be

$$F(t) = A(t)F_{max}\left[\sin\left(\omega_{pump}t - \frac{bt^2}{2}\right) + \sin(\omega_{Stokes}t)\right].$$

Here $F_{max} = 0.01068$ a.u., $b = 3.18 \times 10^{-5}$ eV/fs is the chirp rate, and the pulse shape A(t) has a 1.7 ps flat central part and a $\sin^2(\pi t/T_r)$ rise and fall with $T_r = 17.7$ fs. The time-dependent Schrödinger equation for two coupled vibrational (nuclear) wavefunctions for the ground and excited electronic states $\chi_{1,2}(R, t)$ was numerically integrated (without the RWA) using the split-operator grid method:

$$\left(i\hbar\frac{\partial}{\partial t} + \frac{\hbar^2}{2m}\frac{\partial^2}{\partial R^2}\right)\bar{\bar{I}}\begin{pmatrix}\chi_1(R,t)\\ \chi_2(R,t)\end{pmatrix} = \begin{bmatrix}V_1(R) & V_{12}(R,t)\\ V_{12}(R,t) & V_2(R)\end{bmatrix}\begin{pmatrix}\chi_1(R,t)\\ \chi_2(R,t)\end{pmatrix};$$

see further Section 9.2. Here $V_{1,2}(R)$ are the Morse potentials and $V_{12}(R)$ is $\mu(R)F(t)$, where the dipole function $\mu(R)$ is known for H_2. For a small chirp just large enough to drive the first Λ transition in the sequence, $X'\Sigma_g^+(v = 0 \to v = 1)$, it was numerically demonstrated that the evolution is almost adiabatic and produces close to complete inversion, the results for some vibrational excitation steps being reminiscent of the intermediate situation shown in Fig. 9.17. For the full-range chirp from 3% below to 3% above the central wavelength, the results shown in Fig. 9.23 are the time evolutions of the ground-state vibrational sublevel occupation probabilities and of the dissociation probability. Clean stepwise behavior up the vibrational ladder is strikingly evident. The peaks are smooth, indicating adiabatic transitions. Dissociation is significant, about 25%. Experiments to directly confirm

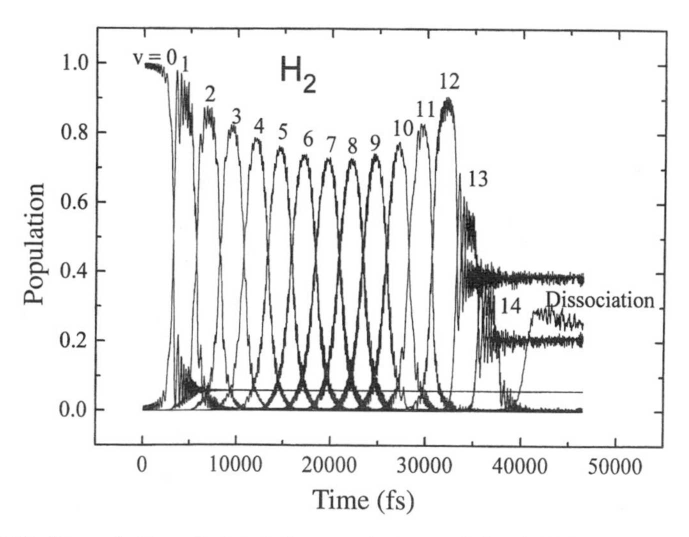

Figure 9.23. Numerically calculated time evolutions of the individual vibrational state probabilities and the dissociation probability of a frequency-chirped sinusoidally driven Morse oscillator. The Morse oscillator parameters were chosen to model the H_2 molecule. During the laser chirp a sequence of Λ three-level population inversions occurs, with an excited electronic state being the intermediate state for each inversion. The oscillator probability distribution is efficiently and selectively driven up to the 13th and 14th vibrational energy levels, which then are efficiently dissociated. From Chelkowski and Gibson (1995). Copyright 1995 by the American Physical Society.

these results for H_2 would require 3 fs pulses to achieve the necessary bandwidth for the needed chirping range. However, for molecules with smaller anharmonicity, such as HCl, this approach for efficient molecular dissociation appears promising.

9.3c. Optimal Control of Quantum Transport (Advanced Topic)

In general, optimal control theory is concerned with finding the best choice of parametric functions for systems of equations, under necessary constraints and with selected objectives. This theory has had well-developed mathematical foundations for over two decades (Bryson and Ho, 1975; Luenberger, 1979). Like other ideas used in this chapter, such as two-state quantum adiabatic evolution and the vector model, early quantum applications were developed in the field of nuclear magnetic resonance (NMR) (Conolly et al., 1986). The temporal design of radio-frequency fields for NMR is now very sophisticated and important for magnetic resonance imaging in modern medicine.

The mathematical problem fundamental to optimal quantum control is an inverse problem, which is to say that we are not solving Schrödinger's equation for a well-defined problem, but rather sorting through an infinite set of "ill-defined" problems in search of one or more of them that achieve desired objectives. Inverse problems are notoriously difficult to solve (Tihkonov and Arsenin, 1977), yet the potential value of their solutions can, in principle, provide enormous assistance for experimental progress. In the set of problems we consider here, the issue is finding the time-dependent optical electric field that optimally makes the time evolution of a driven quantum system produce particular desired results.

In our example we will be concerned with functionals, that is, functions of functions. We want to find those particular functions—such as for the time dependence of a laser field—that make the functional extremal (e.g., minimal). The minimization of a functional with respect to sets of functions can be achieved using the calculus of variations (Morse and Feshbach, 1953, Section 3.1; Wan, 1995). Let us first consider an example: the determination of the resonant driving laser field F(t) that could dissociate a 1-d classical Morse oscillator with minimal laser pulse energy (Krempl et al., 1992). We can discern a number of steps in the procedure:

- *Specification of the Physical Constraints.* Among these constraints we must include the equation(s) of motion that the system must satisfy. For our example we have one Hamilton's equation, the Newtonian equation of motion,

$$\begin{aligned}\frac{dp}{dt} &= -\frac{dV^{\text{Morse}}}{dx} + q'F(t) \\ &= -2aD[\exp(-ax) - \exp(-2ax)] + q'F(t).\end{aligned} \tag{9.31}$$

  Here D and a are the dissociation energy and range parameter of the Morse potential $V^{\text{Morse}}(x)$, q' is the effective charge (the dipole gradient), and these three parameters are chosen to model the HF molecule (Walker and Preston, 1977). A second constraint is that a fixed amount of optical interaction energy (assumed adequate for dissociation) should be absorbed by the molecule within the fixed laser pulse time T_p:

$$\int_0^{T_p} dt\, F(t)p = \text{constant}.$$

- *Identification of the Adjustable Functions.* Here this is just the laser field strength function F(t).
- *Specification of the Physical Objectives.* Our example has just one objective—that the total laser pulse energy be minimal,

$$\int_0^{T_p} dt\, F^2(t) = \text{minimum}.$$

- *Construction of the Functional J to be Minimized.* The variational procedure involves the introduction of Lagrangian multiplier functions $\lambda_i(t)$ that are to be varied to find the optimal solution. Our functional becomes

$$J = \int_0^{T_p} dt \left\{ F^2(t) + \lambda_1 F(t)p + \lambda_2(t)\left[\frac{dp}{dt} + \frac{dV^{Morse}}{dx} - q'F(t)\right]\right\},$$

where the term in square brackets reflects the constraint that the equation of motion must continuously be satisfied. Since our objective is not to reach a distinct point in (phase) space but to gain a given energy transfer, a variable end-point variational calculus is used.

- *Derivation of the Euler Differential Equations and Their Solution.* The variational requirement $\delta J = 0$ in the case of our example was shown (Krempl et al., 1992) to lead to a set of differential equations—called the Euler equations—to be simultaneously satisfied along with appropriate boundary conditions:

$$\frac{dp}{dt} + \frac{dV^{Morse}}{dx} - q'F(t) = 0,$$

$$m\frac{d^2F}{dt^2} + F(t)\frac{d^2V^{Morse}}{dx^2} = 0,$$

$$\delta\left(\frac{p^2}{2m} + V^{Morse}\right) = 0, \quad \text{at } t = 0,$$

$$F(t)\frac{dV^{Morse}}{dx} - \frac{dF}{dt}p = 0, \quad \text{at } t = T_p.$$

The electric field was found to satisfy $F(t) = 2p/e(t + t_0)$, where $t_0 = -2(dV/dx)_{x_0}/e(dF/dt)_0$ and e is the charge of the electron. Inserting this result for F(t) into the equation of motion (9.31) gave an ordinary differential equation that was numerically integrated for given initial conditions. The time-dependent functions x(t), p(t), and F(t) are then determined; see Fig. 9.24a for a typical result. The field F(t) oscillates almost at constant amplitude, but, as we expect, the field is chirped with decreasing frequency.

Quantum results were also obtained by Krempl et al. (1992). The optical classical field F(t) of Fig. 9.24a was used in the Hamiltonian, and an FFT grid integration was employed (Koonin and Meredith, 1990). In Fig. 9.24b the quantum evolution of the expectation values $\langle x \rangle$ appears as the solid curve. At short times the evolution is close to the classical one, while deviations increasingly occur at later times, as expected.

As an example of a fully quantum optimal control problem, we now consider the selective laser excitation of a Morse oscillator from the ground energy eigenstate $|n = 0\rangle$ to the targeted excited energy eigenstate $|n_{target} \equiv n_T = 2\rangle$. Now the equation of motion is Schrödinger's equation rather than Newton's equation, and we wish to maximize the squared wavefunction overlap

$$|\langle n_T|\psi(t)\rangle|^2_{t=T_p} = \langle\psi(t)|n_T\rangle\langle n_T|\psi(t)\rangle_{t=T_p}.$$

We note that the usual (Franck–Condon) laser pulse excitation of the Morse oscillator by itself produces a time-evolving excited-state wavefunction, which is not the presently targeted energy eigenstate. The formulation of our fully quantum

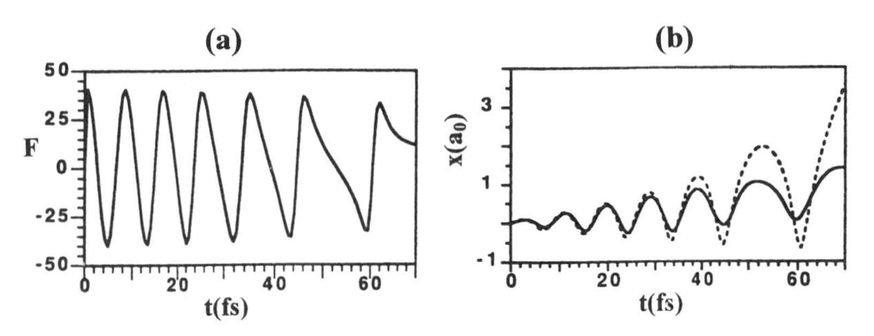

Figure 9.24. The classical and quantum time evolutions of a driven Morse oscillator, optimized to produce the absorption of a fixed amount of classical laser interaction energy during a fixed laser pulse time, using a minimum amount of laser pulse energy. The Morse oscillator parameters are chosen to model the HF molecule. (a) The optimal time evolution of the laser electric field, calculated for the classical problem. (b) The resulting time evolution of the classical (dashed line) and quantum (solid line) mean separation of the atoms in the molecule, using the results shown in panel (a). From Krempl et al. (1992). Copyright 1992 by the American Physical Society.

control problem is too lengthy to include here; see Shi and Rabitz (1990). We remark that the functional J to be minimized is

$$J = \int_0^{T_p} dt\{w_t\langle\psi(t)|(1 - |n_T\rangle\langle n_T|)\psi(t)\rangle + \tfrac{1}{2}w_F F^2(t)\}$$
$$+ w_{T_p}\langle\psi(T_p)|(1 - |n_T\rangle\langle n_T|)\psi(T_p)\rangle.$$

Here w_t, w_F, and w_{T_p} are fixed weighting factors for the three objectives, which, respectively, are (1) that the time integral of the population into the target state be maximal; (2) that the total laser pulse energy be minimal; and (3) that the final deviation of the population from that of the target state be minimal. For the case $w_t = w_{T_p} = w_F = 1$, plus a reasonable set of Morse oscillator parameters and

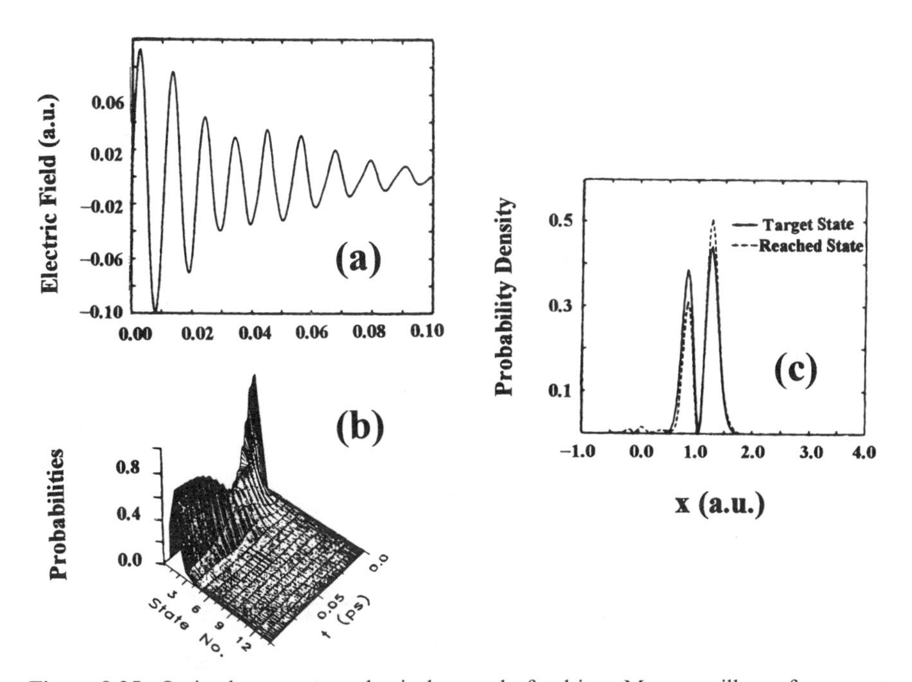

Figure 9.25. Optimal quantum mechanical control of a driven Morse oscillator, for converting the ground vibrational state of the undriven oscillator into the target second vibrational state after a fixed laser pulse time T_p. The time integral of the population flow is maximized, the total laser pulse energy is minimized, and the deviation is minimized between the probability distributions of the reached state and of the target state. (a) The optimal instantaneous laser electric field decreases in magnitude with time and has a downward frequency chirp. (b) The probability distribution over the undriven oscillator states is shown as a function of time. (c) The optimal state reached at T_p is compared with the target state. Reprinted with permission from Shi and Rabitz (1990). Copyright 1990 American Institute of Physics.

$T_p = 0.1$ ps, Fig. 9.25a shows the optimal field, and Fig. 9.25b shows the evolution of the system in terms of probabilities for members of the Morse oscillator energy eigenstates. The field F(t) is not so intuitive, being more complicated that just combined pulsing and linear frequency chirping. Figure 9.25c compares the final probability distribution $|\psi(T_p)|^2$ of the reached state with that of the n = 2 target state; although not perfect, the small difference is encouraging. It should be noted that these numerical results are based on split-operator FFT techniques and thus account for coupling between all of the Morse oscillator basis states.

Quantum optimal control techniques have also addressed the issue of selectively breaking one bond versus another, using a model linear triatomic molecular system consisting of two Morse oscillators coupled via a kinetic energy coupling (Rabitz and Shi, 1991). The magnitudes of the dipole moments D of the two oscillators were taken equal, and the oscillator fundamental frequencies were taken almost the same. The two oscillator dissociation energies were chosen to be quite different, however,

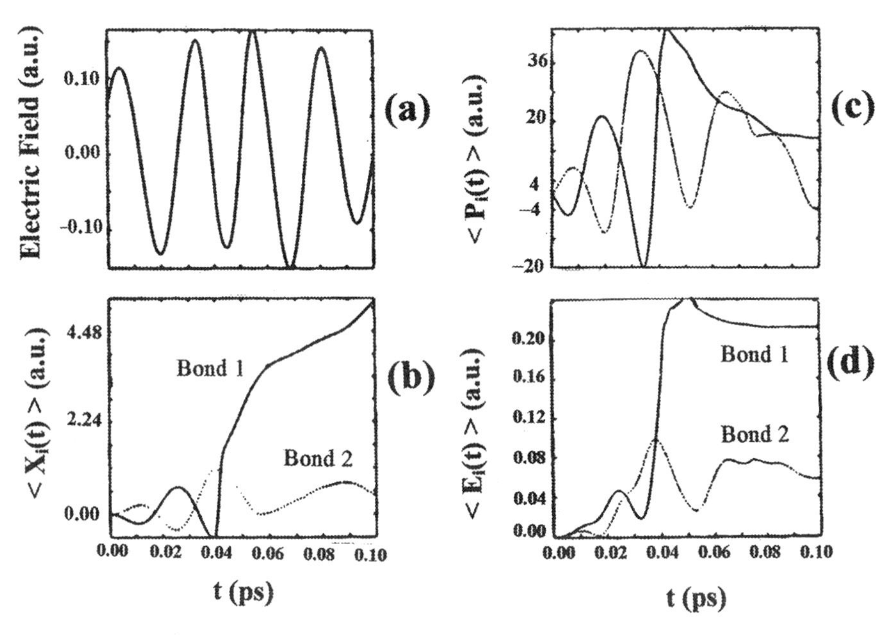

Figure 9.26. Optimized selective laser dissociation of a model quantum linear triatomic molecule. The dissociation energies of bonds 1 and 2 are 0.1976 and 0.137 a.u., respectively. Bond 1 is represented by solid lines, bond 2 by dashed lines. The objective is to dissociate the stronger bond 1, while minimally disturbing bond 2. Starting at the top, (a) the evolution of the optimal laser electric field; (b) the evolutions of the expectation values of the stretch of the two bonds; (c) the evolutions of the two conjugate momenta; and (d) the evolutions of the two bond energies. At the end of the evolutions, the stronger bond has about four times the energy of the weaker bond and is stretched five times more than the weaker bond. From Rabitz and Shi (1991). Copyright 1991 JAI Press.

so that one bond in the triatomic molecule was stronger than the other bond. The principal objective was to break (dissociate) the stronger bond, while leaving the weaker bond intact. Figure 9.26 shows the results for a controlling time $T_p = 0.1$ ps. Panel (a) shows the optimal field F(t). This field produces the quantum evolutions of the two bond stretches (b) and bond energies (d). At the final time T_p, the stretch and energy of the *strong* bond are much larger than for the weak bond, indicating that the stronger bond is dissociating while the weaker bond is not.

APPENDIX A

NOTES ON NUMERICAL METHODS FOR NONLINEAR CLASSICAL DYNAMICS

A.1. LOCAL SCIENTIFIC COMPUTING

As personal computers (PCs) become more powerful and scientific workstations less expensive, it is increasingly practical to do numerical calculations on your own computer. However, some of the calculations that we need to carry out—especially quantum mechanical ones—require days, weeks, or more on a fast modern PC with 128 MB of memory. As an alternative, "supercomputers" operating at gigaflop computing speeds can be formed from a cluster of workstations linked by multiple fast (100 Mbits/s) Ethernet lines and switches. The basic objective of concurrent computing on a cluster of N computers is to achieve a computing speed N times that of one of the computers. Message-passing software systems are crucial, such as parallel virtual machine (PVM) and message-passing interface (MPI) software. Such software can also be used on relatively inexpensive networks of fast PCs (Ballüder et al., 1996; Sterling, 1996). For information on one such "parallel" cluster of PCs, see http://www.beowulf.org. Be aware that some codes written for such clusters have required four PCs to match the computing speed of one optimized PC (Van de Velde, 1994) before the incremental gain in speed proportional to the number of PCs sets in. On the other hand, if one has the expertise to assemble PCs from purchased parts, then a 16-PC system can be constructed for perhaps $20,000.

One important and rapidly developing area is the writing and compiling of "parallelized" computer codes that take full advantage of distributed parallel computing systems. Unfortunately, at present these codes vary with the hardware and message-passing software employed. Writing "parallel-ready" code for concurrent scientific computing has been discussed for many of the codes used in numerical quantum and classical dynamics (Van de Velde, 1994). Some parallel-ready code is already available (Press et al., 1996).

At the moment, most inexpensive local scientific computing uses either some version of Unix, such as the freeware Linux (Templon, 1996) or the Windows NT 4.0 operating system.

Although program compilers for the program language C are used at times, most scientific computer users still employ compilers for Fortran 77 or the newer Fortran 90 because extensive collections of packaged Fortran scientific subroutines have been utilized for decades and hence are thoroughly debugged. These collections include the commercial NAG and Visual Numerics (formerly IMSL) libraries. There also are the large sets of algorithms contained in sources such as Slatec, CMLIB, and TOMS (ACM Transactions on Mathematical Software). Also available are the algorithms accompanying the text and reference book *Numerical Recipes* (Press et al., 1992). A widely used source of software is Netlib, which can be accessed on the World Wide Web at http://www.netlib.org.

In any case, optimizing Fortran codes requires some care (Schofield, 1989). If a computing job is not too extensive, one could avoid the Fortran programming altogether and have more fun using an interactive system such as Mathcad or Mathematica.

In this appendix we assume that we need not discuss numerical implementation of basic mathematical operations, such as evaluation of functions, interpolation and extrapolation, simple differentiation and integration, matrix computations, location of extrema of functions, and sorting. Codes for such operations are discussed in standard reference books, such as Press et al. (1992) and Lederman (1981).

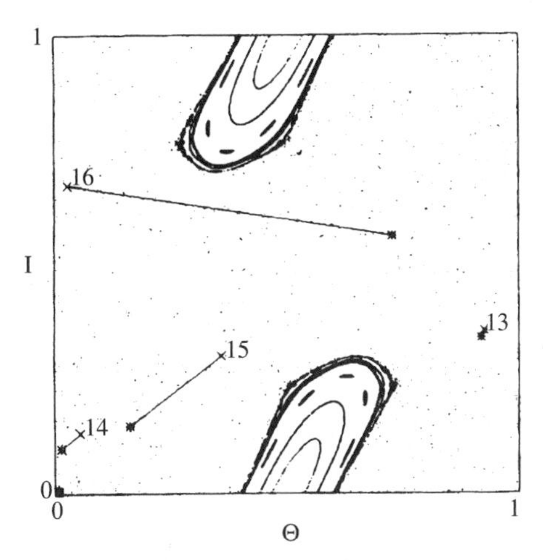

Figure A.1. Points in action-angle space (I, θ) for a chaotic trajectory determined by a 1-d discrete-time map (the standard map). The points mimic those in a stroboscopic Poincaré section for a continuous time evolution, which, however, in 1-d would not have chaotic trajectories. The initial condition for the trajectory is located at $(I_0, \theta_0) = (0.08, 0.01)$. The 13th through 16th points are shown, computed in double precision $(*)$ and in single precision $(\times)$. Straight lines are used to make the correspondences of points for the two calculations. At the 13th iteration, there is reasonable agreement for the two calculations, whereas after the 14th iteration the trajectory in single precision is grossly in error. From Grebogi et al. (1990). Copyright 1990 by the American Physical Society.

In numerical calculations for dynamical problems, high numerical accuracy can be essential. Double-precision Fortran programming and/or compiling are often required; see the interations of a chaotic trajectory in Fig. A.1 for an example.

A.2. ITERATING HAMILTON'S EQUATIONS

Hamilton's equations (1.7) for a Hamiltonian system constitute a coupled set of 2N ordinary differential equations (ODEs). We often want to calculate and display the time evolution of one or more trajectories, trajectories that are numerical solutions of the differential equations for sets of initial conditions. Of course we'll have to settle for solutions at a finite set of values of time, which should be chosen to capture all significant aspects of the evolution. One can discuss much about the numerical treatment of coupled ODEs by considering only one degree of freedom, $N = 1$:

$$\frac{dx}{dt} = f(x, p, t),$$
$$\frac{dp}{dt} = g(x, p, t).$$

This is a coupled pair of ODEs. Here the *derivative functions* f and g are known functions of the position, momentum, and time variables x, p, t. This time evolution problem is an *initial-value problem*, where we know the values x_0 of position and p_0 of momentum at the initial value t_0 of time, and we simply wish to calculate other values of x, p at times $t > t_0$. To numerically iterate Hamilton's equations (i.e., to propagate them forward in time), the basic approach is to replace the differentials dx, dp, dt with small finite steps (discrete changes) Δx, Δp, Δt, and then to multiply the equations by Δt. This *discretization* of the ODEs gives algebraic formulas for the changes Δx, Δp in the dependent variables x, p when the independent variable t is "stepped" forward by an amount equal to one "stepsize" Δt.

A.2a. One-Step Methods for the Evolution of an ODE

Before considering sets of ODEs, it helps to simplify to just one ODE. We can still learn a good deal about numerical iteration approaches, which are easily generalized to sets of coupled ODEs.

Let us consider the equation

$$\dot{x} \equiv \frac{dx}{dt} = f(x, t). \tag{A.1}$$

Using Taylor's theorem, the known derivative function $f(x, t)$ can be expanded about a point x_0, t_0 to obtain

$$\frac{dx}{dt} = f(x, t) = f(x_0, t_0) + (t - t_0)\frac{\partial f(x_0, t_0)}{\partial t} + \frac{(t - t_0)^2}{2}\frac{\partial^2 f(x_0, t_0)}{\partial t^2} + \mathcal{O}(t - t_0)^3, \tag{A.2}$$

where $\partial f(x_0, t_0)/\partial t$ denotes the derivative $\partial f(x, t)/\partial t$ evaluated at $x = x_0$, $t = t_0$.

Discretizing Eq. (A.2) to lowest order in $t - t_0 \equiv \Delta t$ will immediately give $\Delta x = (\Delta t) f(x_0, t_0)$ for a first step. Generalizing to the nth step of an iteration, we then have

$$x_{n+1} = x_n + f(x_n, t_n)(t_{n+1} - t_n). \tag{A.3}$$

Equation (A.3) explicitly and immediately advances the solution from time t_n to time t_{n+1} and defines *Euler's method* for evolving the solution of our ODE (A.1). As the solution moves from t_n to t_{n+1}, the change is in the direction of the local gradient evaluated at t_n. Note that in Euler's method the move is asymmetric in the sense that we only use information about the derivative at the beginning of the interval Δt. A *one-step method*, such as Euler's, therefore has a most useful property: we can find the value of x_{n+1} at each step by using only the previous value of x_n and the differential equation (A.1). One-step methods can generally use evaluations of $f(x, t)$ at several points within the interval $[t_0 + n\,\Delta t,\ t_0 + (n+1)\,\Delta t]$. These methods, however, do not use the previously calculated values of $x_{n-1}, x_{n-2}, \ldots$, nor do they use the values $f_{n-1}, f_{n-2}, \ldots$ of the derivative—facts that distinguish one-step methods from the multistep methods discussed in the next section.

From Eq. (A.2) it is clear that Euler's method has an error proportional to $(\Delta t)^2$, only one power of Δt higher than the term $f\,\Delta t$ used for the evolution. We say that a *method* is *of order n* if its error term is proportional to $(\Delta t)^{n+1}$. Thus Euler's method is a first-order method.

A.2b. Runge–Kutta Methods

Improving on Euler's method, the class of Runge–Kutta methods utilizes a weighted mean of the derivative function evaluated at a set of points in space-time—a set of points being located in the neighborhood of each of the points (x_n, t_n). For instance, in the *second-order midpoint method*, we first step time from t_n to $t_n + \Delta t/2$ and set $k = f(x_n, t_n)\,\Delta t/2$. We then use values for x and t assigned to the midpoint in order to compute the step across the full interval

$$x_{n+1} = x_n + f\left(x_n + \frac{k}{2}, t_n + \frac{\Delta t}{2}\right)\Delta t + \mathcal{O}(\Delta t)^3. \tag{A.4}$$

This equation can be generalized to an arbitrary intermediate point by introducing constants $\alpha, \beta, \gamma, \delta$ such that

$$\begin{aligned} x_{n+1} &= x_n + \alpha k_1 + \beta k_2 + \mathcal{O}(\Delta t)^3, \\ k_1 &\equiv f(x_n, t_n)\,\Delta t, \\ k_2 &\equiv f(x_n + \delta k_1, t_n + \gamma\,\Delta t). \end{aligned} \tag{A.5}$$

Equation (A.5) can be forced to agree with the second-order Taylor series expansion (A.1) if $\alpha + \beta = 1$ and $\beta\delta = \beta\gamma = \frac{1}{2}$ (Lederman, 1981). The next term of order $(\Delta t)^3$

is minimized when $\alpha = \frac{1}{3}$, $\beta = \frac{2}{3}$, and $\delta = \gamma = \frac{3}{4}$. This produces the *optimal second-order Runge–Kutta method.*

One can further increase the order of the error by again introducing weighted values of the derivative function f, with f evaluated, however, at more than one intermediate point. The most commonly used Runge–Kutta scheme is fourth order and involves the introduction of three intermediate points and ten constants as parameters:

$$x_{n+1} = x_n + \frac{k_1}{6} + \frac{k_2}{3} + \frac{k_3}{3} + \frac{k_4}{6} + \mathcal{O}(\Delta t)^5, \tag{A.6}$$

where

$$\begin{aligned}
k_1 &\equiv f(x_n, t_n)\,\Delta t,\\
k_2 &\equiv f\left(x_n + \frac{k_1}{2}, t_n + \frac{\Delta t}{2}\right)\Delta t,\\
k_3 &\equiv f\left(x_n + \frac{k_2}{2}, t_n + \frac{\Delta t}{2}\right)\Delta t,\\
k_4 &\equiv f(x_n + k_3, t_n + \Delta t).
\end{aligned}$$

Often, but not always, the fourth-order Runge–Kutta method enables at least twice the stepsize of the midpoint method, while achieving the same accuracy. Pushing to a still higher-order method will usually, but not always, give little improvement. High order does not always mean high accuracy, nor higher efficiency.

It is important, however, to modify a Runge–Kutta code to utilize the adaptive stepsize-control technique developed by Fehlberg (1968), who presents results for fourth through eighth order. His approach involves estimating the leading truncation error term at each step using the difference in results of calculations in n^{th} order compared to those of order $n + 1$. If the error is too large, then the stepsize is reduced until the accuracy becomes a little less than the acceptable value. This stepsize is then used to start the next step. For discussions of Runge–Kutta–Fehlberg (RKF) algorithms, see Press (1992) or De Vries (1994). For the regularized driven Kepler problem, we note that the eighth-order RKF method was found to be an optimal mix of speed and accuracy (Leopold and Richards, 1985). Note also that while Runge–Kutta methods converge as the stepsize Δt goes to zero, for some differential equations these methods can be unstable (Lederman, 1981).

A.2c. Methods for Higher Precision (Advanced Topic)

When the ultimate accuracy is required, multistep methods can be far more efficient (in computer time) than the optimized fourth-order Runge–Kutta–Fehlberg method. *Multistep methods* use equations that relate x_{n+1} not only to x_n, but also to points

$x_{n-1}, x_{n-2}, \ldots$ "in the past"—points that already have been computed. Thus the equations take the form

$$\sum_{\ell=-1}^{m} \alpha_\ell x_{n-\ell} = \Delta t \sum_{\ell=-1}^{m} \beta_\ell f_{n-\ell}, \tag{A.7}$$

where, as before, $x_{n-\ell} \equiv x(t_{n-\ell})$ and $f_{n-\ell} \equiv f(x_{n-\ell}, t_{n-\ell})$. The quantities α_ℓ, β_ℓ are suitably chosen constants, and m—the number of points in the past that are used—is often called the *order of the multistep method*.

Equations of the form (A.7) are in one of two classes, depending on whether or not $\beta_{-1} = 0$. If $\beta_{-1} = 0$, the right-hand side of Eq. (A.7) will not depend on f_{n+1}, enabling us to solve this equation for x_{n+1}. Such equations with $\beta_{-1} = 0$ are called *predictor formulas*. When $\beta_{-1} \neq 0$, we have *corrector formulas*; now x_{n+1} is present on both sides of Eq. (A.7), and the equation usually cannot be solved explicitly for x_{n+1}. The solution of corrector formulas then require *implicit methods* that involve iterative procedures or other similar techniques.

Predictor–corrector methods utilize a two-stage procedure involving a pair of multistep equations of the form (A.7). The first stage of the procedure uses the predictor method, which employs extrapolation based on a fitted polynomial function of the values of the derivatives $f_{n-\ell}$ to obtain an approximate (predicted) value of x_{n+1}. A popular procedure is to extrapolate f by a cubic polynomial fitted to $f_n, f_{n-1}, f_{n-2}, f_{n-3}$, to obtain the *Adams–Bashforth four-step predictor formula*

$$x_{n+1} = x_n + \frac{\Delta t}{24}(55f_n - 59f_{n-1} + 37f_{n-2} - 9f_{n-3}) + \mathcal{O}(\Delta t)^5. \tag{A.8}$$

Note that here we have a recursion relation obtained by taking all the α_ℓ to be zero, except for $\alpha_{-1} = 1$ and $\alpha_0 = -1$.

Since Eq. (A.8) involves values of (x, t) from several previous steps, in order to start a predictor–corrector calculation we must accurately compute the needed starting steps using some other method, such as the Runge–Kutta method discussed above, or a Taylor series method based on Eq. (A.2). One must take care that the accuracy of the entire predictor–corrector calculation is not limited by the accuracy of these starting steps.

The second stage of the predictor–corrector method starts by evaluating the derivative function f(x, t) using the predicted value of x_{n+1}—from Eq. (A.8)—to obtain a value for the derivative function f_{n+1}. This value is then inserted into a corrector formula in order to obtain a more accurate value of x_{n+1}, the corrected value. Commonly used is the *Adams–Moulton three-step corrector formula*, which is obtained by a cubic polynomial interpolation:

$$x_{n+1} = x_n + \frac{\Delta t}{24}(9f_{n+1} + 19f_n - 5f_{n-1} + f_{n-2}) + \mathcal{O}(\Delta t)^5. \tag{A.9}$$

Having final values for both x_{n+1} and f_{n+1} at $t_{n+1} = t_n + \Delta t$, we can now proceed to take another step by advancing the step-label n by unity. The degree of accuracy p indicated by the residual terms $\mathcal{O}(\Delta t)^{p+1}$ in Eqs. (A.8) and (A.9) has been established by Dahlquist (1956). He also addressed the question of the stability of the predictor–corrector method: for stable iterations at sufficiently small Δt, one should adjust the integer coefficients in Eqs. (A.8) and (A.9) such that $p = m + 1$, where m is the order of the multistep method.

Predictor–corrector methods are awkward when it comes to changing stepsize, since increasing stepsize involves more points from the past than one might have saved. To get around this requires bookkeeping. Also, as the integration proceeds, the best predictor–corrector codes adaptively change the order of the method as well as the stepsize. In addition, these codes employ elaborate techniques utilizing unequal spacing of steps. All of this leads to complicated codes that are roughly twice as long as comparable Runge–Kutta routines and that are difficult to program (Gear, 1971). Fortunately, standard and well-tested subroutine packages exist for Gear's methods.

There are other high-accuracy ODE integrators competitive with, or even better than, the predictor–corrector method. We mention the more recent Bulirsch–Stoer extrapolation method (Press et al., 1992). We also note that for the problem of two oscillators coupled by a quartic term in the Hamiltonian, a Taylor series integration algorithm has been found to be much better than the Runge–Kutta or predictor–corrector algorithms (Meyer, 1986). Also, a long-time high-accuracy calculation for the driven pendulum employed a seventh-order Taylor series method with an explicit truncation-error formula (Grebogi et al., 1990). Thus we should make our choice of algorithm according to the problem at hand.

A.2d. Methods for Stiff Sets of ODEs (Advanced Topic)

We now can consider a system of first-order ODEs, written in vector form as

$$\frac{d\mathbf{x}}{dt} = \mathbf{f}(\mathbf{x}, t), \tag{A.10}$$

where the explicit first-order approximation is—see Eq. (A.3)—

$$\mathbf{x}_{n+1} = \mathbf{x}_n + \mathbf{f}(\mathbf{x}_n, t_n)\,\Delta t. \tag{A.11}$$

The set of ODEs (A.10) introduces a new factor: the different dependent variables x_i may evolve on very different highest-frequency time scales. For numerical accuracy, we might have to follow the variations in the evolution on the shortest time scale, even when the desired time resolution would allow a much larger stepsize. Then the set of ODEs, along with their initial conditions, may produce evolution that is not numerically "well-behaved"—meaning that while the stepsize is short enough to make accurate evolution possible for one dependent variable, another dependent variable that is much more sensitive to initial round-off error (truncation error) dramatically compounds error because of the multitude of steps. Since all the dependent scalar variables in Eq. (A.10) usually are coupled, the whole evolution can be unstable. When the potential for such instability exists, we say we have a set of *stiff ODEs*.

The evolution of Hamilton's equations often involves very different simultaneous time scales. However, for physical reasons, variables in many cases exhibit no unusual sensitivity to initial error or truncation error. Regular evolution is such a case: we usually can find an independent set of variables such that none of the equations of motion possess terms that have singularities that are approached during the evolution. Then the Runge–Kutta and other methods are easily generalized to evolve Eq. (A.10).

For chaotic evolution, however, there is strong sensitivity to initial conditions and the equations can be stiff. For the Runge–Kutta and predictor–corrector methods described above, this sensitivity can be dealt with (Gear, 1971; Stoer and Bulirsch, 1980). The good schemes are quite complicated. Thus we will consider the basic idea of *implicit differencing*, but only in first and second orders. The idea is to directly involve the end point of a step in evaluating the derivatives in the difference equations. Such a procedure greatly reduces sensitivity to early errors when the stepsize is relatively large; it does, however, sacrifice accuracy to some extent.

We define the implicit first-order approximation as one that uses the new point $(\mathbf{x}_{n+1}, t_{n+1})$ in evaluating the derivative function $\mathbf{f}$ in Eq. (A.11):

$$\mathbf{x}_{n+1} = \mathbf{x}_n + \mathbf{f}(\mathbf{x}_{n+1}, t_{n+1})\,\Delta t. \tag{A.12}$$

Generally, Eq. (A.12) is a set of nonlinear scalar equations that must be solved iteratively at each step. For small enough stepsize Δt, we can linearize Eq. (A.12) using a generalization of Eq. (A.2) to first order:

$$\mathbf{x}_{n+1} = \mathbf{x}_n + \left[\mathbf{f}(\mathbf{x}_n, t_{n+1}) + \frac{\partial \mathbf{f}}{\partial \mathbf{x}}\bigg|_{\mathbf{x}_n, t_n} \cdot (\mathbf{x}_{n+1} - \mathbf{x}_n)\right]\Delta t. \tag{A.13}$$

Note that $t_{n+1} = t_n + \Delta t$ is known since t_n is known; the implicit differencing is only in $\mathbf{x}$. The implicit Eq. (A.13) is solved using iterative or other techniques. The procedure is usually (but not always) stable because the local behavior is similar to the case where $\partial\mathbf{f}/\partial\mathbf{x}$ is a constant matrix.

A second form of Eq. (A.13) is obtained by replacing just t_{n+1} with t_n in the square brackets of Eq. (A.13). We can then construct a useful second-order method by taking the average of the two forms for Eq. (A.13):

$$\mathbf{x}_{n+1} = \mathbf{x}_n + \frac{\Delta t}{2}\left[\mathbf{f}(\mathbf{x}_n, t_{n+1}) + \mathbf{f}(\mathbf{x}_n, t_n) + 2\frac{\partial \mathbf{f}}{\partial \mathbf{x}}\bigg|_{t_n} \cdot (\mathbf{x}_{n+1} - \mathbf{x}_n)\right]. \tag{A.14}$$

A.3. AN ALGORITHM FOR LYAPUNOV EXPONENTS

Section 5.2a introduced the set of Lyapunov exponents of a trajectory in phase space. Here we discuss aspects important for the accurate numerical computation of these exponents. We consider an N-d autonomous system, so that $n \equiv 2N$ is the dimension of phase space. Because of the digital nature of today's computers, we utilize logarithms to the base two in the definition Eq. (5.4) for the ith Lyapunov exponent; this exponent is defined in terms of the length $r_i(t)$ of the ith principal axis of the evolving ellipsoid in phase space. The ellipsoid is generated from an infinitesimal n-spheroid of initial conditions $r_i(0)$, $i = 1, \ldots, n$, so that Eq. (5.4) becomes

$$\lambda_i = \lim_{t \to \infty} \log_2 \frac{r_i(t)}{r_i(0)}.$$

We order the λ_i—with λ_1 the largest and λ_n the smallest. For numerical purposes, it is important to realize several facts: (1) that the longest ("linear") extent of the ellipsoid grows as $2^{\lambda_1 t}$; (2) that the area defined by the largest two principal axes grows as $2^{(\lambda_1+\lambda_2)t}$; (3) that the volume defined by the first three principal axes grows as $2^{(\lambda_1+\lambda_2+\lambda_3)t}$; and so on (Shimada and Nagashima, 1979; Benettin et al., 1980). These properties yield an alternative definition of the set (the spectrum) of exponents: the sum of the first k exponents is determined by the long-term averaged exponential growth of a k-volume element of initial conditions in phase space.

Next, we must realize that computing positive (chaotic) values of λ_i to an accuracy of a few percent or better usually means that we cannot simply follow two initially close trajectories. Because of the trajectories' exponential growth in their separation, computer limitations won't allow the initial separation to be small enough. This exponential-divergence numerical problem carries over to the numerical evolution of an n-D sphere of initial conditions. The problem can be avoided, however, by first working with (1) a fiducial (guiding) trajectory that starts at the center of the sphere and (2) an evolving ellipsoid defined by the evolution of the linearized equations of motion for starting points infinitesimally separated from the guiding trajectory. The principal axes of the ellipsoid are then determined by the evolution, via the linearized equations of an initially orthonormal vector frame anchored to the guiding trajectory. Thus we simultaneously integrate the exact equations of motion for the guiding trajectory along with the linearized equations of motion for n different initial conditions, conditions that define an arbitrarily oriented initial frame of n orthonormal vectors $\{\mathbf{V}_1, \mathbf{V}_2, \ldots, \mathbf{V}_n\}$. In time, each of these vectors $\mathbf{V}_i(t)$ will diverge in magnitude, and each will also collapse onto the common local direction of most rapid growth determined by λ_1.

Key to eliminating the exponential-divergence numerical problem is the Gram–Schmidt reorthonormalization (GSR) of the vector frame—see Wolf et al. (1985) for details and an often-used Fortran code. The way repeated GSR works is suggested by Fig. 5.4 and proceeds as follows. By choosing an elapsed time Δt well before any of the $\mathbf{V}_i(t)$ have grown or collapsed beyond computer limitations, we stop the evolution after a time Δt and compute a new orthonormal set $(\mathbf{V}'_1, \mathbf{V}'_2, \ldots, \mathbf{V}'_n)$:

$$\mathbf{V}'_1 = \frac{\mathbf{V}_1}{|\mathbf{V}_1|}, \tag{A.15}$$

$$\mathbf{V}'_2 = \frac{\mathbf{V}_2 - (\mathbf{V}_2 \cdot \mathbf{V}'_1)\mathbf{V}'_1}{|\mathbf{V}_2 - (\mathbf{V}_2 \cdot \mathbf{V}'_1)\mathbf{V}'_1|}, \tag{A.16}$$

$$\vdots$$

$$\mathbf{V}'_n = \frac{\mathbf{V}_n - (\mathbf{V}_n \cdot \mathbf{V}'_{n-1})\mathbf{V}'_{n-1} - \cdots - (\mathbf{V}_n \cdot \mathbf{V}'_1)\mathbf{V}'_1}{|\mathbf{V}_n - (\mathbf{V}_n \cdot \mathbf{V}'_{n-1})\mathbf{V}'_{n-1} - \cdots - (\mathbf{V}_n \cdot \mathbf{V}'_1)\mathbf{V}'_1|}.$$

Note the interplay of GSR with the different degrees of growth determined by the different values of the λ_i. Repeated GSR clearly doesn't change the direction of $\mathbf{V}_1$—see Eq. (A.15)—so this vector tends to track the most rapidly growing direction. The magnitude of this vector is just repeatedly renormalized to unity, with each normalization factor determining an estimate of $2^{\lambda_1(\Delta t)}$. During repeated GSR, we accumulate an average of such estimates, which provides an accurate value of λ_1. We turn to $\mathbf{V}'_2$—given by Eq. (A.16)—which before normalization is $\mathbf{V}_2$ with its projection onto $\mathbf{V}'_1$ subtracted out. The projection removes the tendency of the second vector to track the most rapidly growing direction. But $\mathbf{V}'_1$ and $\mathbf{V}'_2$ span the same two-dimensional subspace as do $\mathbf{V}_1$ and $\mathbf{V}_2$—see Eqs. (A.15) and (A.16)—and this subspace tends to become the 2-D subspace that is most rapidly growing. Thus the area defined by these two vectors is proportional to $2^{(\lambda_1+\lambda_2)\Delta t}$. Because $\mathbf{V}'_1$ and $\mathbf{V}'_2$ are orthogonal, we can determine an accurate value of λ_2 directly from the rate of growth of $\mathbf{V}_2 \cdot \mathbf{V}'_2$, again averaged over repeated GSR. All remaining vectors are treated the same as the second vector: (1) the subspace spanned by the first k vectors is unaffected by GSR; (2) the average long-term evolution of the k-volume determined by these k vectors gives 2^{μ}, $\mu \equiv \sum_{i=1}^{k} \lambda_i(\Delta t)$; (3) projections of evolved vectors onto new orthonormal frames correctly update the rates of growth of each of the first k principal axes; and (4) these growth rates, in turn, lead to accurate values of the k largest Lyapunov exponents λ_i.

A.4. LOCATING PERIODIC ORBITS

A periodic orbit is a time-periodic solution of Hamilton's equations (1.7). The solution repeats itself in time after some period T. Thus the equations to be solved have the form

$$\dot{\mathbf{Z}} \equiv \frac{d\mathbf{Z}}{dt} = \mathbf{f}(\mathbf{Z}(t), \lambda), \qquad \mathbf{Z}(t+T) = \mathbf{Z}(t).$$

The quantity λ denotes a parameter in the Hamiltonian that we might wish to vary. Note that initially we may not know the period T of the orbit that we seek. For the next few pages it is assumed this difficulty has already been resolved; then we will return to address it. Our present objective is to see how we can locate a periodic orbit in phase space, once an adequate approximate value of T is available.

We begin by choosing the zero of time, which is equivalent to choosing an initial phase condition for the periodic evolution. A reasonable approach is to select from the vector $\mathbf{f}$ the scalar component f_k that has the simplest structure, and then to impose an initial condition such as $f_k(\mathbf{Z}(0), \lambda) = 0$. We then have the augmented boundary-value problem

$$\dot{\mathbf{Z}} = \mathbf{f}(\mathbf{Z}(t), \lambda), \qquad \mathbf{g} \equiv \begin{pmatrix} \mathbf{Z}(0) - \mathbf{Z}(T) \\ f_k(\mathbf{Z}(0), \lambda) \end{pmatrix} = \mathbf{0}. \tag{A.17}$$

Equations (A.17) constitute a *two-point boundary-value problem*

$$\dot{\mathbf{Z}} = \mathbf{f}(\mathbf{Z}(t), \lambda), \qquad \mathbf{g}(\mathbf{Z}(0), \mathbf{Z}(T) = \mathbf{0}, \quad 0 \leq t \leq T. \tag{A.18}$$

Largely because of their global nature, two-point boundary-value problems are more difficult to deal with than initial-value problems. We should be aware that general theorems for the existence and uniqueness of solutions are not presently available for two-point boundary-value problems.

Nevertheless, to develop numerical solutions of Eqs. (A.18), we associate these equations with the initial-value problem

$$\dot{\mathbf{u}} = \mathbf{f}(\mathbf{u}, t), \qquad \mathbf{u}(0) = \mathbf{s}, \quad 0 \leq t \leq T, \tag{A.19}$$

where $\mathbf{s}$ is a "trial" initial vector. Let us assume $\mathbf{u}(\mathbf{s}, t)$ is a solution of Eqs. (A.19) that exists on $[0, T]$. Then solving Eqs. (A.18) is equivalent to solving the system of nonlinear equations

$$\mathbf{g}(\mathbf{s}, \mathbf{u}(\mathbf{s}, T)) = 0. \tag{A.20}$$

Equation (A.20) has as many distinct solutions as Eqs. (A.18).

To solve Eq. (A.20), we will use a *shooting method* (Plybon, 1992; Press et al., 1992; Gautschi, 1997). Numerically evolving the initial-value problem (A.19) is now called "shooting." Satisfying the second boundary condition—by solving Eq. (A.20)—is the "target" of the shooting. What we need now is a numerical method for readjusting the "aim" or trial initial condition $\mathbf{s}$, based on the amount the target has been missed.

Some shooting methods are generalizations of the Newton method of finding zeros of a scalar function $f(x)$, a method sometimes encountered in a calculus course. In Newton's method we assume that x_0 is an estimate of a root r of the equation $f(x) = 0$, and that the function f is differentiable in some interval I that contains r and x_0. The mean value theorem for derivatives tells us that for points x and x_0 in I, $f(x) - f(x_0) = f'(\xi)(x - x_0)$, where $f' \equiv df/dx$ and ξ lies between x and x_0. Then setting $x = r$, we obtain $-f(x_0) = f'(\xi)(r - x_0)$, where ξ lies between r and x. Hence $r \cong x_0 - f(x_0)/f'(\xi)$. If x_0 is sufficiently close to r, then we expect that $r \cong x_1 \equiv x_0 - f(x_0)/f'(x_0)$, if $f'(x_0)$ is not zero. The quantity x_1 probably is not exactly r, but x_1 can be taken as a new and better estimate of r; and x_1, in our argument, can be used in place of x_0 to carry out the procedure a second time to obtain a better estimate x_2. We continue this iteratively until the sequence of estimates has converged to the desired accuracy. This is called the *Newton–Raphson root-search method.*

To find the solution of the set of nonlinear scalar equations represented by the vector equation (A.20), we can generalize the Newton–Raphson method. This involves multidimensional calculus. The Jacobian $\partial \mathbf{g}(\mathbf{s})/\partial \mathbf{s}$ appears in place of the derivative $f'(x_0)$. The basic tool is "linearization at the current approximation," which we stop to explain. If we wish to solve a set of nonlinear equations, $\mathbf{f}(\mathbf{x}) = \mathbf{0}$, and if we already have an approximation $\mathbf{x}_0$, we then make a Taylor expansion and truncate after the linear term, $\mathbf{f}(\mathbf{x}_0) + \partial \mathbf{f}(\mathbf{x})/\partial \mathbf{x}|_{\mathbf{x}_0} (\mathbf{x} - \mathbf{x}_0) \cong \mathbf{0}$. The shooting sequence of approximations generated is then $\mathbf{x}_{n+1} = \mathbf{x}_n + \Delta_n$, where Δ_n is the numerical solution of the set of linear equations $\partial \mathbf{f}(\mathbf{x})/\partial \mathbf{x}|_{\mathbf{x}_n} \Delta_n = -\mathbf{f}(\mathbf{x}_n)$.

Thus we see that by satisfying both Eqs. (A.19) and (A.20), the solution of Eqs. (A.18) amounts to (1) making an initial choice $\mathbf{s}_0$ for $\mathbf{s}$, (2) integrating the initial-value problem (A.19) adjoined with that of its Jacobian—the results being $\mathbf{g}(\mathbf{s}_0, T)$ and $\partial \mathbf{g}(\mathbf{s}, T)/\partial \mathbf{s}|_{\mathbf{s}_0}$, then (3) applying the Newton–Raphson formula $\mathbf{s}_1 = \mathbf{s}_0 + \Delta_0 = \mathbf{s}_0 - [\partial \mathbf{g}(\mathbf{s}, T)/\partial \mathbf{s}|_{\mathbf{s}_0}]^{-1} \mathbf{g}(\mathbf{s}_0, T)$ to obtain a better value $\mathbf{s}_1$ for $\mathbf{s}$. We then use $\mathbf{s}_1$ to restart the iterative shooting toward an $\mathbf{s}_n$, where $\mathbf{s}_n$ will accurately describe a point on the desired periodic orbit.

Although this procedure is formally straightforward, it can be difficult to implement. One issue is that we may not be able to integrate our system of equations over the entire time interval to get to the final point T, because some component of $\mathbf{g}$ may diverge first. This problem is usually dealt with using *multiple shooting*—which means dividing the interval $[0, T]$ into N subintervals and applying shooting concurrently on each subinterval. The hope is that if the subintervals are sufficiently small, not only does the appropriate boundary-value problem have a unique solution, but this solution is also not given a chance to grow excessively. Note that the multiple-shooting procedure will include continuity conditions at the interior subdivision points. The most effective multiple-shooting codes dynamically generate the interval subdivision, appropriately changing the subdivision to force the solution to never grow beyond some chosen factor. Putting all these features into a code, we gain the potential for obtaining accurate solutions to some very difficult problems.

A second issue has to do with the convergence of the Newton–Raphson method, for the method typically requires $\mathbf{s}_0$ to be very close to a true initial vector for an exact solution of the boundary-value problem. Initial random shooting over a range of $\mathbf{s}$ is an obvious approach for finding a good starting vector $\mathbf{s}_0$.

Since we are finding periodic orbits, it is possible to work with the monodromy matrix $\bar{\bar{U}} \equiv \mathbf{U}$—Eqs. (3.4) and (3.5)—instead of working with the Jacobian. This choice produces the *monodromy matrix method*, an extension of the Newton–Raphson method. The monodromy matrix method works with the Poincaré section. The method can handle unstable periodic orbits, unlike some other methods that are restricted to stable periodic orbits.

To proceed with the monodromy matrix method, we must first calculate enough points on a Poincaré section both to determine the neighborhood of the periodic orbit of interest and to obtain an initial approximate value for the orbit period T. (This parallels the need to be in the neighborhood of a root of an equation to make efficient use of the Newton–Raphson root-search method.) We will begin our shooting method by using the monodromy matrix $\mathbf{U}$ of an approximate orbit—an orbit in the neighborhood of the periodic orbit as determined by the first return to the surface of section. The use of $\mathbf{U}$ will be analogous to the use of the derivative function in the Newton–Raphson method. Then we shoot and obtain an improved estimate for the location of the periodic orbit, and continue the shooting to "zero in" on that orbit. By combining a fine-enough grid on the surface of section, we can in principle locate all the periodic orbits using this method.

In the monodromy matrix method, a shooting integration back to the surface of section in the time T yields both a point that is an approximation for the fixed point $\mathbf{Z}^{\mathrm{FP}}$ and an approximation for the monodromy matrix $\mathbf{U}$ of that fixed point. The matrix equation to be iterated to the fixed point is

$$\mathbf{U}[\mathbf{Z}(0) - \mathbf{Z}^{\mathrm{FP}}(0)] = \mathbf{Z}(\mathrm{T}) - \mathbf{Z}^{\mathrm{FP}}(\mathrm{T}) \approx \mathbf{Z}(\mathrm{T}) - \mathbf{Z}^{\mathrm{FP}}(0). \tag{A.21}$$

The last step is an approximation, since T is not precisely the period of the periodic orbit we seek. After subtracting $\mathbf{Z}(0)$ from both sides of Eq. (A.21), we rearrange the equation to obtain the basic equation of the monodromy matrix method:

$$\mathbf{Z}(\mathrm{T}) - \mathbf{Z}(0) = (\mathbf{I} - \mathbf{U})[\mathbf{Z}^{\mathrm{FP}}(0) - \mathbf{Z}(0)]. \tag{A.22}$$

Solving this equation for the new value $\mathbf{Z}^{\mathrm{FP}}(0)$ obtained from one shooting iteration, we find

$$\mathbf{Z}^{\mathrm{FP}}(0) = \mathbf{Z}(0) + (\mathbf{I} - \mathbf{U})^{-1}[\mathbf{Z}(\mathrm{T}) - \mathbf{Z}(0)]. \tag{A.23}$$

This quantity is then used for $\mathbf{Z}(0)$ in the next iteration.

The direct inversion of Eq. (A.22) to obtain the result Eq. (A.23) has a potential problem in that there are two or more directions in phase space that lead to unit eigenvalues s_i of $\mathbf{U}$. In these directions the inverse of the matrix $\mathbf{I} - \mathbf{U}$ does not exist—that is, the matrix is singular. Eigenvalues of $\mathbf{I} - \mathbf{U}$ that are approximately zero arise from several constraints: (1) the fact that the exact orbit and the approximate orbit both start and end on the same surface of section; (2) the presence of energy conservation; and (3) the presence of other conserved dynamical quantities, such as angular momentum. A method for treating these constraints—analogous to the introduction of constraints by use of Lagrange multipliers—should be used to remove those components present within Eq. (A.22) that are responsible for the near-zero eigenvalues. This will reduce the dimension of the matrix **U** and will make **U** nonsingular (Baranger et al., 1988; Marcinek and Pollak, 1994). As an alternative to this treatment of constraints, one can employ finite-difference methods that altogether avoid the calculation of the matrix **U** (Marcinek and Pollak, 1994). These finite-difference methods involve following a set of orbits suitably displaced from the desired periodic orbit, rather than following only the one displaced orbit of the monodromy matrix method.

The dependence of the solutions of Eq. (A.17) on the parameter λ is very important for an understanding of a dynamical system, as well as for parametric control of such systems (Section 9.1). Consider some property of the solution $\mathbf{Z}(t, \lambda)$ plotted versus λ. Individual smooth curves in such plots are called "branches," and the plots are called *branching diagrams*. Individual branches can connect at bifurcation points. Numerical branch-tracing (path-following) is based on *continuation methods* in the parameter λ, rather than in the time variable. Nevertheless, continuation in λ can be carried out using the predictor–corrector method utilized for time evolution (Section A.2c). The continuation method generates a chain of solutions at different values of λ. As we go stepping along in λ, it is desirable to adaptively calculate candidates for the stepsize $\Delta\lambda$. We now also need a scheme for determining when a bifurcation point is close by or when it has been passed. We usually need to check the stability of the periodic orbits, and to switch from one branch to another in a methodical way. This becomes an extensive subject; see the book by Seydel (1988).

A.5. SPECTRAL ANALYSIS

In quantum mechanics, as in optics and electronics, we often improve our understanding of complicated time evolution by inspecting the values of frequency, amplitude, and relative phase of the frequency-components of the evolution. This remark applies for the evolution of any observable property of the physical system of interest. The present section deals with numerical procedures for *spectral analysis*—the decomposition of a time evolution into its frequency components. A glance at Eqs. (A.24) through (A.27) below reminds us of the basic analytical equations connecting the time and frequency domains.

A number of applications of spectral analysis appear in this book. Such applications include (1) the calculation of theoretical spectra to be compared with experimental spectroscopic data (Figs. 5.27 and 7.8); (2) the spectral analysis of wavepacket evolution to ascertain whether the evolution is regular or irregular (Fig. 5.21); and (3) the spectral analysis of the results of basis-set expansion calculations for comparison with the spectral results of periodic orbit theory (Fig. 7.10). The reverse analysis—passing from the frequency domain back to the time domain—is also useful. For example, in order to establish that quantum particle evolution in semiconductor devices is ballistic (Figs. 7.18 and 7.30), reverse analysis is used to determine orbit-period spectra from experimental frequency, energy, or related spectra. Furthermore, there are important spectral analysis applications not discussed in this book, such as the location of Arnold webs in phase space (Fig. 5.8), and the estimation of their local rates of transport; see Dumas and Lasker, 1993.

A very different and important use of transformations between a variable (such as time) and its conjugate variable (such as frequency) arises in some modern approaches for numerical quantum evolution. Some of the most powerful numerical techniques involve changing quantum mechanical representations during each basic timestep. Switching from the position representation to the momentum representation, and later switching back, is at the heart of the pseudospectral methods discussed in Sections B.1b and B.2e.

To numerically compute the Fourier transform Eq. (A.25), it's usually best to avoid using standard direct integration techniques to carry out the integrals for many values of the frequency. The fast Fourier transform (FFT) technique—described in Section A.5c below—is computationally much faster. In practice the FFT is much preferred, if not essential, for most applications.

A.5a. Fourier Analysis

Let the continuous time evolution of a scalar physical quantity be represented by the continous real function f(t). Once properly normalized, the two sets of harmonic functions $\{\sin \omega t\}$, $\{\cos \omega t\}$ together form a complete orthonormal basis set, if the frequency variable ω is allowed to take on the values of all real numbers. We can handle an expansion of f(t) in both sets of trigonometric functions by introducing a complex-number representation based on De Moivre's theorem $e^{i\omega t} = \cos \omega t + i \sin \omega t$. The set of expansion coefficients then becomes a continuous function $g(\omega)$ such that

$$f(t) = \frac{1}{\sqrt{2\pi}} \int_{-\infty}^{\infty} g(\omega) e^{-i\omega t}\, d\omega. \tag{A.24}$$

The orthonormality of our basis set means that

$$g(\omega) = \frac{1}{\sqrt{2\pi}} \int_{-\infty}^{\infty} f(t) e^{+i\omega t}\, dt. \tag{A.25}$$

(We note that there exist pathological functions f(t) for which $g(\omega)$ does not exist, but we do not consider such cases here.) We call $g(\omega)$ the *Fourier transform* $\mathbb{F}$ of f(t),

$$\mathbb{F}[f(t)] \equiv g(\omega),$$

and f(t) we call the inverse Fourier transform $\mathbb{F}^{-1}$,

$$\mathbb{F}^{-1}[g(\omega)] \equiv f(t).$$

Sometimes a function f(t) has a Fourier transform that can be evaluated analytically; see the collection of tables in Oberhettinger (1973).

The set of discrete frequencies $\omega_n = 2\pi n/T$ determines a complete orthonormal basis set for expanding a function f(t) that is periodic with period $T = 2\pi/\omega$. Then Eq. (A.24) reduces to the *Fourier series*

$$f(t) = \frac{1}{\sqrt{2\pi}} \sum_{n=-\infty}^{\infty} \frac{\pi}{T} g\left(\frac{2\pi n}{T}\right) e^{-i2\pi nt/T}, \tag{A.26}$$

and we have for the coefficients

$$g\left(\frac{2\pi n}{T}\right) = \frac{1}{\sqrt{2\pi}} \int_{-T/2}^{T/2} f(t) e^{+i2\pi nt/T}\, dt. \tag{A.27}$$

There are quite a few useful elementary properties of the Fourier transform, which are listed in De Vries (1994) and Press et al. (1992). As an example, if the time correlation of two functions $f_1(t)$, $f_2(t)$ is defined as

$$\int_{-\infty}^{\infty} f_1(\tau + t) f_2(\tau)\, d\tau, \tag{A.28}$$

then this time correlation corresponds in the frequency domain to the composite Fourier transform $g_1(\omega)g_2(-\omega)$. If we want to calculate the time autocorrelation function of a wavepacket $f_1(t)$, we set $f_2(t) = f_1(t)$. We can then either do the integral (A.28) numerically, or compute $g_1(\pm\omega)$ and form the composite Fourier transform and lastly transform from the frequency domain back to the time domain—whichever is easier.

For conjugate variables other than time and frequency, we note that Fourier transforms may be multidimensional. We know, for instance, that in quantum mechanics the particle position vector $\mathbf{r}$ and particle momentum vector $\mathbf{p} \equiv \hbar \mathbf{k}$ are conjugate variables. Then we may utilize the 3-d Fourier transform of a function $f(\mathbf{r})$ given by

$$\mathbb{F}[f(\mathbf{r})] \equiv g(\mathbf{k}) = \left(\frac{1}{2\pi}\right)^{3/2} \int_{-\infty}^{\infty} \int_{-\infty}^{\infty} \int_{-\infty}^{\infty} f(\mathbf{r}) e^{+i\mathbf{k}\cdot\mathbf{r}}\, d\mathbf{r}. \tag{A.29}$$

A.5b. The Discrete Fourier Transform

To develop a simple approximate Fourier transform that can be evaluated numerically, let us begin by assuming that we have N consecutive sampled values of a function f(t):

$$f_k \equiv f(t_k), \qquad l_k \equiv k\Delta, \quad k \equiv 0, 1, 2, \ldots, N-1. \tag{A.30}$$

The range of the label k in Eq. (A.30) is chosen to be consistent with most numerical codes, although this range differs from the symmetric labeling used for the Fourier series (A.26). Again for consistency with codes, we take the range of frequencies to be symmetric—$[-\omega_c, \omega_c]$—and seek estimates of the transform only at the discrete values

$$\omega_n \equiv \frac{2\pi n}{N\Delta}, \quad n = -\frac{N}{2}, \ldots, \frac{N}{2}. \tag{A.31}$$

Using Eqs. (A.30) and (A.31), we can now approximate the integral (A.25) by the discrete sum

$$g(\omega_n) = \frac{1}{\sqrt{2\pi}} \int_{-\infty}^{\infty} f(t) e^{+i\omega_n t}\, dt \approx \sum_{k=0}^{N-1} f_k e^{i\omega_n t_k} \Delta = \Delta \sum_{k=0}^{N-1} f_k e^{2\pi i k n/N}. \tag{A.32}$$

The final summation in Eq. (A.32) is called the *discrete Fourier transform* (DFT) of the N data points f_k. To symmetrize the notation, we write $g(\omega_n)/\Delta \equiv g_n$. Normally the codes are written with n and k both ranging from 0 to $N - 1$, which leads to the array labeling shown in Fig. 12.2.2 of Press et al. (1992). The formula for the inverse DFT is then

$$f_k = \frac{1}{N} \sum_{n=0}^{N-1} g_n e^{-2\pi i k n/N}. \tag{A.33}$$

A.5c. The Fast Fourier Transform

Since we are planning to use the DFT in numerical work, our first question is how much data can be included in the calculation. The larger the number N, the more precise our computations should be. In a direct approach to implementing the DFT, the issue is how many basic numerical operations—addition, multiplication, and the like—we need to carry out for a given N. Letting $W \equiv e^{2\pi i/N}$, the DFT (A.32) can be written as

$$g_n = \sum_{k=0}^{N-1} W^{nk} f_k,$$

which is an expression stating that the desired output vector g_n is the product of a matrix W^{nk} times the input vector f_k of data. The elements of the matrix are n-times-k powers of the complex constant W and hence are all nonzero complex numbers. Each g_n is then a linear combination of all the f_k—that is, a combination that requires N multiplications and $N - 1$ additions for a total number of operations of order N. But since we want all N of the g_n, the full order of the calculation is N^2. If $N = 10^6$ and if the central processor of our computer does one numerical operation per microsecond, then a single Fourier transform would take several days of computing. If we try to do this for the 3-d DFT analogous to Eq. (A.29), the computational effort is of order $N^6 = 10^{36}$ operations, which is out of the question.

Fortunately, the situation is much more favorable because of an ingenious numerical procedure that is qualitatively more efficient, with the computational effort of order $N \log_2 N$ instead of N^2 in 1-d. This procedure is called the *fast Fourier transform* (FFT). It is built on two ideas: the *Danielson–Lanczos decomposition* and *bit-reversed data ordering* (Press et al., 1992).

Let us assume that N is an exact power of 2. Danielson and Lanczos showed that a DFT vector of "length N" (the number of data points) can be rewritten as the sum of two DFT vectors g_n^e, g_n^o, each of length N/2:

$$g_n = g_n^e + W^n g_n^o. \tag{A.34}$$

The DFT vector g_n^e is formed from the even number of data points f_k having k even, while g_n^o uses the remaining odd points. We can again apply Eq. (A.34) to each of g_n^e and g_n^o, to obtain equations for four DFTs g_n^{ee}, g_n^{eo}, g_n^{oe}, g_n^{oo} of order N/4, and so on. Thus we can recursively subdivide all the data down to DFT decomposition equations of length one; each of these equations must copy one input f_k into one output g_n. Thus for every complete superscript pattern of e's and o's there is a one-to-one correspondence between an f_k and a g_n. From the formula for combinations, there are $\log_2 N$ such binary patterns, so the overall order of the full Danielson–Lanczos decomposition process is $N \log_2 N$.

Bit-reversed data ordering tells us which e–o binary pattern ultimately connects a given f_k with its g_n. We take each member of the original vector of data, write its k as a binary number, and then switch all the zeros in this number to ones and all the ones to zeros. This gives us the integer n as a binary number. The value of f_k then becomes the value of the nth component of a bit-reversed input vector for insertion into the set of Danielson–Lanczos equations at the lowest (length-one) level. The newly ordered initial data points have become the one-point transforms; adjacent pairs of data points are combined to give two-point transforms, adjacent pairs of adjacent pairs become four-point transforms, and so on, until the first and second halves of the data are combined—using Eq. (A.34)—to give the final transform. The structure of a code implementing this approach—called the Cooley–Tukey algorithm—can be found in Press et al. (1992), along with a Fortran 77 implementation.

APPENDIX B

NOTES ON COMPUTATIONAL QUANTUM DYNAMICS

B.1. NUMERICAL QUANTUM TIME EVOLUTION

A time-dependent quantum numerical approach must be used when a quantal Hamiltonian depends explicitly on time (e.g., when an external electromagnetic field or outside solvent forces are present) or when an initial quantum state is nonstationary (e.g., when the state is produced using excitation by a short laser pulse). We have an initial-value problem (in time) for the numerical quantum evolution of an initial state—a situation analogous to numerical classical evolution. For numerical reasons the initial quantum wavefunction will be defined on a finite set of spatial starting points—a spatial grid—and we will see that efficient quantum calculations require the selection of special grid points. We call such a wavefunction a *discretized wavefunction*.

Accurate and efficient numerical schemes are available for evaluating the action of the Hamiltonian operator on a discretized wavefunction. There exist numerically efficient and adequately accurate quantum time evolution operators. For these reasons, there is increasing interest in the numerical evolution of quantum systems.

The quantum evolution problems encountered in this book are of the simplest type, for they focus on just a few elementary particles and the electromagnetic interaction. In this appendix we focus on the more popular numerical approaches for such small quantum systems, where the approaches can be accurate for relatively long times—meaning hundreds of characteristic periods or longer. Note that there is keen interest in extending numerical work on spatial grids to more than three spatial degrees of freedom (3-d), but given the power of present computers, this remains quite a challenge. For these bigger systems, some calculations can be carried out for short times using other algorithms; for example, see the Lanczos algorithm approach of Park and Light (1986).

TABLE B.1. Comparison of the Numerical Algorithms

Name of Method	CN	AC	PR	SPO	FSOD	CH	SIL
Type of method	Finite difference	Finite difference	Finite difference	FFT grid	FFT grid	FFT grid	Iterative
Wavefunction norm	Preserved	Not preserved	Formally preserved	Preserved	Formally preserved	Not preserved	Preserved
Conservation of energy	Preserved			Not preserved	Preserved	Not preserved	Preserved
Stability	Unconditional		Unconditional	Unconditional	Conditional	Conditional	Unconditional
Error scaling with Δt	Quadratic	Quadratic	Quadratic	Quadratic	Quadratic	Exponential	High order
Hamiltonian limitations	No restriction	No restriction	No restriction	No mixed terms $f(x)g(p)$	Strictly Hermitian	Time independent	No restriction

Table B.1 is a partial summary of the principal established approaches to numerical quantum evolution. We find three finite-difference methods, three pseudo-spectral (FFT-grid) methods, and the short iterative Lanczos (SIL) method that utilizes the Lanczos algorithm. The SIL method avoids directly evolving Schrödinger's equation and thus calculating a Hamiltonian matrix; this fact enhances its usefulness for problems beyond 2-d, where the matrix becomes huge. The price paid is that the wavefunction is not calculated; only certain derived quantities are calculated, such as the time autocorrelation function.

The Chebyshev (CH) and the SIL methods are global evolution methods, where we work with functions that are each defined on the entire spatial grid. The other methods in Table B.1 are local or semilocal, where time evolution calculations are made sequentially for each grid point by using information about the Hamiltonian matrix at nearby grid points.

Note that most methods in Table B.1 have one or more drawbacks, which are indicated in the table. The drawbacks may or may not prove significant for a particular calculation on a particular quantum system. Numerical stability is an overriding concern. In addition, worrisome numerical errors can arise, including significant nonpreservation of total probability (the norm), of the energy $\langle\psi(t)|H|\psi(t)\rangle$ or quasienergy, or of the uncertainty relations. However, one can monitor such errors and use them as indicators of a calculation's accuracy.

B.1a. Finite-Difference Methods

B.1a.1. Finite Differencing the Hamiltonian on a Spatial Grid. In this section we focus on the problem of a particle in a potential $V(\mathbf{r})$, where we can locally assign a value $V(\mathbf{r}_j)$ to each point $\mathbf{r}_j$ on the spatial grid. Finite-differencing methods use a semilocal representation of the Laplacian operator on the grid. As we will see, this representation is given in terms of a polynomial function of the wavefunction's values at a grid point and at its neighboring grid points. The convergence of such methods, with respect to decreasing grid spacing, will then follow a power law.

Quantum mechanics is a nonlocal theory. Thus as a time-dependent numerical calculation using semilocal approximations proceeds, one expects there will eventually be a noticeable numerical violation of the uncertainty principle. Often this error does not appear right away, and an adequately accurate calculation can still be made.

As an example, we now consider the finite differencing of the motion of the electron in a hydrogen atom. Let us first consider the choice of spherical coordinates r, θ, φ. For zero angular momentum, the radial wavefunction R(r)—which is some linear combination of $\ell = 0$ energy eigenstates—satisfies the differential equation

$$i\frac{\partial}{\partial t}R = \left(-\frac{1}{2r^2}\frac{\partial}{\partial r}r^2\frac{\partial}{\partial r} - \frac{1}{r}\right)R. \tag{B.1}$$

If we set $\psi(r) \equiv rR(r)$, then the Hamiltonian for the hydrogen atom changes to

$$H = K + V = -\frac{1}{2}\frac{\partial^2}{\partial r^2} - \frac{1}{r}. \tag{B.2}$$

Let us adopt the notation that the value $\psi(r_j)$ of the wavefunction at the jth grid point r_j is denoted by ψ_j, $\psi_j \equiv \psi(r_j)$. If the radial spatial grid spacing is Δr, then a three-point finite-difference formula that locally approximates the action of the second derivative in Eq. (B.2) on ψ_j is

$$\frac{\partial^2}{\partial r^2}\psi_j \approx \frac{\psi_{j+1} - 2\psi_j + \psi_{j-1}}{(\Delta r)^2} + \mathcal{O}(\Delta r)^2. \tag{B.3}$$

The dominant error term can be shown to be

$$-\frac{(\Delta r)^2}{12}\frac{\partial^4 \psi_j}{\partial r^4} \tag{B.4}$$

(DeVries, 1994). Equation (B.3) is the standard general formula for *three-point spatial differencing* of the Laplacian operator in 1-d.

However, because $V(r) = -1/r$ for the Coulomb problem, our choice of the spherical radial coordinate leads to numerical difficulties. We need to change to modified cylindrical coordinates ξ, z, ϕ such that $r = (\xi^3 + z^2)^{1/2}$ and then work out the corresponding 2-d generalization of Eq. (B.3). In these new coordinates, the leading term of the ground-state hydrogen atom wavefunction becomes linear in ξ near the origin, so that the three-point finite-differencing formula will then be accurate even near the atomic nucleus. The new finite differencing is consistent with Schrödinger's equation, with the dominant error becoming

$$-\frac{(\Delta z)^2}{24}\frac{\partial^4\psi(\xi, z)}{\partial z^4} - \frac{(\Delta\xi)^2}{54\xi^3}\left(\xi^2\frac{\partial^4\psi(\xi, z)}{\partial\xi^4} - 2\xi\frac{\partial^3\psi(\xi, z)}{\partial\xi^3}\right)$$

(Kono et al., 1997). Thus the choice of the spatial coordinate system can be crucial for accurate finite differencing.

Of course, finite grids generally have boundaries, where the differencing formula (B.3) will need revision. This situation is indicated in Fig. B.1 for a 2-d rectangular grid. At the internal grid point "A", Eq. (B.3) can be used for each coordinate and the two terms added. Thus we use values of ψ at the five indicated points. At point "B" on the boundary, there is no right-hand grid point located at the symbol $\otimes$. At the symbol we should take $\psi = 0$, promising ourselves not to let our localized initial quantum state grow to reach the grid boundaries.

There is general consensus that the additional effort of refining the three-point differencing formula Eq. (B.3) is usually not justified. For fine grids, an extra level of

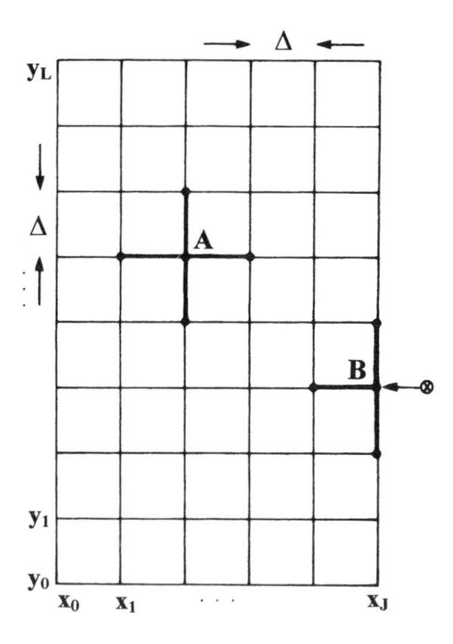

Figure B.1. Points used for three-point finite differencing of the Laplacian operator on a two-dimensional spatial grid. The second spatial derivatives at the point A are evaluated using point A and the four points shown connected to it by the heavy straight lines. The second derivatives at point B are evaluated using the connected points plus "right-hand side" boundary information, shown schematically by the symbol ⊗. From Press et al. (1986). Copyright 1986 Numerical Recipes Software.

accuracy can be obtained more readily by using Richardson extrapolation (DeVries, 1994) to refine the coarse results.

B.1a.2. The Crank–Nicholson Method for Time Propagation. The Crank–Nicholson (CN) finite-difference method is a standard method for numerical integration of the time-dependent Schrödinger equation. Its appealing features are threefold: the norm of the wavefunction is preserved during the integration process; the energy is preserved; and the method is unconditionally stable. For one small timestep Δt, the error is proportional to $(\Delta t)^3$.

We now define the Crank–Nicholson time propagator and combine it with spatial finite differencing to construct the space-time differencing equation (B.8) of the CN method. The first step in developing the CN method—or any other method for iterating the time-dependent Schrödinger equation—is to make an approximation for the short-time time evolution operator $U(\Delta t) = \exp(-iH\,\Delta t)$. The standard approach is to select some rational approximation $R(x)$ for $\exp(x) \equiv 1 + x + xx/2 + xxx/6 + \cdots$ and then replace x by $-iH\,\Delta t$. The CN method corresponds to the choice $R(x) = (1 + x/2)(1 - x/2)^{-1}$.

To further develop the CN method, we note that the time evolution operator $U(\Delta t)$ has the property that $\psi_n = [U(\Delta t)]^n \psi_0$. In particular, we can then write

$$\psi_{n+1} = \exp(-iH\,\Delta t)\psi_n \quad \text{and} \quad \psi_{n-1} = \exp(iH\,\Delta t)\psi_n. \qquad \text{(B.5)}$$

Eliminating ψ_n from these two equations yields the identity $\exp(iH\,\Delta t)\psi_{n+1} = \exp(-iH\,\Delta t)\psi_{n-1}$. We then discretize in time by taking the leading terms of the Taylor series expansion of $\exp(\mp iH\,\Delta t)$:

$$(1 + iH\,\Delta t)\psi_{n+1} = (1 - iH\,\Delta t)\psi_{n-1}. \tag{B.6}$$

Introducing the three-point spatial differencing, Eq. (B.3), and setting $\psi_n(x_j) \equiv \psi_{j,n}$, we obtain the action of the Hamiltonian on the wavefunction:

$$H\psi_{j,n\pm1} = \frac{1}{2}\frac{\psi_{j+1,n\pm1} - 2\psi_{j,n\pm1} + \psi_{j-1,n\pm1})}{(\Delta x)^2} + V_j\psi_{j,n\pm1}. \tag{B.7}$$

Then we insert Eq. (B.7) into Eq. (B.6), which yields the CN differencing equation

$$\begin{aligned}[1 + i(2\alpha + V_j\,\Delta t)]\psi_{j,n+1} - i\alpha[\psi_{j+1,n+1} + \psi_{j-1,n+1}]\\ = [1 - i(2\alpha + V_j\,\Delta t)]\psi_{j,n-1} + i\alpha[\psi_{j+1,n-1} + \psi_{j-1,n-1}],\end{aligned} \tag{B.8}$$

where $\alpha \equiv \frac{1}{2}[\Delta t/(\Delta x)^2]$. If we define $\boldsymbol{\Psi}_{n\pm1}$ as the vectors with components $\psi_{j,n\pm1}$, then Eq. (B.8) becomes a matrix equation

$$\bar{\bar{A}}\boldsymbol{\Psi}_{n+1} = \bar{\bar{\mathbf{B}}}\boldsymbol{\Psi}_{n-1}, \tag{B.9}$$

where $\bar{\bar{B}} = \bar{\bar{A}}^*$ and the elements of $\bar{\bar{A}}$ and $\bar{\bar{B}}$ can be picked out of Eq. (B.8).

The time differencing in Eq. (B.8) is *implicit*. Thus for each timestep, one must solve a tridiagonal system of linear algebraic equations by inverting the matrix $\bar{\bar{A}}$. This is a serious drawback from the point of view of computer time, as the number of arithmetic operations required to invert a matrix increases approximately as the cube of the size of the matrix.

For 1-d, 2-d, and 3-d, the CN scheme Eq. (B.9) reads, respectively,

$$\begin{aligned}A_{jk}\psi_{k,n+1} &= B_{jk}\psi_{k,n-1},\\ A_{mnjk}\psi_{jk,n+1} &= B_{mnjk}\psi_{jk,n-1},\\ A_{mnpjk\ell}\psi_{jk\ell,n+1} &= B_{mnpjk\ell}\psi_{jk\ell,n-1}.\end{aligned} \tag{B.10}$$

The 1-d case is not difficult numerically, especially since the matrices A_{jk} and B_{jk} are banded matrices of width 3. In the form of Eq. (B.10), the algebraic equations for the 2-d case with the four-dimensional matrices A_{mnjk} and B_{mnjk} are almost unmanageable. Nevertheless, the CN method was first used by McCullough and Wyatt (1969, 1971) to solve a 2-d $H + H_2$ scattering problem. They found a contraction in the indices from four dimensions down to two. However, the contracted matrices still were large, and moreover were no longer thin-banded. Given the computers available then, the numerical effort required was formidable.

B.1a.3. The Askar–Cakmak Propagation Method. An *explicit* finite-differencing time-propagation scheme is used in the *Askar–Cakmak (AC) method*. Like the CN, this is a *finite-difference second-order differencing method*. Unlike the CN method, we now keep ψ_n in the pair of Eqs. (B.5) and subtract one equation from the other to obtain the identity

$$\psi_{n+1} - \psi_{n-1} = [\exp(-iH\,\Delta t) - \exp(iH\,\Delta t)]\psi_n. \tag{B.11}$$

Again expanding $\exp(\pm iH\,\Delta t)$ into a Taylor series, we obtain approximately

$$\psi_{n+1} = -2iH\,\Delta t\,\psi_n + \psi_{n-1}. \tag{B.12}$$

When combined with Eq. (B.12), a finite-difference expression analogous to Eq. (B.3) produces

$$\psi_{j,n+1} = -2i[(2\alpha + V_j\,\Delta t)\psi_{j,n} - \alpha(\psi_{j+1,n} + \psi_{j-1,n})] + \psi_{j,n-1}, \tag{B.13}$$

which has the matrix form

$$\boldsymbol{\Psi}_{n+1} = \bar{\bar{A}}\boldsymbol{\Psi}_n + \boldsymbol{\Psi}_{n-1}. \tag{B.14}$$

Equation (B.14) explicitly gives the state of the system at the time $n+1$ in terms of the states at past times. No inversion of a matrix equation is needed at each timestep. This greatly reduces the computing time and makes 2-d calculations quite feasible. For the Schrödinger equation, the AC method is stable and has the same order of magnitude accuracy as the CN scheme. The AC method preserves the norm to order $(\Delta t)^4$, which often is acceptable.

B.1a.4. The Peaceman–Rachford Propagation Method. The CN and AC methods work well for 1-d problems, but even the AC method is relatively inefficient for 2-d problems. To improve efficiency for the 2-d case, we use a powerful operator-splitting and time-splitting scheme called the *alternating-direction implicit (ADI) method* (Press et al., 1992). The idea is to divide each timestep Δt into two substeps of size $\Delta t/2$. We also need operators in exponents to be sums of operators that depend on different coordinates, such as $H(x, y) = H_x(x) + H_y(y)$. Then the ADI propagator is written as

$$\exp[-i(H_x + H_y)\Delta t] \approx \frac{1}{1 + iH_x\,\Delta t/2}\frac{1 - iH_y\,\Delta t/2}{1 + iH_y\,\Delta t/2}(1 - iH_x\,\Delta t/2). \tag{B.15}$$

The operation of (B.15) on the wavefunction is separated into two implicit schemes by introducing an "artificial" intermediate state $\psi_{n+1/2}$ to obtain the propagation equations for the two substeps

$$(1 + iH_y\, \Delta t/2)\psi_{n+1/2}(x, y) = (1 - iH_x\, \Delta t/2)\psi_n(x, y), \tag{B.16a}$$

$$(1 + iH_x\, \Delta t/2)\psi_{n+1}(x, y) = (1 - iH_y\, \Delta t/2)\psi_{n+1/2}(x, y). \tag{B.16b}$$

This is called the *Peaceman–Rachford (PR) method* (Koonin et al., 1977; Galbraith et al., 1984). We can again use the three-point spatial finite-difference formula (B.3) for the operation of the differential operators on the wavefunction in Eqs. (B.16). The PR method then reduces our problem to two simple tridiagonal systems of algebraic equations, one for each substep. For Eq. (B.16a), a tridiagonal system of equations connecting the probability amplitudes at grid points $\{y_j\}$ is obtained at each x-grid point, while for Eq. (B.16b) a tridiagonal system of equations connecting the probability amplitudes at grid points $\{x_j\}$ is obtained at each y-grid point. Thus given $\psi_n(x, y)$, we solve Eq. (B.16a) for $\psi_{n+1/2}(x, y)$, we substitute this result into the right-hand side of Eq. (B.16b), and then we solve Eq. (B.16b). We see that a one-step 2-d propagation has been reduced to two half-step propagations, one in each spatial dimension. Each half-step propagation is like that of the implicit CN Eq. (B.8). Since the CN method works well in 1-d, we are in business. For instance, the dynamics of the $m = 0$ 3-d hydrogen atom in a pulsed strong linearly polarized laser field has been propagated by the PR method (Kono et al., 1997).

B.1b. FFT Pseudospectral Spatial-Grid Methods

A *spectral method* is a method that involves transforming a problem from its original description into a problem involving the Fourier transform of the solution (Section A.5). Sometimes the partial differential equation (PDE) of evolution for the transformed solution is simpler than the original PDE, and more easily solved. After finding the transformed solution, one applies the inverse Fourier transform to find the solution of the original PDE.

In the case of the time-dependent Schrödinger equation for a particle—even in a 1-d potential—the simple spectral method doesn't work because the kinetic energy is a function of momentum. The kinetic energy operator is a partial derivative with respect to position, while the potential energy operator depends on particle position. The trick is now to treat the kinetic energy operator in the Fourier transform (momentum) space, while the potential energy operator is treated in position space. Such an approach is called a *pseudospectral method*.

Because of the fast Fourier transform (Section A.5c), we have a very fast numerical method of transforming back and forth between position and momentum space. Thus codes utilizing FFT pseudospectral methods are usually very efficient. They also converge more rapidly than do the finite-difference methods (Section B.1a).

In practice, a numerical calculation on a digital computer involves the evaluation of variables at discrete sets of values. Thus the numerical application of pseudospectral methods to Schrödinger's equation is carried out on a grid of points in position space—so these methods are called *FFT grid methods*. We begin by discussing how to represent a wavefunction on a spatial grid. In Section B.1b.2 we will return to pseudospectral time propagation of such a wavefunction.

B.1b.1. The Fourier Global Spatial-Grid Representation. On a spatial grid, a wavefunction $\psi(x)$ is represented by its values ψ_j on the grid points x_j, $\psi_j \equiv \psi(x_j)$. We use a *collocation method* to find an approximation to $\psi(x)$ by (1) expanding in a set of analytic basis functions $g_k(x)$, and then (2) finding the expansion coefficients by matching the approximate and true solutions on the set of N_g grid points. The grid points are also called *collocation points*, and the basis functions sometimes are called *collocating functions*. We initially write the expansion as

$$\psi(x) \approx \bar{\psi}(x) \equiv \sum_{n=0}^{N_g-1} a_n g_n(x) \tag{B.17}$$

and the matching condition then becomes

$$\psi(x_j) \equiv \bar{\psi}(x_j) = \sum_{n=0}^{N_g-1} a_n g_n(x_j). \tag{B.18}$$

If the number of grid points equals the number of expansion functions, Eq. (B.18) will lead to a set of N_g linear equations for the a_n; see Eqs. (B.21) below. The relationship (B.18) between grid points and expansion coefficients is called the *collocation relation*.

It is very useful, both theoretically and numerically, to introduce a special combination of grid points and orthogonal collocating functions, such that the functions are also orthogonal when summed over the grid points

$$\sum_{j=0}^{N_g-1} g_n(x_j) g_\ell(x_j) = \delta_{n\ell}. \tag{B.19}$$

We then have a *global spatial-grid representation* of the wavefunction $\psi(x)$.

To obtain the *Fourier global spatial-grid representation*, we first choose the N_g spatial grid points to be equally spaced, $x_j = (j-1)\,\Delta x$, $j = 1, 2, \ldots, N_g$. In addition, the collocating functions $g_k(x)$ are taken to be the complex exponentials, which are both global and orthogonal in the continuous variable x,

$$g_n(x) = \exp\left(\frac{i2\pi n x}{L}\right), \quad n = -\frac{N_g}{2}, \ldots, 0, \ldots, \frac{N_g}{2} - 1, \tag{B.20}$$

where $L = (N_g - 1)\,\Delta x$ is the length of the spatial grid. Since we are representing the wavefunction of a quantum particle on the spatial grid, physically these complex exponentials are plane matter waves $\exp(ip_n x/\hbar)$ defined on an equally spaced grid in momentum space, $p_n = n(h/L)$. The orthogonality (B.19) of the collocating functions $g_n(x)$ allows us to invert the relation (B.18) to obtain

$$a_n = \frac{1}{N_g}\sum_{j=1}^{N_g} \psi(x_j)\exp\left(-\frac{i2\pi n x_j}{L}\right) = \frac{1}{N_g}\sum_{j=1}^{N_g} \psi(x_j)\exp\left(-\frac{i2\pi n(j-1)}{N_g - 1}\right). \qquad \text{(B.21)}$$

These a_n constitute a discrete representation of the wavefunction in momentum space.

To represent a physical problem at energy E, the full grid in phase space should enclose at least the classically accessible phase-space area Ω. The phase-space area of our rectangular grid is $(p_{max} - p_{min})L = (N_g - 1)h$, and therefore Ω sets a minimum value for the number N_g of spatial grid points. This requirement by itself may be almost enough for an adequate representation, because of the exponential decay of the wavefunction $\psi(x)$ in the classically forbidden region, and because of the exponential convergence with decreasing Δx of the best numerical time-propagation methods that use the Fourier grid representation (Kosloff, 1993).

The Fourier grid representation method may be extended by appropriately changing the spatial coordinate system and then placing a rectangular grid on the spatial part of the new phase space. By increasing the overlap of the grid in phase space with the classically allowed region Ω in phase space, this extension can enhance the efficiency of the grid representation (Fattal et al., 1997). For example, in the case of Kepler problems, the usual radial coordinate r can be transformed to a new coordinate Q according to $r = Q - A\tan^{-1}(\beta Q)$, where A and β are appropriately chosen constants. By effectively including a large number of grid points at small values of r, this transformation greatly reduces the difficulties associated with the singularity of the Coulomb potential.

The x–p uncertainty principle is preserved by Fourier spatial-grid representations (Kosloff and Kosloff, 1983). This preservation helps enable an accurate global representation of the Laplacian on the grid.

B.1b.2. Split-Operator Pseudospectral Time Propagation in the Fourier Global Spatial-Grid Representation. A common element in several numerical methods for the Schrödinger equation is the pseudospectral spatial-grid approach: it appears in the split-operator (SPO) method, the Fourier second-order-differencing (FSOD) method, the Chebyshev (CH) method, and the short iterated Lanczos time-propagation (SIL) method; see Table B.1. The first of these methods—the SPO—is very popular and is often called the *split-operator FFT grid method*, or just the *FFT grid method*.

Based on the general ideas of operator-splitting methods (Press et al., 1992), the SPO method was introduced for the Schrödinger equation by Feit et al. (1982). One begins by partitioning the time evolution operator into a product of short-time propagators

$$U(\mathrm{t}) = U(\mathrm{t}, \mathrm{t} - \Delta\mathrm{t}) \cdots U(2\,\Delta\mathrm{t}, \Delta\mathrm{t}) U(\Delta\mathrm{t}, 0). \qquad \text{(B.22)}$$

The time increment $\Delta\mathrm{t}$ is chosen short enough so that the (possibly time-dependent) Hamiltonian $H(\mathrm{t}) = K(\mathrm{t}) + V(\mathrm{t})$ is both approximately constant over the time increment and sufficiently short to permit use of the Trotter product formula (Trotter, 1959):

$$U(\mathrm{t}, \mathrm{t} - \Delta\mathrm{t}) = \exp\left(-\frac{\mathrm{i}H\,\Delta\mathrm{t}}{\hbar}\right) \approx \exp\left(-\frac{\mathrm{i}K\,\Delta\mathrm{t}}{2\hbar}\right) \exp\left(-\frac{\mathrm{i}V(\mathbf{r}, \mathrm{t})\,\Delta\mathrm{t}}{\hbar}\right) \exp\left(-\frac{\mathrm{i}K\,\Delta\mathrm{t}}{2\hbar}\right). \qquad \text{(B.23)}$$

This symmetric split-operator form of the propagator over a small time interval divides the kinetic energy term into two symmetric parts. We obtain an expression accurate to order $(\Delta\mathrm{t})^3$, as opposed to order $(\Delta\mathrm{t})^2$. We can see this increased accuracy by considering the Baker–Campbell–Hausdorff identity, which states that for three operators A, B, C,

$$\exp A \exp B = \exp C \quad \text{if and only if} \quad C = A + B + \tfrac{1}{2}[A, B] + \cdots. \qquad \text{(B.24)}$$

The identity (B.24) can be extended to any order by using the definition $\exp x = 1 + x + (1/2!)xx + (1/3!)xxx + \cdots$ for $x = A$ and $x = B$. Thus this equation is very useful for developing approximations of known error. If we take the operator C to be $-\mathrm{i}H\,\Delta\mathrm{t}$ with $K = p^2/2\mathrm{m}$, then the term in Eq. (B.24) that is proportional to $(\Delta\mathrm{t})^2[p^2, V]$ cancels out and leaves a lowest-order error of approximately

$$\text{error} \approx \frac{1}{\hbar^3} \max\left(\left\{-\mathrm{i}\frac{(\Delta\mathrm{t})^3}{16\mathrm{m}}[V, [V, p^2]]\right\}, \left\{-\mathrm{i}\frac{(\Delta\mathrm{t})^3}{32\mathrm{m}^2}[p^2, [p^2, V]]\right\}\right) \qquad \text{(B.25)}$$

(Kosloff, 1988). This error accumulates in both the phase and the energy of an evolving wavefunction. We note that the error will be smaller in several cases: for a small timestep $\Delta\mathrm{t}$; for a large mass m; and for a potential with small first-order (and higher-order) spatial derivatives.

The symmetrical split-operator form (B.23) has been used in calculations for the 1-d driven Kepler and Morse problems (Bardsley et al., 1988; Tanner and Maricq, 1989; respectively). An alternative to Eq. (B.23) is to divide the potential energy term into two symmetric parts:

$$\exp\left(-\mathrm{i}\frac{(K + V)\,\Delta\mathrm{t}}{\hbar}\right) \approx \exp\left(-\frac{\mathrm{i}V\,\Delta\mathrm{t}}{2\hbar}\right) \exp\left(-\frac{\mathrm{i}K\,\Delta\mathrm{t}}{\hbar}\right) \exp\left(-\frac{\mathrm{i}V\,\Delta\mathrm{t}}{2\hbar}\right). \qquad \text{(B.26)}$$

For applications of this alternative expression, see DeVries (1994) and Braun et al. (1996). The general SPO approach also works in phase-space representations (Torres-Vega and Frederick, 1991).

Often the initial state for an SPO calculation is obtained numerically, using the Fourier global spatial-grid representation (Section B.1b.1) or other means. In such cases it is wise to find the numerical initial state on a spatial grid with the grid spacing considerably smaller than the spacing used later during the time propagation. This procedure reduces the likelihood of the initial-state error growing during the time propagation to become the dominant source of error in the results at later times.

A numerical calculation of the first step in the time evolution of an initial state then proceeds as follows. We consider the case of Eq. (B.23). The initial state is expanded in a basis of momentum eigenstates. The result is then scalar-multiplied by the right-hand kinetic energy propagator of Eq. (B.23) for each momentum grid point $p_n = -h/(2\,\Delta x) + (n-1)h/(N_g\,\Delta x)$, $n = 1, \ldots, N_g$. The result is then inverse Fourier-transformed into position space and scalar-multiplied by the potential energy propagator, for each of the N_g coordinate grid points. To complete the evolution over one time increment Δt, we Fourier-transform back into momentum space and scalar-multiply by the remaining kinetic energy propagator, for each of the momentum grid points. Repeating this sequence of steps N_t times yields the quantum evolution over a time $t = N_t\,\Delta t$.

The full numerical implementation of this FFT grid method requires discretization of both the position and momentum spaces (Section B.1b.1) in addition to the discretization of the uniformly spaced time grid that was introduced in Eq. (B.22). Thus not only is the stepsize Δx in space governed by $\Delta k\,\Delta x \leq 1$, but the stepsize Δt in time is also limited by $\Delta E\,\Delta t \leq 1$. The spacings Δx and Δt can be chosen significantly larger than those required for the Crank–Nicholson (CN) method (Section B.1a.2).

As Table B.1 indicates, the SPO method is stable when the stepsizes satisfy the conditions just mentioned. Because only unitary operators are involved, this propagation scheme is formally norm-preserving. Energy is not preserved, but the commutator error terms—expression (B.25)—usually limit the accuracy. Thus the SPO introduces phase errors in the wavefunction, errors that accumulate with time.

How serious are the accumulated phase errors for the SPO method? We expect this to depend on the problem at hand, so no general statements can be made. The issue of errors can be addressed when we can compare SPO results with those obtained using a full "exact" energy eigenfunction (time-dependent) basis-set expansion. (See Section B.2b for the time-independent case). For wavepackets evolving within a 1-d Morse-oscillator potential well, the comparison has been carried out by Braun et al. (1996). Evolutions were carried out to a time of 10 ns, about 9000 vibrational periods for the packets used. For a timestep $\Delta t = 5$ a.u. (which means 10^8 propagation steps were taken!), the imaginary part of the two-packet autocorrelation function $\int dr\, \psi^*_{exact}(r, t)\psi_{SPO}(r, t)$ remained zero, indicating that phase error was negligible. For the larger stepsize $\Delta t = 50$ a.u., these results did not occur.

An excellent way to become acquainted with the split-operator pseudospectral time-propagation method is to follow the parallel development of the equations and of a computer code found in De Vries (1994). The code can be used to recalculate (much more quickly and accurately) the early results of Goldberg et al. (1967) for a Gaussian wavepacket that is incident on either a rectangular potential barrier or a rectangular potential well; see Fig. 6.2.

B.1b.3. Other Methods for Pseudospectral Time Propagation. For autonomous systems, one can use *global propagators*, which are polynomial expansions of U. The main idea is to write

$$U(\mathrm{t}) = \exp\left(-\frac{\mathrm{i}H\mathrm{t}}{\hbar}\right) \approx \sum_{\mathrm{n}=0}^{\mathrm{N_g}} \mathrm{a_n P_n}\left(-\frac{\mathrm{i}H\mathrm{t}}{\hbar}\right), \tag{B.27}$$

where the $\mathrm{P_n}$ are polynomials and the $\mathrm{a_n}$ are expansion coefficients. For a scalar function $\mathrm{f(x)}$ defined on the interval $[-1, 1]$, it is known that the Chebyshev polynomials $\mathrm{T_n(x)} \equiv \cos(\mathrm{n}\cos^{-1}\mathrm{x})$ are optimal for such an expansion, because the maximum error of the approximation is minimal compared to almost all possible approximations (Press et al., 1992). This bound on the error is a powerful property of the *Chebyshev-expansion time-propagation (CH) method.* In addition, the CH method leads to high accuracy because of exponential convergence with decreasing $\Delta\mathrm{x}$ and $\Delta\mathrm{t}$. Complex Chebyshev polynomials $\phi_\mathrm{n}(X)$ are used, where the scalar function x is replaced by an operator X, which for us is the exponent $-\mathrm{i}H\mathrm{t}$ in the time propagator (B.27). The range of definition of these polynomials is from $-\mathrm{i}$ to $+\mathrm{i}$, meaning that the Hamiltonian operator has to be renormalized to become H_{norm}. This renormalization of H is achieved by dividing H by the full span of real energy eigenvalues $\Delta\mathrm{E} \equiv \mathrm{E_{max}} - \mathrm{E_{min}}$, which must be estimated. If $\Delta\mathrm{E}$ is chosen too large, then the algorithm loses efficiency. If the estimated $\Delta\mathrm{E}$ is smaller than the true value, then the calculation will diverge.

As an example, consider wavepacket evolution in 2-d for a potential that depends on the cylindrical coordinates ρ, z. The evolved wavefunction after an elapsed time τ takes the form

$$\Psi(\rho, \mathrm{z}, \tau) = \exp(-\alpha) \sum_{\mathrm{n}=1}^{\mathrm{N_g}} \mathrm{a_n}(\alpha)\phi_\mathrm{n}(-\mathrm{i}H_{\mathrm{norm}})\Psi(\rho, \mathrm{z}, 0),$$

(Tal-Ezer and Kosloff, 1984; LeForestier et al., 1991). Here $\alpha \equiv \Delta\mathrm{E}\,\tau/2\hbar$, $\mathrm{a_0} = \mathrm{J_0}(\alpha)$, and $\mathrm{a_n} = 2\mathrm{J_n}(\alpha)$ for $\mathrm{n} \geq 1$. Quantities ϕ_n and $\mathrm{J_n}$ denote, respectively, the $\mathrm{n^{th}}$ order Chebyshev polynomials and the Bessel functions of the first kind. We calculate the actions of the operators $\phi_\mathrm{n}(-\mathrm{i}H_{\mathrm{norm}})$ on the initial-state wavefunction $\psi(\rho, \mathrm{z}, 0)$ by using the recurrence relation for the Chebyshev polynomials

$$\chi_{\mathrm{n}+1} = -2\mathrm{i}H_{\mathrm{norm}}\chi_\mathrm{n} + \chi_{\mathrm{n}-1},$$

with

$$\chi_n \equiv \phi_n(-iH_{\text{norm}})\Psi(\rho, z, 0).$$

The expansion coefficients $a_n(\alpha)$ drop off very quickly for $n > \alpha$. We use the split-operator FFT grid technique to evaluate the operation of the kinetic energy and potential energy operators. This evaluation must be done over the entire 2-d spatial grid; the particular procedure depends on how the Hamiltonian varies with the spatial coordinates. The Chebyshev (CH) method is not unitary (the norm of the spatial probability distribution is not conserved), but because of the CH method's high accuracy, the deviation of the norm from unity can be used as an accuracy check.

A drawback of the Chebyshev (and other global evolution methods) is that results at intermediate times are not obtained. We can, however, split the propagation into a set of propagations over subintervals. Another drawback is that the CH method is designed for the propagation of autonomous systems only. Nevertheless, the CH method is now a standard technique for the design of (undriven) semiconductor solid state heterostructure devices (Pandey et al., 1994). When such devices are modeled by sequences of rectangular potential barriers and wells, the CH method can handle the discontinuities in the potential V(x) more easily than can the SPO method.

The FSOD method (see Table B.1) is a good method, but not as popular as the SPO and CH methods. The SIL method is relatively new but very promising; see Fig. B.2. The SIL method has been routinely used to check Floquet-state predictions

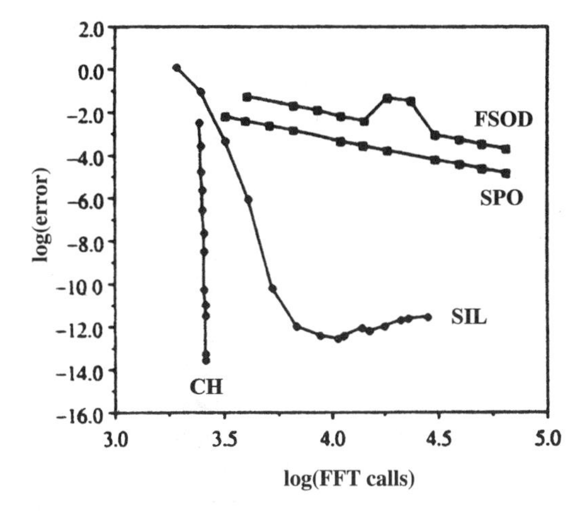

Figure B.2. A log–log plot of the accumulated phase error in the final-state wavefunction for a wavepacket moving in a 1-d Morse potential well, at a fixed elapsed time of 2400 atomic units. The horizontal axis is a measure of numerical effort, the number of FFT calls used. More calls means the use of a smaller timestep. The four pseudospectral time-propagation methods—CH, SIL, SPO, and FSOD—are discussed in the text. From LeForestier et al. (1991). Copyright 1991 Academic Press, Inc.

for highly controlled chirped-pulse driven systems (Guérin, 1997). Descriptions of the FSOD and SIL methods can be found in LeForestier et al. (1991).

B.1c. Numerical Computation of Floquet States and Quasienergies

Computation of (time-decaying) Floquet states can be accomplished using the split-operator (SPO) method (Bardsley et al., 1988). For the usual Floquet states, the amplitude of the driving field is held constant (Section 8.1). All we need do is propagate in time for one period T of the driving field, taking $t = T$ in Eq. (B.22), and then find the eigenvectors of this one-period (Floquet) time propagator $U_F = U(T, 0)$; see Eq. (8.12). To account for probability flow into the continuum of the system being driven, one must either impose outgoing boundary conditions or use the complex coordinate rotation method described in Section B.2c. (The complex coordinate rotation method is often preferred since the wavefunctions for both bound states and decaying (resonance) states vanish for large distances x from the initial system being driven.)

A common numerical procedure for the propagation is to utilize the Fourier global spatial-grid method—Eqs. (B.17)–(B.21)—which expands a Floquet state in plane waves, $\psi_F(x, t) = \sum_n c_n(t) \exp(i\pi nx/x_{max})$. In this grid representation, the one-period propagator U_F becomes a matrix $\bar{\bar{U}}_{mn}$. This matrix is obtained by numerically propagating each plane wave state in turn, using the SPO method with either Eq. (B.23) or Eq. (B.26). When $c_n(0)$ is taken to be the delta function δ_{nj}, the quantity U_{mj} is given by the value of $c_m(t)$ at time T. The Floquet states are then the eigenstates of the matrix $\bar{\bar{U}}_{mn}$ and are obtained by matrix diagonalization; see Section B.2b. The real quasienergies ϵ_k and the decay rates Γ_k are computed from the complex eigenvalues of $\bar{\bar{U}}_{mn}$, since after diagonalization $\bar{\bar{U}}_{mn}$ becomes $\exp[i(\epsilon - i\Gamma/2)T]$.

B.1d. The Calculation of Evolution of Energy-Level Crossing Systems

Split-operator techniques have been used to evolve wavepackets on two or more coupled potential energy surfaces. The numerical technique of choice depends on the number N_s of such surfaces. For $N_s = 2$, the SPO is most efficiently carried out in the diabatic (molecular) representation (Section 9.2b). The reason for the efficiency is that the nondiagonal part of the SPO time propagator (the diabatic potential energy) can be written exactly as a matrix polynomial. There is an identity for the exponential of a symmetric 2×2 square matrix. If the matrix is written as

$$\bar{\bar{A}}_2 \equiv \begin{bmatrix} a & b \\ b & c \end{bmatrix},$$

then the polynomial expansion is

$$\exp(-i\bar{\bar{A}}_2) = \exp(-i\alpha_0)\left[\bar{\bar{I}}_2 \cos\alpha - i\frac{\sin\alpha}{\alpha}(\bar{\bar{A}}_2 - \alpha_0\bar{\bar{I}}_2)\right],$$

with

$$\alpha = \left[\left(\frac{a-c}{2}\right)^2 + b^2\right]^{1/2} \quad \text{and} \quad \alpha_0 = \frac{a+c}{2} = \frac{1}{2}\mathrm{Tr}[\bar{\bar{A}}_2]$$

(Broeckhove et al., 1990; Péoux et al., 1996). For $N_s = 3$, an exact polynomial expression for a fully symmetrix 3×3 square matrix can be used once again, the expression having the form

$$\exp(-i\bar{\bar{A}}_3) = \exp(-i\beta)\{\bar{\bar{I}}_3 + \gamma_1[\bar{\bar{A}}_3 - \beta\bar{\bar{I}}_3] + \gamma_2[\bar{\bar{A}}_3 - \beta\bar{\bar{I}}_3]^2\}.$$

Here $\bar{\bar{I}}_3$ is the identity matrix and β, γ_1, and γ_2 are complex coefficients determined by the elements of the matrix $\bar{\bar{A}}_3$ (Péoux et al., 1996). Using this $N_s = 3$ approach, this group carried out efficient SPO calculations of the photodissociation of the HBr molecule. The calculations involved an initial wavepacket produced on the potential energy curve for the lowest $^1\Pi$ state. This state was coupled to two other dissociative excited states to produce an $N_s = 3$ problem.

In the *polynomial-propagator SPO method* just described, the numerical effort is determined by the matrix multiplications in the polynomial terms and therefore scales as $N_g^{N_g}$, where N_g is the number of SPO grid points. Fortunately, there is another approach that has the numerical effort scaling as N_g^3. This second approach uses split-operator techniques that combine adiabatic–diabatic molecular representation switching with the usual SPO method. For the HBr photodissociation problem, this second approach has been found to be half as fast as the polynomial-propagator SPO method (Péoux et al., 1996). For $N_s > 3$, however, adiabatic–diabatic representation switching may be useful. The idea is to propagate the kinetic energy in the diabatic-and-momentum representation, and to propagate the potential energy in the adiabatic-and-position representation (Almeida and Metiu, 1988). Of course the adiabatic–diabatic switching procedure requires a representation of the transformation matrix $\bar{\bar{L}}$ from the adiabatic molecular representation to the diabatic molecular representation. A numerical procedure for efficently calculating $\bar{\bar{L}}$ is proposed in Fernández and Micha (1992).

Now we return to the general problem of multiple spatial dimensions. Currently, accurate numerical calculations of wavepacket evolution on two coupled potential surfaces can be carried out for at most only $N = 3$ or 4 spatial degrees of freedom, as is the case for one-surface problems. There is, however, a *Gaussian wavepacket path integral* (GWD-PI) *method* that is accurate for $N = 2$ and that may also be accurate for surface-crossing problems with much larger N (Coalson, 1996).

The path integral methodology in the GWD-PI method parallels the methodology for the exact solution of the spin-boson model. This model can be thought of as a pair of multidimensional, linearly displaced harmonic oscillator potential surfaces that are coupled by a constant nonadiabatic surface-hopping matrix element. We are interested here in considering the evolution of an initial wavepacket. To do this we need to consider the simultaneous propagation of nuclear coordinate wavepackets on each of the two surfaces—for all possible paths determined by all possible sequences of surface-hopping times—and then coherently sum up amplitude contributions for all the paths. For the spin-boson problem, the nuclear coordinate propagation can be done analytically for each path, reducing the numerical work to summing amplitude contributions obtained from all allowed spin paths. The summing can be carried out to useful times (Evans and Coalson, 1995).

To extend the spin-boson methodology to general coupled anharmonic potential surfaces, one uses an approximate single-surface propagation method that is rapid and multidimensional, namely, the *Gaussian wavepacket dynamics* (GWD) *algorithm* (Heller, 1981; Neria and Nitzan, 1993). The procedure is similar to the semiclassical wavepacket time-propagation method of Section 2.1c. The key feature of this algorithm is that the single-surface evolutions are obtained by integrating a set of first-order equations. These equations prescribe the evolution of the time-dependent parameters that define the evolving Gaussian wavepacket.

Results of the GWD-PI method have been compared with accurate SPO calculations for a special problem: two potential surfaces are chosen to have the form $V_j(x, y) \equiv A_j \exp(\alpha_j x - \beta_j y) + k_j x^2/2 + V_j^0$, $j = 1, 2$, and the coupling function is chosen as $g(x, y) = g_0 \exp(-\gamma y)$. These potential surfaces produce direct dissociation along the y-coordinate and vibrational evolution along the x-coordinate. The comparisons in Fig. B.3 are encouraging. For problems having possibly

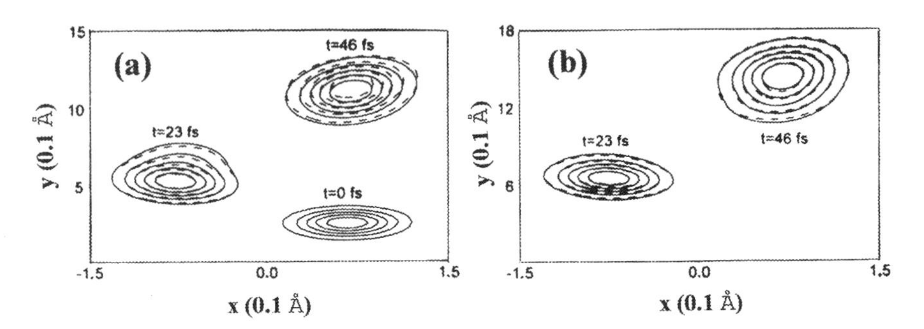

Figure B.3. Contour plots showing the time evolution of wavepacket nuclear coordinate probability on coupled (a) radiatively bright and (b) radiatively dark 2-d potential energy surfaces, at three different times. Because of the initial condition, at $t = 0$ there is no probability in panel (b). The solid contours are accurate results obtained by the split-operator FFT grid method (SPO), while the dashed contours are results obtained in the GWD-PI approximation. From Cárdenas and Coalson (1997). Copyright 1997, with permission from Elsevier Science.

hundreds of spatial degrees of freedom, the GWD-PI method has the potential for accuracy at short times.

B.2. THE NUMERICAL CALCULATION OF STATIONARY STATES AND THEIR PROPERTIES

Since stationary states of the Schrödinger equation are time-independent (static) solutions, we are faced here with a PDE spatial boundary-value problem. For a 1-d problem, the PDE reduces to an ODE, and there are many methods that give good results. A general approach for higher-dimensional systems is to use the time-dependent numerical techniques of Section B.1, followed by the utilization of theory and more numerical work to extract energy eigenvalues and eigenstates from the numerical time-dependent wavefunction (Section B.2e). This procedure works well for particle evolution in nonsingular potentials. However, in the case of certain special problems—such as those involving the Coulomb potential—time-independent techniques have generally been more useful, at least until very recently.

B.2a. Application of Shooting Methods

The time-independent 1-d Schrödinger equation is just a second-order ODE, which can be solved using shooting methods (Section A.4). One need not reduce the equation to two first-order ODEs (Koonin and Meredith, 1990). For 1-d boundary-value problems, the shooting method is almost always adequate for finding the solution. A straightforward generalization to more dimensions is not available, however, as we would need to shoot *not* to a point in space, but to some higher-dimension object, such as a continuous curve or surface. So we stay in 1-d and consider a particle in a potential well, such as that shown in Fig. 1.3. Our approach to finding energy eigenvalues and eigenstates is to shoot to hit an intermediate fitting point, say, the right-hand turning point r_{out} in Fig. 1.3. We begin with a trial value for the energy $E < 0$ and start the shooting integration far enough to the left of the left-hand turning point r_{in} so that the starting wavefuntion is very close to zero (Fig. 1.4). We then integrate outward from there to the fitting point. We proceed in a similar manner to shoot again to the fitting point, now starting to the right of r_{out} and integrating inward. The iterative shooting procedure of Section A.4 now uses the energy E as the adjustable parameter, and the target condition is to match the derivatives of the two partial wavefunctions at the fitting point. Available shooting programs for a particle in a 1-d potential well include SHOOTF (Press et al., 1992) and EXAMPLE3.FOR (Koonin and Meredith, 1990).

Can the shooting method provide any help in solving a 2-d energy eigenstate problem? At times it can, for we can often do a 2-d basis-set expansion calculation—see the next subsection—where the basis set is the direct product of 1-d basis states for each of the two spatial coordinates. This is useful when the Hamiltonian has the form $H(x, y) = H_x(x) + H_y(y) + V(x, y)$, where the two sets of 1-d basis states are sets of energy eigenstates of H_x and of H_y. Then the shooting method may be used to find an adequate number of 1-d eigenstates for each 1-d basis. As an example, refer to Section 7.3 and the treatment of the nanotechnology problem of a particle within a semiconductor quantum well and in a tilted magnetic field. Shooting methods have been used to find the 1-d basis energy eigenfunctions for the particle motion along the z-coordinate that is perpendicular to the parallel planes of the heterostructure (Muratov et al., 1997).

B.2b. Basis-Set Expansion Calculations Using Natural Basis States

Section 1.2a presented the ideas behind basis-set expansion calculations of energy eigenvalues and eigenstates. A separation of the Hamiltonian $H = H_0 + V$ was the starting point. The key ideas were that (1) the partial Hamiltonian H_0 should be analytically solvable (Section 1.2a), thereby providing the basis set $\{\phi_k\}$ for an expansion of the full wavefunction $\Psi_E = \sum_{k=1}^{N} c_k \phi_k$; and (2) in practice the matrix elements $\langle \phi_j | V | \phi_k \rangle$ should be evaluated analytically, or numerically if relatively few of the matrix elements are nonzero. When both (1) and (2) are true, we call the basis set $\{\phi_k\}$ a *natural basis set.*

Also in Section 1.2a, when the basis set is orthogonal, we found the matrix equation $(\bar{\bar{H}}_0 + \bar{\bar{V}} - E\bar{\bar{I}})\bar{C} = \bar{0}$, where $\bar{\bar{I}}$ is the identity matrix and the energy eigenvalues E_n and corresponding energy eigenfunctions $\bar{C}_n$ remain to be found. Standard numerical direct matrix diagonalization algorithms can solve the related matrix equation $\bar{\bar{H}}\bar{\bar{C}} = \bar{\bar{C}}\bar{\bar{E}}$, where $\bar{\bar{E}}$ is the diagonal energy eigenvalue matrix and $\bar{\bar{C}}$ is the $N \times N$ matrix of energy eigenvectors. We will see that the direct diagonalization algorithms actually use a finite sequence of transformations to generate $\bar{\bar{C}}$ and $\bar{\bar{E}}$. The *n*th column of $\bar{\bar{C}}$,

$$\bar{C}_n = \begin{pmatrix} C_{1n} \\ C_{2n} \\ \vdots \\ C_{Nn} \end{pmatrix},$$

then gives the expansion of the *n*th energy eigenvector in terms of the basis set.

One numerical approach for direct matrix diagonalization proceeds as follows (Koonin and Meredith, 1990; Press et al., 1992). Matrix theory tells us that the symmetric matrix $\bar{\bar{H}}$ is similar (in the matrix theory sense) to a real diagonal matrix. For any real square matrix $\bar{\bar{H}}$, the Householder algorithm prescribes how to find a matrix $\bar{\bar{P}}$ such that $\bar{\bar{P}}\bar{\bar{H}}\bar{\bar{P}} \equiv \bar{\bar{T}}$ is a tridiagonal matrix. The QR theorem states that there is a unique upper triangular matrix $\bar{\bar{R}}$ and an orthogonal matrix $\bar{\bar{Q}}$ such that $\bar{\bar{T}} = \bar{\bar{Q}}\bar{\bar{R}}$. So we use a QR algorithm to convert $\bar{\bar{T}}$ to $\bar{\bar{R}}$, with the diagonal elements of $\bar{\bar{R}}$ then being the energy eigenvalues. Inverse vector iteration then gives the matrix $\bar{\bar{C}}$. A discussion of various "canned" Fortran routines for carrying out this procedure, including the popular EISPACK routines, can be found in Press et al. (1992).

It is important to realize that one can find natural basis sets that are not eigenstates of a partial Hamiltonian—but yet the sets span the appropriate Hilbert space. For example, in the case of 3-d perturbed hydrogen atom problems, we can expand in Sturmian radial functions (and in the usual spherical harmonics), rather than using the ordinary hydrogen atom radial wavefunctions. By definition, the *spherical Sturmian (radial) functions* $S_k(r) \equiv S_{n\ell}(r)$ satisfy the differential equation (in atomic units)

$$\left(-\frac{1}{2}\frac{d^2}{dr^2} + \frac{\ell(\ell+1)}{2r^2} - \frac{\alpha_k}{r}\right)S_k(r) = ES_k(r). \tag{B.28}$$

Here the boundary conditions are (1) $S_k(0) = 0$, and (2) $S_k(r)$ decays exponentially for sufficiently large r (Rotenberg, 1962; Halley et al., 1993). Equation (B.28) is similar to the radial Schrödinger equation, but the energy E appears as a fixed parameter. Instead of E, the effective charge α_k acts as the eigenvalue. The spherical Sturmian functions are not solutions of the radial Schrödinger equation, yet we can expand in these functions as a step in a numerical Hamiltonian matrix diagonalization algorithm. The Sturmians do constitute a natural basis (Buchleitner et al., 1994, 1995).

The primary difference between the hydrogen atom radial functions and the Sturmian radial functions lies in the scaling of the radial variable r. The scaled radial variable for the atom functions is $2r/n$, while for the Sturmian functions it is ηr, with $\eta \equiv \sqrt{(-8E)}$. Thus the scaled variable for the atom functions scales with the principal quantum number n, while the Sturmian functions have fixed radial scaling. The parameter η may assume any single positive value without affecting the completeness and orthogonality properties of the Sturmians. The value $\eta = 2$ $(E = -\frac{1}{2})$ was chosen by Rotenberg, but for calculations requiring many basis states, η should be chosen for convergence and will often be considerably smaller than unity (Clark and Taylor, 1982).

The Sturmian functions are valuable because they constitute an infinite, discrete and complete set of states, a basis set that contains no continuum! In the Sturmian basis, each Sturmian state contains a part of the atomic continuum (Clark and Taylor, 1982). This fact is very advantageous for basis-set expansion calculations. The Sturmian states $|k\rangle$ are "orthogonal"—but with the potential energy r^{-1} as a weighting function

$$\langle k|r^{-1}|k'\rangle = \frac{\alpha_{k'}}{n^2}\delta_{kk'}.$$

If we expand in Sturmian functions and if there is a perturbing Hamiltonian that is polynomial in the coordinates, then analytically calculable banded Hamiltonian matrices will follow. Thus the diamagnetic Kepler problem (Section 7.1) and the sinusoidally driven Kepler problem (Section 8.3) are often treated using Sturmian basis-set expansion calculations (Buchleitner et al., 1994, 1995).

We know that the hydrogen atom problem can be separated and solved using several different spatial coordinate systems—spherical, cylindrical, parabolic, and semiparabolic (Section 7.1a). Similarly, several different useful Sturmian basis sets have been explored (Delande and Gay, 1987). If one chooses a Sturmian basis tailored to the particular wavefunction of interest, computer time can be reduced by a factor of 20, thus enabling calculations to be carried out on a personal computer rather than a supercomputer (Delande and Gay, 1986). Wunner et al. (1986) used 70 radial Sturmian states and 70 spherical harmonics to treat the diamagnetic Kepler problem. Comparisons with calculations carried out other ways indicated that the energy eigenvalues were accurate to eight significant figures, and electric dipole matrix elements can be good to order 1%.

Lastly, considering the diamagnetic Kepler problem once again—but now in semiparabolic coordinates, Wintgen and Friedrich (1986) showed that different scaled Hamiltonians prove useful for the three different energy regions that are below, near, and above the ionization limit of the hydrogen atom. As alternatives to the Sturmian basis-set expansion calculations, their calculations utilized harmonic oscillator basis sets; see also Merani et al. (1995).

B.2c. The Complex Coordinate Rotation Method (Advanced Topic)

Many physical systems have a range of energies (or quasienergies) where the system can come apart as time evolves onward. In such a regime, there is a continuous energy spectrum associated with the unbound motion represented by continuum-energy eigenstates. Often there exist values of energy where the unbound motion is delayed by a transient tendency for the system to stay together. Then one sees resonance peaks in the energy spectrum with definite heights and widths. A *resonance state* is characterized by a complex energy $E_c = E + i\Gamma/2$, where Γ is the decay rate of the state.

For the calculation of resonances, the *complex coordinate rotation (CCR) method* has proved to be a useful numerical tool. This method defines a rotated non-Hermitian Hamiltonian $H(\theta)$—obtained from the original Hermitian Hamiltonian H—by a "dilatation" analytic transformation. As an example, consider the diamagnetic Kepler problem, where H is given by Eq. (7.2). The *complex rotated Hamiltonian* $H(\theta)$ is obtained by making both the position and momentum complex,

$$\mathbf{r} \to \mathbf{r}\exp(i\theta), \qquad \mathbf{p} \to \mathbf{p}\exp(-i\theta), \tag{B.29}$$

where θ is a real parameter called the "rotation angle." Thus we have

$$H(\theta) = \tfrac{1}{2}p^2\exp(-2i\theta) - \frac{\exp(-i\theta)}{r} + \tfrac{1}{2}\gamma L_z + \tfrac{1}{8}\gamma^2(x^2 + y^2)\exp(2i\theta). \tag{B.30}$$

The quantity $H(\theta)$ can formally be written in terms of the complex rotation operator $R(\theta)$:

$$R(\theta) = \exp\left(-\theta\frac{(\mathbf{r}\cdot\boldsymbol{p}+\boldsymbol{p}\cdot\mathbf{r})}{2}\right),$$
$$H(\theta) = R(\theta)HR(-\theta). \tag{B.31}$$

In this transformation, the canonical commutation relations between components of $\mathbf{r}$ and $\boldsymbol{p}$ are preserved. The key properties of $H(\theta)$ are (Reinhardt, 1982; Ho, 1983; Buchleitner et al., 1994):

- The bound (discrete) spectra of $H(\theta)$ and H coincide below the continuum (i.e., below the first ionization threshold).
- The continua are rotated by the angle 2θ into the lower half of the complex energy plane, about branching points.
- The resonances of H coincide with the complex eigenvalues of $H(\theta)$. The quantities E and Γ are independent of θ, *provided* that the rotation of the continua has uncovered the resonances.

Because of this third property, matrix diagonalization methods using natural basis sets can be transferred directly to states in the continuum, the differences being that the basis functions now depend on complex-rotated coordinates, and that the Hamiltonian matrix becomes a complex-valued symmetric matrix. Figure B.4 shows an example of a calculated partial spectrum of the rotated Hamiltonian (B.30). For these results, the matrix to be diagonalized was constructed using the spherical Sturmian radial basis states that satisfy Eq. (B.28). For the 3-d hydrogen atom, the rotated continua are degenerate—because of the angular momentum ℓ. This degeneracy is removed in a truncated basis, and instead of just a single line, one finds several approximately straight-line segments of complex eigenvalues. An alternative to the Sturmian approach is to employ the CCR method in semiparabolic coordinates and introduce the harmonic oscillator states (Merani et al., 1995).

If one carries out the matrix diagonalization of $H(\theta)$ for at least a few different values of θ in the range $0 \leq \theta \leq \pi/4$, the different resonances can be clearly identified. Note also that because of basis-set truncation, unphysical amplitudes at the top of the basis must be carefully truncated away (Buchleitner et al., 1994). This truncation is necessary because the diagonal matrix elements $\langle n\ell m|R(\theta)|n\ell m\rangle$ scale roughly as $\exp(n\theta)$, a consequence of using exponentially decreasing Sturmian functions to represent resonance wavefunctions that cannot be normalized.

The CCR method can be used to obtain Floquet states of the driven Kepler problem (Buchleitner et al., 1995). The time variable should not be complexified.

Speaking more generally, only for certain classes of Hamiltonians are there theorems that rigorously justify using the CCR method. Before using the method with confidence, we must consult the mathematical literature (Balslev and Combes, 1971; Simon, 1973; Yajima, 1982; Graffi et al., 1985).

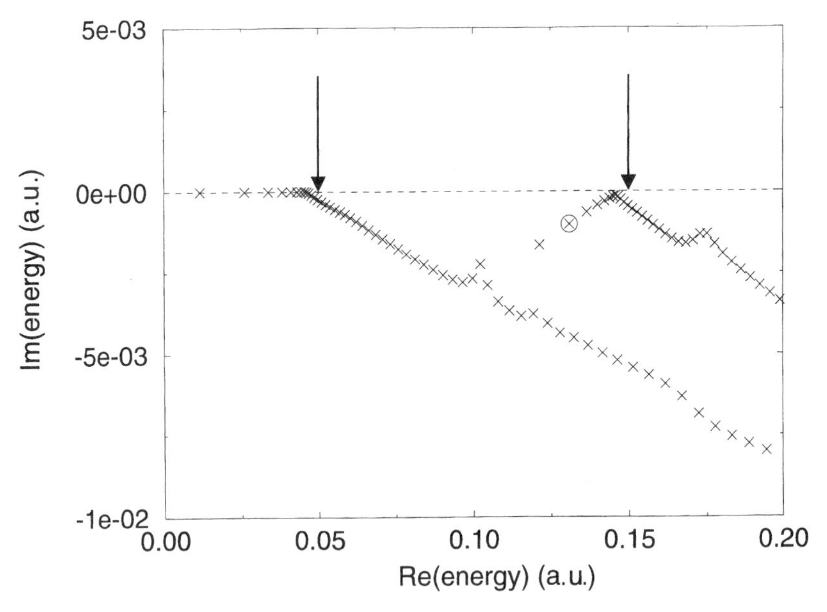

Figure B.4. A partial spectrum of a complex rotated Hamiltonian $H(\theta)$ for the diamagnetic Kepler problem; see Eq. (B.30). The Hamiltonian matrix is truncated and diagonalized in a finite basis, so that the continuum appears as sets of discrete eigenvalues in the complex energy plane—sets that lie approximately on straight lines rotated about threshold points on the positive real energy axis. Other points represent resonances as complex energy eigenvalues. From Buchleitner et al. (1994). Copyright 1994 Institute of Physics Publishing.

B.2d. The Discrete-Variable Global Spatial-Grid Representation (Advanced Topic)

Closely related to the Fourier spatial-grid method (Section B.1b.1) is a method for representing certain Hamiltonian matrices—a method called the *discrete-variable spatial-grid representation* (DVR). The DVR is an approximate discretized coordinate-space representation, where the potential energy matrix $\bar{\bar{\text{V}}}$ is diagonal and the kinetic energy matrix $\bar{\bar{\text{K}}}$ is sparse. A natural basis set must be available, a set that is built up from direct products of solutions of 1-d differential equations that separately involve the kinetic energy terms for each coordinate. The basic assumption of a DVR is that the matrix elements are integrals that can be approximated by quadrature formulas. For discussions of such formulas, see Press et al. (1992) and Section 25.4 of Abramowitz and Stegun (1964).

The basis-set expansion uses a discretized version of the natural basis (rather than the complex exponential basis of the Fourier method), where the discretized basis again is defined to be an orthonormal basis on a set of grid points along each dimension of coordinate space, Eqs. (B.17)–(B.19). The grid points are not equally spaced, but instead are the quadrature points associated with the natural basis. In addition to being orthonormal, the discretized natural basis functions have the notable property that they are zero at all quadrature points except one. The matrix elements of the potential energy become just the values of the potential energy at each grid point; that is, the potential energy matrix is diagonal in the DVR.

For the diamagnetic Kepler problem of Section 7.1b, setting up the DVR is described in Grozdanov et al. (1997). Energy eigenvalues obtained using the DVR Hamiltonian matrix can be accurate to eight significant figures, and dipole matrix elements can be good to at least 0.1% (Grozdanov et al., 1997). For further discussion of the DVR method, see Groenenboom and Colbert (1993).

B.2e. Energies and Eigenfunctions from Time-Dependent Pseudospectral Methods

The split-operator pseudospectral (SPO) technique (Section B.1b.2) was originally proposed as part of a numerical scheme for energies and eigenfunctions (Feit et al., 1982). Actually, the SPO technique was used before that as a numerical propagating-beam method for computing electric fields of normal-mode eigenfunctions in optical waveguides (Feit and Fleck, 1980). In the paraxial-ray approximation, the 2-d Schrödinger equation describes the wavefunction in the spatial dimensions perpendicular to the direction along the waveguide. Suppose we use the SPO method to obtain the time evolution of an initial (wavepacket) quantum state, out to a time T. We then compute the time autocorrelation function—Eq. (2.14)—and Fourier analyze this function to obtain all the energy eigenvalues that we can. The number we get depends on T; the longer T is, the more eigenvalues we get. What the Fourier transform actually gives us, of course, is a set of hopefully sharp (resonant) spectral lines centered near energies E_n, but having finite linewidths. Using lineshape fitting techniques, both the positions and heights of the resonances can be determined accurately. The heights determine the weights of the stationary states that compose the initial state.

Once the eigenvalues are known, the corresponding eigenfunctions are computed by numerically evaluating the integrals

$$\psi(\mathbf{r}, E_n) = \int_0^T \psi(\mathbf{r}, t) w(t) \exp(iE_n t)\, dt, \tag{B.32}$$

where w(t) is an appropriate window function that is needed because of the finite time T over which $\psi(\mathbf{r}, t)$ is known.

BIBLIOGRAPHY

Abramowitz, M., and I. A. Stegun (1964). *Handbook of Mathematical Functions*, National Bureau of Standards, Washington, DC. Reprinted 1968 by Dover Publications, New York.

Agam, O., and S. Fishman (1994). *Phys. Rev. Lett.* **73**, 806.

Alekseev, K. N., and G. P. Berman (1994). *JETP* **78**, 296.

Almeida, R., and H. Metiu (1988). *Chem. Phys. Lett.* **146**, 47.

Anderson, P. W., B. I. Halpern, and C. M. Varma (1972). *Philos. Mag.* **25**, 1.

Andreev, A. V., O. Agam, B. D. Simons, and B. L. Altshuler (1996). *Phys. Rev. Lett.* **76**, 3947.

Aranson, I., B. Meerson, and T. Tajima (1992). *Phys. Rev. A* **45**, 7500.

Arnold, V. I. (1967). *Funct. Anal. Appl.* **1**, 1.

Arnold, V. I., and A. Avez (1968). *Ergodic Problems of Classical Mechanics*, Benjamin, New York.

Arranz, F. J., F. Borondo, and R. M. Benito (1996). *Phys. Rev. E* **54**, 2458.

Auerbach, A., and S. Kivelson (1985). *Nucl. Phys. B* **257**, 799.

Autler, S. H., and C. H. Townes (1955). *Phys. Rev.* **100**, 703.

Averbukh, I. Sh., and N. F. Perel'man (1990). *Sov. Phys. JETP* **69**, 464.

Avron, J. E., R. Seiler, and L. G. Yaffe (1987). *Commun. Math. Phys.* **110**, 33.

Ballüder, K., J. A. Scales, C. Schroter, and M. L. Smith (1996). *Comput. Phys.* **10**, 17.

Balslev, E., and J. M. Combes (1971). *Commun. Math. Phys.* **22**, 280.

Ban, M. (1996). *Int. J. Theoret. Phys.* **35**, 1947.

Band, Y. B., and O. Magnes (1994). *Phys. Rev. A* **50**, 584.

Baranger, M., K. T. R. Davies, and J. H. Mahoney (1988). *Ann. Phys. (NY)* **186**, 95.

Bardsley, J. N., B. Sundaram, L. A. Pinnaduwage, and J. E. Bayfield (1986). *Phys. Rev. Lett.* **56**, 1007.

Bardsley, J. N., A. Szöke, and M. J. Comella (1988). *J. Phys. B* **21**, 3899.

Bayfield, J. E. (1987). In *Quantum Measurement and Chaos*, E. R. Pike and S. Sarkar (eds.), Plenum, New York, p. 1.

Bayfield, J. E., and D. W. Sokol (1988). *Phys. Rev. Lett.* **61**, 2007.

Bayfield, J. E., E. E. Nikitin, and A. I. Reznikov (1973). *Chem. Phys. Lett.* **19**, 471.

Bayfield, J. E., L. D. Gardner, Y. Z. Gulkok, and S. D. Sharma (1981). *Phys. Rev. A* **24**, 138.

Bayfield, J. E., S. Y. Luie, L. C. Perotti, and M. P. Skrzypkowski (1995). *Physica D* **83**, 46.

Bayfield, J. E., S. Y. Luie, and L. C. Perotti (1998). Unpublished.

Beenacker, C. W. J. (1991). *Phys. Rev. B* **44**, 1646.

Bellerman, M. R. W., P. M. Koch, D. R. Mariani, and D. Richards (1996). *Phys. Rev. Lett.* **76**, 892.

Bender, C. M., F. Cooper, J. E. O'Dell, and L. M. Simmons, Jr. (1985). *Phys. Rev. Lett.* **55**, 901.

Benettin, G., L. Galgani, and J.-M. Strelcyn (1976). *Phys. Rev. A* **14**, 2338.

Benettin, G., L. Galgani, A. Giorgilli, and J.-M. Strelcyn (1980). *Meccanica* **15**, 9.

Benkadda, S., Y. Elskens, B. Ragot, and J. T. Mendoça (1994). *Phys. Rev. Lett.* **72**, 2859.

Bensch, F., H. J. Korsch, B. Mirbach, and N. Ben-Tal (1992). *J. Phys. A* **25**, 6761.

Benvenuto, F., G. Casati, and D. L. Shepelyansky (1997). *Phys. Rev. A* **55**, 1732.

Berman, G. P., O. F. Vlasova, and F. M. Izrailev (1987). *Sov. Phys. JETP* **66**, 269.

Berry, M. V. (1978). In *Topics in Nonlinear Dynamics*, S. Jorna (ed.), Amer. Inst. Phys. Conf. Proc. **46**, p. 16.

Berry, M. V. (1981). *Eur. J. Phys.* **2**, 91.

Berry, M. V. (1981). In *Singularities in Waves and Rays*, R. Balian, M. Kleman, and J. P. Poirier (eds.), North Holland, Amsterdam, p. 453.

Berry, M. V. (1985). *Proc. R. Soc. London A* **400**, 229.

Berry, M. V., and K. E. Mount (1972). *Rep. Prog. Phys.* **35**, 315.

Berry, M. V., and M. Tabor (1976). *Proc. R. Soc. London A* **349**, 101.

Berry, M. V., and M. Tabor (1977a). *Proc. R. Soc. London A* **356**, 375.

Berry, M. V., and M. Tabor (1977b). *J. Phys. A* **10**, 371.

Bersons, I. Ya. (1983). *Sov. Phys. JETP* **58**, 40.

Bethe, H. A., and E. E. Salpeter (1977). *Quantum Mechanics of One- and Two-Electron Atoms*, Plenum, New York.

Bird, J. P., K. Ishibashi, D. K. Ferry, Y. Ochiai, Y. Aoyagi, and T. Sugano (1995). *Phys. Rev. B* **51**, 18037.

Birkhoff, G. D. (1927). *Dynamical Systems*, American Mathematical Society, Providence, RI.

Bixon, M., and J. Jortner (1968). *J. Chem. Phys.* **48**, 715.

Blumel, R., and U. Smilansky (1989). *Phys. Scr.* **40**, 386.

Bogoliubov, N. N., and Y. A. Mitropolsky (1961). *Asymptotic Methods in the Theory of Nonlinear Oscillations*, Gordon and Breach, New York.

Bogomolny, E. B. (1988). *Physica D* **31**, 169.

Bohigas, O., S. Tomsovic, and D. Ullmo (1993). *Phys. Rep.* **223**, 43.

Bohm, D. (1951). *Quantum Theory*, Prentice-Hall, Englewood Cliffs, NJ.

Bohr, N. (1961). *Atomic Theory and the Description of Nature*, The University Press, Cambridge, England.

Bonvalet, A., J. Nagle, V. Berger, A. Migus, J.-L. Martin, and M. Joffre (1996). *Phys. Rev. Lett.* **76**, 4392.

Boris, S. D., S. Brandt, H. D. Dahmen, T. Stroh, and M. L. Larson (1993). *Phys. Rev. A* **48**, 2574.

Born, M. (1960). *The Mechanics of the Atom*, Ungar, New York.

Brandt, S., and H. D. Dahmen (1994). *Quantum Mechanics on the Personal Computer*, 3rd ed., Springer-Verlag, Berlin.

Brandt, S., and H. D. Dahmen (1995). *The Picture Book of Quantum Mechanics*, 2nd ed., Springer-Verlag, New York.

Braun, M., C. Meier, and V. Engel (1996). *Comput. Phys. Commun.* **93**, 152.

Breuer, H. P., and M. Holthaus (1989). *Phys. Lett. A* **140**, 507.

Breuer, H. P., and M. Holthaus (1991). *Ann. Phys. (NY)* **211**, 249.

Broeckhove, J., B. Feyen, L. Lathouwers, F. Arickx, and P. Van Leuven (1990). *Chem. Phys. Lett.* **174**, 504.

Broeckhove, J., B. Feyen, L. Lathouwers, and P. Van Leuven (1992). In *Time-Dependent Quantum Molecular Dynamics*, J. Broeckhove and L. Lathouwers (eds.), Plenum, New York, p. 391.

Broers, B., H. B. van Linden van den Heuvell, and L. D. Noordam (1992). *Phys. Rev. Lett,* **69**, 2062.

Bryson, A., and Y. Ho (1975, 1988). *Applied Optimal Control Theory*, Hemisphere Publishing, Washington, DC.

Buchleitner, A., and D. Delande (1995). *Chaos, Solitons & Fractals* **5**, 1125.

Buchleitner, A., and D. Delande (1997). *Phys. Rev. A* **55**, R1585.

Buchleitner, A., B. Grémaud, and D. Delande (1994). *J. Phys. B* **27**, 2663.

Buchleitner, A., D. Delande, and J. C. Gay (1995). *J. Opt. Soc. Am. B.* **12**, 505.

Burgnies, L., O. Vanbésien, and D. Lippens (1997). *Appl. Phys. Lett.* **71**, 803.

Campolieti, G., and P. Brumer (1994). *Phys. Rev. A* **50**, 997.

Cárdenas, A. E., and R. D. Coalson (1997). *Chem. Phys. Lett.* **265**, 71.

Carnegie, A., and I. C. Percival (1984). *J. Phys. A* **17**, 801.

Cary, J. R., and R. T. Skodje (1988). *Phys. Rev. Lett.* **61**, 1795.

Cary, J. R., and R. T. Skokje (1989). *Physica D* **36**, 287.

Cary, J. R., D. F. Escande, and J. L. Tennyson (1986). *Phys. Rev. A* **34**, 4256.

Cary, J. R., P. Rusu, and R. T. Skodje (1987). *Phys. Rev. Lett.* **58**, 292.

Casati, G., and B. Chirikov, eds. (1995). *Quantum Chaos: Between Order and Disorder*, Cambridge University Press, Cambridge, England.

Casati, G., B. V. Chirikov, D. L. Shepelyanski, and I. Guarneri (1987). *Phys. Rep.* **154**, 77.

Chang, A. M., H. U. Baranger, L. N. Pfeiffer, and K. W. West (1994). *Phys. Rev. Lett.* **73**, 2111.

Chang, A. M., H. U. Baranger, L. N. Pfeiffer, K. W. West, and T. Y. Chang (1996). *Phys. Rev. Lett.* **76**, 1695.

Chelkowski, S., and A. D. Bandrauk (1993). *J. Chem. Phys.* **99**, 4279.

Cheklowski, S., and G. N. Gibson (1995). *Phys. Rev. A* **52**, R3417.

Cheklowski, S., A. D. Bandrauk, and P. B. Corkum (1990). *Phys. Rev. Lett.* **65**, 2355.

Chirikov, B. V. (1979). *Phys. Rep.* **52**, 265.

Christoffel, K. M., and P. Brumer (1986). *Phys. Rev. A* **33**, 1309.

Chu, S.-I., and T.-F. Jiang (1991). *Comp. Phys. Commun.* **63**, 482.

Cimiraglia, R. (1992). In *Time-Dependent Quantum Molecular Dynamics*, J. Broeckhove and L. Lathouwers (eds.), Plenum, New York, p. 11.

Clark, C. W., and K. T. Taylor (1982). *J. Phys. B* **15**, 1175.

Clarke, R. M., I. H. Chan, C. M. Marcus, C. I. Durüoz, J. S. Harris, Jr., K. Campman, and A. C. Gossard (1995). *Phys. Rev. B* **52**, 2656.

Coalson, Rob D. (1996). *J. Phys. Chem.* **100**, 7896.

Cohen, G., and B. Meerson (1993). *Phys. Rev. E* **47**, 967.

Cohen-Tannoudji, C., B. Diu, and F. Laloë (1977). *Quantum Mechanics*, Wiley, New York, p. 420.

Conner, J. N. L. (1969). *Chem. Phys. Lett.* **4**, 419.

Conolly, S., D. Nishimara, and Q. Maeovski (1986). *IEEE Med. Im.* **5**, 106.

Courtney, M., N. Spellmeyer, H. Jiae, and D. Kleppner (1995). *Phys. Rev. A* **51**, 3604.

Crespi, B., G. Perez, and S.-J. Chang (1989). Preprint No. ILL-(TH)-89-#56; B. Crespi, Ph. D. thesis, University of Illinois at Urbana-Champaign, 1990, available from the University of Illinois Library.

Crespi, B., G. Perez, and S.-J. Chang (1993). *Phys. Rev. E* **47**, 986.

Cycon, H. L., R. G. Froese, W. Kirsch, and B. Simon (1987). *Schrödinger Operators, with Application to Quantum Mechanics and Global Symmetry*, Springer-Verlag, Berlin.

Dahlquist, G. (1956). *Math. Scand.* **4**, 33.

Davis, M. J., and E. J. Heller (1979). *J. Chem. Phys.* **71**, 3383.

Davis, M. J., and E. J. Heller (1981a). *J. Chem. Phys.* **75**, 246.

Davis, M. J., and E. J. Heller (1981b). *J. Chem. Phys.* **75**, 3916.

Davis, M. J., E. B. Stechel, and E. J. Heller (1980). *Chem. Phys. Lett.* **76**, 21.

Davydov, A. S. (1976). *Quantum Mechanics*, 2nd ed., Pergamon, New York.

Delande, D., and J. C. Gay (1986). *J. Phys. B* **19**, L173.

Delande, D., and J. C. Gay (1987). *Phys. Rev. Lett.* **59**, 1809.

Delos, J. B., S. K. Knudson, and D. W. Noid (1983a). *Phys. Rev. A* **28**, 7.

Delos, J. B., S. K. Knudson, and D. W. Noid (1983b). *Phys. Rev. Lett.* **50**, 579.

Delos, J. B., S. K. Knudson, and D. W. Noid (1984). *Phys. Rev. A* **30**, 1208.

Demkov, Yu. N., and V. I. Osherov (1968). *Sov. Phys. JETP* **26**, 916.

DeVogelaere, R. (1958). In *Contributions to the Theory of Nonlinear Oscillations*, S. Lefschotz (ed.), Princeton University Press, Princeton, NJ, p. 54.

DeVries, P. L. (1994). *A First Course in Computational Physics*, Wiley, New York.

Dietz, K., J. Henkel, and M. Holthaus (1992). *Phys. Rev. A* **45**, 4960.

Dittrich, T., and U. Smilansky (1995). In *Quantum Chaos: Between Order and Disorder*, G. Casati and B. Chirikov (eds.), Cambridge University Press, Cambridge, England, p. 605.

Dittrich, W., and M. Reuter (1994). *Classical and Quantum Dynamics: From Classical Paths to Path Integrals*, 2nd ed., Springer-Verlag, New York.

Doron, E., and S. D. Frischat (1995). *Phys. Rev. Lett.* **75**, 3661.

Dumas, H. S., and J. Laskar (1993). *Phys. Rev. Lett.* **70**, 2975.

Eaves, L. (1990). In *Electronic Properties of Multilayers and Low-Dimensional Semiconductor Structures*, J. M. Chamberland, L. Eaves, and J.-C. Portal (eds.), Plenum, New York, p. 243.

Eberly, J., N. Narozhny, and J. Sanchez-Mondragon (1980). *Phys. Rev. Lett.* **44**, 1323.

Eckhardt, B. (1993). In *Quantum Chaos*, G. Casati, I. Guarneri, and U. Smilansky (eds.), North Holland, Amsterdam, p. 77.

Eckhardt, B., G. Hose, and E. Pollak (1989). *Phys. Rev. A* **39**, 3776.

Eckhardt, B., S. Fishman, K. Müller, and D. Wintgen (1992). *Phys. Rev. A* **45**, 3531.

Ehrenfest, P. (1917). *Philos. Mag.* **33**, 500.

Ehrenfest, P. (1923). *Naturwissenshaftung* **11**, 543.

Einstein, A. (1917). *Vehr. Deutsch. Phys. Geis.* **19**, 82.

Eiselt, J., and H. Risken (1991). *Phys. Rev. A* **43**, 346.

El Ghafar, M., P. Törmä, V. Savichev, E. Mayr, A. Zeiler, and W. P. Schleich (1997). *Phys. Rev. Lett.* **78**, 4181.

Evans, D. G., and R. D. Coalson (1995). *J. Chem. Phys.* **104**, 3598.

Farrelly, D. (1986). *J. Chem. Phys.* **85**, 2119.

Fattal, E., R. Baer, and R. Kosloff (1997). *Phys. Rev. E* **53**, 1217.

Fehlberg, H. (1968). *NASA TR R-287*, Huntsville, Alabama.

Feingold, M. (1985). *Phys. Rev. Lett.* **55**, 2626.

Feit, M. D., and J. A. Fleck, Jr. (1980). *Appl. Opt.* **19**, 1154.

Feit, M. D., J. A. Fleck, Jr., and A. Steiger (1982). *J. Comput. Phys.* **47**, 412.

Feldmann, J., T. Meier, G. von Plessen, M. Koch, E. O. Göbel, P. Thomas, G. Bacher, C. Hartmann, H. Schweizer, W. Schafer, and H. Nickel (1993). *Phys. Rev. Lett.* **70**, 3027.

Fernández, F. M., and D. A. Micha (1992). *J. Chem. Phys.* **97**, 8173.

Ferry, D. K., and S. M. Goodnick (1997). *Transport in Nanostructures*, Cambridge University Press, Cambridge, England.

Feynman, R. P., and A. R. Hibbs (1965). *Quantum Mechanics and Path Integrals*, McGraw-Hill, New York.

Feynman, R. P., F. L. Vernon, Jr., and R. W. Hellwarth (1957). *J. Appl. Phys.* **28**, 49.

Floquet, M. G. (1883), *Ann. Ecole Norm. Sup.* **12**, 47.

Fogarty, A., T. M. Fromhold, L. Eaves, F. W. Sheard, M. Henini, T. J. Foster, P. C. Main, and G. Hill (1994). *Superlattices and Microstructures* **15**, 287.

Folk, J. A., S. R. Patel, S. F. Godijn, A. G. Huibers, S. M. Cronenwett, C. M. Marcus, K. Campman, and A. C. Gossard (1996). *Phys. Rev. Lett.* **76**, 1699.

Ford, J., S. D. Stoddard, and J. S. Turner (1973). *Prog. Theor. Phys.* **50**, 1574.

Frakenfelder, H., F. Parak, and R. D. Young (1988). *Annu. Rev. Biophys. Chem.* **17**, 451.

Frederick, J. H., E. J. Heller, J. L. Ozment, and D. W. Pratt (1988). *J. Chem. Phys.* **88**, 2169.

Friedland, L. (1997). *Phys. Rev. E* **55**, 1929.

Friedrich, H., and J. Trost (1996). *Phys. Rev. Lett.* **76**, 4869.

Friedrich, H., and D. Wintgen (1989). *Phys. Rep.* **183**, 37.

Fröman, N., and P. O. Fröman (1965). *JWKB Approximation*, North Holland, Amsterdam.

Fromhold, T. M., P. B. Wilkinson, F. W. Sheard, L. Eaves, J. Miao, and G. Edwards (1995). *Phys. Rev. Lett.* **75**, 1142.

Gaeta, Z. D., M. W. Noel, and C. R. Stroud, Jr. (1994). *Phys. Rev. Lett.* **73**, 636.

Galbraith, I., Y. S. Ching, and E. Abraham (1984). *Am. J. Phys.* **52**, 60, see the appendix.

Galvez, E. J., B. E. Sauer, L. Moorman, P. M. Koch, and D. Richards (1988). *Phys. Rev. Lett.* **61**, 2011.

Gao, J., and J. B. Delos (1994). *Phys. Rev. A* **49**, 869.

Gaspard, P., S. A. Rice, and K. Nakamura (1989). *Phys. Rev. Lett.* **63**, 930.

Gaubatz, U., P. Rudecki, S. Schiemann, and K. Bergmann (1990). *J. Chem. Phys.* **92**, 5363.

Gautschi, W. (1997). *Numerical Analysis: An Introduction*, Birkhäuser, Boston.

Gear, C. W. (1971). *Numerical Initial Value Problems in Ordinary Differential Equations*, Prentice-Hall, Englewood Cliffs, NJ.

Geerligs, L. J., C. J. M. P. Harmans, and L. P. Kouwenhoven, eds. (1993). *The Physics of Few Electron Nanostructures*, North Holland, Amsterdam.

Gentile, T. R., B. J. Hughes, D. Kleppner, and T. W. Ducas (1989). *Phys. Rev. A* **40**, 5103.

Goldberg, A., H. M. Schey, and J. L. Schwartz (1967). *Am. J. Phys.* **35**, 177.

Goldstein, H. (1980). *Classical Mechanics*, 2nd ed., Addison-Wesley, Reading, MA.

Grabert, H., and H. Wipf (1990). In *Festkörperprobleme—Advances in Solid State Physics*, Vol. 30, Vieweg, Braunschweig, Germany, p. 1.

Gräf, H.-D., H. L. Harney, H. Lengeler, C. H. Lewenkopf, C. Rangacharyulu, A. Richter, P. Schardt, and H. A. Weidenmüller (1992). *Phys. Rev. Lett.* **69**, 1296.

Graffi, S., V. Grecchi, and H. J. Silverstone (1985). *Ann. Inst. H. Poincaré* ***42***, 215.

Graham, R., and M. Höhnerbach (1984). *Z. Phys. B* **57**, 233.

Graham, R., and M. Höhnerbach (1992). *Phys. Rev. A* **45**, 5078.

Gray, S. K. (1983). *Chem. Phys.* **75**, 67.

Grebogi, C., S. M. Hammel, J. A. Yorke, and T. Sauer (1990). *Phys. Rev. Lett.* **65**, 1527.

Greenberg, W. R., A. Klein, and C.-T. Li (1995). *Phys. Rev. Lett.* **75**, 1244.

Greene, J. M. (1968). *J. Math. Phys.* **9**, 760.

Greene, J. M. (1979a). *J. Math. Phys.* **20**, 1183.

Greene, J. M. (1979b). In *Nonlinear Dynamics and the Beam-Beam Interaction*, M. Month and J. C. Herrera (eds.), AIP Conf. Proc. No. 57, American Institute of Physics, New York, p. 257.

Greiner, W. (1992). *Quantum Mechanics: An Introduction*, 2nd ed., Springer-Verlag, New York.

Groenenboom, G. C., and D. T. Colbert (1993). *J. Chem. Phys.* **99**, 9681.

Grosdanov, T. P., L. Andric, C. Manescu, and R. McCarroll (1997). *Phys. Rev. A* **56**, 1865.

Grossmann, F., P. Jung, T. Dittrich, and P. Hänggi (1991a). *Z. Phys. B* **84**, 315.

Grossmann, F., T. Dittrich, P. Jung, and P. Hänggi (1991b). *Phys. Rev. Lett.* **67**, 516.

Guérin, S. (1997). *Phys. Rev. A* **56**, 1458.

Gutzwiller, M. C. (1967). *J. Math. Phys.* **8**, 1979.

Gutzwiller, M. C. (1969). *J. Math. Phys.* **10**, 1004.

Gutzwiller, M. C. (1970). *J. Math. Phys.* **11**, 1791.

Gutzwiller, M. C. (1971). *J. Math. Phys.* **12**, 343.

Gutzwiller, M. C. (1982). *Physica D* **5**, 183.

Gutzwiller, M. C. (1990). *Chaos in Classical and Quantum Mechanics*, Springer-Verlag, New York.

Haake, F. (1991). *Quantum Signatures of Chaos*, Springer-Verlag, Berlin.

Halley, M. H., D. Delande, and K. T. Taylor (1993). *J. Phys. B* **26**, 1775.

Harriman, J. E. (1994). *J. Chem. Phys.* **100**, 3651.

Hasegawa, H., M. Robnik, and G. Wunner (1989). *Prog. Theor. Phys. Suppl.* **98**, 198.

Heller, E. J. (1981). *Acc. Chem. Res.* **14**, 368.

Heller, E. J. (1984). *Phys. Rev. Lett.* **53**, 1515.

Heller, E. J. (1995). *J. Phys. Chem.* **99**, 2625.

Heller, E. J., P. W. O'Connor, and J. Gehlen (1989). *Phys. Scr.* **40**, 354.

Henkel, J., and M. Holthaus (1992). *Phys. Rev. A* **45**, 1978.

Hénon, M., and C. Heiles (1964). *Astron. J.* **69**, 73.

Herring, C. (1962). *Rev. Mod. Phys.* **34**, 631.

Hilliary, M., R. F. O'Connell, M. O. Scully, and E. P. Wigner (1984). *Phys. Rep.* **106**, 121.

Ho, Y. K. (1983). *Phys. Rep.* **99**, 1.

Holle, A., G. Wiebusch, J. Main, B. Hager, H. Rottke, and K. H. Welge (1986). *Phys. Rev. Lett.* **56**, 2594.

Holle, A., G. Wiebusch, J. Main, K. H. Welge, G. Zeller, G. Wunner, T. Ertl, and H. Ruder (1987). *Z. Phys. D* **5**, 279.

Holle, A., J. Main, G. Wiebusch, H. Rottke, and K. H. Welge (1988). *Phys. Rev. Lett.* **61**, 161.

Holthaus, M. (1994). *Prog. Theor. Phys. Suppl.* **116**, 417.

Holthaus, M. (1995). *Chaos, Solitons & Fractals* **5**, 1143.

Holthaus, M., and M. E. Flatté (1994). *Phys. Lett. A* **187**, 151.

Holthaus, M., and B. Just (1994). *Phys. Rev. A* **49**, 1950.

Holtzclaw, K. W., and D. W. Pratt (1986). *J. Chem. Phys.* **84**, 4713.

Howland, J. S. (1980). In *Mathematical Methods and Applications of Scattering Theory*, J. A. DeSanto, A. W. Saenz, and W. W. Zachary (eds.), Springer-Verlag, Berlin, p. 163.

Huang, Z. H., T. E. Feuchtwang, P. H. Cutler, and E. Kazes (1990). *Phys. Rev. A* **41**, 32.

Hund, F. (1927). *Z. Phys.* **43**, 803.

Hutchinson, J. S. (1996). In *Dynamics of Molecules and Chemical Reactions*, R. E. Wyatt and J. Z. H. Zhang (eds.), Marcel Dekker, New York, p. 561.

Jaynes, E. T., and F. W. Cummings (1963). *Proc. IEEE* **51**, 126.

Jordan, D. W., and P. Smith (1987). *Nonlinear Ordinary Differential Equations*, 2nd ed., Clarendon Press, Oxford, p. 245.

Kastner, M. A. (1993). *Rev. Mod. Phys.* **64**, 849.

Kato, T. (1950). *J. Phys. Soc. Jpn.* **5**, 435.

Kay, K. G. (1994). *J. Chem. Phys.* **101**, 2250.

Keating, J. P., and M. V. Berry (1987). *J. Phys. A* **20**, L1139.

Keller, J. B. (1958a). *Ann. Phys. (NY)* **4**, 180.

Keller, J. B. (1958b). In *Calculus of Variations and Its Applications*, McGraw-Hill, New York; also *Proc. Symp. Appl. Math.* **8**, 27.

Kemble, F. C. (1937). *Fundamental Principles of Quantum Mechanics*, McGraw-Hill, New York. Reprinted 1958 by Dover Publications, New York.

Kilin, S. Ya., P. R. Berman, and T. M. Maevskaya (1996). *Phys. Rev. Lett.* **76**, 3297.

Klafter, J., G. Zumofen, and M. F. Schlesinger (1995). In *Chaos—The Interplay Between Stochastic and Deterministic Behavior*, P. Garbaczewski, M. Wolf, and A. Weron (eds.), Springer-Verlag, Berlin, p. 183.

Klauder, J. R. (1987a). *Phys. Rev. Lett.* **59**, 748.

Klauder, J. R. (1987b). *Ann. Phys. (NY)* **180**, 108.

Klauder, J. R., and B. S. Skagerstam (1985). *Coherent States*, World Scientific, Singapore.

Kolos, W. (1970). *Adv. Quant. Chem.* **5**, 99.

Kono, H., A. Kita, Y. Ohtsuki, and Y. Fujimura (1997). *J. Comput. Phys.* **130**, 148.

Koonin, S. E., and D. C. Meredith (1990). *Computational Physics—Fortran Version*, Addison-Wesley, Reading, MA.

Koonin, S. E., K. T. R. Davies, V. Maruhn-Rezwani, H. Feldmeier, S. J. Krieger, and J. W. Negele (1977). *Phys. Rev. C* **15**, 1359.

Kosloff, R. (1988). *J. Phys. Chem.* **92**, 2087.

Kosloff, R. (1993). In *Numerical Grid Methods and Their Application to Schrödinger's Equation*, C. Cerjan (ed.), Kluwer Academic, Dordrecht, The Netherlands, p. 175.

Kosloff, D., and R. Kosloff (1983). *J. Comput. Phys.* **52**, 35.

Kramers, H. A. (1957). *Quantum Mechanics*, North Holland, Amsterdam, and Interscience, New York.

Kravchenko, Yu. P., M. A. Liberman, and B. Johansson (1996). *Phys. Rev. Lett.* **77**, 619.

Krempl, S., T. Eisenhammer, A. Hübler, G. Mayer-Kress, and P. W. Milonni (1992). *Phys. Rev. Lett.* **69**, 430.

Kurchan, J., P. Leboeuf, and M. Saraceno (1989). *Phys. Rev. A* **40**, 6800.

Kus, M., F. Haake, and D. Delande (1993). *Phys. Rev. Lett.* **71**, 2167.

Landau, L. D., and E. M. Lifshitz (1958). *Quantum Mechanics: Non-Relativistic Theory*, Pergamon, London; and (1977), 3rd ed., Pergamon, Oxford.

Langer, R. E. (1937). *Phys. Rev.* **51**, 669.

Latka, M., P. Grigolini, and B. J. West (1993). *Phys. Rev. A* **47**, 4649.

Latka, M., P. Grigolini, and B. J. West (1994). *Phys. Rev. A* **50**, 1071.

Leboeuf, P. (1991). *J. Phys. A* **24**, 4575.

Leboeuf, P., and A. Voros (1990). *J. Phys. A* **23**, 1765.

Lederman, W., ed., (1981). *Handbook of Applicable Mathematics: Vol. 3, Numerical Methods*, Wiley, Chichester, UK.

Lee, B. (1993). *Superlattices and Microstructures* **14**, 295.

LeForestier, C., R. H. Bisseling, C. Cerjan, M. D. Feit, R. Friesner, A. Guldberg, A. Hammerich, G. Jolicard, W. Karrlein, H.-D. Meyer, N. Lipkin, O. Roncero, and R. Kosloff (1991). *J. Comput. Phys.* **94**, 59.

Leichtle, C., I. Sh. Averbukh, and W. P. Schleich (1996). *Phys. Rev. Lett.* **77**, 3999.

Leopold, J. G., and D. Richards (1985). *J. Phys. B* **18**, 3369.

Levit, S., and U. Smilansky (1977). *Ann. Phys. (NY)* **108**, 165.

Lichtenberg, A. J. (1969). *Phase Space Dynamics of Particles*, Wiley, New York.

Lichtenberg, A. J., and M. A. Lieberman (1992). *Regular and Chaotic Dynamics*, 2nd ed., Springer-Verlag, New York.

Liu, W.-K., B. Wu, and J.-M. Yuan (1995). *Phys. Rev. Lett.* **75**, 1292.

Loudon, R. (1983). *The Quantum Theory of Light*, 2nd ed., Clarendon Press, Oxford.

Luenberger, D. (1979). *Introduction to Dynamical Systems, Theory, Models and Applications*, Wiley, New York.

MacKay, R. S., J. D. Meiss, and I. C. Percival (1984). *Physica D* **55**, 340.

Main, J., A. Holle, G. Wiebusch, and K. H. Welge (1987). *Z. Phys. D* **6**, 295.

Main, J., G. Wiebusch, K. Welge, J. Shaw, and J. B. Delos (1994). *Phys. Rev. A* **49**, 847.

Mao, J.-M., and J. B. Delos (1992). *Phys. Rev. A* **45**, 1746.

Marcinek, R., and E. Pollak (1994). *J. Chem. Phys.* **100**, 5894.

Maslov, V. P., and M. V. Fedoriuk (1981). *Semi-Classical Approximations in Quantum Mechanics*, D. Reidel, Dordrecht, The Netherlands.

McCullough, E. A., and R. E. Wyatt (1969). *J. Chem. Phys.* **51**, 1253.

McCullough, E. A., and R. E. Wyatt (1971). *J. Chem. Phys.* **54**, 3578.

McDonald, S. W., and A. N. Kaufman (1988). *Phys. Rev. A* **37**, 3067.

McLaughlin, D. W. (1972). *J. Math. Phys.* **13**, 1099.

McWeeny, R., and C. Coulson (1948). *Proc. Camb. Philos. Soc.* **44**, 413.

Meerson, B., and L. Friedland (1990). *Phys. Rev. A* **41**, 5233.

Meerson, B. I., E. A. Oks, and P. V. Sasarov (1982). *J. Phys. B* **15**, 3599.

Melinger, J. S., A. Hariharan, S. R. Gandhi, and W. S. Warren (1991). *J. Chem. Phys.* **95**, 2210.

Melinger, J. S., S. R. Gandhi, A. Hariharan, D. Goswami, and W. S. Warren (1994). *J. Chem. Phys.* **101**, 6439.

Merani, N., J. Main, and G. Wunner (1995). *Astron. Astrophys.* **298**, 193.

Merzbacher, E. (1961). *Quantum Mechanics*, Wiley, New York.

Messiah, A. (1962). *Quantum Mechanics*, North Holland, Amsterdam, p. 749.

Meyer, H.-D. (1986). *J. Chem. Phys.* **84**, 3147.

Miller, W. H. (1968). *J. Chem. Phys.* **48**, 1651.

Miller, W. H. (1970). *J. Chem. Phys.* **53**, 3578.

Miller, W. H. (1974). *Adv. in Chem. Phys.* **25**, 69.

Miller, W. H. (1975). *J. Chem. Phys.* **63**, 996.

Miller, W. H., and T. F. George (1972). *J. Chem. Phys.* **56**, 5668.

Moerner, W. E., and T. Basche (1993). *Angew. Chem. Int. Ed. Engl.* **32**, 457.

Moiseyev, N., and A. Peres (1983). *J. Chem. Phys.* **79**, 5945.

Moiseyev, N., H. J. Korsch, and B. Mirbach (1994). *Z. Phys. D* **29**, 125.

Møller, K. B., T. G. Jørgensen, and Go. Torres-Vega (1997). *J. Chem. Phys.* **106**, 7228.

Moore, F. L., J. C. Robinson, C. Bharucha, P. E. Williams, and M. G. Raizen (1994). *Phys. Rev. Lett.* **73**, 2974.

Morse, P. M., and H. Feshbach (1953). *Methods of Theoretical Physics*, McGraw-Hill, New York, 2 vols.

Müller, G., G. S. Boebinger, H. Mathur, L. N. Pfeiffer, and K. W. West (1995). *Phys. Rev. Lett.* **75**, 2875.

Muratov, L. S., M. I. Stockman, L. N. Pandey, T. F. George, W. J. Li, B. D. McCombe, J. P. Kaminski, S. J. Allen, and W. J. Schaff (1997). *Superlattices and Microstructures* **21**, 501.

Narozhny, N. B., J. J. Sanchez-Mondragon, and J. Eberly (1981). *Phys. Rev. A* **23**, 236.

Nauenberg, M. (1989). *Phys. Rev. A* **40**, 1113.

Neria, E., and A. Nitzan (1993). *J. Chem. Phys.* **99**, 1109.

Nikitin, E. E., and S. Ya. Umanskii (1984). *Theory of Slow Atomic Collisions*, Springer-Verlag, Berlin.

Noid, D. W., and R. A. Marcus (1977). *J. Chem. Phys.* **67**, 559.

Noid, D. W., M. L. Koszykowski, and R. A. Marcus (1979). *J. Chem. Phys.* **71**, 2864.

Noid, D. W., M. L. Koszykowski, M. Tabor, and R. A. Marcus (1980). *J. Chem. Phys.* **72**, 6169.

Oberhettinger, F. (1973). *Fourier Transforms of Distributions and Their Inverses: A Collection of Tables*, Academic, New York.

Ohanian, H. C. (1990). *Principles of Quantum Mechanics*, Prentice-Hall, Englewood Cliffs, NJ.

Oosterkamp, T. H., L. P. Kouwenhoven, A. E. A. Koolen, N. C. van der Vaart, and C. J. P. M. Harmans (1997). *Phys. Rev. Lett.* **78**, 1537.

Oppenländer, A., Ch. Rambaud, H. P. Trommsdorff, and J. C. Vial (1989). *Phys. Rev. Lett.* **63**, 1432.

Ortigoso, J. (1996). *Phys. Rev. A* **54**, R2521.

Oseledec, V. I. (1968). *Trans. Moscow Math. Soc.* **19**, 197.

Ozorio de Almeida, A. M. (1984). *J. Phys. Chem.* **88**, 6139.

Ozorio de Almeida, A. M. (1988). *Hamiltonian Systems: Chaos and Quantization*, Cambridge University Press, Cambridge, England.

Ozorio de Almeida, A. M., and J. H. Hannay (1982). *Ann. Phys. (NY)* **138**, 115.

Pandey, L. N., L. S. Muratov, M. I. Stockman, and T. F. George (1994). *Phys. Status Solidi* **185**, 151.

Park, D. A. (1974). *Introduction to Quantum Theory*, McGraw-Hill, New York.

Park, T. J., and J. C. Light (1986). *J. Chem. Phys.* **85**, 5870.

Parker, J., and C. R. Stroud, Jr. (1986). *Phys. Rev. Lett.* **56**, 716.

Peliti, L., ed., (1992). *Biologically Inspired Physics*, Plenum, New York.

Péoux, G., M. Monnerville, and J.-P. Flament (1996). *J. Phys. B* **29**, 6031.

Percival, I., and D. Richards (1982). *Introduction to Dynamics*, Cambridge University Press, Cambridge, England

Perelomov, A. (1986). *Generalized Coherent States and Their Applications*, Springer-Verlag, Berlin.

Perotti, L. C., M. Skrzypkowski, and J. E. Bayfield (1998). Unpublished.

Peterson, G. L., and C. D. Cantrell (1985). *Phys. Rev. A* **31**, 807.

Plybon, B. F. (1992). *An Introduction to Applied Numerical Analysis*, PWS-Kent, Boston.

Pollak, E., M. S. Child, and P. Pechukas (1980). *J. Chem. Phys.* **72**, 1669.

Pomphrey, N. (1974). *J. Phys. B* **7**, 1909.

Popov, V. S., B. M. Kamakov, and V. D. Mur (1996). *Phys. Lett. A* **210**, 402.

Porter, C. E. (1965). *Statistical Theories of Spectra: Fluctuations*, Academic, New York.

Press, W. H., B. P. Flannery, S. A. Teukolsky, and W. T. Vetterling (1986). *Numerical Recipes: The Art of Scientific Computing*, Cambridge University Press, Cambridge, England.

Press, W. H., S. A. Teukolsky, W. T. Vetterling, and B. P. Flannery (1992). *Numerical Recipes in Fortran 77*, 2nd ed., Cambridge University Press, Cambridge, England.

Press, W. H., S. A. Teukolsky, W. T. Vetterling, and B. P. Flannery (1996). *Numerical Recipes in Fortran 90: The Art of Parallel Scientific Computing*, 2nd ed., Cambridge University Press, Cambridge, England.

Provost, D., and P. Brumer (1995). *Phys. Rev. Lett.* **74**, 250.

Rabi, I. I. (1932). *Phys. Rev.* **51**, 652.

Rabitz, H., and S. Shi (1991). *Adv. Mol. Vib. Coll. Dynamics* **1A**, 187.

Rasband, S. N. (1990). *Chaotic Dynamics of Nonlinear Systems*, Wiley, New York.

Reichl, L. E. (1992). *The Transition to Chaos in Conservative Classical Systems: Quantum Manifestations*, Springer-Verlag, New York.

Reichl, L. E., and L. Haoming (1990). *Phys. Rev. A* **42**, 4543.

Reinhardt, W. P. (1982). *Annu. Rev. Phys. Chem.* **33**, 223.

Remoissenet, M. (1996). *Waves Called Solitons: Concepts and Experiments*, 2nd ed., Springer-Verlag, Berlin.

Richards, D. (1983). *J. Phys. B* **16**, 749.

Robnik, M., and L. Salasnich (1997). *J. Phys. A* **30**, 1711.

Rotenberg, M. (1962). *Ann. Phys. (NY)* **19**, 262.

Ruder, H., G. Wunner, H. Herold, and F. Geyer (1994). *Atoms in Strong Magnetic Fields*, Springer-Verlag, Berlin.

Sagdeev, R. Z., D. A. Usikov, and G. M. Zaslavsky (1988). *Nonlinear Physics, from the Pendulum to Turbulence and Chaos*, Harwood Academic, Chur, Switzerland.

Sambe, H. (1973). *Phys. Rev. A* **7**, 2203.

Schlautmann, M., and R. Graham (1995). *Phys. Rev. E* **52**, 340.

Schofield, C. F. (1989). *Optimising Fortran Programs*, Halstead, New York.

Schulman, L. S. (1981). *Techniques and Applications of Path Integration*, Wiley, New York.

Schulman, L. S. (1994). *J. Phys. A* **27**, 1703.

Schuster, H. G. (1992). In *Nonlinear Dynamics in Solids*, H. Thomas (ed.), Springer-Verlag, Berlin, p. 22.

Schutte, C. J. H. (1968). *The Wave Mechanics of Atoms, Molecules and Ions*, Edward Arnold, London.

Schweizer, W., R. Niemeier, G. Wunner, and H. Ruder (1993). *Z. Phys. D* **25**, 95.

Seligman, T. H., J. J. M. Verbaarschot, and M. R. Zirnbauer (1984). *Phys. Rev. Lett.* **53**, 215.

Sepúlveda, M. A., and F. Grossman (1996). In *Advances in Chemical Physics* **96**, I. Prigogine and S. A. Rice (eds.), Wiley, New York, p. 191.

Seydel, R. (1988). *From Equilibrium to Chaos*, Elsevier, New York.

Shi, S., and H. Rabitz (1990). *J. Chem. Phys.* **92**, 364.

Shimada, I., and T. Nagashima (1979). *Prog. Theor. Phys.* **61**, 1605.

Shin, J.-Y., and H.-W. Lee (1994). *Phys. Rev. E* **50**, 902.

Shirley, J. H. (1965). *Phys. Rev. B* **138**, 1979.

Shore, B. W. (1990). *The Theory of Coherent Atomic Excitation*, Wiley, New York, 2 vols.

Shore, B. W., and P. L. Knight (1993). *J. Mod. Opt.* **40**, 1195.

Shuryak, E. V. (1976). *Sov. Phys. JETP* **44**, 1070.

Simon, B. (1973). *Ann. Math.* **97**, 247.

Sirko, L., and P. M. Koch (1995). *Appl. Phys. B* **60**, S195.

Skinner, J. L., and T. P. Trommsdorff (1988). *J. Chem. Phys.* **89**, 897.

Skodje, R. T. (1984). *Chem. Phys. Lett.* **109**, 221.

Skodje, R. T. (1989). *J. Chem. Phys.* **90**, 6193.

Skodje, R. T., and F. Borondo (1985). *Chem. Phys. Lett.* **118**, 409.

Skodje, R. T., F. Borondo, and W. P. Reinhardt (1985). *J. Chem. Phys.* **82**, 4611.

Skodje, R. T., H. W. Rohrs, and J. Van Buskirk (1989). *Phys. Rev. A* **40**, 2894.

Smith, F. T. (1969). *Phys. Rev.* **179**, 111.

Solov'ev, E. A. (1982). *Sov. Phys. JETP* **55**, 1017.

Sridhar, S., and A. Kudrolli (1994). *Phys. Rev. Lett.* **72**, 2175.

Stein, J., and H.-J. Stockmann (1992). *Phys. Rev. Lett.* **68**, 2867.

Sterling, T. (1996). *Commun. of the ACM* **39**, 11.

Stoer, J., and R. Bulirsch (1980). *Introduction to Numerical Analysis*, Springer-Verlag, New York.

Suárez Barnes, I. M., M. Nauenberg, M. Nockleby, and S. Tomsovic (1993). *Phys. Rev. Lett.* **71**, 1961.

Suárez Barnes, I. M., M. Nauenberg, M. Nockleby, and S. Tomsovic (1994). *J. Phys. A* **27**, 3299.

Susskind, S. M., and R. V. Jensen (1988). *Phys. Rev. A* **38**, 711.

Tabor, M. (1983). *Physica D* **6**, 195.

Tabor, M. (1989). *Chaos and Integrability in Nonlinear Dynamics*, Wiley, New York.

Takahashi, K. (1989). *Prog. Theor. Phys. Suppl.* **98**, 109.

Takatsuka, K., and A. Inoue (1997). *Phys. Rev. Lett.* **78**, 1404.

Tal-Ezer, H., and R. Kosloff (1984). *J. Chem. Phys.* **81**, 3967.

Tanner, J. J., and M. M. Maricq (1989). *Phys. Rev. A* **40**, 4054.

Teissier, R., J. W. Cockburn, J. J. Finley, M. S. Skolnick, P. D. Buckle, D. J. Mowbray, R. Grey, G. Hill, and M. A. Pate (1994). *Superlattices and Microstructures* **15**, 373.

Templon, J. A. (1996). *Comput. Phys.* **10**, 49.

Tennyson, J. L., J. R. Cary, and D. F. Escande (1986). *Phys. Rev. Lett.* **56**, 2117.

Tesche, C. D. (1990). *Phys. Rev. Lett.* **64**, 2358.

Thachuk, M., and D. M. Wardlaw (1995). *J. Chem. Phys.* **102**, 7462.

Tien, P. K., and J. P. Gordon (1963). *Phys. Rev.* **129**, 647.

Tihkonov, A., and V. Arsenin (1977). *Solutions of Ill-Posed Problems*, Winston/Wiley, Washington, DC.

Ting, J. J.-L. (1994). *J. Phys. B* **27**, 1249.

Tinkham, M. (1964). *Group Theory and Quantum Mechanics*, McGraw-Hill, New York.

Tolman, R. C. (1938). *The Principles of Statistical Mechanics*, The University Press, Oxford.

Tomsovic, S. (1995). In *Chaos—The Interplay Between Stochastic and Deterministic*

Behavior, P. Garbaczewski, M. Wolf, and A. Weron (eds.), Springer-Verlag, Berlin, p. 331.

Tomsovic, S. (1996). *Phys. Rev. Lett.* **77**, 4158.

Tomsovic, S., and E. J. Heller (1991). *Phys. Rev. Lett.* **67**, 664.

Tomsovic, S., and E. J. Heller (1993). *Phys. Rev. Lett.* **70**, 1405.

Tomsovic, S., and J. H. Lefebvre (1997). *Phys. Rev. Lett.* **79**, 3629.

Tomsovic, S., and D. Ullmo (1994). *Phys. Rev. E* **50**, 145.

Tomsovic, S., M. Grinberg, and D. Ullmo (1995). *Phys. Rev. Lett.* **75**, 4346.

Torres-Vega, Go., and J. H. Frederick (1991). *Phys. Rev. Lett.* **67**, 2601.

Torres-Vega, Go., and J. H. Frederick (1993). *J. Chem. Phys.* **98**, 3103.

Trotter, H. F. (1959). *Proc. Am. Math. Soc.* **10**, 545.

Truscott, W. S. (1993). *Phys. Rev. Lett.* **70**, 1900.

Tully, J. C., and R. K. Preston (1971). *J. Chem. Phys.* **55**, 562.

Ueda, M., T. Wakabayashi, and M. Kuwata-Gonokami (1996). *Phys. Rev. Lett.* **76**, 2045.

Ullmo, D., M. Grinberg, and S. Tomsovic (1996). *Phys. Rev. E* **54**, 136.

Utermann, R., T. Dittrich, and P. Hänggi (1994). *Phys. Rev. E* **49**, 273.

Vanbésien, O., and D. Lippens (1994). *Appl. Phys. Lett.* **65**, 2439.

Vanbésien, O., and D. Lippens (1995). *Superlattices and Microstructures* **17**, 197.

Van de Velde, E. F. (1994). *Concurrent Scientific Computing*, Springer-Verlag, New York.

Van Fleck, J. H. (1928). *Proc. Math. Acad. Sci. USA* **14**, 178.

Van Hemmen, J. L., H. Hey, and W. F. Wreszinski (1997). *J. Phys. A* **30**, 6371.

Van Leeuwen, K. A. H., G. V. Oppen, S. Renwick, J. B. Bowlin, P. M. Koch, R. V. Jensen, O. Rath, D. Richards, and J. G. Leopold (1985). *Phys. Rev. Lett.* **55**, 2231.

Voth, G. A. (1986). *J. Phys. Chem.* **90**, 3624.

Wagner, M. (1996). *Phys. Rev. Lett.* **76**, 4010.

Walker, R. B., and R. K. Preston (1977). *J. Chem. Phys.* **67**, 2017.

Wan, F. Y. M. (1995). *Introduction to the Calculus of Variations and Its Applications*, Chapman & Hall, New York.

Watanabe, N., Y. Mori, and H. Kawai (1987). In *Thin Film Growth Techniques for Low-Dimensional Structures*, R. F. C. Farrow, S. S. P. Parkin, P. J. Dobson, J. H. Neave, and A. S. Arrott (eds.), Plenum, New York, p. 115.

Waterland, R. L., J.-M. Yuan, C. C. Martens, R. E. Gillilan, and W. P. Reinhardt (1988). *Phys. Rev. Lett.* **61**, 2733.

Weiper, F. J., J. Ankerhold, H. Grabert, and E. Pollak (1996). *Phys. Rev. Lett.* **77**, 2662.

Weisshaar, A., J. Lary, S. M. Goodnick, and V. K. Tripathi (1991). *J. Appl. Phys.* **70**, 355.

Whittaker, E. T. (1964). *A Treatise on the Analytical Dynamics of Particles and Rigid Bodies*, Cambridge University Press, Cambridge, England.

Wilkie, J., and P. Brumer (1977a). *Phys. Rev. A* **55**, 27.

Wilkie, J., and P. Brumer (1997b). *Phys. Rev. A* **55**, 43.

Wilkinson, M. (1986). *Physica D* **21**, 341.

Wilkinson, M., and J. H. Hannay (1987). *Physica D* **27**, 201.

Wintgen, D. (1987). *J. Phys. B* **20**, L511.

Wintgen, D. (1988). *Phys. Rev. Lett.* **61**, 1803.

Wintgen, D., and H. Friedrich (1986). *J. Phys. B* **19**, 991.

Wolf, A., J. B. Swift, H. L. Swinney, and J. A. Vastano (1985). *Physica D* **16**, 285.

Wu, J. C., M. N. Wybourne, A. Weisshaar, and S. M. Goodnick (1993). *J. Appl. Phys.* **74**, 4590.

Wunner, G., M. Kost, and H. Ruder (1986). *Phys. Rev. A* **33**, 1444.

Wyatt, R. E., G. Hose, and H. S. Taylor (1983). *Phys. Rev. A* **28**, 815.

Xavier, Jr., A. L., and M. A. M. Aguiar (1997). *Phys. Rev. Lett.* **79**, 3323.

Yajima, K. (1982). *Commun. Math. Phys.* **87**, 331.

Yeazell, J. A., M. Malialieu, and C. R. Stroud, Jr. (1990). *Phys. Rev. Lett.* **17**, 2007.

Yuan, J.-M., and W.-K. Liu (1998). *Phys. Rev. A* **57**, 1992.

Zakrzewski, J., and D. Delande (1993). *Phys. Rev. E* **47**, 1650.

Zaslavsky, G. M., and N. N. Filonenko (1968). *Sov. Phys. JETP* **25**, 851.

Zhu, C., and H. Nakamura (1995). *J. Chem. Phys.* **102**, 7448.

INDEX

Bold-face pages indicate primary information about a topic and/or its definition.

RETURN PHYSICS LIBRARY

1-MONTH